GNVQ

Advanced Engineering

Systems, Processes, Materials and Design

Alan Darbyshire and David Taylor

First published 1997 by:
Stanley Thornes Publishers Ltd
Ellenborough House
Wellington Street
Cheltenham
GL50 1YW
UK

97 98 99 00 01 / 10 9 8 7 6 5 4 3 2 1

ISBN 0 7487 2886 4

Typeset by Florencetype Ltd, Stoodleigh, Devon
Printed and bound in Great Britain by Scotprint, Musselburgh

Contents

Acknowledgements

The following figures are reproduced from the publications listed with permission of the copyright holders:

A. Green, *Plane and Solid Geometry*, Stanley Thornes (Publishers) Ltd.
Figures: 1.17, 9.14a and b, 11.7a and b, 11.25, 11.40, 11.41, 11.43, 11.45, 11.46, 11.47, 11.48, 11.49, 11.52, 11.53, 11.54, Tables: 11.7, 11.8

M. Parker and L. Dennis, *Engineering Drawing Fundamentals*, Stanley Thornes (Publishers) Ltd.
Figures: 9.24, 10.7, 11.20, 11.33, 11.38, 11.39, 11.50

M. Parker and F. Pickup, *Engineering Drawing with Worked Examples*, 3rd edition, Stanley Thornes (Publishers) Ltd.
Figures: 11.8, 11.9, 11.10, 11.11, 11.13, 11.15, 11.26, 11.31, 11.32, 11.34, 11.51

E. Ramsden, *Materials Science*, Stanley Thornes (Publishers) Ltd.
Figures: 3.43, 3.46, 6.1, 6.12, 6.13, 6.17, 6.23, 6.24, 6.28, 6.31

Technology of Skilled Processes – Workholding and Toolholding, Removing Materials, Stam Press Ltd.
Figures: 3.3, 3.5, 3.6, 3.10, 3.11, 3.12, 3.13, 3.17, 3.18, 3.20, 3.23, 3.24, 3.25

Technology of Skilled Processes – Soft Soldering, Hard Soldering and Brazing, Stam Press Ltd.
Figures: 3.51, 3.52, 3.53, 3.55

Technology of Skilled Processes – Fusion Welding, Stam Press Ltd.
Figures: 3.56, 3.57, 3.58, 3.59, 3.60, 3.61, 3.62, 3.63, 3.64, 3.95, 3.96

Technology of Skilled Processes – Joining, Stam Press Ltd.
Figures: 3.65, 3.66, 3.67, 3.68, 3.69, 3.70, 3.72, 3.73

Technology of Skilled Processes – Assembly and Dismantling, Stam Press Ltd.
Figure: 3.71

Technology of Skilled Processes – Observing Safe Working Practices and Moving Loads, Stam Press Ltd.
Figure: 3.86, 3.92, 3.93, 4.7

Technology of Skilled Processes – Interpreting Drawings, Specifications and Data, Stam Press Ltd
Figure: 11.1, 11.2, 11.3, 11.4, 11.5, 11.6, 11.14, 11.17, 11.18, 11.19, 11.21, 11.22, 11.23, 11.24, 11.27, 11.28, 11.29, 11.30

The authors and publishers are grateful to the following people and firms for supplying photographs:

Aurora Forgings Ltd. (figure 3.38); British Aerospace Airbus Ltd. (figure 6.6); Brookes & Vernons (figure 1.14); Martyn Chillmaid (figure 6.7); Duffmarine Ltd. (figure 7.2); Dyson Appliances Ltd. (figure 10.10); Ford Motor Co. (figure 10.8); Massey Ferguson (figure 10.1); Microsoft Ltd.(figure 9.13); Nissan (figure 1.3); Orange (figure 6.5); P.A. News (figures 9.10, 9.23, 10.9); Pashley Hand Built Bicycles (figure 10.2); Peugeot (figure 9.11); Pilkington Plc. (figure 6.11); Q.A. Photos Ltd. (figure 9.3); Rover Group (figure 9.12); Science Photo Library (figure 7.5); Siemens Medical Engineering (figure 7.3); Vickers plc. (figure 6.46); and GKN Westland (figure 10.4).

Introduction to GNVQs in Engineering

GNVQs (General National Vocational Qualifications) are a new type of qualification available for students who want to follow a course linking traditional areas of study with the world of work. Those who qualify will be ideally placed to make the choice between progressing to higher education and applying for employment.

GNVQs are available at three levels:

Foundation	normally studied full time for one year
Intermediate	normally studied full time for one year
Advanced	normally studied full time for two years, and often referred to as 'vocational A-levels'.

Each GNVQ is divided into various types of units:

Mandatory units	which every one must study
Optional units	from which a student must choose to study
Additional units	offered by the various awarding bodies in order to allow students to top up their skills and knowledge
Core skills units	covering the essential skills related to numeracy, IT and communication.

Units are clearly structured. The principal components are:

elements	which focus on specific aspects of the unit
performance criteria	which tell you what you must be able to do
range	which tell you what areas you must be able to apply your knowledge to.

Advanced GNVQs in Engineering

This book is intended primarily for students of the Advanced GNVQ in Engineering. This course is divided as follows:

eight **mandatory units** covering all key aspects of engineering as a discipline, namely:

Unit 1: Engineering and commercial functions in business
Unit 2: Engineering systems
Unit 3: Engineering processes
Unit 4: Engineering materials
Unit 5: Design development
Unit 6: Engineering in society and the environment
Unit 7: Science for engineering
Unit 8: Mathematics for engineering

various specialist pathways (e.g. mechanical engineering, electrical and electronic engineering) which can be studied by selecting particular combinations of the **optional units**.

additional units chosen from those offered by the various awarding bodies. Many students will wish to study at least one additional mathematics unit if they intend to study engineering at a higher level some time in the future.

core skills units in the three areas at level three.

Stanley Thornes publishes titles specifically developed to cover all of the mandatory units above. This book covers Units 2, 3, 4 and 5, and integrates core skills throughout. Each part of the book is written to reflect the GNVQ unit structure, thus enabling the student to map his or her progress against the official specifications.

Assessment

GNVQs are assessed using two methods:

Unit tests	one for each unit, taken by everybody.
Portfolio of evidence	developed individually by each student to demonstrate competence and understanding.

The unit tests can be sat on a number of occasions throughout the year, and can be sat again should a student not pass. Any number of attempts can be made to pass, but a pass must be achieved before a student's portfolio will be assessed. A GNVQ cannot be awarded until the student has passed the unit test for each of the mandatory units.

Each unit test consists of 20–25 multiple choice questions, and the pass mark is 70%. A practice paper is included at the end of each unit in this book, to give students an idea of what is expected from them.

Portfolio of evidence

Each student must develop a portfolio (i.e. organised collection) of evidence that they have achieved the required level of skill and understanding for each of the elements of the units they take. The unit specifications say exactly what type of evidence will be acceptable for assessment, but do not dictate the details of any project to be undertaken, nor exactly how the evidence is to be compiled. Written, photographic, video or audio tape records of achievement are all acceptable, as are testimonials from work placements.

Regardless of the media employed to deliver evidence, it is absolutely essential that the content is organised and referenced so that it can be easily and effectively assessed. There is no point in assembling masses of evidence if its presentation makes it impossible to assess or understand. An index should be included, and evidence relevant to more than one element cross-referenced, or duplicated if necessary.

Grading

A GNVQ can be awarded as a pass, a merit or a distinction. To achieve a merit or a distinction a student must demonstrate notable ability in a range of skills related to the development of high-quality portfolio evidence. Your lecturer or teacher will be able to give you detailed guidance on how to develop and demonstrate these skills.

Additional titles developed to support GNVQs in Engineering are listed on the back cover of this book, and further details are available from Stanley Thornes Publishers Customer Services Department on 01242 228888.

Key skills: Summary of units at level 3

This summary shows how the activities and assignments provide opportunities to provide evidence of the key skills. When planning a piece of work, think carefully about the key skills you can provide evidence for. For example, use information technology as often as possible when producing written work, charts and diagrams.

Key Skills	Activities	Assignments
Communication		
3.1 Take part in discussion	3.4, 3.6, 3.7, 3.11, 4.2, 5.1, 5.2, 6.2, 6.3, 6.4, 6.7, 6.8, 7.1, 7.2, 7.3, 7.4, 7.5, 7.6, 10.9	3, 4, 5, 6, 7, 8, 9, 10
3.2 Produce written material	1.2, 1.3, 1.4, 1.5, 1.6, 1.7, 1.8, 1.9, 1.10, 1.11, 1.12, 1.13, 2.1, 2.2, 2.3, 2.4, 2.5, 2.7, 2.8, 2.9, 2.10, 3.4, 3.6, 3.10, 4.1, 4.2, 5.1, 5.2, 6.3, 6.4, 6.5, 6.7, 6.8, 9.1, 9.2, 9.3, 9.4, 9.5, 9.6, 9.7, 9.9, 9.10, 9.12, 9.16, 9.17, 9.18, 9.19	1, 2, 3, 4, 5, 6, 7, 8, 9, 10
3.3 Use images	1.1, 1.8, 1.9, 1.10, 1.11, 1.12, 1.13, 2.1, 2.2, 2.3, 2.4, 2.5, 2.7, 2.8, 2.9, 2.10, 3.1, 3.3, 3.9, 3.10, 4.1, 5.2, 6.3, 6.4, 6.7, 7.3, 7.4, 7.6, 9.7, 9.8, 9.9, 9.10, 9.11, 9.15, 9.16, 10.10, 10.11, 11.2, 11.3, 11.4, 11.5, 11.6, 11.7, 11.8, 11.9, 11.10	1, 2, 3, 4, 5, 6, 7, 8, 9, 10, 11
3.4 Read and respond to written material	3.1, 3.3, 3.5, 3.6, 3.7, 3.8, 3.11, 4.2, 5.2, 6.2, 6.4, 6.5, 7.1, 7.2, 7.3, 7.4, 7.6, 7.7	1, 2, 3, 4, 5, 6, 7, 8, 9, 10
Information Technology		
3.1 Prepare information	3.4, 3.6, 3.10, 4.1, 4.2, 5.1, 5.2, 6.3, 6.4, 6.5, 6.7, 6.8, 7.1, 7.3, 7.4, 7.6, 7.7, 9.14, 11.1, 11.2, 11.4, 11.5, 11.7, 11.8, 11.9, 11.10 + all activities listed under Communication 3.2 & 3.3 if done on computer	1, 2, 3, 4, 5, 6, 7, 8, 9, 10, 11
3.2 Process information	3.4, 3.6, 3.10, 4.1, 4.2, 5.1, 5.2, 6.3, 6.4, 6.5, 6.7, 6.8, 7.1, 7.3, 7.4, 7.6, 7.7, 9.14, 11.1, 11.2, 11.4, 11.5, 11.7, 11.8, 11.9, 11.10 + all activities listed under Communication 3.2 & 3.3 if done on computer	1, 2, 3, 4, 5, 6, 7, 8, 9, 10, 11
3.3 Present information	3.4, 3.6, 3.10, 4.1, 4.2, 5.1, 5.2, 6.3, 6.4, 6.5, 6.7, 6.8, 7.1, 7.3, 7.4, 7.6, 7.7, 9.14, 11.1, 11.2, 11.4, 11.5, 11.7, 11.8, 11.9, 11.10 + all activities listed under Communication 3.2 & 3.3 if done on computer	1, 2, 3, 4, 5, 6, 7, 8, 9, 10, 11
3.4 Evaluate the use of information technology	10.3	

Key Skills	Activities	Assignments
Application of Number		
3.1 Collect and record data	2.6, 2.7, 2.8, 2.9, 2.10, 6.3, 9.10, 9.13, 9.14, 9.22, 10.2, 10.4, 10.5, 10.8, 10.10, 10.11, 10.12, 10.13	2, 4, 5, 9, 10
3.2 Tackle problems	2.6, 2.7, 2.8, 2.9, 2.10, 3.2, 3.9, 6.1, 6.3, 6.6, 9.13, 9.14, 9.22, 10.4, 10.5, 10.8, 10.10, 10.11, 10.12, 10.13	2, 4, 5, 9, 10
3.3 Interpret and present data	2.6, 2.7, 2.8, 2.9, 2.10, 3.2, 3.9, 4.10, 6.1, 6.3, 6.6, 9.13, 9.14, 9.22, 10.4, 10.5, 10.8, 10.10, 10.11, 10.12, 10.13, 11.3	2, 4, 9, 10

PART ONE: ENGINEERING SYSTEMS

A knowledge of engineering systems is the key to understanding how to design, make and maintain devices and machines. All devices are systems that consist of interconnected parts that work together to produce the desired result. For example, an aircraft is a system for moving goods and people through the air. An air traffic control system uses electronic systems to control the aircraft movements. The aircraft and the air traffic control system are subsystems that together make up a complete transportation system. Systems are by their nature multidisciplinary, and a single system can use electrical, electronic, thermodynamic, mechanical, fluidic and chemical elements. This multidisciplinary approach is what makes the study of systems such an interesting and fascinating subject.

The key to an understanding of a system is a knowledge of the inputs to the system and how these inputs are transformed by the system into outputs. Before an engineer can do any design work, the alternative systems must be thoroughly investigated and subjected to analysis. This part of the book gives examples of a wide range of engineering systems and also demonstrates simple ways of comprehending and analysing systems. Whenever any engineering work is done, the correct way to approach it is to consider the basic principles involved and to apply a systems approach to the solving of the problem. This can then lead to a design brief, followed by a design solution with correct choice of materials and processes for the realisation of the design. Proper consideration of systems gives a basic foundation, which then enables optimum solutions to be found to engineering problems.

Chapter 1: Inputs and outputs of engineering systems

This chapter covers:
Element 2.1: Investigate engineering systems in terms of their inputs and outputs.

. . . and is divided into the following sections:

- System terms
- Modelling
- The purposes of engineering systems
- Relationship of output to input.

Before engineering systems can be considered, the concept of systems will be examined. Systems are mentioned in everyday life, such as the:

- London Underground system
- telephone system
- parliamentary system
- education system.

When people speak of a system in this way, they are mostly not aware of what they mean. It is just used as a term that gives a name to whatever system they are using or talking about. As engineers, we have to be much more precise when referring to systems. A system can be described as an organised grouping and interconnection of individual parts. The key here is the organised grouping and interconnection. The individual parts of a car engine heaped together on the floor do not make a system. It is only when they are assembled together correctly that they make up the power system called an internal combustion engine.

Key characteristics of systems are:

- A function is performed by the system, which is of interest or use to a group or a person. Clearly, a system is of no use unless it actually does something for people.
- Individual parts of a system perform a precise function in the system. If they are removed from the system, then their function may change. Another way of putting this is that they are affected by being in the system and are also affected when they are removed from it.

Figure 1.1 The whole is more than the sum of its parts!

Case study

Changing systems

It is instructive to look at what happens when a specifically designed system is changed. This can occur when components or subsystems are put together from different sources. For example, an engine and generator set were designed to work together as a standby power supply system. Later, other generator sets were attached to the engine, but problems of vibration and reliability occurred. Similar problems occurred when the generator set was attached to different engines. Clearly, as the individual subsystems were separated for use in other systems, they behaved differently with the new subsystems.

System terms

Some terms in general use with systems are:

- Inputs are what goes into the system. The input to an electric motor drive system is electrical energy.
- Outputs are the products of the system, both desirable and undesirable. In the case of the electric motor drive system, the desirable output is kinetic energy in the form of a rotating shaft. The undesirable output is the heat generated by the motor.
- Feedback is where the output is connected in some way to the input by a feedback loop. A system with feedback is called a closed-loop system. The desired output is set as an input, and then the output is measured and compared with the input value. A system without feedback is called an open-loop system. An open-loop system is a feedforward system in which a desired output value is input, but no automatic correction is made if an error in output occurs. In an

electric motor drive system with feedback, the output speed can be measured and fed back to be used to vary the input energy, so controlling the speed of the motor at a set value called the set point. For a system to be stable, the feedback must be negative. This means the measured value must be subtracted from the set value.

- Environment means the things outside of the system that can affect it, but that are not themselves normally affected by the system. The electric motor drive system can be affected by ambient temperature changes, but it will not normally have a significant effect on the ambient temperature.
- Boundary is the dividing line between the system and its environment. Anything inside this line is in the system and anything outside is part of the environment.
- Disturbance (or secondary input) is usually a random undesirable change in input or environment. For an electric motor drive system, any large supply voltage fluctuations will be a disturbance and will affect the system. Similarly, environmental disturbances like thunderstorms may affect the system.

Systems are often described using a block diagram, as shown in Figure 1.2.

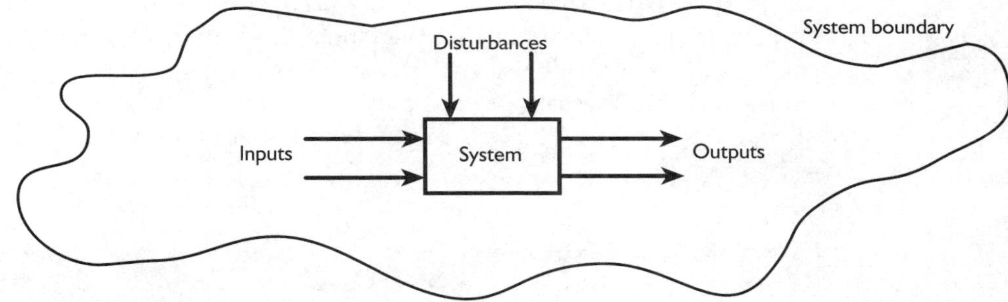

Figure 1.2 General system block diagram

Modelling

Representing systems with block diagrams is an attempt to model the process.

Modelling tries to represent the real world, so that systems can be designed and tested before the real-life system is built.

Case study

Design and testing of systems

Design and testing of systems has been radically changed by the use of simulation techniques. This avoids the need to build actual models to test the behaviour and response of the system. One example is a particular design of internal combustion engine that can be modelled on computer before the real engine is built and tested. Inputs can be changed and outputs measured. The system can also be changed on screen, e.g. valve numbers and sizes, piston shape, etc. can easily be modified. There will be a visual display of the way the gases flow through the engine as well as computed figures of things such as power output and fuel consumption. Also, stresses in vital components can be measured using finite element stress analysis. Therefore, the system is thoroughly tested and proved before any engine needs to be built. Firms such as Rolls Royce and Rover make extensive use of this technology to bring in new designs and systems.

Activity 1.1

Investigate five different types of systems. For each one:

a Find the inputs and outputs.
b Establish whether there is feedback between output and input.
c Draw a block diagram of the system.

The purposes of engineering systems

All systems have a function that is of interest to individuals or groups. Most systems will also have a purpose, which is achieved by the functional activity of the system in converting the inputs into outputs.

Case study

Car factory robots

The purpose of a particular robot in a car factory is to spray paint onto the car body. Its function is to convert the inputs of electrical energy into the outputs of rotation of the robot's waist, shoulder, elbow and wrist. Electric drive motors rotate each moving part of the robot. The spray nozzle is mounted on the wrist of the robot, which follows a programmed path around the car body, as shown in Figure 1.3.

Figure 1.3 Robot spray painting a car body

Manufacture

The purpose of manufacture is to convert materials and components into products. So the purpose of a manufacturing system is to take inputs and use them to produce a useable output, as illustrated in Figure 1.4. However, this system will function by taking inputs of things like energy and converting them into things such as heat, force, motion, etc.

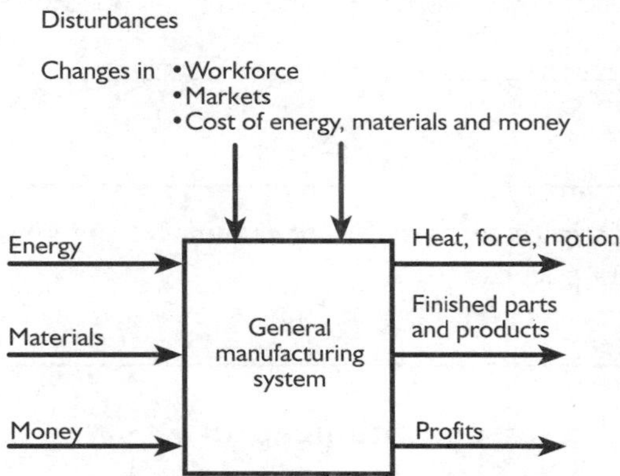

Figure 1.4 Block diagram of manufacturing system

Any complete system, such as a manufacturing system, will usually be made up of many subsystems. It will be useful to study some of these to differentiate clearly between purpose and function. Table 1.1 gives some examples of manufacturing subsystems.

Table 1.1 Examples of manufacturing subsystems

Subsystem	Purpose	Function
Casting of metals	Production of parts by pouring liquid metal into a mould	Conversion of chemical or electrical energy into heat
Machining	Production of parts by cutting material with cutting tools	Conversion of electrical energy into kinetic energy
Automatic insertion of ICs onto PCB	Manufacture of electronic circuit boards	Conversion of electrical energy into kinetic energy

Case study

Manufacturing system in a brewery

A brewery is a good example of a manufacturing system. A typical brewery will produce something like a hundred million gallons (approximately 457 million litres) of beer a year. The inputs are:

- raw materials – barley, hops, yeast, etc.
- energy, labour, etc.

The outputs are:

- beer – in barrels and bottles
- waste and by-products, e.g. yeast, spent hops and barley, etc.

The brewery system has purpose and function as shown in Table 1.2.

Table 1.2 Purpose and function of the brewery system

Purpose	Function
Production of beer from raw materials	Conversion of sugar in barley into alcohol by using yeast for fermentation, which is a chemical process

Activity 1.2

Investigate three different manufacturing subsystems. For each system state:
a its purpose
b its function.

Information and data communication

Information is defined as knowledge communicated by others or obtained by investigation, study or instruction.

Data is the plural of datum, which is defined as detailed information of any kind.

Thus, data are also information, but of a more detailed kind.

Case study

Information and data

It would be purely information to report that the world land speed record had been broken in a turbine-powered car. More data on this could be given in the form of:

- actual average speed achieved and previous record
- specification of the vehicle, e.g. power, thrust, weight.

So a television news programme might state that 'the world land speed record was broken today by A.N. Other in his turbine-powered car'. A more detailed treatment in a quality newspaper might also say, in addition, 'the car was powered by a Rolls Royce Avon gas turbine producing *y* kW of power with a thrust of *x* kg. The average speed of two runs was 1200 km/h. This beat the previous record of 1100 km/h set by Joe Bloggs in 1996'.

It is often difficult to distinguish between information and data. Since data are a type of information, it is not critical which term is used. Generally, if numbers are being communicated, it is called data. However, with digital technologies, as all information is transmitted in the form of numbers, there is really no difference between the two.

Systems that communicate information and data are generally a hybrid of computer and telecommunication systems. These systems are generally referred to as information technology systems.

The purpose of information and data communication systems is to:

- obtain and process information and data
- transmit and display information and data.

The function of information and data communication systems is to:

- convert information and/or data into electronic signals
- send signals down wire or optic fibre, or through the atmosphere by a discrete pulse (digital) or by continuously varying waveform (analogue)
- reconvert signals back to information and/or data.

By definition, the main outputs and inputs of information and data communication systems should be the same.

Case study

A broadcasting system

One example of an information and data communication system is a broadcasting system. A radio listener should hear the same sounds from a concert as are heard by the people in the concert hall. Figure 1.5 is a block diagram of this system. Information in the form of sound waves is input to a digital or analogue broadcasting system. This system converts the sound waves into voltages or pulses and transmits them over the air to the listener's radio. This then reconverts the information back to sound waves.

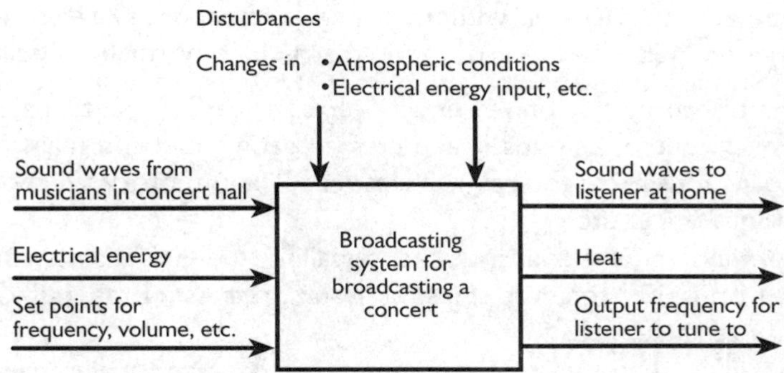

Figure 1.5 Broadcasting from a concert hall

Another example of an information and data communication system is a computer network as shown in Table 1.3.

Table 1.3 Purpose and function of a computer network

System	Purpose	Function
Computer network	Processing and communication of information and data	Conversion of information to digital pulses, processing the pulses, transmitting the pulses and converting them back to information

Activity 1.3

Investigate three different information and/or data communication systems. For each one state:
a its purpose
b its function.

Transportation

The purpose of transportation systems is to arrange physical movement from one place to another. This covers a very wide range and includes:

- movement through land, water or air by road vehicles, trains, ships and aircraft
- transmission of liquids and gases such as oil, water, natural gas, etc. by pipeline
- conveyers, transfer lines and other systems in factories, mines, etc., for moving materials, parts and products.

Case study

Transport and trade

Trade is the lifeblood of modern economies, so industry and agriculture would not be able to carry on without effective transport systems. These systems have a large boundary because of the infrastructure normally required. For example:

- Air transport not only uses aircraft but needs airports, navigation systems, air traffic control systems, weather forecasting, fuel and maintenance, etc.
- Road transport requires vehicles, roads and motorways, police, fuel and maintenance, etc.
- Railways require trains, tracks, signalling, fuel and maintenance, etc.
- Sea transport requires ships, harbours, cranes, roads, railways, fuel, maintenance, etc.
- Pipelines need rights of way, pumps, storage and maintenance, etc.

Some examples of transportation systems are given in Table 1.4.

Table 1.4 Examples of Transportation systems

System	Purpose	Function
Transfer line in car factory	To move cars, bodies and engines between processes	Converts electrical energy into rotary then linear movement, i.e. kinetic energy
Diesel-powered freight lorry	To move goods of all kinds locally and long distance	Converts chemical energy into heat; gas expands converting heat into kinetic energy. Motion is transmitted from piston to wheels

Activity 1.4

Investigate three different transportation systems. For each one state:
a its purpose
b its function.

Environmental control

Human beings can only function satisfactorily between a fairly narrow range of environmental parameters. Research in Britain has established that light factory work requires a temperature of 16–20°C. To maintain temperatures at this level requires the use of systems for:

- heating
- ventilating
- air conditioning.

Environmental control is also used to give specific environmental conditions in industrial and domestic applications such as:

- refrigeration and freezing – mainly used to preserve food
- process control – some processes require special conditions, e.g. printing requires temperatures of around 25°C for ink to flow; precise temperature control of furnaces is necessary for heat treating and melting metals.

To summarise, it is the purpose of environmental control systems to maintain environmental parameters such as temperature and humidity at set levels.

Case study

Environmental control systems

Table 1.5 gives examples of environmental control systems.

Table 1.5 Examples of environmental control systems

System	Purpose	Function
Domestic central heating	Provision of heat to maintain internal temperatures at set levels	Conversion of chemical or electrical energy into thermal energy (heat)
Cold store in frozen food company	Keep food fresh by providing temperatures below 0°C	Extraction of heat from store by converting electrical energy into kinetic energy for pumping heat out

Activity 1.5

Investigate three different environmental control systems. For each one state:
a its purpose
b its function.

Power generation and transmission

This is nowadays taken to mean exclusively electrical power. This is because electricity is the one form of power that can easily be transmitted over long distances. However, other forms of power can be used where it is safer or more convenient. Thus, high-pressure air, water or hydraulic fluid can be generated and transmitted through pipes.

Power is defined as the rate of working.

Work and energy have the same units, which are joules (J). A joule is the amount of energy required to move a force of 1 newton (1 N) through 1 metre (1 m), as shown in Figure 1.6.

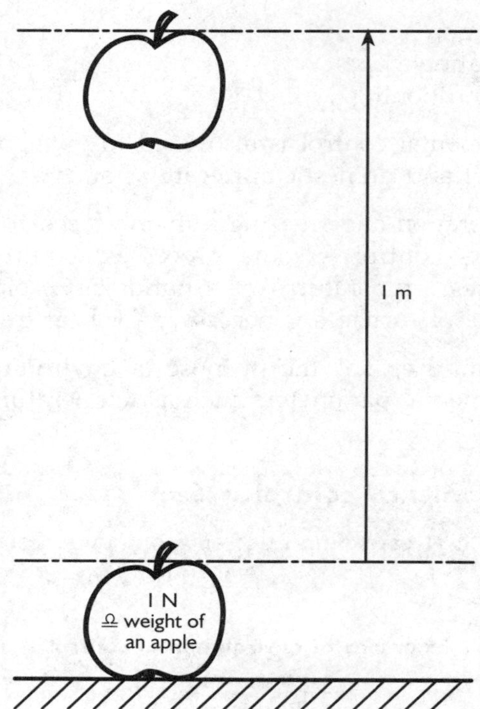

Figure 1.6 A force of 1 newton raised through 1 metre

Power introduces the time factor. If it takes 1 second (1 s) to move the force of 1 N by 1 m, then this is defined as 1 watt (1 W). So 1 W = 1 N m/s = 1 J/s.

In electrical power, the more usual units are the kilowatt (kW), megawatt (MW) and gigawatt (GW) where:

- 1 kW = 1000 W = 10^3 W
- 1 MW = 1000 kW = 1 000 000 W = 10^6 W
- 1 GW = 1000 MW = 1 000 000 000 W = 10^9 W.

Electrical equipment is normally rated in kW, and output from power stations is given in MW.

Case study

Power generation

Modern power stations typically have an output of 600 MW, which is enough power to light up 6 million 100-watt light bulbs! In the year 1993–94, the peak demand for power in England and Wales was 49.4 GW – or, to put it another way, the output from 82 power stations each of 600 MW. Although the electricity companies supply us with power, it is the total amount of energy supplied that is charged for. The charging unit is the kilowatt hour (kW h). This is 1 kW of power supplied for 1 h. Therefore:

1 kW h = (1000 × 60 × 60) J = 3 600 000 J = 3.6 MJ (megajoules).

This is a typical example of an energy calculation:

- A 10-kW electric motor is run at full power for 24 hours. What will be the cost of electricity if it is charged at 5 p/kW h?
- Electricity consumption = power × time = 10 kW × 24 h = 240 kW h
- Cost = consumption × cost/kW h = £(240 kW h × £5/100 kW h) = £12.

The purpose of power generation and transmission systems is to:

- Produce power
- Transmit it to where it is needed.

Table 1.6 gives some examples of power generation and transmission systems.

Table 1.6 Power generation and transmission systems

System	Purpose	Function
Nuclear power station	Generation of electrical power	Energy conversions: nuclear to thermal (steam) to kinetic (turbine) to electrical (alternator)
Local distribution substation	Reduction of electricity voltage to domestic levels	Induction of reduced electromagnetic force (EMF) in transformer

Activity 1.6

Investigate three different power generation and/or transmission systems. In each case state:

a its purpose
b its function.

Progress check

1 Define what is meant by a system.
2 State the two key characteristics of all systems.
3 Explain each of the following systems terms:
   a inputs
   b outputs
   c environment
   d boundary
   e feedback.
4 For a system define:
   a purpose
   b function.
5 An electrically powered robot has an average power consumption of 25 kW. What would be the annual cost of electricity (at 5 pence per kW h) to run it, if it works 20 hours per day, 7 days a week and 50 weeks per year?

The three performance criteria – inputs and outputs of engineering systems; the representation of systems using block diagrams; and the implications of inputs and outputs for systems – are specified separately in the course element 2.1. However, it makes more sense not to separate them, since they are each an essential part of the analysis when investigating systems. They will be considered together for each type of system.

Inputs and outputs have already been defined. What are considered as inputs and outputs will vary according to where the system boundary is drawn.

Case study

Cheese-making equipment

A portable electrically powered hand grinder is used to smooth out welded joints in a factory making cheese-processing equipment.

Considering just the grinder as a system, the input is:

- electrical power.

The outputs are:

- kinetic energy of the rotating grinding wheel
- heat.

This is shown in a block diagram in Figure 1.7.

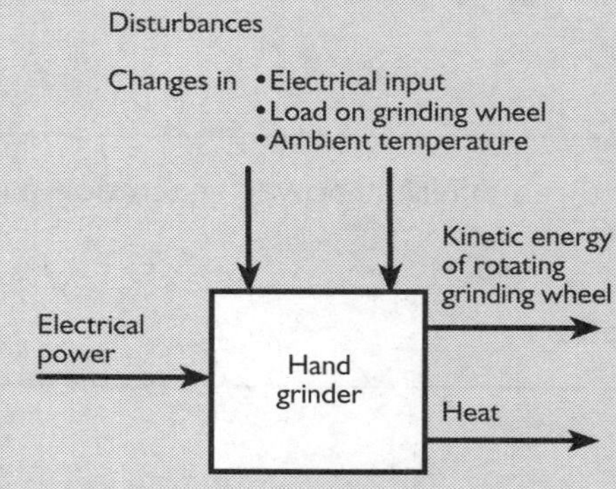

Figure 1.7 Block diagram of a hand grinder

The system boundary can be enlarged so that the grinder becomes part of the manufacturing system for the cheese-making equipment. The inputs are now:

- materials, energy and money.

The outputs are:

- finished parts and products, waste and profits.

Figure 1.8 shows this as a block diagram.

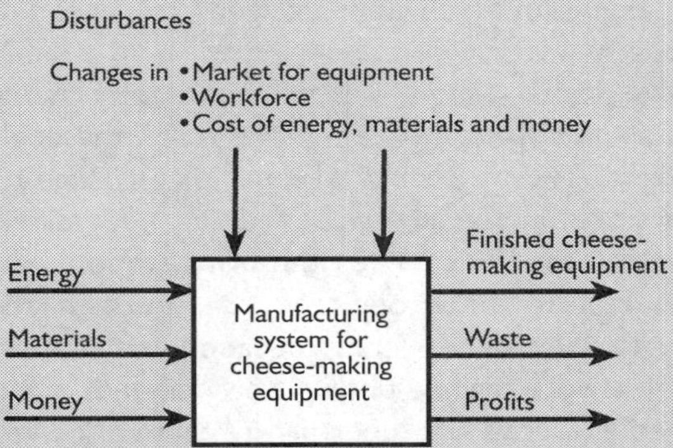

Figure 1.8 Block diagram of a manufacturing system for cheese-making equipment

We can enlarge the boundary even more to include the power station where the electrical energy is generated, cheese making, transport and food retailing. The inputs will now be:

- fuel, milk and money.

Giving outputs of:

- Cheese on a shop or supermarket shelf, by-products, waste and possible pollution of air and water.

Figure 1.9 shows this as a block diagram.

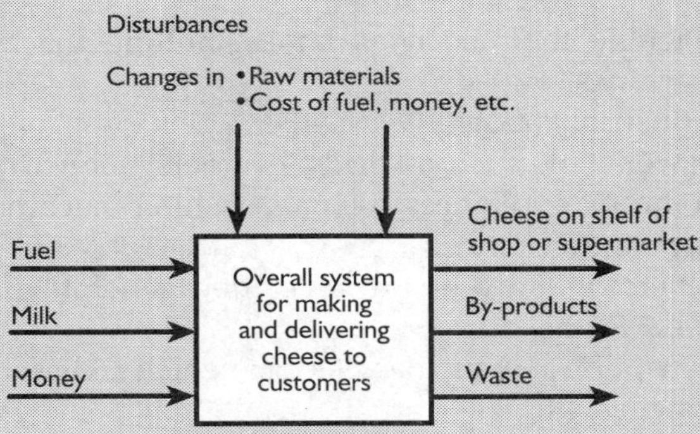

Figure 1.9 Block diagram of electricity generation, cheese-making, delivering and retailing

Relationship of output to input

It is important to realise that inputs and outputs are directly related. The system modifies the inputs to produce the desired output. Hence, a change of input will cause a change of output and could mean that the system will also have to change.

Case study

A car engine

The engine of a vehicle is designed to run on a particular grade of fuel. If the wrong fuel is put in, the engine will give a different output. Power will probably be reduced and emissions will increase. This might be acceptable temporarily, but, if the situation became long term, the engine would have to be modified or changed. Thus, a change of input may change both the output and the system. This situation actually happened at a new supermarket filling station in Yeovil, Somerset, in 1995. When the filling station was built, the petrol pumps were connected to the wrong tanks. So people filled up with leaded petrol when they thought they were getting unleaded and vice versa. Most people realised their engines were malfunctioning but were not sure why. The problem was finally traced and cost the supermarket quite a lot in compensation. Today's car engines are designed very specifically for one type of fuel.

Activity 1.7

Investigate an engineering system. Find the inputs and outputs of the system for three different system boundaries, progressively drawing the boundaries wider.

Electromechanical systems

These systems cover the full range of applications where electricity is used to provide power, or where a mechanical system is used to generate electrical power. Typical examples include:

- machine tools such as lathes, milling machines, machining centres, drilling machines, etc.
- industrial robots
- power tools such as drills, grinders, screwdrivers, saws, etc.
- domestic appliances such as washing machines, food mixers, vacuum cleaners, lawn mowers, etc.
- generators such as power station generators, stand-by generators, alternators, wind farms, etc.
- transportation such as electric vehicles, diesel electric trains, conveyer systems, winches, etc.
- heavy plant and construction equipment such as cranes, lifts, conveyers, etc.
- fluidic systems such as compressors, pumps, wave power, etc.

Outputs and inputs of electromechanical systems

Considering the general case of an electromechanical system, the inputs can be:

- electrical energy
- raw materials
- set points for distance, velocity, power, pressure, flow rate, etc.
- disturbances, e.g. variations in electrical input, temperature, barometric pressure, load, etc.

The outputs can be:

- kinetic energy (used to do work)
- heat (losses in motor and friction)
- change of position
- change of velocity
- processed materials and finished components
- waste materials.

This general model of an electromechanical system is represented by the block diagram shown in Figure 1.10.

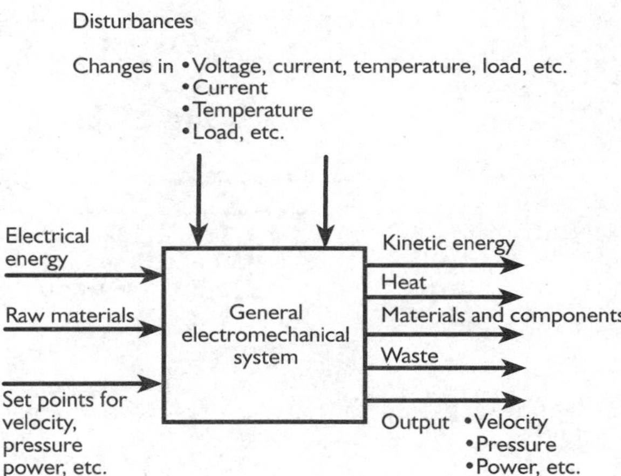

Figure 1.10 Block diagram of a general model of an electromechanical system

Implications of inputs and outputs for electromechanical systems

Some implications, for the system, of the inputs and outputs are:

- Source and type of electricity available. The system may have to be designed to operate from various alternating current (AC) voltages, direct current supply (DC) or batteries.
- Safety – the main hazards are likely to be: risk of electrocution – all systems will need to be fully insulated and earthed, with safety trips on all circuits; and from rotating and reciprocating parts – all moving parts should be balanced and guarded.

- Heat – this will usually be an extraneous output and can be usefully used or dissipated by heat sinks, radiators, fans, etc.
- Waste outputs will have to be disposed of safely without causing pollution.

Case study

CNC lathe

A specific electromechanical system will now be considered. From the large variety of machine tools, a good example is a CNC (computer numerical control) lathe. Note that CNC lathes are sometimes called turning centres. Figure 1.11 shows a CNC lathe in diagrammatic form.

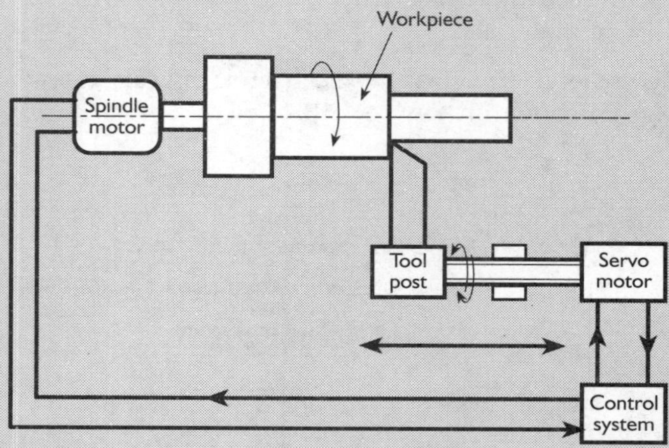

Figure 1.11 CNC lathe (turning centre)

The inputs to the system are:

- raw materials
- electrical energy to motors for main spindle and axes drives
- set points for spindle and axes speeds, spindle rotational position and X and Z linear position
- disturbances such as variations in electrical energy, material, tools, temperature, slide wear and friction, etc.

The outputs of the system are:

- main spindle rotation at set speed
- axes movement to set position at set speed and angle
- finished part (profiled by cutting tool being moved along X and Z axes and material being rotated by main spindle)
- waste material removed from workpiece
- heat (from motors and cutting process).

The system, with its inputs and outputs, is represented by the block diagram in Figure 1.12.

Some implications of the inputs and outputs for the system are:

- Different materials require different system conditions. For example, light alloys can be machined at much higher speeds with a lower power consumption than tougher materials such as stainless steel. Therefore, output power will vary, causing changes in the input of electrical energy.

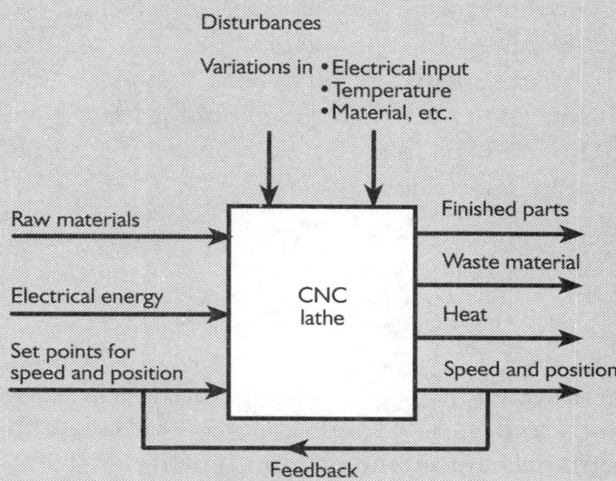

Figure 1.12 Block diagram of CNC lathe system

- The high-velocity output of work rotation and chips of cut material means that the system will have to be guarded. An interlocking device will normally be built in, so that the lathe cannot be operated with the guard open.
- If high accuracy is required, outputs will have to be compared with inputs by a closed-loop negative feedback system. This then takes account of system disturbances. An open-loop system would not do this.
- The high volume of waste output means that the system must have an efficient means of removing it.

Activity 1.8

Investigate a typical electromechanical system and describe:

a the inputs to the system
b the outputs from the system
c the implications for the system of the inputs and outputs.

Represent the inputs and outputs of the system using a block diagram

Progress check

1 Define and describe what is meant by an electromechanical system.
2 Give an example of an electromechanical system from each of these fields:
   a machine tools
   b industrial robots
   c power tools
   d generators
   e transportation
   f heavy plant and construction
   g fluidic systems.
3 For the general case of electromechanical systems describe:
   a all inputs to the system
   b all outputs from the system.
4 Describe two possible implications for the system of the inputs and outputs.
5 Draw a block diagram, with inputs and outputs, of the general model of an electromechanical system.

Fluidic systems

These systems cover anything to do with the movement, processing or use of liquids and gases. There is some overlap with other systems, particularly with electromechanical and thermodynamic systems. Among systems that can be classified as fluidic are:

- Hydraulic control and actuation systems including control and actuation of earth-moving equipment such as JCBs; control and actuation of power presses for manufacturing processes such as deep drawing; hydraulic systems for vehicle brakes, clutches and automatic transmissions.
- Pneumatic control and actuation systems including air tools, such as drills and grinders; air-controlled and -operated devices such as pick and place robots, grippers, clamps, etc.
- Water supply and treatment, which covers the storage, treatment and pumping of water and sewage.
- Hydroelectric systems. This covers the way water is stored, transported and used to power turbines to generate electricity.

Inputs and outputs of fluidic systems

For the general model of a fluidic system, the inputs can be:

- energy, e.g. kinetic, potential, electrical or chemical
- fluid
- disturbances such as impurities, temperature changes, energy changes, etc.
- set points such as pressure, flow rate.

The outputs can be:

- pressure and force
- energy
- linear and rotary motion
- raw materials, e.g. pure water.

The block diagram of this general model is shown in Figure 1.13.

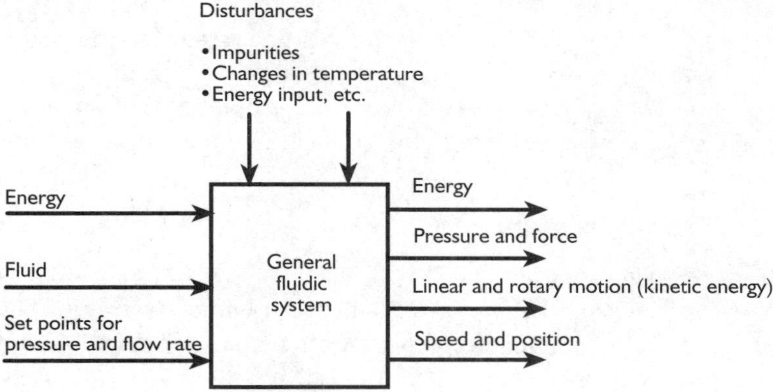

Figure 1.13 Block diagram of a general model of a fluidic system

Implications of inputs and outputs for fluidic systems

Some implications, for the system, of the inputs and outputs are:

- Cavitation – this is likely to occur in fluidic systems and is erosion of materials by fluid or gas flow over a surface. It has to be taken care of in a system by careful design and the use of correct materials and maintenance.
- Environmental effects such as noise, heat, pollution, consumption of resources, etc.
- Fluids may need to be cleaned and filtered for safe and satisfactory operation.

Case study

A hydraulic excavator

A good specific case to examine is a hydraulic excavator, the well-known JCB shown in Figure 1.14 being a good example.

Figure 1.14 JCB excavator

Only the fluidic system will be considered; the internal combustion engine is a separate thermodynamic system. The inputs can be:

- energy from the thermodynamic system for the pump
- low-pressure fluid into the pump
- set points for position from the operator's manual controls
- disturbances from impurities in hydraulic fluid, temperature changes, friction, etc.

The outputs are:

- high-pressure fluid
- linear or rotary movement from kinetic energy of fluid
- force (combined with motion enables work to be done, e.g. digging a trench)
- heat (from pump and friction).

Figure 1.15 shows the system as a block diagram.

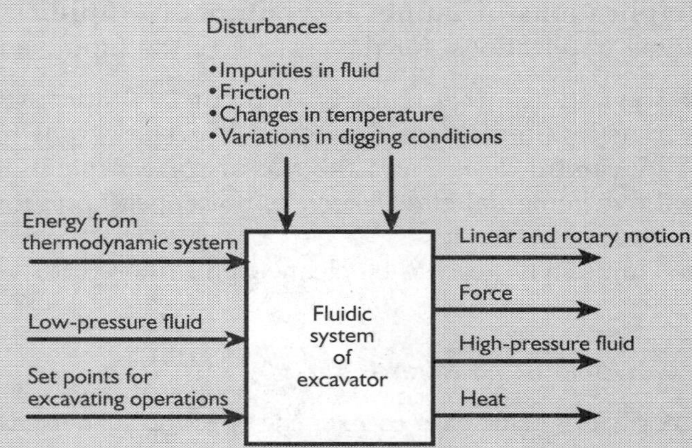

Figure 1.15 Block diagram of the fluidic system of an excavator

Some implications, for the system, of the inputs and outputs are:

- The fluid input must be clean and free from particulates. Any particles above 5 microns can cause problems with the valves. Micronic filtration is essential for satisfactory operation of the system
- Since the pump output is constant and the load demands may vary, a hydraulic accumulator is provided. This stores fluid at high pressure until it is needed
- A fail-safe system must be provided in case of any input or output failure. Thus, a non-return valve would hold the hydraulic ram in position if the supply pressure failed
- A pressure relief valve must be provided to prevent excessive output pressure
- The hydraulic pump churns up the oil, causing it to become oxidised. The oil temperature is also increased by the pump. Therefore, the temperature and condition of the oil needs to be monitored. Cooling and regular replacement may be necessary.

Activity 1.9

Investigate a typical fluidic system and describe:
a the inputs to the system
b the outputs from the system
c the implications for the system of the inputs and outputs.
Represent the inputs and outputs of the system using a block diagram.

Progress check

1 Define and describe what is meant by a fluidic system.
2 Give an example of a fluidic system from each of the following fields:
  a hydraulic actuation and/or control
  b pneumatic actuation and/or control
  c water supply
  d hydroelectricity.
3 For the general case of fluidic systems give:
  a all inputs
  b all outputs.
4 For the general case of fluidic systems state two implications, for the system, of the inputs and outputs.
5 Draw a block diagram model of a general fluidic system, showing the inputs and outputs.

Chemical systems

A chemical system is one in which chemical or electrochemical reactions are used for useful purposes. This occurs mainly in manufacturing systems. Typical examples are:

- Production of materials including polymerisation of molecules, principally of hydrogen and carbon, to produce plastics; metal extraction from ores.
- Electrolysis including electroplating; electrochemical machining (ECM); corrosion and corrosion protection; anodising.
- Chemical etching including the production of tracks on printed circuit board (PCB); production of profiled parts in thin metal; etching of metal samples to reveal grain structure.
- Chemical reaction gluing – two-part glues, which when mixed have a chemical reaction (usually polymerisation) and solidify.
- Combustion – the burning of any fuel is a chemical reaction and so is a chemical system.

Inputs and outputs of chemical systems

For a general model of a chemical system, the inputs can be:

- elements or compounds
- catalysts – these are not part of a chemical reaction but they speed it up

- energy – thermal or electrical
- parts for processing
- disturbances, e.g. variations in temperature, energy, materials, etc.

The outputs can be:

- elements or compounds
- energy – thermal or electrical
- finished parts.

The general model of a chemical system is shown in the block diagram in Figure 1.16.

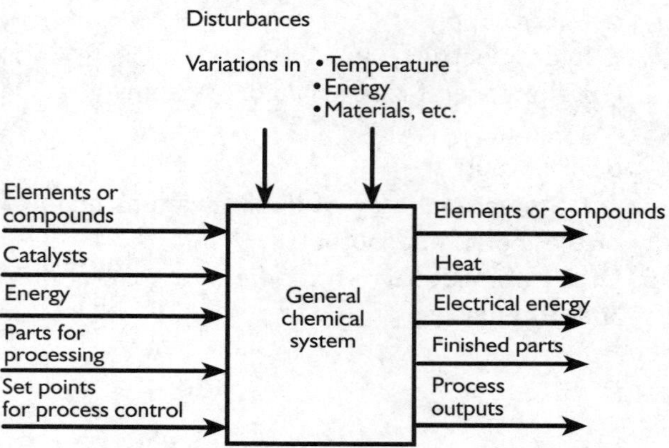

Figure 1.16 General model of a chemical system

Implications of inputs and outputs in a chemical system

Typical implications, for the system, of the inputs and outputs are:

- Some chemical reactions, classified as endothermic, require the input materials to be heated. Therefore, an energy input is also required. The production of some materials requires a large energy input. To produce 1 kg of aluminium requires about 240 MJ of energy (equivalent to 67 kW h). This is why aluminium smelters need to be near a cheap source of electricity.
- Some chemical reactions, classified as exothermic, release a lot of heat. In these cases, energy output is a by-product of the system. A typical exothermic reaction is the combustion of fuel with oxygen. For example, the combustion of 1 litre of fuel oil will generate 41 MJ of energy (equivalent to 11 kW h).
- Catalysts may have to be provided to start a chemical reaction or to keep it going efficiently. An example of this is the catalytic converter in a vehicle. Platinum is used in the exhaust system to ensure that any unburnt fuel reacts with air to provide full combustion. The platinum is non-reactive and can remain in the exhaust system for several years.

Case study

Printed circuit board

A good example of a chemical system is the etching of the circuit tracks on a printed circuit board (PCB). Figure 1.17 shows a typical PCB.

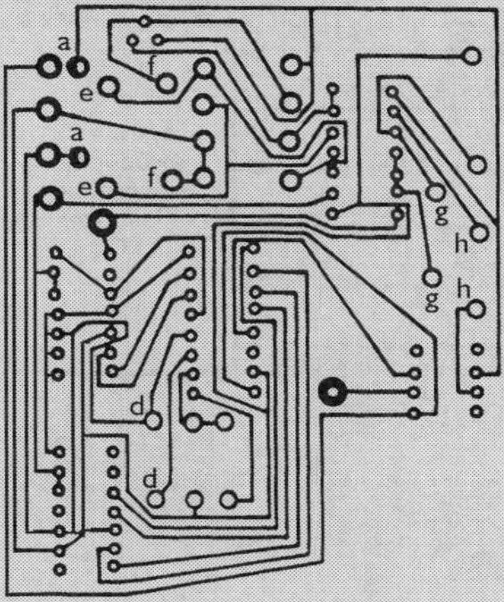

Figure 1.17 Printed circuit board with tracks

The inputs to the system are:

Photo-resistant copper-clad board that has been exposed to UV light in the areas where the copper is to be etched away

- ferric chloride etching solution
- electrical energy – for pump and heater
- low-pressure air from pump
- set points of time, temperature and air pressure
- disturbances such as changes in ambient temperature, variations in materials, etc.

The outputs from the system are:

- finished PCB with etched tracks
- waste chemicals
- heat.

These inputs and outputs are shown in the block diagram in Figure 1.18.

Some implications for the system of the inputs and outputs are:

- The system will need to be inert to the corrosive chemicals used and produced.
- Provision will have to be made for safe handling, storage and disposal of chemicals and waste.

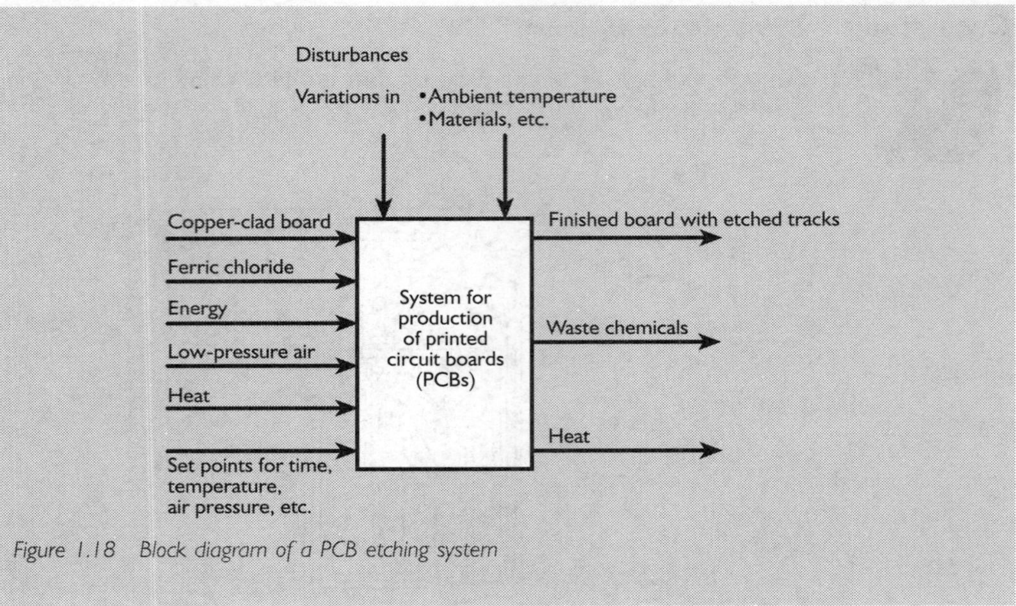

Figure 1.18 Block diagram of a PCB etching system

Activity 1.10

Investigate a typical chemical system and describe:
a the inputs to the system
b the outputs from the system
c the implications for the system of the inputs and outputs.
Represent the inputs and outputs using a block diagram model.

Progress check

1 Define and describe what is meant by a chemical system.
2 Give an example of a chemical system from each of the following fields:
   a production of materials
   b electrolysis
   c chemical etching
   d chemical reaction gluing
   e combustion.
3 For the general case of chemical systems give:
   a all inputs
   b all outputs.
4 For the general case of chemical systems give two implications, for the system, of the inputs and outputs.
5 Draw a block diagram model of general chemical systems, showing the inputs and outputs.

Information and data communication systems

These systems have already been described, in general terms, in the first part of this chapter. An information and/or data communication system processes information or data, transmits it to the recipient and reprocesses it if necessary. Some examples of these systems are:

- Radio and television – conversion of visual images and/or sound into electronic form, transmitting them and converting back to visual images and/or sound.
- Computers and computer networks – processing, display and transmission of data, visual images and sound. The network can link computers locally at one site or can be worldwide, like the Internet, using the telephone network.
- Management information systems – these are systems that inform management of the exact situation of the business. This will include information on orders, work in progress, quality and rejects, deliveries, performance of staff, costs, etc. These systems can be manual, but they are generally implemented by computers and computer networks.
- Programmable logic controllers (PLCs) – dedicated microprocessors with multiple input/output ports for use in process control.
- Telecommunications – conversion of sound into electronic data, transmission of these data and conversion back into sound. Telephone networks can transmit other types of data, including that between computers.
- Telemetry – making, transmitting and processing of measurements from remote sites such as spacecraft, nuclear power stations, etc.
- Sound, visual and computer data recording – conversion of visual images and/or sound into electronic form and copying onto media such as tape or disk.

Inputs and outputs of information and data communication systems

Considering the general model of an information and data communication system, the inputs can be:

- Visual images – input to cameras, scanners, etc.
- Sound – input to microphones
- Analogue signals – input from devices, such as transducers, measuring temperature, velocity, acceleration, flow rate, etc. An analogue signal will be in the form of a continuously varying waveform. It can be converted into a digital signal by an analogue to digital converter (ADC)
- Digital signals – input from devices such as keyboards, digital phones, digital recorders, transducers, etc. A digital signal is in the form of discrete pulses. It can be converted into an analogue signal by a digital to analogue converter (DAC)
- Electrical energy – the usual way of energising information and data communication systems
- Light including transmission in fibre optics systems and reading music compact discs (CDs) and computer CD-ROMs
- Mechanical – little used now but the original method of sound recording used sound vibrations to actuate a stylus, which cut grooves in a revolving disc. This disc was then copied and the sound reproduced from a stylus in contact with the revolving disc
- Disturbances such as 'noise', interference, etc.
- Set points such as the frequency of a particular radio or television station

The outputs can be:

- Visual images – graphics and text on screen or paper
- Sound – output from loudspeaker or headphone
- Analogue signals – signals to control processes, velocity, direction, etc.
- Digital signals – signals to control processes, velocity, direction, etc.
- Heat from the operation of electronic circuits
- Interference to other equipment from unwanted signals, e.g. computers can cause interference to radios and televisions, and mobile phones can interfere with hearing aids and may also be a radiation health risk

Figure 1.19 shows a general model of an information and data communication system.

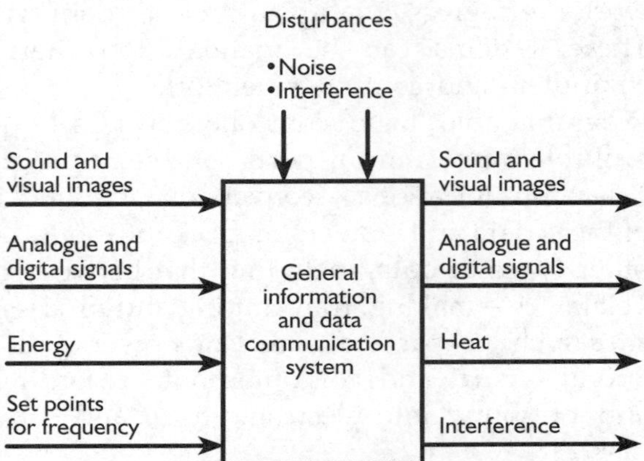

Figure 1.19 Block diagram of a general information and data communication system

Implications of inputs and outputs for information and data communication systems

Some implications for the system of the inputs and outputs are:

- Frequency of input will affect how fast the system will need to work. Input frequency will be in cycles/s (Hz) for an analogue input. For a digital input, the frequency will be in bits/s (baud). The frequency range over which a system can respond to a signal is called the bandwidth. For example, the bandwidth of an analogue telephone system, transmitting through copper wires, is 3000 Hz. It can transmit sound between 30 and 3300 Hz (3.3 kHz). However, the normal range of human speech and hearing is between 50 and 15 000 Hz (15 kHz). Therefore, the full range of the human voice is not transmitted, and a person's voice sounds different over the telephone. Music, however, is very distorted over the telephone, so a high-fidelity music system needs to have a bandwidth of around 15 kHz. But even this is small compared with the bandwidth required for television and telephone by satellite, which require a bandwidth of around 6 000 000 Hz (6 MHz) to transmit all the information at the required speed.
- Noise can distort inputs and outputs and hence affect the way the system functions. Examples of noise are (1) alternate path propagation where a transmitted signal splits into two or more paths and they reach the receiver at different times – this can cause sound echoes or 'ghosting' of screen images;

(2) interference from nearby electrical equipment; (3) electrical storms; (4) errors caused by malfunctioning circuits; (5) thermal and other radiation; (6) natural noise caused by solar radiation and atmospheric electrical generation.

- The implications of noise for systems is that they must be designed to take account of it. Some ways of doing this are by: (1) careful circuit design, including noise filters, to minimise noise – shielding cable and equipment with material impervious to signals to reduce interference; (2) using digital signals – information is transmitted by bits or pulses and so is much less prone to being distorted.
- Security of input and output may be very important. This is particularly true for financial transactions and for military communications. The system must therefore be designed to prevent unauthorised access. Techniques for system design to help ensure security are: (1) personal identification numbers for bank cards and credit cards; (2) passwords for access to a computer system; (3) coding of input (encryption) and decoding of output (decryption).
- Heat generated as a result of the operation of the system will usually have to be removed. This is to prevent an excessive temperature rise from affecting the operation of the system. Typical ways of doing this are: (1) heat sinks – these are large surface areas of a high-conductivity material such as aluminium. Components liable to generate heat are embedded in them and heat is conducted away; (2) air circulation – fans are installed to bring in cool air and remove heat. Computer systems usually have fan cooling.

Case study

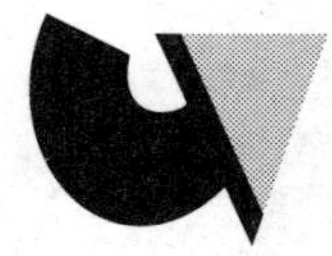

Programmable logic controller

A specific example of an information and data communication system is a programmable logic controller (PLC). These systems are widely used for process control in manufacturing industry. Applications include controlling:

- transfer lines in car production and assembly
- petrochemical plants such as oil refineries
- food plants for mass producing chocolates, biscuits, etc.
- electronic and consumer goods production.

A typical PLC system is shown in Figure 1.20.

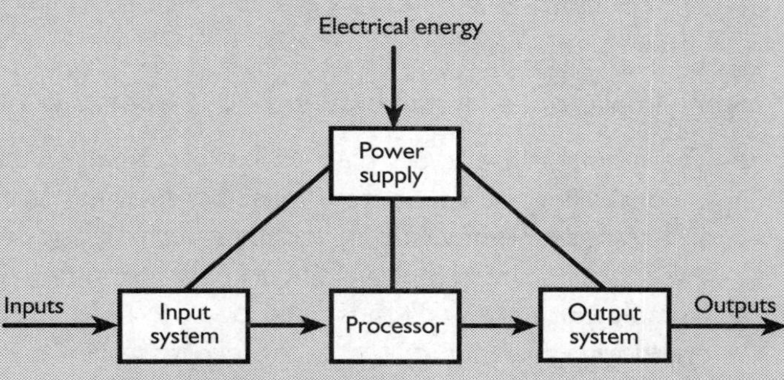

Figure 1.20 Programmable logic controller system

Inputs to the system will be:

- Analogue and digital signals from the process by sensors giving data on the process such as temperature, time, sequence, velocity, pressure, flow rate, etc.
- Energy – electrical energy runs the system
- Set points for variables such as temperature, time, sequence, velocity, position, pressure, flow rate, etc.
- Disturbances such as changes in ambient temperature and pressure, etc.

Outputs from the system include:

- Analogue and digital signals to control processes and achieve set points such as correct temperature, time, sequence, velocity, position, pressure, flow rate, etc.
- Heat – generated by electronic circuits
- On-line process data for operator and management information – continuous information available on screen to assist overall control
- Recorded data in disk and printed form – to give information if required at a later date for analysis.

The block diagram for a system model of a PLC system is shown in Figure 1.21.

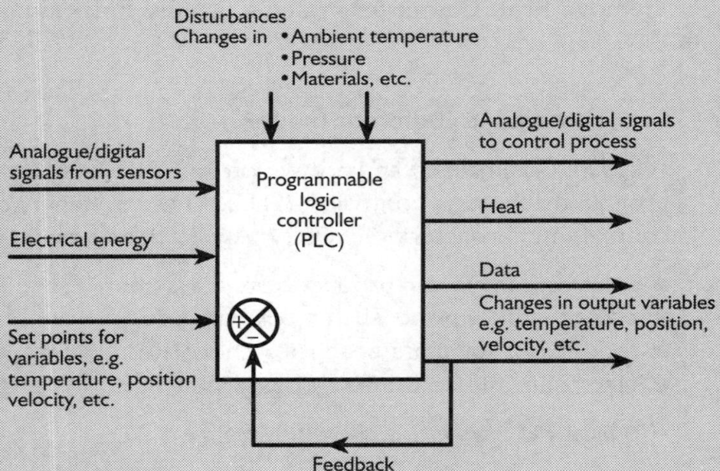

Figure 1.21 Block diagram of a PLC system

Some implications for the system of the inputs and outputs are:

- The inputs and outputs have to be interfaced to the real world that the PLC is controlling. So the system must be designed to handle inputs from sensors and transducers measuring variables such as temperature, pressure, position of a component or a tool, velocity, flow rate, time.
- Suitable output signals must also be supplied to actuators such as motors, cylinders, valves, pumps, etc. and process plant such as furnaces, mixing tanks, etc. As can be seen from the block diagram in Figure 1.21, the system normally needs to be closed loop so that process variables are continuously

monitored and kept to the set point. The system also needs to be versatile so that outputs and inputs can be analogue or digital as required.
- Since a process plant can be dangerous if processes go out of control, the system must be designed to shut down in these circumstances. The system must therefore be designed to be 'fail safe'.

The case study on page 50 in Chapter 2 also deals with the PLC, but in more specific detail.

Activity 1.11

Investigate a typical information and/or data communication system and describe:
a inputs to the system
b outputs from the system
c implications for the system of the inputs and outputs.
Represent the inputs and outputs using a block diagram model.

Progress check

1 Define and describe what is meant by an information and data communication system.
2 Give an example of an information and data communication system from each of these fields:
  a radio and television
  b computers and computer networks
  c telecommunications
  d telemetry
  e recording.
3 Explain these terms used in information and data communication systems, and give an example of each:
  a bandwidth
  b noise.
4 What is the bandwidth of a system that can transmit frequencies between 500 Hz and 5500 Hz?
5 What is the bandwidth of a typical television transmitting system?
6 For a general information and data communication system give:
  a all inputs
  b all outputs.
7 For a general information and data communication system, give two implications of the inputs and outputs.
8 Draw a block diagram model of a general information and data communication system and put in all the inputs and outputs.

Thermodynamic systems

The word thermodynamics is derived from the Greek *therme* meaning heat and *dynamis* meaning force. So a thermodynamic system is one in which heat is used to generate a force to do some work, or in which work is done to produce heat. There are two basic laws that are fundamental to an understanding of thermodynamics:

- The first law of thermodynamics is equivalent to the principle of conservation of energy. That is to say, energy cannot be created or destroyed but only changed in form. This means that heat and mechanical work are just different forms of energy, so that heat can produce work and work can produce heat. However, all the energy produced by heating is not available for doing work, and some heat is just lost. Theoretically, about two-thirds of heat can be used for useful work. However, in practice, other losses occur and most systems can only convert about one-third of the heat output to useful work. The efficiency of the system is defined as:

$$\text{Efficiency } (\eta) = \frac{\text{Output as useful work}}{\text{Input energy}} \times 100\%$$

- The second law of thermodynamics says that heat can only normally flow from a hot source to a cold sink. So a heat exchanger or a radiator works by transferring heat from a hotter to a colder area. To make heat flow the other way requires the use of a heat pump, of which the refrigerator (Figure 1.25) is a good example.

The principle of these two laws is illustrated in Figure 1.22. This shows a heat engine, which is the most common thermodynamic system.

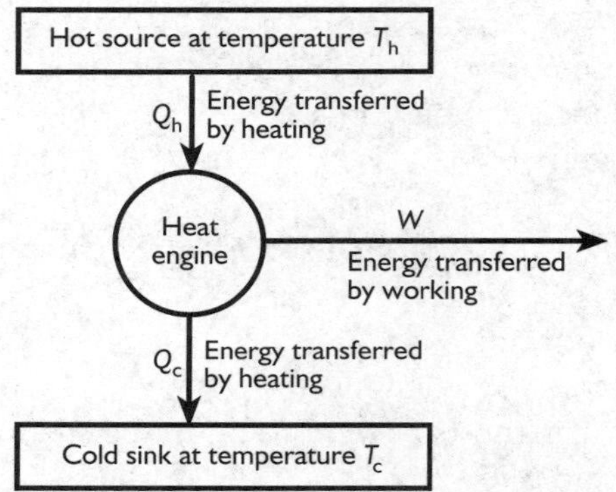

Figure 1.22 A heat engine

The principles illustrated in Figure 1.22 enable the efficiency of a heat engine to be determined:

Efficiency (η) = Energy output transferred by working

Energy input transferred by heating = W/Q_h (1.1)

By the principle of the first law of thermodynamics:

$$Q_h = Q_c + W \qquad (1.2)$$

and rearranging:

$$W = Q_h - Q_c.$$

Substituting this value for W in (1.1) gives:

$$\text{Efficiency } (\eta) = \frac{Q_h - Q_c}{Q_h}$$

and rearranging:

$$\eta = 1 - [Q_c/Q_h].$$

Q_h is proportional to the temperature of the hot source (T_h) in kelvin (K) and Q_c is proportional to the temperature of the cold sink (T_c) in K, where temperature T (K) = temperature (°C) + 273. Therefore:

$$\eta = 1 - (T_c/T_h).$$

For example, steam is supplied to a turbine at 567°C and is returned to a river at 7°C. The maximum theoretical efficiency

$$\eta = 1 - [(7 + 273)/(567 + 273)]$$

$$= 1 - [280/840]$$

$$= 1 - 0.333$$

$$= 0.666 = 0.67 \times 100\% = 67\%.$$

In practice, the actual efficiency would be lower because of other losses.

Two examples of specific thermodynamic systems are:

- The internal combustion engine in which chemical energy is supplied in the form of diesel fuel or petrol. This fuel is combined with air in the combustion chamber where the chemical reaction of combustion produces heat. This heat expands the gas in the cylinder, which provides a force to move the piston. The piston rotates the crankshaft, which is then able to do work, e.g. turn wheels, power a generator, turn a winch, etc. So heat has been produced to do mechanical work. This is shown in Figure 1.23.
- The refrigerator in which a refrigerant that boils at a low temperature is passed, as a vapour, through the cold interior, where it absorbs heat and keeps the temperature at the correct level. This hotter vapour is now compressed, using an electrically powered compressor, to form a liquid. This high-pressure liquid is passed into a heat exchanger at the back of the refrigerator where it gives up the heat to the atmosphere. The compressed fluid is then passed through an expansion valve, where it changes to the colder vapour. So work has been done by the compressor to produce the heat that has been pumped from the inside of the refrigerator to the outside space. This is shown in Figure 1.24.

The main types of thermodynamic systems are:

- Boilers in which the combustion of fuel, or nuclear reaction, is used to generate heat for hot water or hot air for central heating, etc.; steam to power steam turbines for electricity generation, ship propulsion, etc.
- Furnaces and ovens in which the combustion of fuel, or the heating effect of an electric current, is used to generate heat for heating materials for forming and heat treatment; cooking and food processing; pasteurisation and sterilisation.
- Internal combustion engines (as described in the previous example) in which fuel is combusted within the engine to do mechanical work. Types of internal

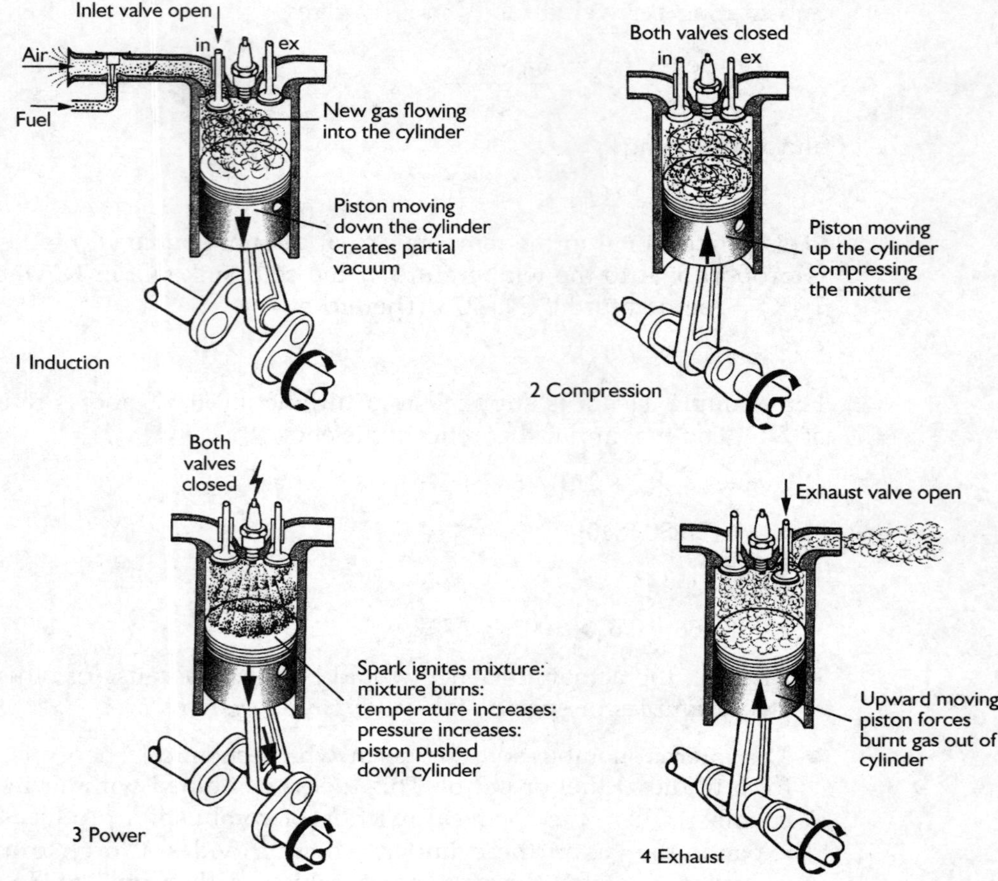

Figure 1.23 Principles of the internal combustion engine

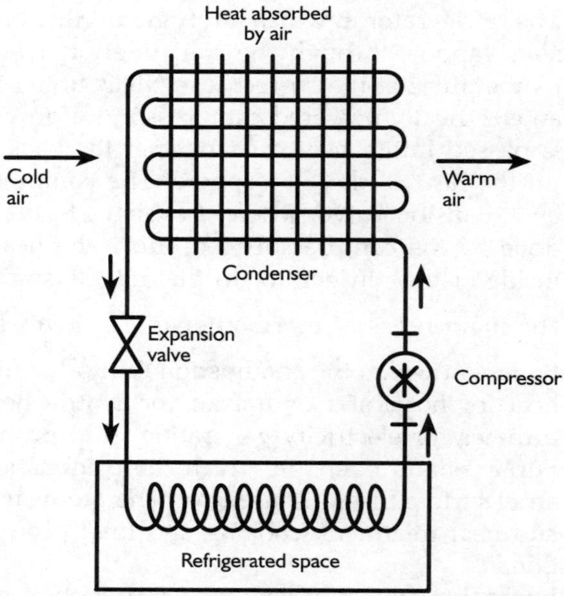

Figure 1.24 Principles of a refrigeration system

combustion engines are reciprocating petrol, gas or diesel engines. Diesel engines are sometimes referred to as compression ignition engines. This is because the fuel ignites owing to the very high compression ratio. There is no need for a spark as with petrol and gas engines. Rotary engines operate by expanding gas turning a rotor rather than moving a piston linearly.

- Gas turbines – ambient air is taken in and compressed; fuel is injected in and combustion occurs. The hot gases produced are then passed through a turbine, which drives the compressor and can also drive an output shaft. The hot gases exit the turbine and can be used to provide thrust. Typical applications are jet engines for aircraft – these can be pure jet, in which the thrust is used for propulsion, or turboprop, in which the turbine rotation is used to turn a propeller; electricity generation. The combined cycle gas turbine (CCGT) is about the most efficient way of generating electricity. As well as using the gas turbine rotation to drive a generator, the waste heat emitted is used to produce steam to drive a steam turbine.
- Refrigeration – refrigerators and deep freezes function by extracting heat from the interior space to keep it cold (see description above).
- Air conditioning – combined heating and refrigeration systems. Air is filtered and circulated within buildings, aircraft and vehicles at the correct temperature.
- Solar energy systems – solar cells use the radiation heat energy of the sun to produce electricity; solar energy is collected by focused reflectors or black-surfaced collectors. This can then be used to heat water or for space heating; buildings can be designed to act as passive collectors of solar energy through south-facing windows, etc.

Inputs and outputs of thermodynamic systems

For a general model of a thermodynamic system, the inputs can be:

- Energy, e.g. chemical energy in fuels such as coal, oil, etc.; electrical energy to power refrigeration systems, etc.; solar energy
- Water – input raw material for heat distribution, steam generation or for cooling
- Gas or vapour including air to provide oxygen for combustion processes; refrigerant vapour in refrigeration and air-conditioning systems; carbon dioxide used as a heat exchanger in some nuclear power plants
- Disturbances – variations in ambient temperature, atmospheric pressure, air quality, quality of fuel supplied, etc.
- Set points – desired outputs of temperature, pressure, power, velocity, etc.

The outputs can be:

- Energy – thermal energy that can be converted to kinetic energy to do work such as the rotation of a shaft for propelling vehicles, generating electricity, etc. or can increase the temperature of materials and components, buildings, etc.
- Waste heat – heat that cannot usefully be used and has to be returned to the environment, e.g. heat going up a boiler chimney, hot water from a power station being returned to the river or sea, heat emitted from a gas turbine, internal combustion engine, etc.
- Waste materials, e.g. solid or liquid residues from fuel burning, disposal of refrigerants, etc.; gases released into the atmosphere, such as the products

of combustion including carbon dioxide, carbon monoxide, sulphur dioxide, nitrous oxide, etc.; solids released into the atmosphere such as solid particulates from power stations, diesel engines, etc.

- Sound (noise) – most combustion processes generate extraneous sound.

The general model of a thermodynamic system with inputs and outputs can be represented by the block diagram shown in Figure 1.25.

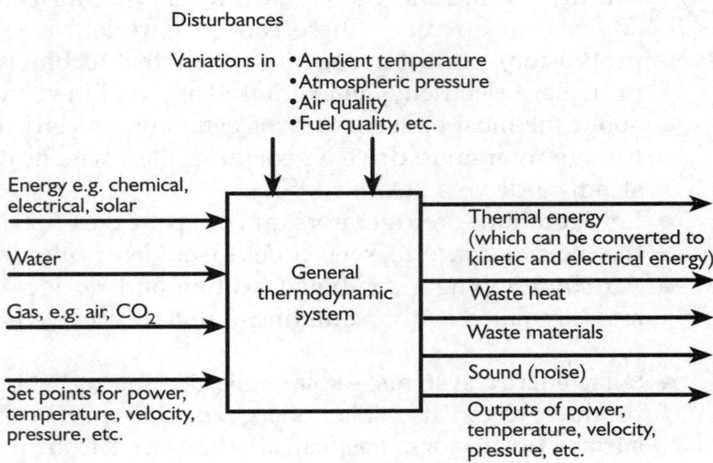

Figure 1.25 Thermodynamic system block diagram

Implications of inputs and outputs for thermodynamic systems

Some implications for the system of the inputs and outputs are:

- The chemical system of combustion generates very high temperatures and pressures and often creates toxic and corrosive by-products. This means that any materials used must be capable of resisting heat and chemical attack, engines must be designed to resist stresses caused by high pressures, pollution must be minimised by the use of catalytic converters, which improve the efficiency of combustion, filters and precipitators. Precipitators are used in power station flues to remove sulphur dioxide, which causes acid rain. The gaseous sulphur dioxide is precipitated to a solid and removed. It is a useful by-product for the manufacture of sulphuric acid.
- Heat produced which is not usefully used must be dissipated, e.g. by radiation, as in a vehicle radiator or air cooling fins, gas discharge to the atmosphere as in a gas turbine, steam from a turbine is passed into a condenser, which turns it into hot water; this is then cooled in a cooling tower or discharged to the sea or a river.
- Excess sound (noise) produced must be dealt with by exhaust silencers for internal combustion engines, designs to minimise sound, e.g. turbofan jet engines, multivalve and fully balanced internal combustion engines, etc., soundproofing, enclosure and provision of ear protectors.

Case study

Combined cycle gas turbine (CCGT)

A good specific example to consider is the combined cycle gas turbine (CCGT), which is now (1997) being brought into service extensively to provide the energy to generate electricity. This is shown in schematic form in Figure 1.26.

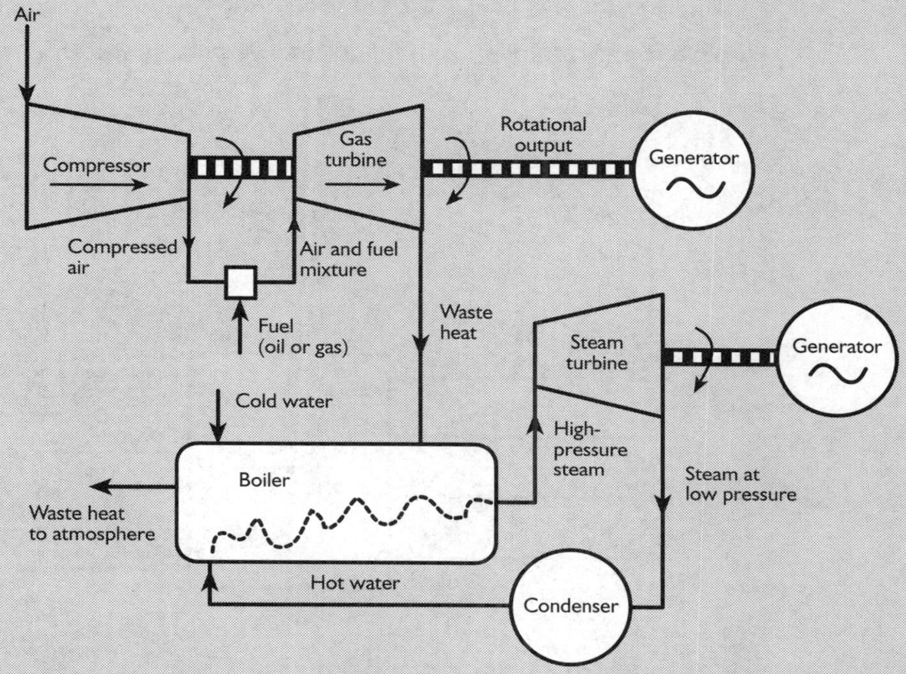

Figure 1.26 Schematic diagram of a combined cycle gas turbine plant

The inputs to the system are:

- fuel – oil or natural gas
- air – mixed with fuel for combustion, after it has been compressed
- water – input to boiler for steam generation
- set points – rotational speed of compressor and gas turbine; rotational speed of steam turbine
- disturbances – variations in calorific value of fuel, temperature of input air and water, output load

The outputs from the system are:

- Energy – thermal energy in the form of the hot gas and steam, which is converted to kinetic energy by the rotation of a turbine. This is converted to electrical energy by the rotation of the generator
- Waste heat – the condensed steam from the steam turbine is returned to the environment as hot water
- Waste materials – there will be some emissions, but the use of natural gas and low-sulphur oil minimises these. A CCGT only emits around 0.35 kg of carbon dioxide per kilowatt hour compared with around 0.8 kg for a

coal-fired power station. Also, compared with a coal-fired station, there are no solid residues

- Sound (noise) – turbines are noisy and can produce sound levels up to 120 decibels, which compares with the 60 decibels of normal traffic noise. However, because decibels are measured on a logarithmic scale, the turbine is a million times louder than the traffic noise!

A block diagram model, with inputs and outputs, is shown in Figure 1.27.

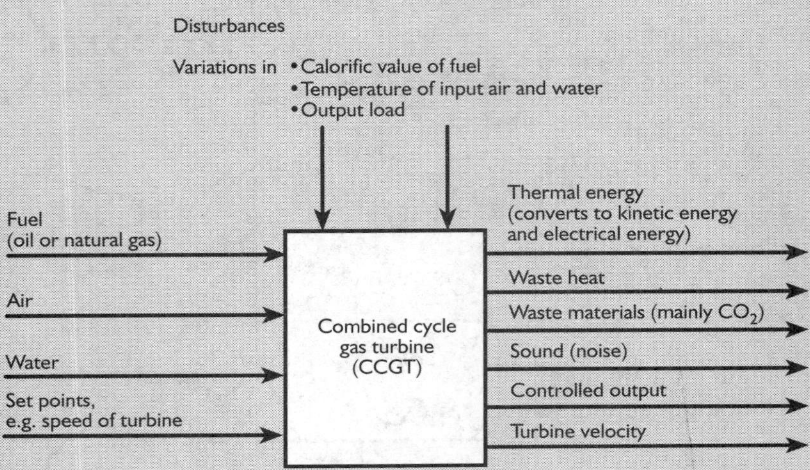

Figure 1.27 Block diagram of a CCGT system

Some implications for the system of the inputs and outputs are:

- High temperatures – temperatures in the gas turbine are up to 1100 K and in the steam turbine up to 550 K. This means that the materials used must be resistant to heat and, in the case of the gas turbine, the products of combustion. Suitable materials include nickel alloys, titanium alloys and ceramics.
- High pressures – the pressures in gas turbines are up to 25 atmospheres (2.8 MPa) and in steam turbines up to 150 atmospheres (17 MPa). Enclosures and components have to be designed to withstand this pressure.
- High rotational velocities – with rotational velocities of several thousand revolutions per minute, the turbines have to have all blades individually balanced, in addition to the whole assembly being balanced. Bearing design and lubrication is also very important.
- Waste heat – although the waste heat of the gas turbine is used, there will still be the waste heat coming out of the steam turbine. Cooling water will have to be used to remove this.
- Loud noise – in view of the noise levels being up to 120 decibels, the turbines and the turbine halls will need to be soundproofed. Operators will normally be expected to wear ear protection.

Activity 1.12

Investigate a typical thermodynamic system and describe:
a the inputs to the system
b the outputs from the system
c the implications for the system of the inputs and outputs.
Represent the inputs and outputs using a block diagram model.

Progress check

1 State the first two laws of thermodynamics.
2 How is thermal efficiency defined?
3 What is the thermal efficiency of a system that supplies 500 MJ of thermal energy to give a useful work output of 150 MJ?
4 Calculate the maximum theoretical efficiency of a heat engine that is supplied with thermal energy from a hot source at 500°C and returns it to the environment cold sink at 10°C.
5 Give an example of a thermodynamic system from each of these fields:
   a boilers and steam engines/turbines
   b furnaces and ovens
   c internal combustion engines
   d gas turbines
   e refrigeration
   f solar energy.
5 Give all the inputs and outputs for a general thermodynamic system.
6 What are the implications, of the inputs and outputs, for a general thermodynamic system?
7 Draw a block diagram model for a general thermodynamic system, showing all the inputs and outputs.

Activity 1.13

Generation of electricity requires these systems:

- Chemical: production of all the materials required; combustion of fuels
- Thermodynamic: boilers; steam turbines; gas turbines
- Electromechanical: generating sets
- Fluidic: water-handling systems; control systems for valve operation, etc.
- Information and data communication: process control information; links to national and international power networks; management information systems.
   a Identify the inputs and outputs for each subsystem and for the complete system.
   b Draw a block diagram showing how the subsystems interface. Put all the relevant inputs and outputs on your diagram.
   c State the implications for the system of the inputs and outputs.

Assignment 1

This assignment provides evidence for:
Element 2.1: Investigate engineering systems in terms of their inputs and outputs
and the following key skills:
Communication: 3.2, 3.3, 3.4
Information Technology: 3.1, 3.2. 3.3

Your tasks

1 Investigate an engineering system in each of the categories:
   a electromechanical
   b fluidic
   c chemical
   d information and data communication
   e thermodynamics.
2 For each system produce a report which:
   a describes the purpose of the system
   b identifies and describes the inputs to the system
   c identifies and describes the outputs from the system
   d explains the implications of the inputs and outputs for the system.

Block diagram models for each system should be included showing inputs and outputs. The report should be written in a clear and understandable style and word processed, which can also give core skills credits in communication and information technology.

Chapter 2: Investigate the operation of engineering systems

This chapter covers:
Element 2.2: Investigate the operation of engineering systems.

. . . and is divided into the following sections:

- Components, function and inter-relationship of engineering systems and subsystems
- System control strategies and techniques
- The evaluation of engineering systems.

Chapter 1 has introduced the basic concepts of engineering systems. This chapter will be less descriptive and more analytical. This means a concentration on:

- Relationships between systems and subsystems
- Systems control strategies and techniques
- Measurement and evaluation of performance of systems.

Relationships between systems and subsystems

There is a well-known piece of humorous verse by Augustus de Morgan that goes:

'Big fleas have little fleas upon their backs to bite 'em
And little fleas have lesser fleas and so ad infinitum.'

As with fleas, so with systems. Within systems, there can be subsystems and within each subsystem further subsystems can exist.

For example, a motor vehicle is:

- a system for transporting goods and people
- the engine is the motive power subsystem of the vehicle
- the fuel injection system is the energy input subsystem of the engine
- the injection timing system is a subsystem of the fuel injection system.

In block diagram terms, the relationship between systems and subsystems can be represented by having one inside the boundary of another, as shown in Figure 2.1.

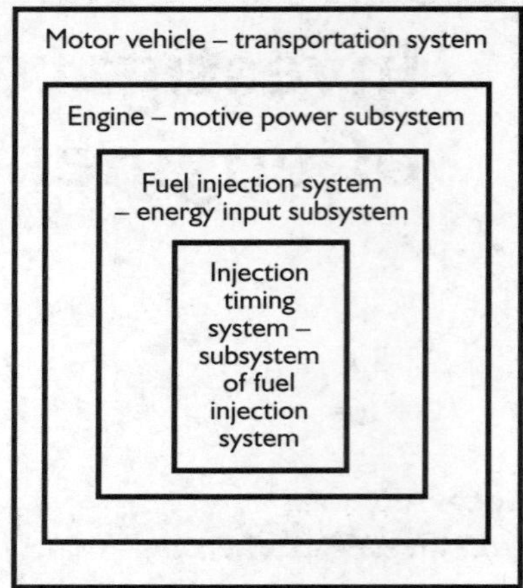

Figure 2.1 Vehicle systems, subsystems and system boundaries

An alternative way of looking at it is that each subsystem is an input to the next system up the hierarchy. This is shown in Figure 2.2.

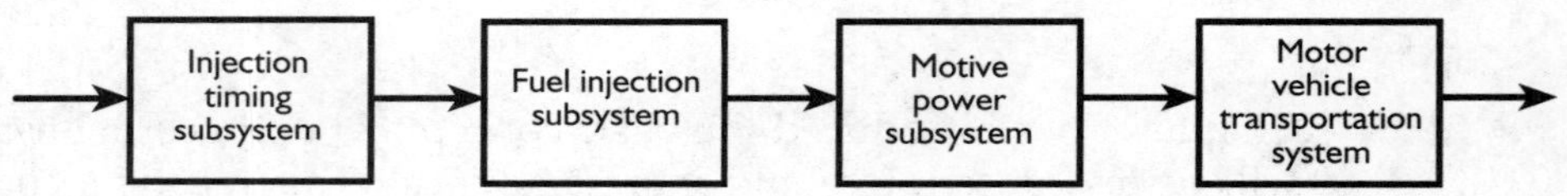

Figure 2.2 Vehicle systems and subsystems showing inputs to and outputs from each

Systems control strategies and techniques

A system is a means of modifying inputs to produce a desired output. Therefore, the mechanisms by which the correct output can be achieved must be understood. The system can then be controlled. The vast majority of systems are:

- Controlled by an actuating mechanism that is powered separately from the input signal that controls it. Therefore, this input signal should not be affected by any changes in loading on the actuating mechanism. This input signal (often called the error signal because it indicates that the desired output has not yet been reached) usually has to be amplified so that the actuating mechanism can be driven.
- Powered by an energy input to the actuating mechanism that is generally proportional to the difference between the actual and desired values of the variable value being controlled. Systems of this type are called closed-loop feedback systems. This is because the output is constantly monitored, fed back and compared with the desired value (set point). Thus, the system has feedforward of desired value and feedback of actual value. Systems that have only feedforward are called open-loop systems.

Figure 2.3 is a block diagram model of a general closed-loop control system with feedback. An open-loop system would not have the feedback path.

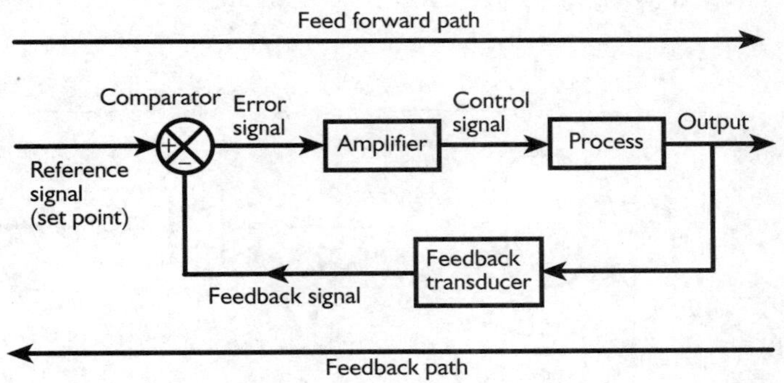

Figure 2.3 Block diagram model of a general closed-loop control system with feedback

Measurement and evaluation of performance of systems

Systems need to be evaluated in terms of their performance. This can often be done in the design stage by building mathematical models and testing the way the system behaves. Measurements can also be taken of the real-world system to monitor the way it performs in service. Typically, evaluations are done of things such as:

- response time to an input change – this could be a set point change or a disturbance
- accuracy of achievement of desired value
- stability.

Components, function and inter-relationships of engineering systems and subsystems

The work will build on that already done in Chapter 1, since the same kinds of systems are covered. However, in this chapter the emphasis will be on subsystems and components rather than just on inputs and outputs.

Electromechanical systems

A particular case will be studied to identify and describe the subsystems and components.

Case study

Steel rolling mill measurement and control system

Rolling mills in steel plants have automatic measurement and control systems, which ensure that the thickness of the finished strip is kept within acceptable limits. Figure 2.4 shows a possible arrangement of this system.

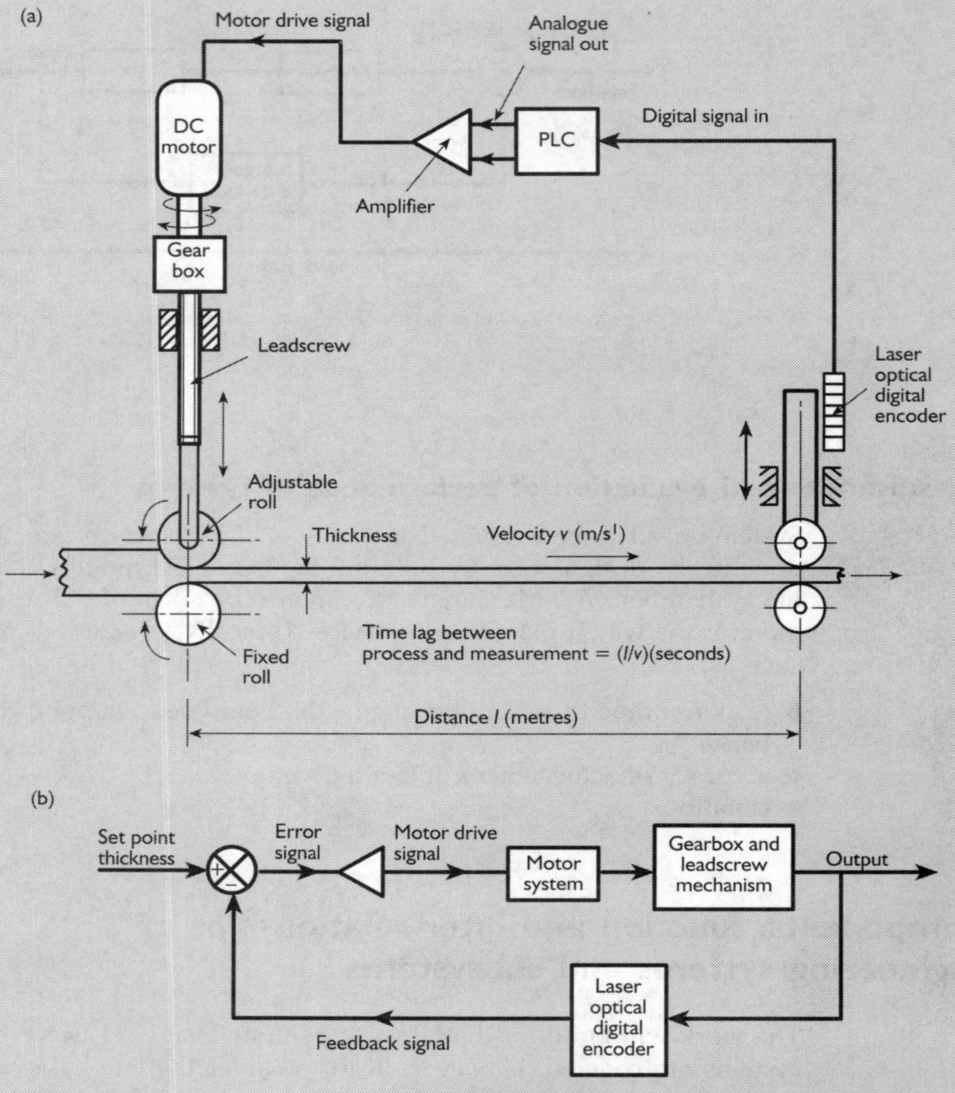

Figure 2.4 Steel rolling mill measurement and control system

This closed-loop system has these operational characteristics and inter-relationships:

- The laser optical encoder on the thickness-measuring probes sends a digital feedback signal of actual thickness to the programmable logic controller (PLC) – see case study in Chapter 1 for a full description of a PLC system

- This digital feedback signal is compared, in the PLC, to the set point for thickness, and any error, plus or minus, is converted by the PLC's internal digital to analogue converter (DAC) to a DC voltage, which is amplified to drive the DC motor of the variable height roll on the rolling mill. This DC motor is connected to a gearbox and a leadscrew, which drives the variable height roll up or down to correct the thickness
- There is a lag, or time delay (*l/v*), between the measurement and the correction being applied to the roll position, and this has to be built into the system
- Mains electrical supply is connected to motors driving the rolls
- Mains electrical supply is transformed (changed in voltage), rectified (converted to DC) and connected to the DC motor of the roll adjuster, encoder and PLC
- The encoder is connected to the PLC
- The amplified signal from the PLC is connected to the armature of the DC motor.

This complete thickness control system comprises the following components and subsystems:

- Laser optical encoder. The function of this system is to measure the thickness of the steel plate and convert this linear measurement to a digital value. It uses graduated glass scales and a laser to give a digital measurement. It is connected to the PLC, to which it feeds back the measured thickness.
- PLC incorporating DAC. The function of the PLC is to produce the signals that control the height of the roll and so control the sheet thickness. The digital signal from the encoder comes in to the digital input port of the PLC and is compared with the set point value. Any difference from the set point value (the error signal) is converted to an analogue value in the DAC.
- Amplifier – the function of the amplifier is to increase the power of the error signal and output it to drive the DC motor controlling the roll. The error signal is input to the amplifier, which increases the 'gain' of the signal. 'Gain' in this context is the ratio of output to input, e.g. gain = output/input.
- DC motor – the function of the motor is to convert the electrical energy of the signal from the PLC into kinetic energy of rotation. The amplified error signal is fed to this motor, which rotates in the correct direction to adjust the height of the roll.
- Gearbox – the function of the gearbox is to reduce the speed of rotation of the shaft from the motor, and to increase the torque to drive the leadscrew. It does this by using different-sized gears to provide the correct output speed/input speed ratio.
- Leadscrew – this has the function of converting the rotary motion of the output shaft of the gearbox into a linear motion. The end of the leadscrew is attached to the roll, which can then be driven up or down.
- Roll – this is the part of the system that actually produces the steel sheet. The two rolls are powered by electric motors, which convert electrical energy into kinetic energy of rotation. The rolls grip the sheet by friction; then the sheet moves forward and reduces in thickness.

Activity 2.1

Select a typical electromechanical system. The same system as chosen for Activity 1.8 in Chapter 1 can be used again or a new system selected. For this system:

a draw schematic and block diagrams
b describe how the system operates
c identify and describe the function of each subsystem
d show the inter-relationships between each subsystem, e.g. show the connections within and between each subsystem.

Progress check

1 Define a typical electromechanical system.
2 Describe the type of functions performed by subsystems in a general electromechanical system.
3 What are the typical inter-relationships between subsystems in a general electromechanical system.
4 Give an example of the use of signal paths and feedback between subsystems in a general electromechanical system.

Fluidic systems

These systems usually have to incorporate electrical/electronic systems so that they can be controlled. Again, a particular case will be studied to show how the system is built up from subsystems and components.

Case study

Cylinder actuation and control

Linear actuators, in the form of hydraulic or pneumatic cylinders, are controlled by valves. These valves can themselves be controlled by electrical relays (solenoids). In turn, these relays can be controlled by a microprocessor or PLC. A system for controlling a cylinder is shown in Figure 2.5.

This open-loop system has the following inter-relationships and operational characteristics:

- Air or oil under pressure is supplied to a solenoid-operated valve. This solenoid valve controls the valve spool allowing fluid to flow to and from each end of the cylinder, which moves the piston in the cylinder backwards or forwards.
- The PLC program switches the solenoid at the correct time and for the right duration to move the cylinder a set distance. The PLC uses computer logic signals of 5 V and 0.1 A and outputs a control signal of up to 200 V and 5 A to the solenoid.
- The PLC is connected to the mains electrical supply and to the solenoids of the valve, and piping connects the valve to the cylinder and to the supply pump and exhaust (air) or reservoir (oil).
- The hydraulic pump or air compressor is connected to the mains electrical supply.

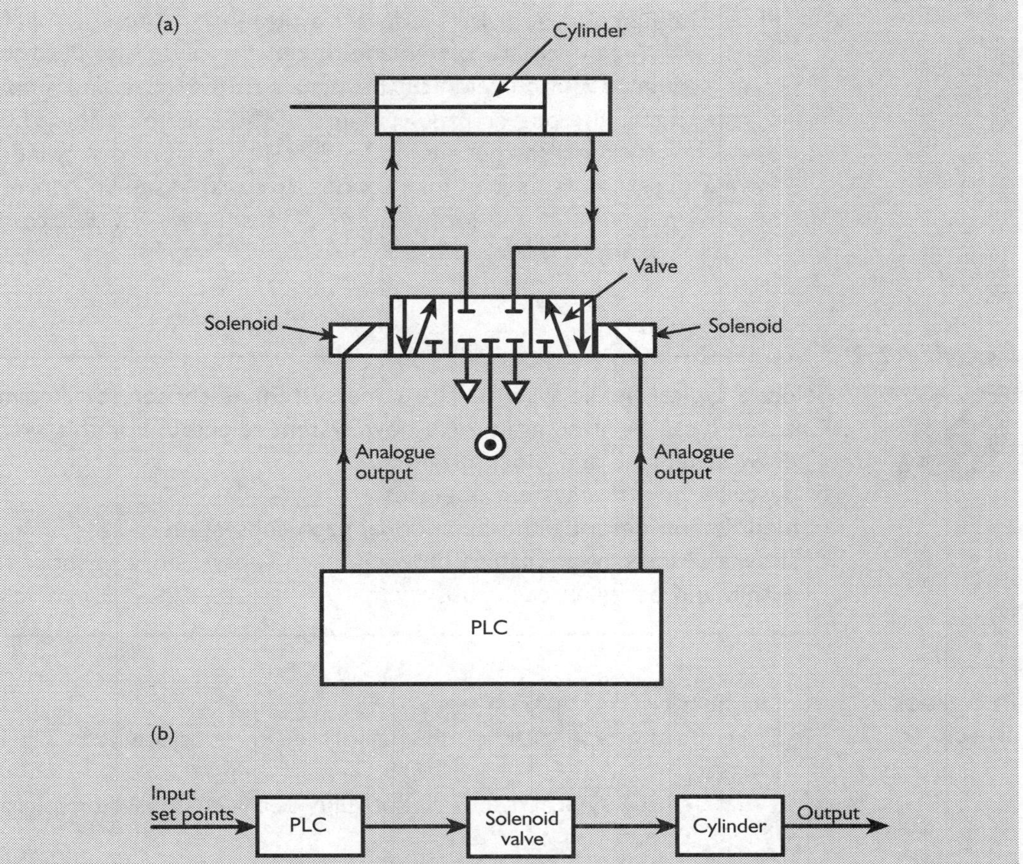

Figure 2.5 Open-loop system for controlling a cylinder

The complete system is made up of the subsystems:

- Air compressor or hydraulic pump to supply fluid under pressure. The function of this is to convert electrical energy into the potential energy of high-pressure fluid. This potential energy can be converted into kinetic energy by opening a valve.
- Solenoid-operated spool valve to control the rate and direction of fluid flow. This has the function of converting the potential energy of the fluid into kinetic energy by allowing it to flow. The solenoids (one at each end of the valve) are operated by a signal from the PLC. This signal moves the solenoid, so converting electrical energy into kinetic energy. The solenoid is attached to the valve spool, which moves sideways and uncovers the input/output ports of the valve. This allows fluid to flow to and from the cylinder according to which solenoid is operated.
- Linear actuator in the form of a piston and cylinder assembly. The function of this is to convert a flow of fluid under pressure into a linear movement. The fluid is simultaneously admitted to one end of the cylinder and allowed to leave the other end, according to which direction it is required to move in. This is achieved by connecting each end to the spool valve.

- PLC. The function here is to convert a digital signal to an amplified analogue signal, which can actuate the solenoids on the valve. The PLC controls the time, sequence and duration of the signals to the valve and amplifies these signals to suit the power requirements of the solenoid. This PLC is part of an open-loop system, so there is no feedback to it from the cylinder. It works from a programmed set of instructions. It could easily be converted to become part of a closed-loop system, if sensors were fitted to the cylinder.

Activity 2.2

Select a typical fluidic system. The same system as chosen for Activity 1.9 in Chapter 1 can be used again or a new system selected. For this system:

a draw schematic and block diagrams
b describe how the system operates
c identify and describe the function of each subsystem
d show the inter-relationships between each subsystem, e.g. show the connections within and between each subsystem.

Progress check

1 Define a typical fluidic system.
2 Describe the type of functions performed by subsystems in a general fluidic system.
3 What are the typical inter-relationships between subsystems in a general fluidic system.
4 Give an example of the use of signal paths and feedback between subsystems in a general fluidic system.

Chemical systems

As already mentioned in Chapter 1, these systems cover a wide variety of processes. A case study is used again to illustrate the basic principles in a general chemical process.

Case study

Chemical processing system

Many chemical processes involve the bringing together of two or more constituents to form a mixture or a compound. Figure 2.6 shows a typical system for such a chemical process.

This closed-loop system has operational and inter-relational features as follows:

- Constituent X is pumped into the reaction tank at a constant, programmed rate. Constituent Y is pumped at a constant rate but is also being regulated by a control valve.

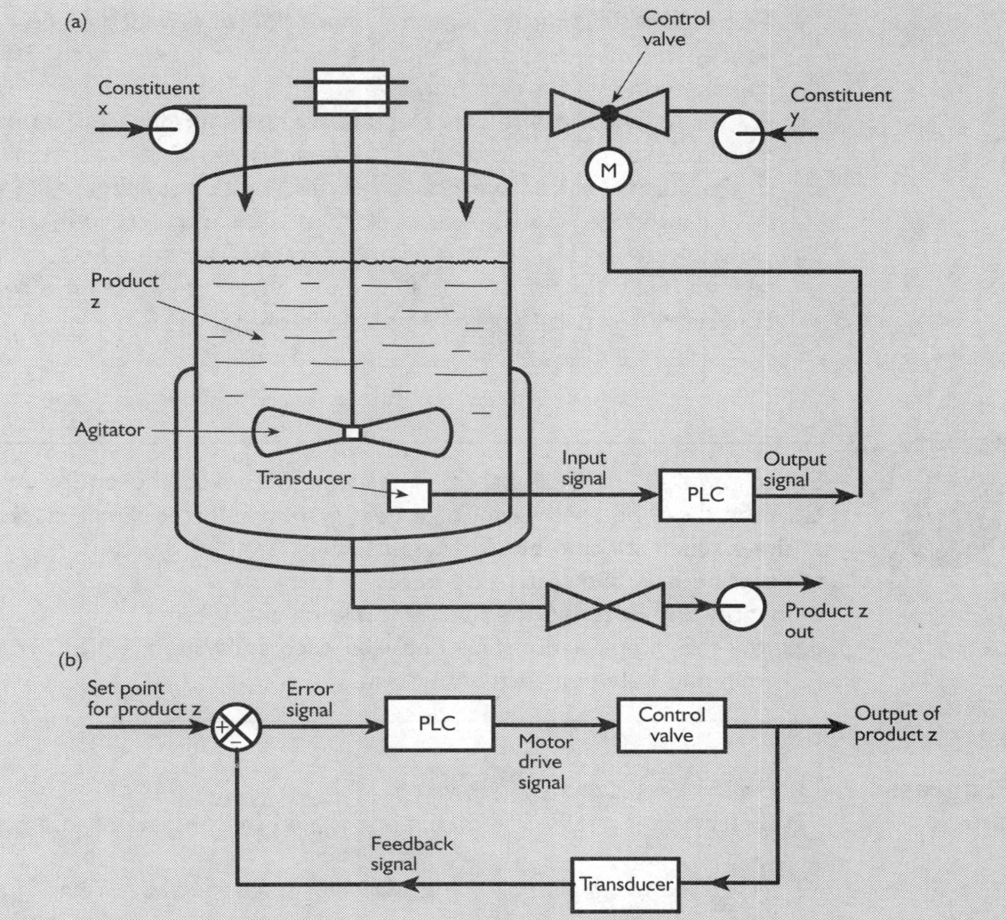

Figure 2.6 Typical closed-loop system for a chemical process

- The composition of the product Z produced by components X and Y is measured by a transducer and a voltage signal from the transducer, corresponding to the composition, is fed back to the PLC where the set point and composition voltages are compared. If there is an error, an amplified signal is sent to the control valve regulating the flow of constituent Y.
- The PLC, pumps and agitators are connected to the mains electrical supply.
- The PLC is connected to the composition transducer and the control valve.

The complete system is made up of the components and subsystems:

- Pumping and piping systems. These pump in constituents X and Y and pump out the finished product Z. The pumps are normally driven by electric motors and have the function of converting electrical energy into kinetic energy of rotation. This pump rotation then pressurises the constituents and causes them to flow along the pipes.
- Reaction tank and agitator. The function here is to admit and contain the constituents and allow the mixing or reaction to take place. The tank needs to be of a material that is inert to the constituents and the product.

- Transducer. This measures the composition of product Z and converts it to a signal, which is sent to the PLC. Its function is therefore to change a chemical composition into an electrical signal.
- PLC. The function of this is to compare the transducer signal with the set point to produce an error signal. The error signal is then amplified and output to the control valve. The PLC does this with a comparator, an amplifier and a power interface with an output port to the control valve motor.
- Control valve. This regulates the flow rate of constituent Y. Its function is to take an input signal from the PLC and convert it into a valve movement, so converting electrical energy into kinetic energy.

Activity 2.3

Select a typical chemical system. The same system as chosen for Activity 1.10 in Chapter 1 can be used again, or a new system selected. For this system:

a draw schematic and block diagrams
b describe how the system operates
c identify and describe the function of each subsystem
d show the inter-relationships between each subsystem, e.g. show the connections within and between each subsystem.

Progress check

1 Define a typical chemical system.
2 Describe the type of functions performed by subsystems in a general chemical system.
3 What are the typical inter-relationships between subsystems in a general chemical system.
4 Give an example of the use of signal paths and feedback between subsystems in a general chemical system.

Information and data communication systems

Chapter 1 should be referred to for general information about these systems. The general function of these systems is to transmit, process and display information. The data and information system can also be used as a subsystem to control other subsystems within a complete system. As discussed in Chapter 1, there are a wide variety of information and data communication systems. One system will be illustrated as a case study.

Case study

Programmable logic controller (PLC)

Since the programmable logic controller (PLC) has been used in the other case studies, it is worth studying in its own right as a data communication system. The PLC is essentially a microcomputer but with built-in interfaces to connect it with industrial machines and devices. Figure 2.7 shows a typical PLC system.

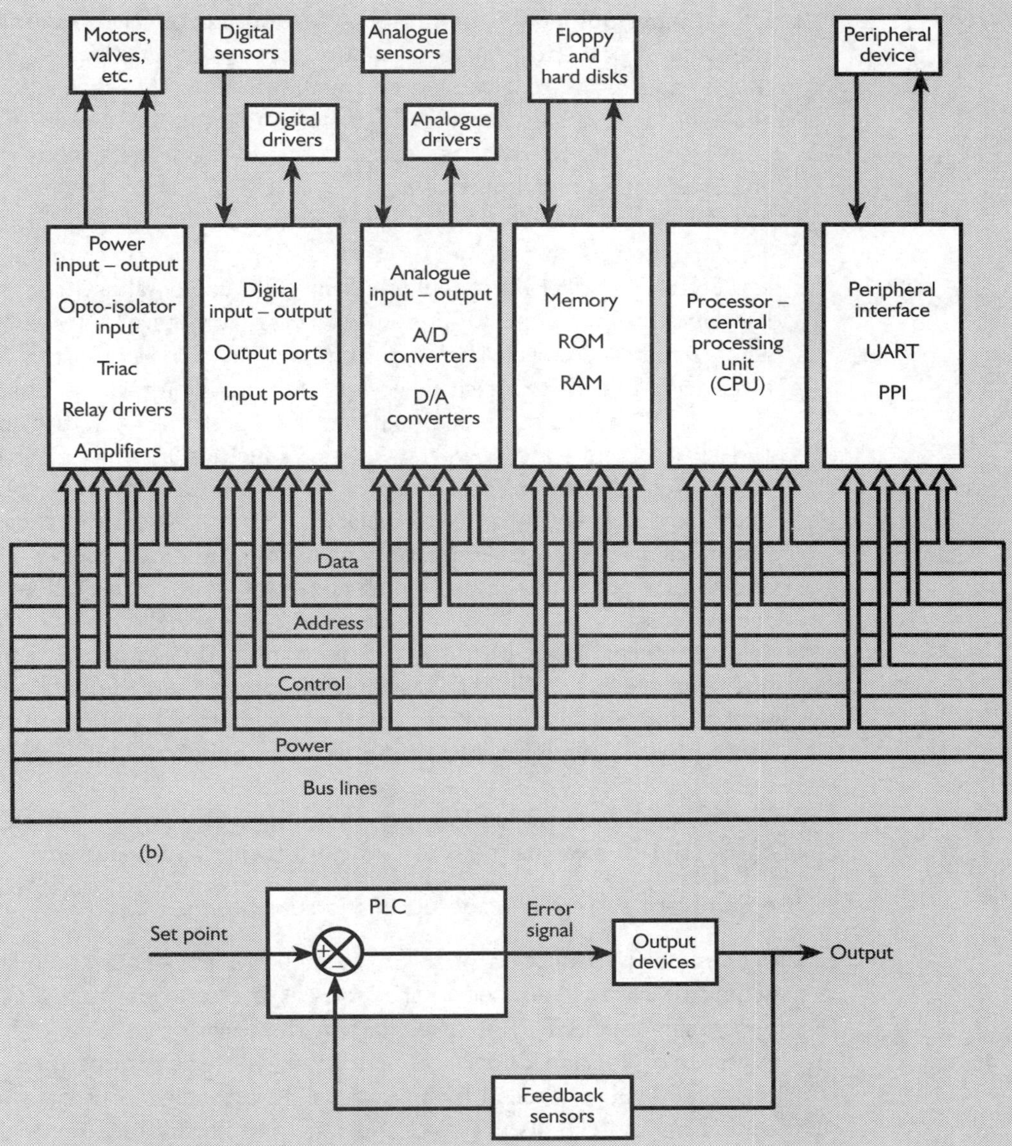

Figure 2.7 A typical PLC system

The essential operational and inter-relational features of this PLC are:

- Processing of information/data by the processor. The processor can do all the normal arithmetic and logic processing to digital data. Since it has a built-in clock, it can also transmit data in a programmed time sequence.
- Temporary storage of data in the random access memory (RAM). This holds data that is currently being worked on, or is being received or transmitted.
- Permanent storage of data in the hard or floppy disk. Both hard and floppy disks can have data written onto or read from them. Data is also stored permanently in the read only memory (ROM). This holds data that is required to start up and run the system.

- Digital input/output (I/O) with input and output ports. This receives or transmits digital data. Analogue input/output with analogue to digital converter (ADC) and digital to analogue converter (DAC). This receives or transmits analogue data. It can also take in analogue data from sensors/transducers and convert it to digital data, as well as taking in digital data, converting it to analogue data and outputting it.
- Interfacing the PLC to devices like monitor, keyboard, printer, telephone, etc. The programmable peripheral interface (PPI) inputs or outputs parallel data, e.g. 8, 16, 32 or 64 bits simultaneously. The universal asynchronous receiver/transmitter (UART) converts data from parallel to serial form, e.g. a stream of single bits for devices that require data in this form.
- Interfacing the PLC to real devices with a power interface. The output part consists of devices such as amplifiers (solid state devices that amplify a signal) and triacs (solid-state power switches), which can output electrical current at the power required to run a motor or turn on a relay or a solenoid. The input part is generally an opto-isolator. This opto-isolator takes in a high-power signal, which switches on a light-emitting diode (LED). This then energises a phototransistor, which allows current to pass as an input signal to the processor.
- All the internal parts of the PLC are connected together by a bus line. The bus lines are the wires that transmit the power and the data to each device as well as addressing the correct locations in memory and controlling sequence and timing. They also allow all the parts to be in communication with each other.
- The digital I/O, analogue I/O, peripheral interface and power I/O are all connected to external devices and can receive and transmit.

The complete system is made up of the subsystems:

- Processor. The function of this is to take in binary data and manipulate it according to programmed instructions. It is often referred to as the central processing unit (CPU). It takes data in the form of streams of bits (binary digits 0s or 1s) and does logic operations and calculations (processing) on this data. It is controlled by an internal clock, which controls the speed of operation and can also control the time and duration of signal output. The processor connects to all the other subsystems of the PLC.
- Random access memory (RAM). This has the function of storing data. It consists of discrete memory locations where bits of data can be temporarily stored during processing. Data can flow in each direction between the processor, RAM and the other subsystems.
- Digital I/O. This functions as an interface for digital data. It can input or output digital data in the form of bits to and from external devices and the other subsystems.
- Analogue I/O. This functions as an interface for analogue data. It can input or output analogue data. It can also convert analogue data to digital data with an ADC, and digital data to analogue data with a DAC. It can also communicate with the other subsystems using the digital data bus.

- Programmable peripheral interface. This functions as an interface for the peripheral devices. It is a buffer between the processor and the peripheral devices. It inputs or outputs data in a stream of parallel bits to and from peripherals such as the keyboard and monitor.
- Universal asynchronous receiver/transmitter (UART). This functions as another type of interface for peripheral devices. It converts a parallel stream of bits into a single serial stream of bits for transmission to peripherals that can only accept data in this form.
- Power interface. The main function here is of amplification of low-power signals and switching of high-power supplies by a low-power signal. It allows the PLC to control and run real industrial devices at whatever voltages and currents are required, using devices such as amplifiers, opto-isolators and triacs. This interface communicates to the other subsystems using the normal computer logic signals of 5 V and 0.1 A.

Activity 2.4

Select a typical information and data communication system. The same system as chosen for Activity 1.11 in Chapter 1 can be used again, or a new system selected. For this system:

a draw schematic and block diagrams
b describe how the system operates
c identify and describe the function of each subsystem
d show the inter-relationships between each subsystem, e.g. show the connections within and between each subsystem.

Progress check

1 Define a typical information and data communication system.
2 Describe the type of functions performed by subsystems in a general information and data communication system.
3 What are the typical inter-relationships between subsystems in a general information and data communication systems.
4 Give an example of the use of signal paths and feedback between subsystems in a general information and data communication system.

Thermodynamic systems

Chapter 1 describes thermodynamic systems in general terms. A case study will be used to demonstrate some principles of thermodynamic systems in terms of subsystems and components.

Case study

Steam turbine and a DC electrical generating system

Figure 2.8 shows a steam turbine and a DC electrical generation system.

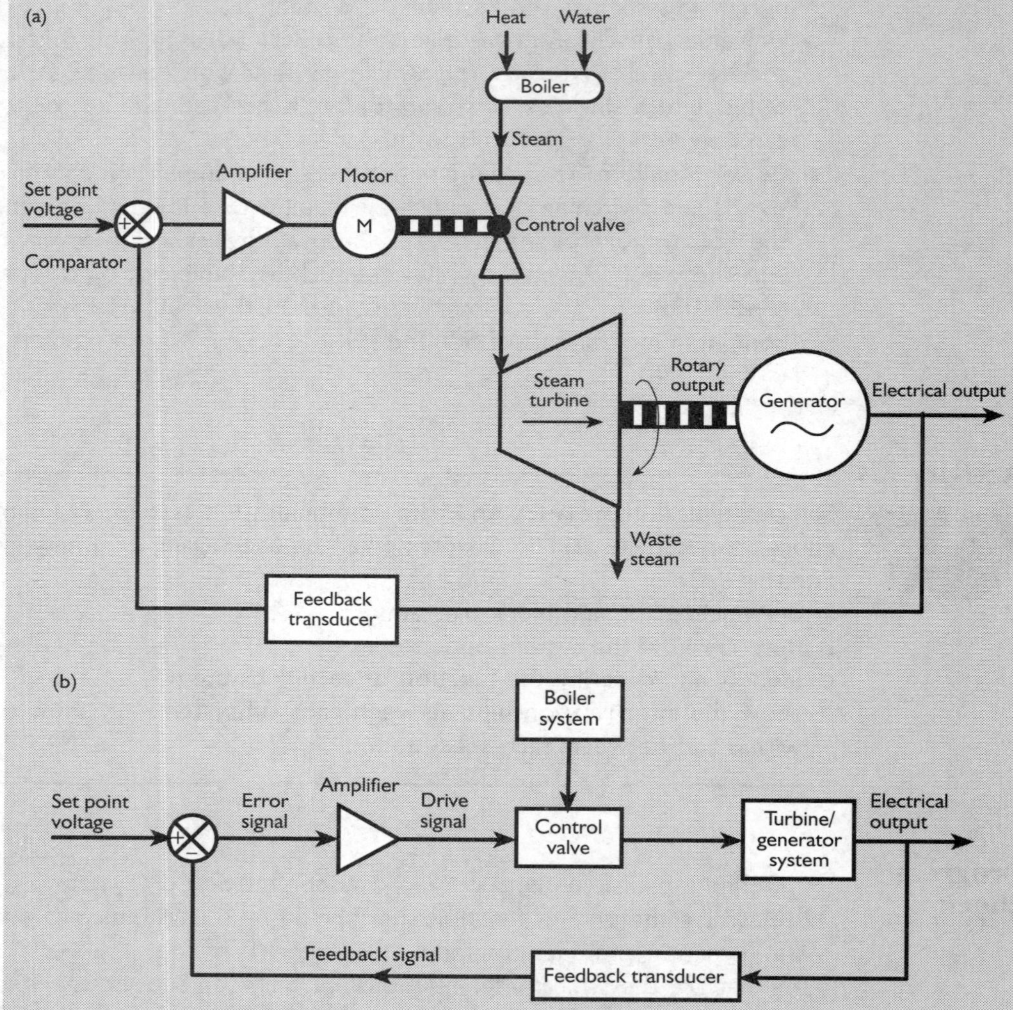

Figure 2.8 A steam turbine and a DC electrical generation system

This closed-loop system has these operational characteristics and inter-relationships:

- High-pressure steam is produced in a boiler by combustion of fuel. This high-pressure steam is piped through an input valve to a steam turbine. The steam is admitted to the turbine and the turbine rotor rotates. The rotating turbine is connected to the generator, which also rotates and outputs electricity.
- The output voltage of the generator is measured and converted by a feedback transducer to a digital signal. This signal is fed back and compared with a set point voltage where any error is converted to an analogue signal, amplified and used to drive a motor controlling the input valve. This input valve changes

the flow of steam to the turbine, so changing the speed, which changes the electrical output of the generator.

The complete system is made up of the subsystems:

- Boiler. The primary function here is the conversion of chemical energy in the fuel to heat energy. It produces high-pressure steam by combustion of fuel. The heat produced by combustion passes into heat exchangers, which transfer the heat into the water. The water is contained in a pressure vessel, which produces dry superheated steam at temperatures up to 600°C and pressures up to 330 bar.
- Input valve. This functions by converting an input signal of electrical energy into kinetic energy of movement. It consists of a motorised valve that varies the diameter of the input pipe to the turbine. Steam is piped from the boiler to this valve. The valve is controlled by a motor, which is driven by an error signal from the comparator unit. Opening or closing the valve varies the flow of steam to the turbine. The function here is to convert the potential energy of high-pressure steam into kinetic energy of rotation. The turbine is driven by the high-pressure steam from the input valve acting on the turbine blades, which rotate the shaft. The shaft is connected to the generator.
- Generator. This functions as a converter of kinetic energy of rotation into electrical energy. The rotor of the generator is turned by the turbine shaft and produces an output of electricity. This output is measured by the feedback transducer and fed back to the comparator, which compares the set point voltage with the feedback voltage. Any error is output to the amplifier, which amplifies the error signal and sends this amplified signal to the motor controlling the input valve.

Activity 2.5

Select a typical thermodynamic system. The same system as chosen for Activity 1.12 in Chapter 1 can be used again, or a new system selected. For this system:

a draw schematic and block diagrams
b describe how the system operates
c identify and describe the function of each subsystem
d show the inter-relationships between each subsystem, e.g. show the connections within and between each subsystem.

Progress check

1 Define a typical thermodynamic system.
2 Describe the type of functions performed by subsystems in a general thermodynamic system.
3 What are the typical inter-relationships between subsystems in a general thermodynamic system.
4 Give an example of the use of signal paths and feedback between subsystems in a general thermodynamic system.

System control strategies and techniques

This section looks at methods of implementing control systems and the way a system responds to a disturbance or change of set point. The two main ways of changing the set point are:

- step change – this is an instantaneous change of input
- ramp change – this is a varying change of input.

Figure 2.9 shows these two types of input change. It can be seen that, in a step change, the desired value goes instantaneously from set point R_0 to set point R_1. In a ramp change, the desired value increases with time.

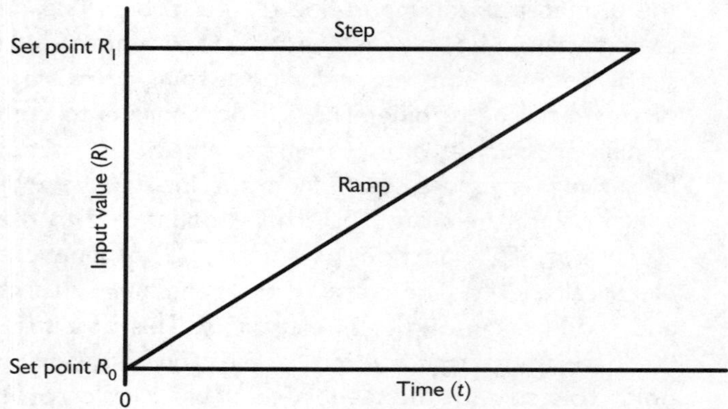

Figure 2.9 Step and ramp input changes

In order to discuss system design, it is necessary to define all the terms used in the analysis of systems. Figure 2.10 shows a typical response of a system to a step change of set point.

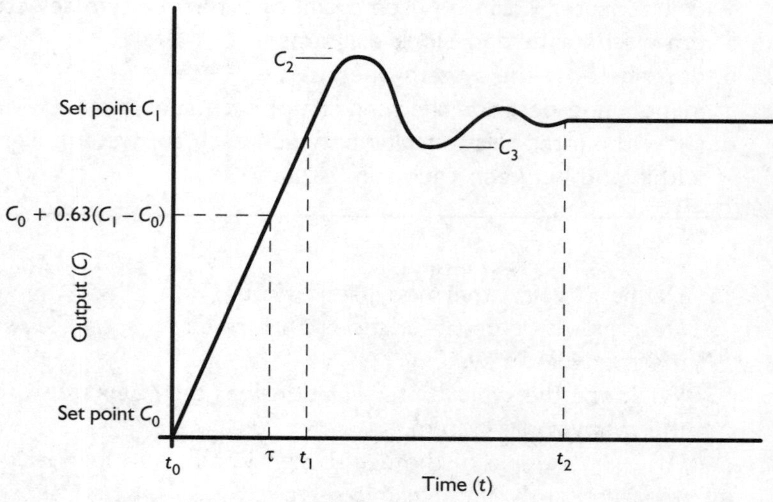

Figure 2.10 System response to step input change

It can be seen in Figure 2.10 that, when set point C_0 is changed to set point C_1 by a step change of input R, the system responds by moving to C_1 in time t_1. It then oscillates above to C_2 and below to C_3 before settling to level C_1 in time t_2. Note that it reaches a value of $C_0 + 0.63\,(C_1 - C_0)$ in time τ. The time τ is called the time constant of the system.

The value of C at the time constant = $C_0 + (C_1 - C_0)\,(1 - e^{-1})$.

The value of C at any time t between t_0 and $t_1 = C_0 + (C_1 - C_0)\,(1 - e^{-t/\tau})$.

Referring to Figure 2.10:

- **Response rate** is the speed the system operates at. It is equal to the change in set point divided by the time taken, which equals $(C_1 - C_0)/(t_1 - t_0)$. The time constant τ is also a very useful measure of system response rate.
- **Overshoot** is the amount by which the set point is exceeded. The first overshoot is equal to $C_2 - C_1$.
- **Undershoot** is the amount by which the value goes under the set point. Here, the first undershoot is $C_3 - C_2$.
- **Settling time** is the time taken for the system to stabilise after a change in set point or a disturbance. Here, the settling time is $t_2 - t_0$.
- **Stability** is the ability of the system to remain at the same value without oscillating above or below it.
- **Accuracy** is the difference between the desired set point and the actual value achieved. It is a permanent error, unlike undershoot or overshoot, which, except in an oscillating system, are temporary. Any error in a control system is called a residual error and in proportional process control systems is called 'offset' (see page 75).
- **Repeatability** is the ability of a system to hit the same value over many cycles of changes. It should not be confused with accuracy. A system can be repeatable but to the same inaccurate value.
- **Sensitivity** is another way of expressing gain and is the ratio between change in output and the corresponding change in input. Therefore, sensitivity = change in output/corresponding change in input.
- **Safety** is self-explanatory and means that the system must be analysed to ensure that no aspect of the system can cause a hazard.
- **Efficiency** is the ratio of input energy to output of heat or work and is simply defined as output/input.

The typical response of a system to a ramp input is shown in Figure 2.11.

It can be seen that the system response lags behind the ramp input by a steady-state error equal to the time constant τ.

Analogue and digital techniques

Although these two techniques are classified as separate techniques, in practice they are frequently used together in 'hybrid' systems. Since devices such as analogue to digital converters (ADCs) and digital to analogue converters (DACs) are readily available, the two techniques can readily be combined. A PLC combines both techniques. The operation of the PLC is covered in case studies in this and the previous chapter.

Analogue control techniques

By definition, an analogue control technique uses:

Continuously varying signals such as voltage, current, pressure, etc.

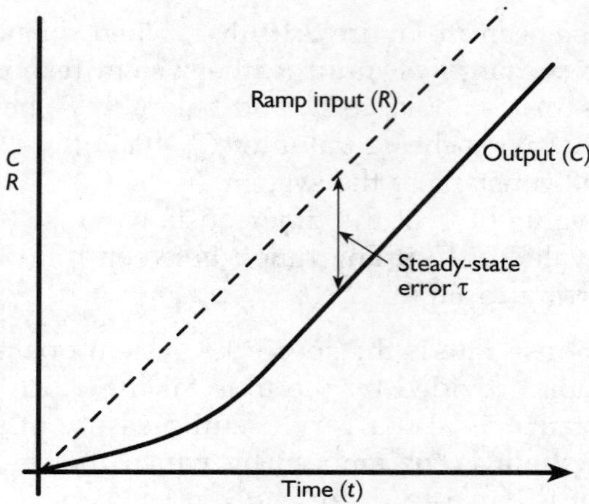

Figure 2.11 System response to ramp input change

Figure 2.12a shows an example of an analogue signal such as the output from a microphone.

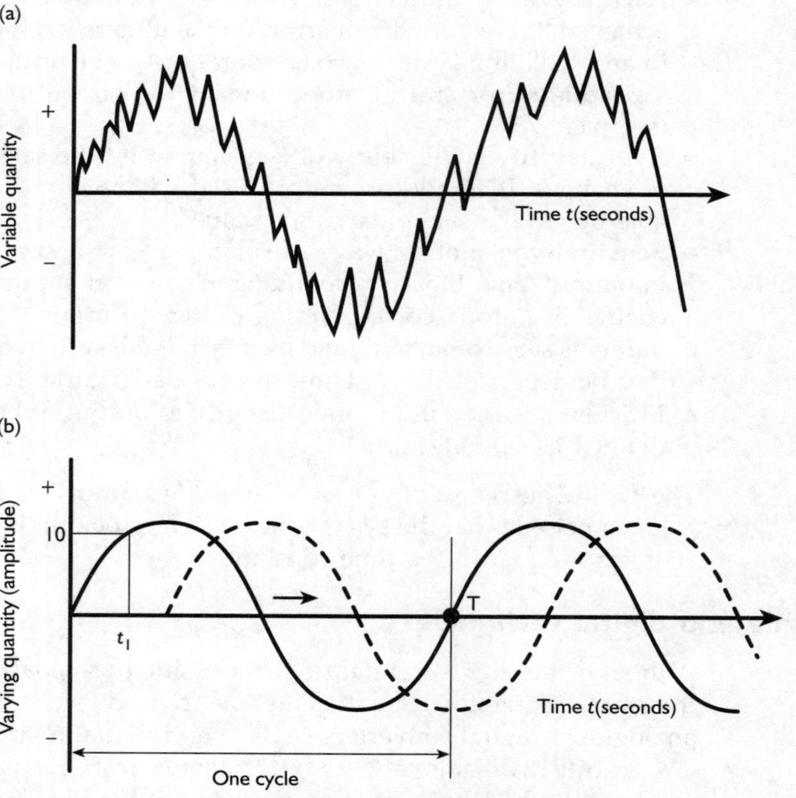

Figure 2.12 Analogue signal. Example of an analogue signal such as the output from a microphone (a) and basic definitions associated with an analogue waveform (b)

Figure 2.12b shows the basic definitions associated with an analogue waveform:

- Amplitude is the value of the varying quantity at a particular time t – in this case, the amplitude is equal to 10 at time t_1.
- Frequency is the number of cycles per second, which equals $1/T$.
- Phase – the dotted line shows a wave that is out of phase with the first one – in this case by around 90°.

Analogue amplification with operational amplifier

Analogue signals from systems components such as transducers are frequently of low power. In order that these signals can be used to control actuators and other equipment, the signals require amplification. This amplification is frequently done with an operational amplifier, which is a semiconductor component. Figure 2.13 shows:

a a circuit symbol for an operational amplifier (op-amp)
b a typical op-amp IC.

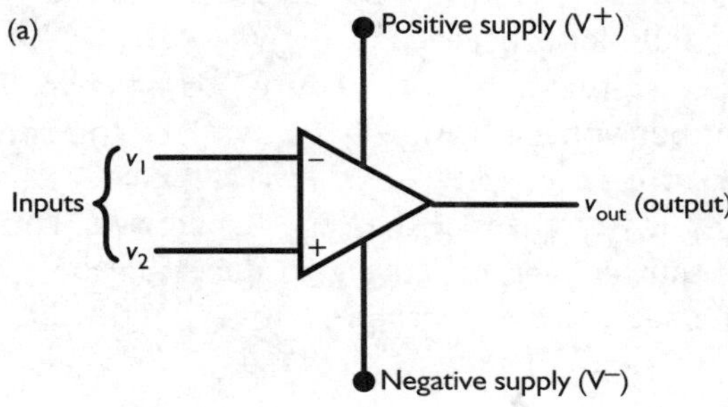

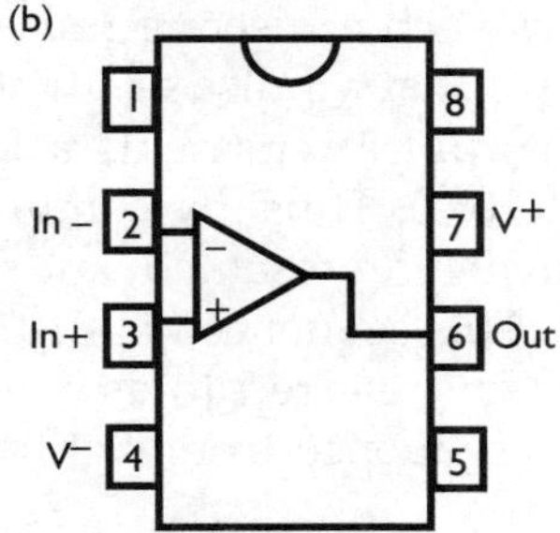

Figure 2.13 Circuit symbol for an operational amplifier (op-amp) (a) and a typical op-amp IC (b)

The gain (or amplification) is defined as:

$$v_{out}/(v_2 - v_1).$$

The gain, in this case, is on the amplitude of the voltage and so is the amplitude of the output signal/amplitude of the input signal.

A typical value of gain is 10^5. Gain is more frequently expressed in decibels (dB), where gain in dB is 20 $\log_{10}$gain, so a gain of

$10^5 = (20 \times 5)$ dB $= 100$ dB.

Case study

Use of a comparator

As well as being used to amplify signals, the op-amp can be used for other functions. One of these is as a comparator. There are many control applications where an input signal has to be compared with a set point. This could be control of things such as:

- speed
- fluid level.

Referring to Figure 2.13a, v_1 is the reference (set point) voltage and v_2 is the input signal from the measuring transducer. Note that the input signal must be in volts, irrespective of whether speed, fluid level or any other variable is being measured. The transducer converts the variable into a voltage. The op-amp works as a comparator as follows:

- If the two voltages are equal v_{out} is zero. In this case, the input signal is equal to the set point, so no output is required. This is the case when the set speed or fluid level is reached.
- If v_2 is greater than v_1, then v_{out} is positive. This positive signal amplifies the output voltage positively to increase the speed or the fluid level until the set point is reached.
- If v_2 is less than v_1, then v_{out} is negative. This negative signal amplifies the output voltage negatively to decrease the speed or reduce the fluid level until the set point is reached.

Digital control techniques

By definition, a digital control technique uses discrete pulses – that is to say, a voltage that lasts for a very short time.

Figure 2.14 shows such a discrete pulse.

Figure 2.14a shows that a pulse of 5 volts is present for a time of 10^{-6} seconds. The presence of this pulse represents a logic state of 1. The absence of a pulse implies a logic state of 0. Thus, by combining the presence or absence of pulses, a binary value can be represented and/or transmitted. Figure 2.14b shows a sequence of pulses. This sequence of 10101001 is what is called eight binary digits (bits), which is usually referred to as a byte of information. Each bit represents a binary number in sequence from right to left as follows:

$$2^0 = 1:\ 2^1 = 2:\ 2^2 = 4:\ 2^3 = 8:\ 2^4 = 16:\ 2^5 = 32:\ 2^6 = 64:\ 2^7 = 128.$$

Therefore, a 1 represents one of these numbers, while a 0 indicates an absence of the number. So, the sequence of 10101001 above is equal to the denary number:

$$1 + 0 + 0 + 8 + 0 + 32 + 0 + 128 = 169.$$

Therefore, the bits can represent numbers, enabling measurement or control of varying values. The 0 and 1 logic states can also be used as on–off switches in control systems.

Conversion can be made between analogue and digital signals by using an analogue to digital converter (ADC). This device has to sample the analogue signal at a rate higher than the signal frequency. The sampled values are then output as a constant stream of bits. The principle of this is shown in Figure 2.15.

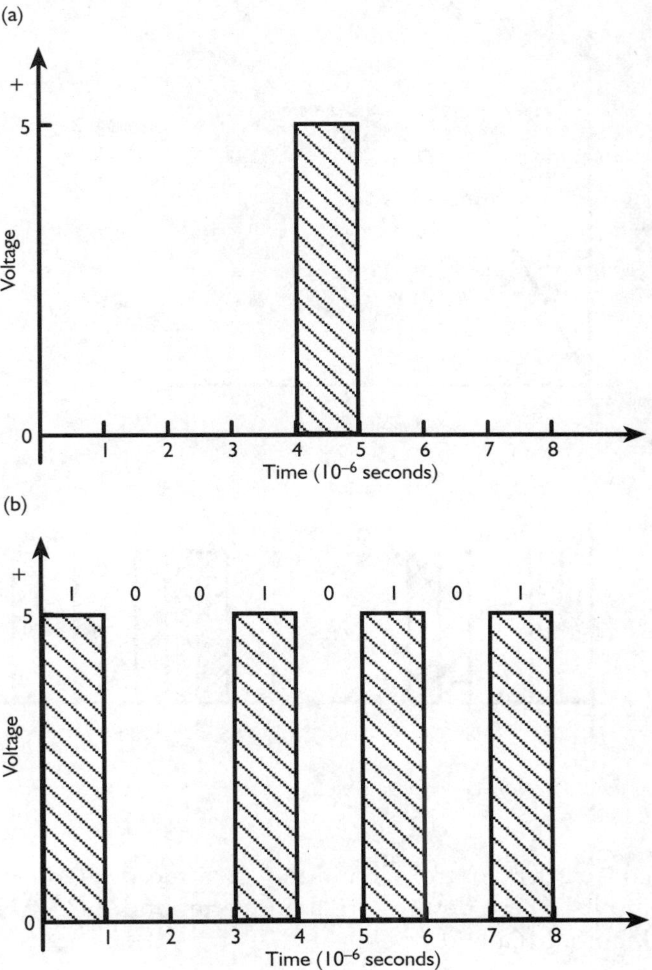

Figure 2.14 *A discrete pulse: (a) single pulse and (b) sequence of pulses.*

Figure 2.15a shows an analogue signal that has been sampled at time $T/12$, $T/6$ and $T/4$, where T is the time for one complete waveform. Assuming the waveform is a sine wave, the values are 50 volts at $T/12$, 87 volts at $T/6$ and 100 volts at $T/4$. These values are converted into the bytes 00110010 (50), 01010111 (87) and 01100100 (100). These are shown diagramatically as pulses in Figure 2.15b. Note that the pulses are read in sequence from left to right on the diagram, but are placed in ascending value from right to left, i.e. reversed.

Conversion between analogue and digital signals can also be made using a digital to analogue converter (DAC). This works in exactly the reverse way to the ADC. The stream of bits is now output as a continuously varying value. Look at Figure 2.15b and just visualise the digital numbers being converted back to the waveform values of Figure 2.15a.

Digital logic control

A digital circuit can be in the logic LOW (off) state (digital 0) or in the logic HIGH (on) state (digital 1). The inputs and outputs to a circuit can be 0 or 1 and so can be used to control devices. Two digital logic devices can be used to

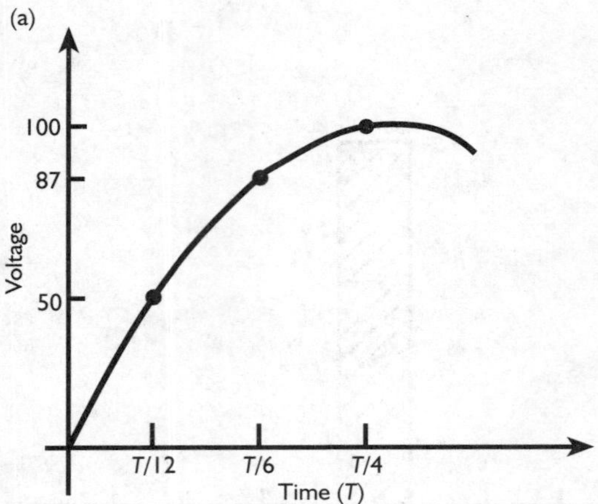

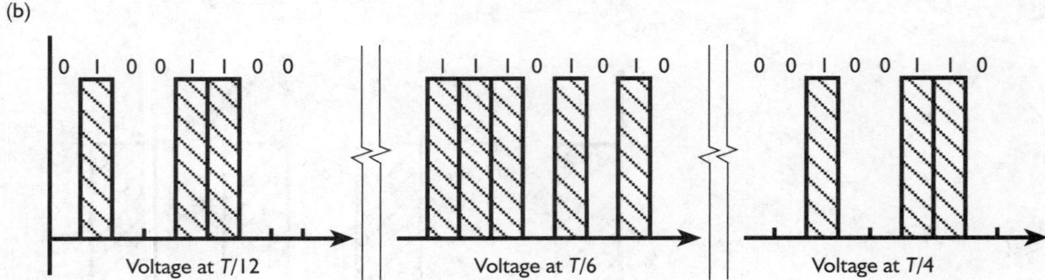

Figure 2.15 (a) An analogue signal sampled at various timepoints and shown diagrammatically as pulses in (b)

construct a control circuit and give the desired outputs for any combination of inputs. These devices are the inverter and the NAND gate, shown as symbols in Figure 2.16a and b.

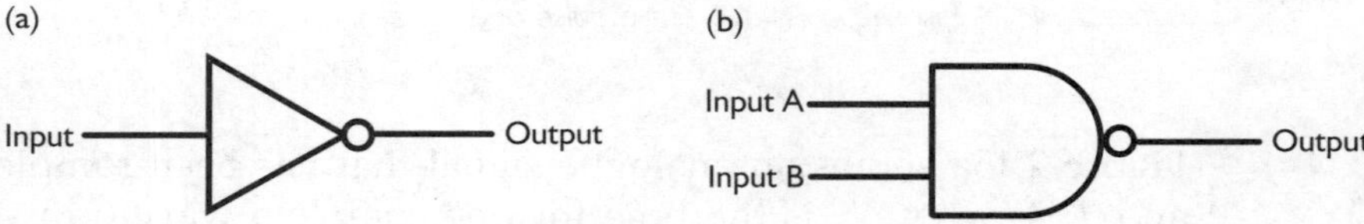

Figure 2.16 Inverter (a) and NAND gate (b)

The inverter works by outputting the inverse of the input. Table 2.1 shows the input–output table and truth table (where false = LOW and true = HIGH) for an inverter.

Table 2.1 Input–output table and truth table for an inverter

Input–output table		Truth table	
Input	**Output**	**Input**	**Output**
0	1	False	True
1	0	True	False

The NAND gate works by outputting 1 when the state of the two inputs is anything but two 1s. When the input is two 1s, a logic 0 is output. Table 2.2 shows the input–output table and the truth table for a NAND gate.

Table 2.2 Input–output table and truth table for a NAND gate

Input–output table			Truth table		
Input A	**Input B**	**Output**	**Input A**	**Input B**	**Output**
0	0	1	False	False	True
0	1	1	False	True	True
1	0	1	True	False	True
1	1	0	True	True	False

A NAND gate can be converted to an equivalent AND gate by connecting an inverter to its output, as shown in Figure 2.17a. The symbol for an AND gate is shown in Figure 2.17b.

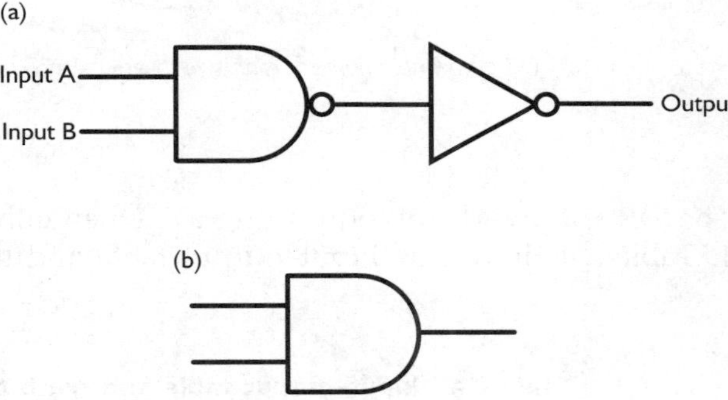

Figure 2.17 (a) AND gate implemented using a NAND gate and an inverter. (b) Symbol for an AND gate

The AND gate works by only outputting a 1 when both input A and input B are 1s. At all other input states, the output is 0. Table 2.3 shows the input–output table and the truth table for an AND gate.

Table 2.3 Input–output table and truth table for an AND gate

Input–output table			Truth table		
Input A	**Input B**	**Output**	**Input A**	**Input B**	**Output**
0	0	0	False	False	False
0	1	0	False	True	False
1	0	0	True	False	False
1	1	1	True	True	True

An equivalent OR gate can be made by connecting inverters to each input of the NAND gate, as shown in Figure 2.18a. The symbol for an OR gate is shown in Figure 2.18b.

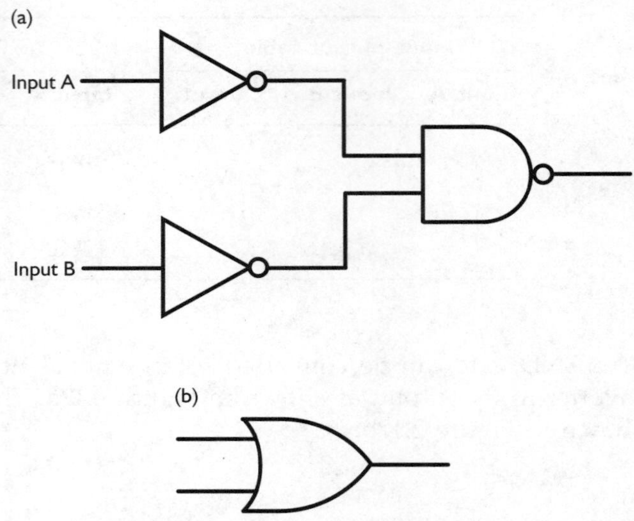

Figure 2.18 (a) OR gate implemented using two inverters and a NAND gate. (b) Symbol for an OR gate

The OR gate works by outputting a 1 when either or both of the two inputs is a 1. Table 2.4 shows the input–output table and the truth table for an OR gate.

Table 2.4 Input–output table and truth table for an OR gate

Input–output table			Truth table		
Input A	**Input B**	**Output**	**Input A**	**Input B**	**Output**
0	0	0	False	False	False
0	1	1	False	True	True
1	0	1	True	False	True
1	1	1	True	True	True

To summarise the main types of logic gates:

- NAND – output is 0 when all inputs are 1 – output is 1 at all other input states
- INVERTER – output is 0 when input is 1 – output is 1 when input is 0
- AND – output is 1 when all inputs are 1 – output is 0 at all other input states
- OR – output is 1 when any input is 1 – output is 0 only if all inputs are 0.

Case study

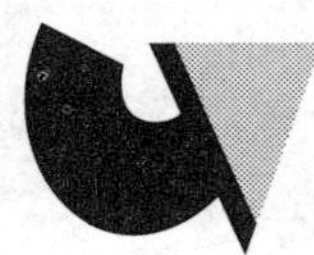

Automatic drilling operation

A good example of digital logic control is in the control of an automatic dr operation. Before the drilling operation can start, the workpiece must be clamped in the fixture, the guard put in place and the operator must press the start button. Thus, before a control signal can be sent to start the operation, the three requirements of fixturing, guarding and operator action must be satisfied. The signals to do this can be provided by a pressure sensor in the fixture, a proximity sensor on the guard and an on–off switch for the operator. Only when a logic 1 (high = true) is received from all three sources will the operation start. Also, if a logic 0 (low = false) is received from any of the sources after the start of the operation, it will stop. So the logic conditions required are:

- output 1 only if all three inputs are 1
- output 0 if any input is 0.

So what is required to implement this logic is a three-input AND gate. The input–output and truth tables for a three-input AND gate are shown in Table 2.5.

Table 2.5 Input–output table and truth table for a three-input AND gate

Input–output table				Truth Table			
Input A	**Input B**	**Input C**	**Output**	**Input A**	**Input B**	**Input C**	**Output**
0	0	0	0	False	False	False	False
0	0	1	0	False	False	True	False
0	1	0	0	False	True	False	False
0	1	1	0	False	True	True	False
1	0	0	0	True	False	False	False
1	0	1	0	True	False	True	False
1	1	0	0	True	True	False	False
1	1	1	1	True	True	True	True

This logic can be implemented by the use of two NAND gates and two inverters, as shown in Figure 2.19.

Figure 2.19b shows suitable ICs for implementing the three-input AND gate. The IC pin connections are:

- pin 1 on 7400 to input A
- pin 2 on 7400 to input B
- pin 3 on 7400 to pin 1 on 7404
- pin 2 on 7404 to pin 4 on 7400
- pin 5 on 7400 to input C
- pin 6 on 7400 to pin 3 on 7404
- pin 4 on 7404 to output.

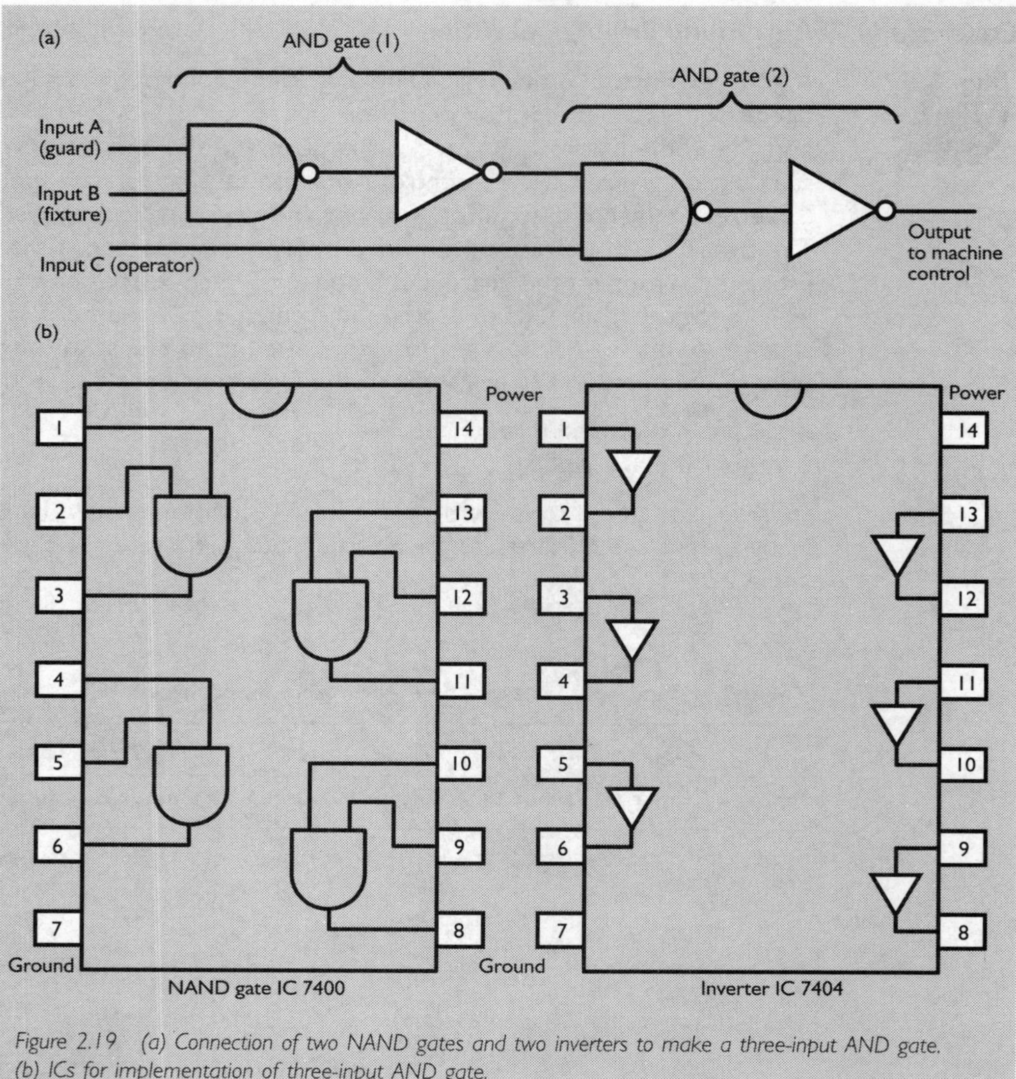

Figure 2.19 (a) Connection of two NAND gates and two inverters to make a three-input AND gate. (b) ICs for implementation of three-input AND gate.

Open- and closed-loop control strategies

As discussed earlier in the introduction to this chapter, two basic control strategies are open loop and closed loop. These strategies are analysed by considering the relationships between the control processes, the inputs and the outputs.

The following terms are used when evaluating control strategies:

- B = feedback signal, which is a function of the controlled variable
- C = controlled variable, which is the output of the system
- E = actuating or error signal, which is the difference between the reference input and the feedback signal

- G = transfer function, which converts input into output and can consist of both control and actuating elements
- H = feedback elements, which change the value of the controlled variable C to a feedback signal
- R = the reference signal input (set point)
- U = disturbance input, which is the total of all the uncontrolled signals that may affect the system. As was seen in Chapter 1, disturbances can be temperature changes, friction, loading changes, etc.

The use of the block diagram and mathematical transfer function enable systems and subsystems to be evaluated.

Open-loop system

Figure 2.20 shows the basic block diagram for an open-loop system.

Figure 2.20 Block diagram of open-loop system

Here, there is no feedback from output to input so:

$$\text{Transfer function } G = \text{output } C/\text{input } R$$

$$G = C/R$$

or, alternatively:

$$\text{Output } C = \text{input } R \times \text{transfer function } G \qquad (2.1)$$

$$C = RG.$$

Because there is no feedback, an open-loop system can give errors if there is a disturbance or a change in external loading. Examples of open-loop systems are:

- Electric motor drives in applications such as lifts, milk floats, machine tool spindles, etc. – any excessive load will cause the motor to slow down.
- Hydraulic and pneumatic cylinders – there is normally no indication of distance travelled by the cylinders.

To summarise, open-loop systems:

- are cheap, simple and very stable
- can be prone to errors because of lack of feedback.

Closed-loop system

Figure 2.21a shows the basic block diagram for a closed-loop system with negative feedback.

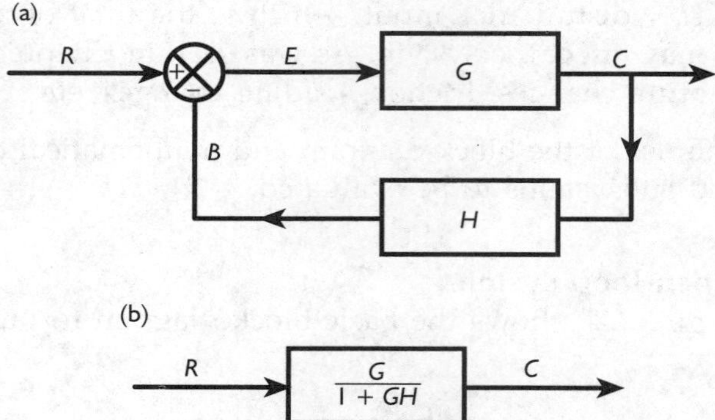

Figure 2.21 Block diagram of closed-loop system with negative feedback (a) and its equivalent transfer function (b)

In this case, the difference is that the actuating (sometimes called error) signal E is obtained by subtracting the feedback signal B from the reference input R so:

Transfer function G = (output C)/(actuating signal E)

$$G = C/E \tag{2.2}$$

or

Output C = actuating signal $E \times$ transfer function G

$$C = EG. \tag{2.3}$$

Because the feedback signal B is subtracted from the reference input R, these systems are referred to as closed-loop negative feedback systems and are self-correcting.

Using the transfer functions in this system gives:

$$E = R - B \tag{2.4}$$

$$C = EG \tag{2.5}$$

$$B = CH \tag{2.6}$$

where H is the transfer function of feedback element.

Substituting (2.4) into (2.5) gives:

$$C = G\ (R - B). \tag{2.7}$$

Substituting (2.6) into (2.7) gives:

$$C = G\ (R - CH). \tag{2.8}$$

Multiplying out gives:

$$C = GR - CGH. \tag{2.9}$$

Subtracting C and GR from both sides gives:

$$GR = C + CGH. \qquad (2.10)$$

So:

$$GR = C\ (1 + GH).$$

Therefore, dividing both sides by (1 + *GH*) gives:

$$C = GR/(1 + GH). \qquad (2.11)$$

So the equivalent transfer function for a system with feedback:

$$= C/R$$

$$= G/(1 + GH). \qquad (2.12)$$

This equivalent transfer function is shown in Figure 2.21b.

Examples of closed-loop negative feedback systems are:

- accurate control of position on CNC machine tools
- automatic control of sound and vision on television receivers.

To summarise, closed-loop systems:

- are accurate and take account of changing conditions
- are more expensive and can be unstable.

Case study

Controlling the speed of an electric motor

In a simple closed-loop system to control the speed of an electric motor:

$G = 10\ \omega/V$ (ω = speed in rad/s, V = voltage)
$H = 0.1\ V/\omega$.

The equivalent transfer function will be:

$$G/(1 + GH) = [10/(1 + (10\ \omega/V \times 0.1\ V/\omega))]\ \omega/V$$

Since $\omega/\omega = 1$ and $V/V = 1$,

$$10\ \omega/V \times 0.1\ V/\omega = 10 \times 0.1 = 1.$$

Therefore:

$$G/(1 + GH) = [10/(1 + 1)]\ \omega/V = [10/2]\ \omega/V = 5\ \omega/V.$$

So this transfer function means that every volt of input gives an output of 5 rad/s compared with the value of 10 rad/s for the transfer function *G*.

Note that the feedback signal *B* produced by the feedback element *H* must be compatible with the input signal *R*.

Combining systems and subsystems

Systems in series (sequential)

Figure 2.22a shows two system blocks in series.

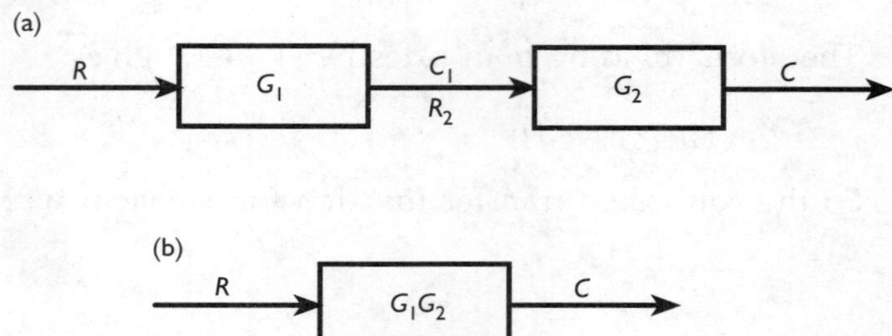

Figure 2.22 Two system blocks in series (a) and the equivalent transfer function (b)

Using the individual transfer functions gives:

$$C_1 = RG_1 \tag{2.12}$$

$$C = R_2G_2 \tag{2.13}$$

since by inspection $R_2 = C_1$.

Therefore, substituting (2.12) into (2.13) gives:

$$C = RG_1G_2.$$

Thus, the equivalent transfer function for two systems (or subsystems) G_1 and G_2 in series is G_1G_2.

This equivalent transfer function is shown in Figure 2.22b.

Systems in parallel (combinational)

Figure 2.23a shows two system blocks in parallel.

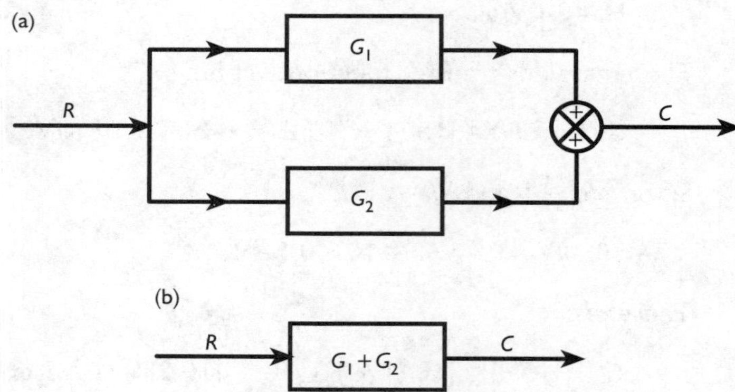

Figure 2.23 Two system blocks in parallel (a) and equivalent transfer function (b)

Using the individual transfer functions gives:

$$C = RG_1 + RG_2 = R(G_1 + G_2).$$

Thus, the equivalent transfer function for two systems in parallel is $G_1 + G_2$.

So the parallel blocks can be replaced by an equivalent single block, as shown in Figure 2.23b.

Activity 2.6

Find a closed-loop system in which the open-loop transfer function of the controller G and the feedback block H is known. Measure the output C and verify that the equivalent transfer function of the system is $G/(1 + GH)$.

Sequential control

This type of control is used when devices have to be switched in sequence. A typical example of this is the control of operations such as punching a hole in a metal strip with a press. A typical sequence would be: switch on the motor to move metal under press; switch off motor; receive signal from guard; wait five seconds; switch control valve that moves press ram down; switch control valve to move press up; wait five seconds; receive signal from guard.

The programmable logic controller (PLC) described on pages 50–3 is ideal for sequence control. This is because it can send and receive analogue and digital signals in any kind of timed sequence.

On–off control

As the name implies, on–off control is the simplest type of control action for a system. The system being controlled can only be in one of two states, either on or off. There are no intermediate states, so on–off control is termed discontinuous. Typical examples of on–off control include heating or cooling systems with closed-loop thermostatic control, such as electric and gas cookers, furnaces, central heating, refrigerators, freezers, etc. When the temperature is above the set point in a cooling system or below the set point in a heating system, the system switches on. When the temperature reaches the set point, the system switches off.

The way on–off control works is shown in Figure 2.24.

It can be seen that the system responds by switching on, because the initial value is less than the set point. It switches off when the set point is reached but, because of the system inertia (sometimes called hysteresis), the value rises above the set point (overshoots). The value then falls to just below the set point when the system switches on. The value still continues to fall, however, because of the system inertia (undershoots). The system inertia acts as damping for the system to prevent rapid oscillations. Figure 2.25 shows what happens in a low-inertia system.

Although the value has less overshoot or undershoot, it can be clearly seen that the system is unstable with rapid oscillations and consequent rapid switching on–off. This would cause problems in a central heating system with the boiler constantly switching on and off. A mechanical equivalent to this is the behaviour of a suspension spring on a vehicle with worn shock absorbers. Normally, the shock absorbers act as dampers to prevent the spring oscillating. However, if the shock absorbers are worn, the spring will oscillate and the vehicle will constantly bounce up and down.

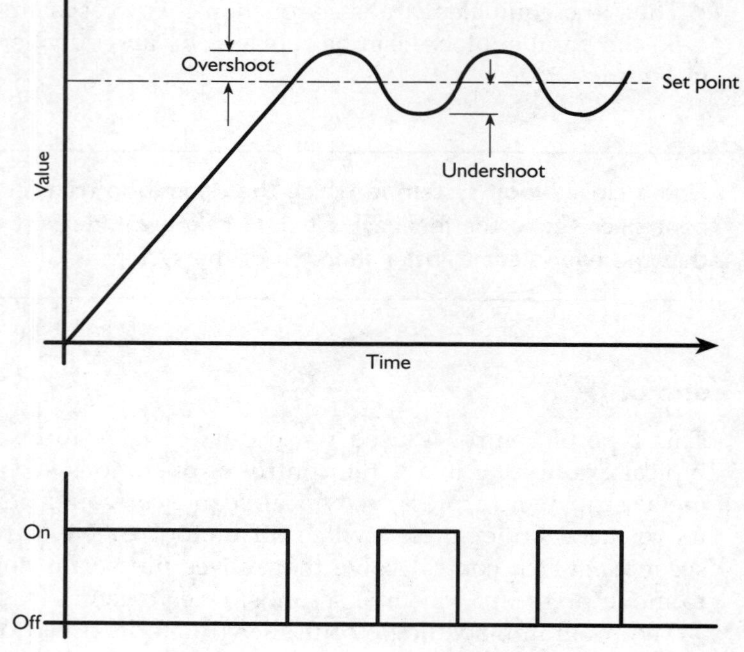

Figure 2.24 Set point on–off control

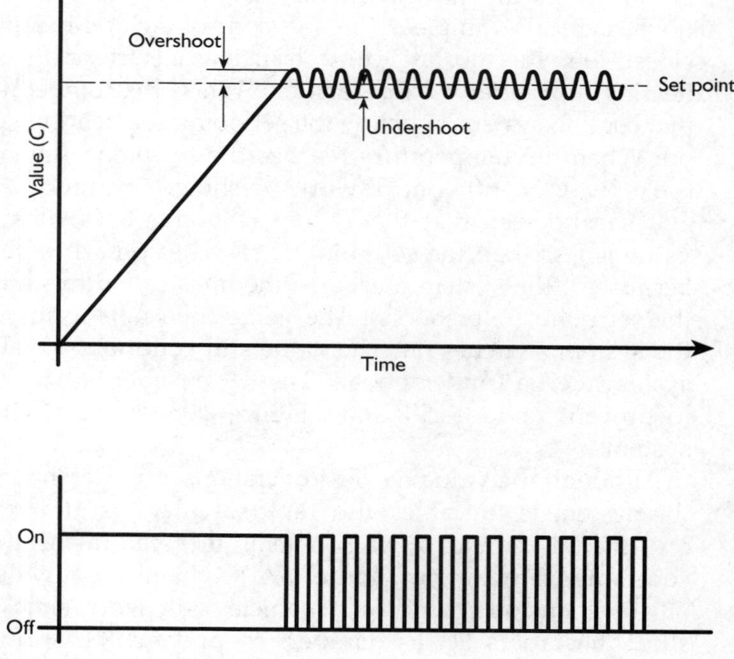

Figure 2.25 A low-inertia system

On–off control systems can be summarised as:

- subject to cycling with undershoot and overshoot about set point
- unsuitable for very precise control.

Activity 2.7

Investigate an on–off control in a heating system such as a furnace, a cooker or a heater. A suggested method of doing this is:

- start up the system and allow it to reach an initial set point thermostat value
- measure this value with a thermometer
- change the set point to a higher thermostat value and note the time
- measure the temperature at regular intervals both up to reaching the set point and for an extended period afterwards
- change the set point to a lower value and measure the temperature at regular intervals as above
- plot all temperature values (*y*-axis) against time (*x*-axis) on a graph.

Comment on the system behaviour with regard to response rate, overshoot/undershoot, oscillation, hysteresis, differences and reasons for differences between going up to a set point and coming down to a set point, safety, efficiency, accuracy, repeatability, stability, settling time.

If time allows, the investigation can be repeated, but with a disturbance being introduced into the system before the set point is reached, e.g. parts being put into the furnace, food being put into the cooker or refrigerator, a window being opened near the heater, a reduction in supply voltage/gas pressure, etc.

Proportional control

Proportional control overcomes some of the defects in a simple on–off system by using an error signal that is proportional to the difference between the set point and the actual value of the variable. This is expressed mathematically as:

$$C = EG$$

where C is the control action (or output of the controller), E is the error (or actuating signal), which is the difference between the actual value and the set point ($R - B$), where R is the set point and $B = CH$. G is the transfer function (gain) of the controller (Figure 2.21 – closed-loop system).

Figure 2.26 shows an idealised representation of proportional control.

Figure 2.27 shows typical responses of different types of proportional control systems.

In Figure 2.27a, it can be seen that the control action progressively reduces until the set point is reached, and this will usually prevent overshoot. A system that behaves like this is said to be overdamped and would generally be somewhat sluggish in operation.

Figure 2.27b shows the ideal type of response combining a reasonably fast response without overshoot or oscillation. This is called a critically damped system.

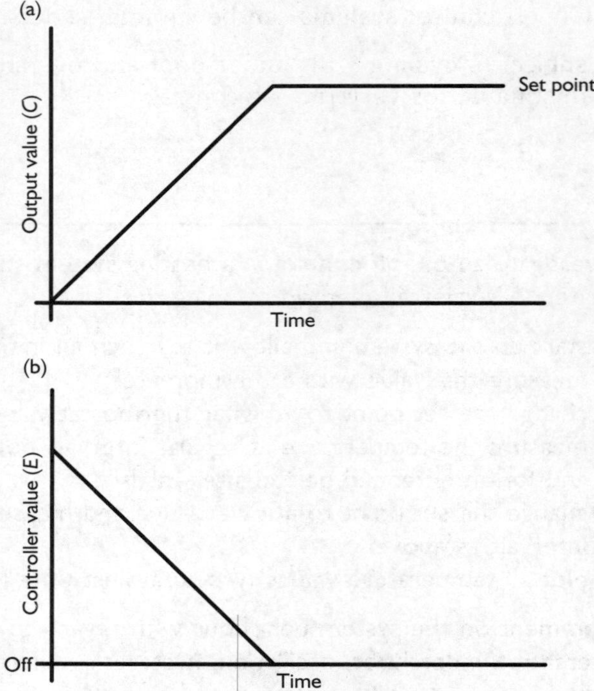

Figure 2.26 *Idealised representation of proportional control*

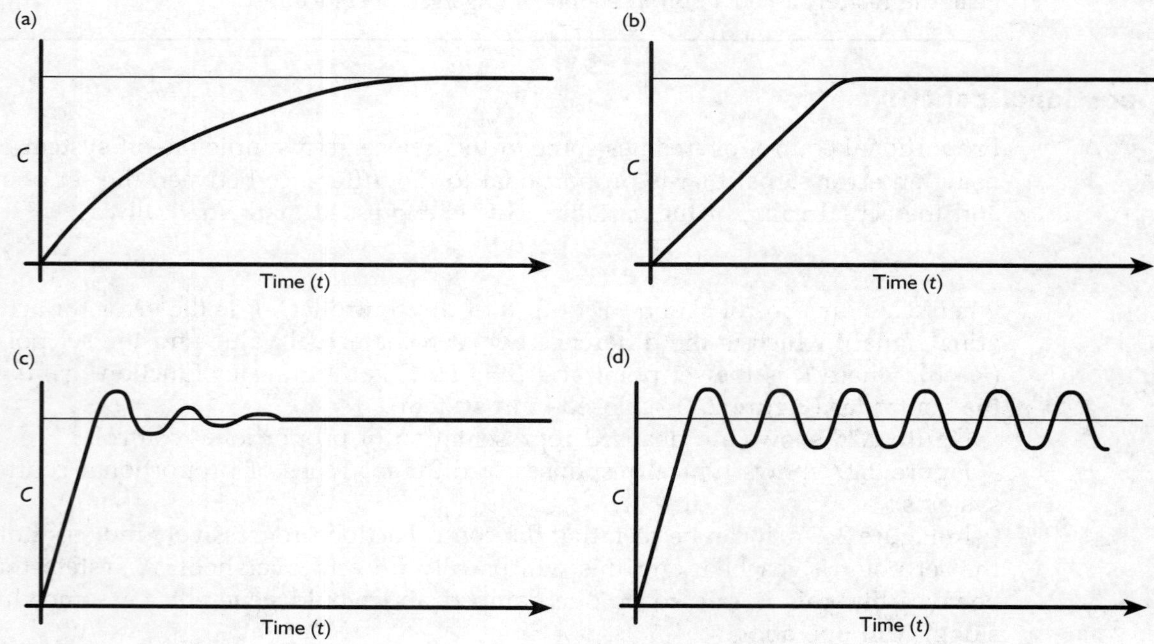

Figure 2.27 Typical responses of different types of proportional control systems. (a) Overdamped system; (b) critically damped system; (c) underdamped system; (d) unstable oscillating system

Figure 2.27c shows the response of a system in which overshoot occurs with the level settling after a few oscillations. This type of system is underdamped. It combines rapid response with some oscillation.

Figure 2.27d shows an unstable oscillating system in which there is insufficient damping to prevent oscillation.

Offset

In many proportional control systems, e.g. in a heating control system, an error will always be present because of the load (in the case of a heating system, it will be heat loss). This is because the system response lags the controller action. This steady-state error is known as 'offset'. It can be prevented by adding a constant K to the error E, so that even when the error is zero, there is still an output of K to the controller. The problem with this is that, as the load changes, the value of K will have to be changed. The equation for proportional control to eliminate 'offset' is:

$$C = EG + K_p.$$

Figure 2.28a shows a response to a step input with offset, and Figure 2.28b shows the response with the offset constant K_p added.

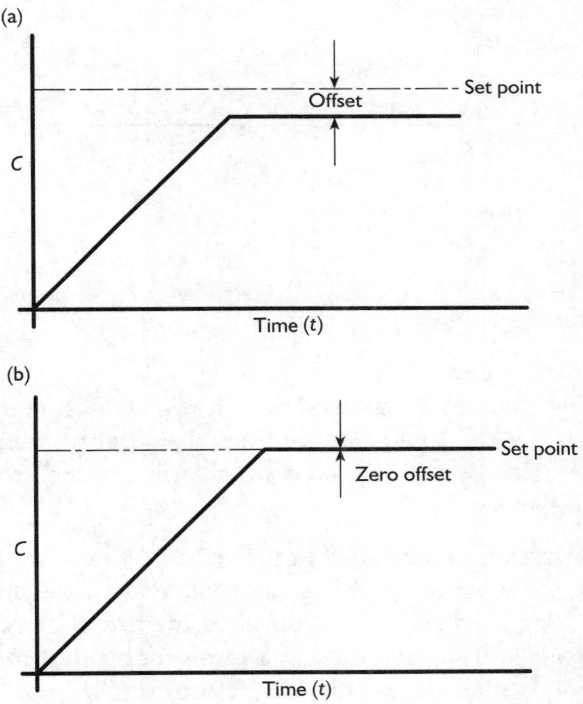

Figure 2.28 (a) Response to a step input with offset. (b) Response with offset constant K_p added

Case study

Liquid level in a tank

A good example of a simple proportional controller is the float control for controlling liquid level in a tank such as a toilet, a main water tank in the roof of a building, a tank in a chemical process, etc. This is shown schematically in Figure 2.29.

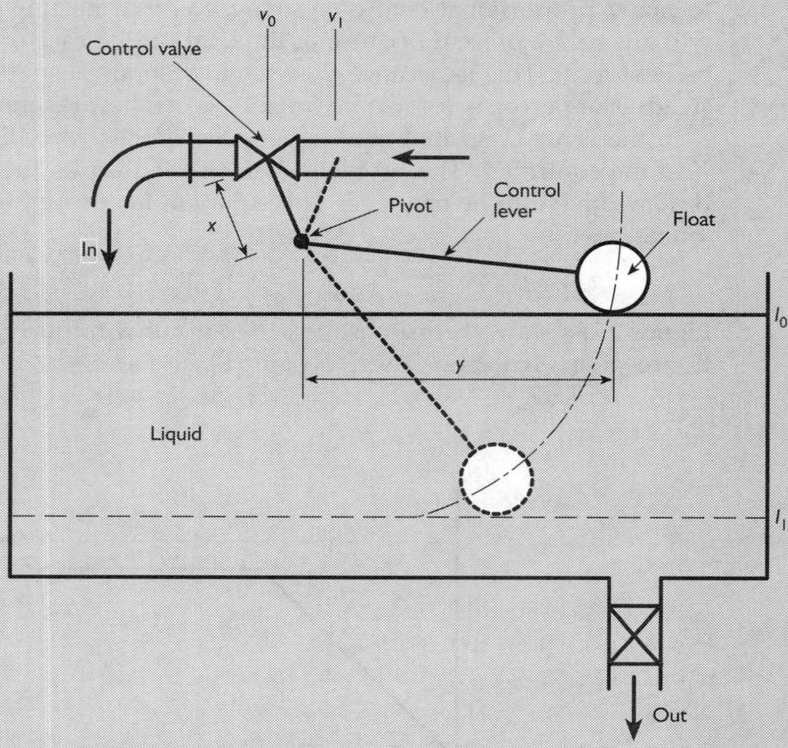

Figure 2.29 Diagram of float control in a tank

The value to be controlled is the liquid level in the tank. The float measures the level of the liquid and transmits this level to the control valve. This valve is actuated by the pivoted control lever attached to the float and the valve. The system works by:

- the valve being held in the totally off position (v_0) when the liquid level is at l_0
- the valve being opened as soon as the level drops below l_0, with the valve opening being proportional to the drop in level
- liquid being admitted at a rate proportional to the difference in level until level l_0 is reached and the valve closes totally.

As shown in Figure 2.29, if the distance from pivot to float is y and from pivot to valve is x, the valve is at position v_0 when the level is at l_0 and at position v_1 when the level is at l_1, then neglecting the effect of the pivot arm radius, it can be assumed and deduced that:

$$(v_1 - v_0)/x = (l_0 - l_1)/y$$

Error signal $E = l_0 - l_1$
$(v_1 - v_0)/x = E/y$
$v_1 - v_0 = Ex/y$
Valve position $v_1 = E\,(x/y) + v_0$ (2.14)

Gain (or transfer function) of the proportional controller (see page 67 on analogue amplification) is equal to the change in valve position/change in fluid level.

Gain $= (v_0 - v_1)/(h_0 - h_1) = x/y$.

Activity 2.8

Obtain a water tank with a float control system. Modify it so that the float can be adjusted to vary the gain. Modify the water outlet so that the level can be changed at differing rates. Investigate the system, setting the gain at the:

a maximum value possible
b minimum value possible
c midway between the maximum and minimum.

A suggested method for the investigation is to:

1 Fill the tank to level l_0 so that the inlet supply is off.
2 Reduce the liquid level to a new level l_1:
   a as fast as possible
   b at a very slow rate
   c at an intermediate rate.
3 For each case, measure the changes in liquid level and valve position at frequent intervals. Plot these changes on two graphs similar to those in Figure 2.26.

From an analysis of your results and graphs comment on:

- response rate (time to reach set point)
- stability
- repeatability
- accuracy
- undershoot/overshoot
- effect of changes in gain (sensitivity) and rate of change of level
- settling time
- safety
- efficiency.

Proportional, integral and derivative control

While a simple proportional controller is adequate for controlling liquid level, there are many systems for which it is not suitable. A proportional controller can be improved by adding other control methods.

Proportional, integral and derivative (PID) control

To eliminate the problems inherent in a simple proportional controller, it is usual to add integral and derivative control.

Integral control can be added by integrating the error signal. Mathematically, this can be expressed as:

$$C = K_i \Sigma \, Edt$$

where K_i is the integral control constant.

The addition of integral control will normally eliminate 'offset'.

Derivative or anticipatory control can be added by deriving a corrective action proportional to the rate of change of error. Mathematically, this can be expressed as:

$$C = K_d \, dE/dt$$

where K_d is the derivative control constant.

The addition of derivative control allows a much higher gain to be used without the system overshooting.

It is usual to implement all three terms in PID control together and, for this reason, it is also called three-term control. The full PID algorithm is:

$$C = K_pE + K_i \Sigma \, Edt + K_d dE/dt.$$

A block diagram of a three-term controller is shown in Figure 2.30a. These three parallel terms can be combined to give Figure 2.30b.

A three-term controller can be implemented by either analogue or digital means. A programmable logic controller (PLC) can be used to implement PID control by either means.

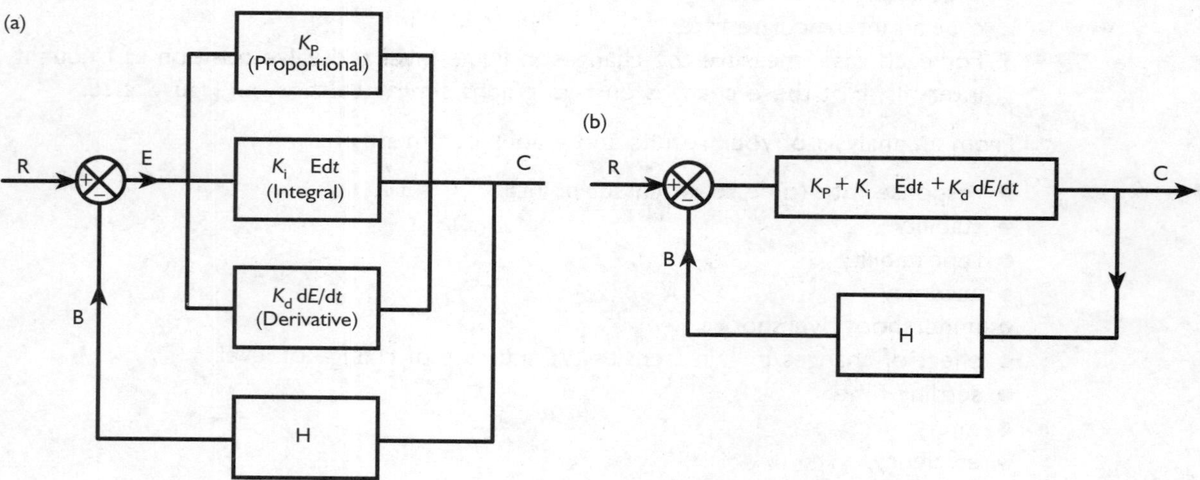

Figure 2.30 PID control. Block diagram (a) of three-term controller combined to give (b)

Case study

CNC lathe

A CNC lathe as used in an earlier case study (see pages 18–19) is a good example of where PID control would be used. PID control would be used because the system must have a rapid response and be very accurate. Overshoot is not desirable because this would cause errors to the workpiece. The slides of the lathe are moved by leadscrews, which are rotated by electric servomotors. The output of distance moved is measured by transducers on the slideways. The derivative term in the controller is obtained directly from velocity transducers on the motors rather than by differentiating the error E. This is possible because velocity is the derivative of distance and time. The schematic and block diagrams are shown in Figure 2.31a and b. A change in the set point R causes the controller to turn the servomotor in the correct direction. Signals are fed back from the tachogenerator, which measures velocity, and the optical encoder, which measures distance.

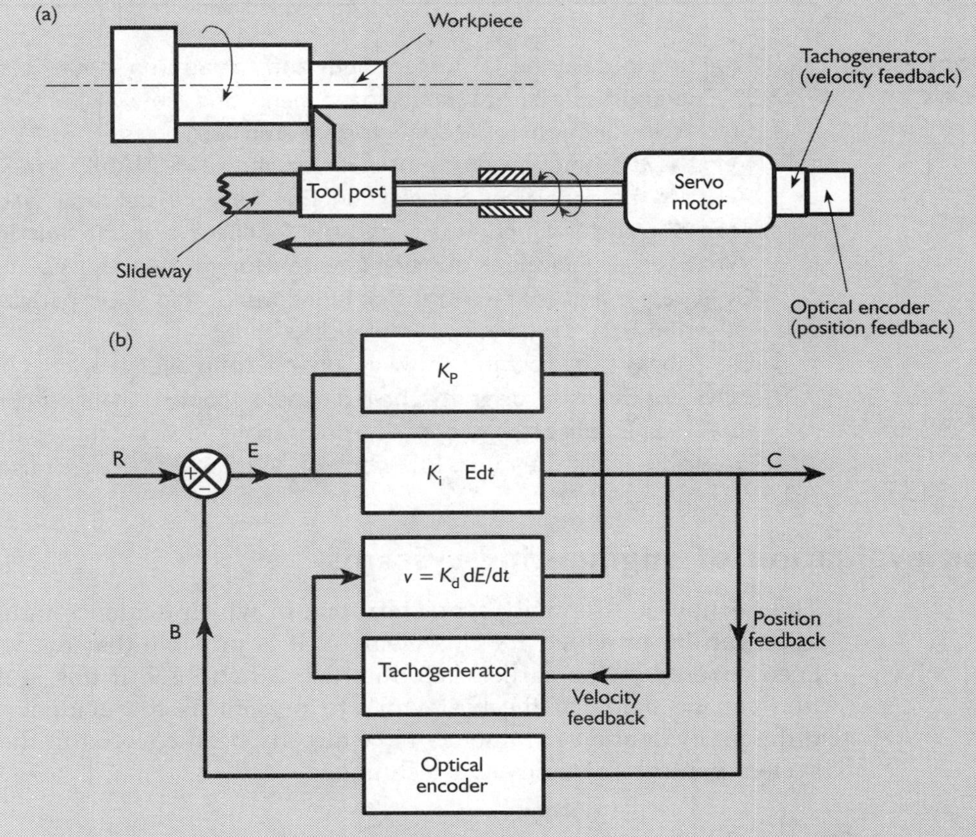

Figure 2.31 CNC lathe. Schematic (a) and block (b) diagrams

Activity 2.9

Find a CNC lathe (or any CNC machine tool) that has the slides driven by servo-motor and has both velocity and positional feedback. Investigate the system by:

1 Programming the slide to move a reasonable distance, say 200 mm, along the slideways at five different feed rates from the slowest possible to the fastest possible.
2 Initiating the slide movement and plotting distance moved against time. Repeat this three or four times to check accuracy and repeatability. If the necessary equipment for measuring time and distance is not available, it can be done indirectly by machining a continuously varying workpiece like a taper. The taper diameter can then be measured at regular intervals along the workpiece.

Present the results as a graph of output distance against time similar to the graphs in Figure 2.27. Comment on: accuracy, repeatability, stability, overshoot/undershoot, sensitivity, rate of response,; settling time, safety and efficiency.

Progress check

1 Define and describe (a) a step input and (b) a ramp input.
2 Define and describe (a) an analogue signal and (b) a digital signal.
3 State two functions of an operational amplifier.
4 What is the function of (a) an inverter and (b) a NAND gate?
5 What is the equivalent transfer function for a closed-loop system if the control transfer function is G and the feedback transfer function is H?
6 What is the equivalent transfer function for (a) two transfer functions G_1 and G_2 in series and (b) two transfer functions G_1 and G_2 in parallel?
7 Describe how sequence control works.
8 Describe, using diagrams, how an on–off controller works.
9 (a) Describe, using diagrams, how a simple proportional control system works. (b) What is 'offset' in proportional control and what causes it?
10 Describe, using diagrams, PID control. Why is PID control used?

The evaluation of engineering systems

This section is essentially a practical one in which evidence indicators on evaluation can be provided for this element. It is possible that the work has already been covered in the earlier Activities 2.7, 2.8 and 2.9. If this is the case, a selection can be made of the best work to present as the evidence indicators. The different evaluation parameters have already been covered in the introduction to system control strategies and techniques.

Activity 2.10

Select a control system from any of the categories of system covered earlier. Evaluate this system in a similar way to those suggested in Activities 2.7, 2.8 and 2.9. Use the ten evaluation parameters as described on page 57. It is important that this evaluation is carried out on a real system. If a real system is not available, then a computer simulation could be used instead.

Progress check

Define and describe the following terms used in the evaluation of control systems:

1 safety
2 accuracy
3 efficiency
4 repeatability
5 stability
6 overshoot and undershoot
7 settling time
8 sensitivity
9 response rate
10 time constant.

Assignment 2

This assignment provides evidence for:
Element 2.2: Investigate the operation of engineering systems
and the followig key skills:
Communication: 3.2, 3.3, 3.4
Information Technology: 3.1, 3.2, 3.3
Application of Number: 3.1, 3.2, 3.3

Your tasks

This assignment should be presented as a well-written word-processed document. You should also, if possible, use spreadsheets, graphs, drawings and formulae. The use of all these techniques will also give credits in the core skills of communication, IT and numeracy. This assignment is essentially to ensure that all the evidence indicators for the unit have been covered. This may already have been done in earlier activities, but it will all need pulling together into one portfolio. You can also use the systems that you will need to evaluate for your design evaluation assignment at the end of Chapter 10.

Draw block diagrams of one engineering system from each of the categories of:

- electromechanical
- fluidic
- chemical
- information and data communication
- thermodynamic.

For each system, represent the constituent subsystems and components and:

- describe all subsystem functions
- identify and describe the subsystem inter-relationships
- identify and describe the system control strategies and techniques.

Choose one of the systems and evaluate its performance as described on page 57. Draw the block diagram of the system that has been evaluated showing clearly all feedback loops and control equations.

Unit test 1: Engineering systems

1 A programmable logic controller (PLC) is used to control a flexible manufacturing cell. The PLC is
   A a piece part manufacturing system
   B a data communication system
   C an eletromechanical system
   D a machine tool system
2 The PLC in question 1 is connected to a hydraulic cylinder. Which type of system is the hydraulic cylinder?
   A electromechanical
   B thermodynamic
   C electronic
   D fluidic
3 Which is a chemical system?
   A etching of tracks on a printed circuit board
   B supply of water to industrial customers
   C a steam turbine
   D grinding of high-precision shafts
4 Test figure 1 shows the block diagrams of three different systems 1, 2 and 3. Match each of the block diagrams to one of the system purposes (A, B, C or D).
   A power generation
   B power transmission
   C environmental control
   D manufacture

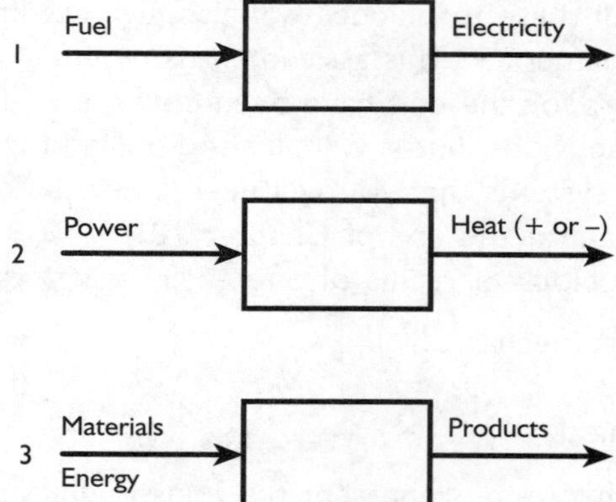

Test Figure 1

5 Typical inputs/outputs of different engineering systems are:
   1 output of finished components
   2 input of fuel oil
   3 output of polymer material

Match each of these inputs/outputs to the correct system (A, B, C or D).

A fluidic system
B manufacturing system
C thermodynamic system
D chemical system

6 Which output could apply to a thermodynamic system?
A fossil fuel
B nuclear radiation
C processed materials
D hot gas

7 Which input could apply to an information and data communication system?
A an analogue signal from a transducer
B a movement of parts in a factory
C a supply of raw materials
D piped natural gas

8 Identify a normal output of a fluidic system
A toxic materials
B rotary motion
C electromagnetic radiation
D useful heat

9 Which subsystem functions would be undertaken by the engine management computer system of a car?
A vehicle guidance
B monitoring and control
C energy conversion
D road safety

10 Which system requires a waste treatment and/or waste discharge subsystem?
A information and data communication
B electromechanical
C pneumatic/fluidic
D thermodynamic

11 A signal from a transducer subsystem to a control system indicating the value of a variable is called
A feedback
B calibration
C hysteresis
D oscillation

12 Which is the relationship between an aircraft computer subsystem and hydraulic actuators attached to surfaces, such as flaps and rudder, to which it sends signals?
A control of the subsystem by the actuators
B monitoring of the position of the surfaces
C control by the subsystem of the actuators
D monitoring of the stresses in the surfaces

13 Three typical engineering control systems are
1 control of the speed of an electric motor by continuously varying the voltage applied to it

2 control of a process, e.g. a washing machine, by just switching devices on and off at the correct time
3 sound control in a compact disc player using a laser beam directed onto the discretely stepped surface of the disc

Match each of the control systems to the most appropriate control system technique (A, B, C or D)

A digital
B analogue
C sequential
D integral

14 Different engineering systems and their required performances are
1 a positioning system in a robot laser system used to attach retinas in operations on human eyes
2 a control system for identifying reject parts at the rate of 1000/minute
3 a process temperature control system where a stated temperature must not be exceeded

Match each system to the most important criteria (A, B, C or D) to be selected when evaluating the system

A response rate
B overshoot
C efficiency
D accuracy

PART TWO: ENGINEERING PROCESSES

Chapter 3: Engineering processes for electromechanical products
Chapter 4: Making electromechanical products to specification
Chapter 5: Engineering services

Production and design engineers often work closely together to ensure that a product will be fit for its intended purpose and profitable to manufacture. The production engineers must evaluate the suitability of any alternative production process and, if necessary, recommend changes to the design that will avoid the use of expensive techniques or the need to purchase new equipment. At the same time, however, they must be aware of advances in manufacturing technology and be prepared to recommend the adoption of new methods and the purchase of new equipment if it will prove to be economical in the long term.

There also needs to be close co-operation between production engineers, supervisors, technicians and process operators. Education and training is generally required before new products or new equipment can be introduced, particularly with regard to safe working practices. Production processes can only operate efficiently and safely if the plant, tools and equipment are maintained in a good state of repair. Major items of plant and equipment should be serviced regularly as part of a scheduled maintenance programme. An engineering firm may have its own maintenance engineers and technicians or it may contract the servicing to outside specialists.

Chapter 3 Engineering processes for electromechanical products

This chapter covers:
Element 3.1: Select processes to make electromechanical engineered products.

. . . and is divided into the following sections:
- Engineering processes and process techniques
- Evaluation and selection of processes and process techniques
- Safety procedures and equipment.

Some of the processes used to make engineered products date back to ancient times. In particular, the casting and forging of metals can be traced back to the Bronze and Iron Ages. Metal and woodworking techniques have been developed and refined by blacksmiths, armourers, shipwrights and carpenters down the centuries. During the late Middle Ages, the need arose for accurate clocks and scientific instruments. New skills and techniques were developed by their makers who were perhaps the first precision engineers.

The Industrial Revolution saw the mechanisation of many processes that had previously been carried out by hand. There were developments in the techniques of iron smelting, and steam power was developed to drive machinery. In recent years, many new materials and processing methods have been developed, and many of the older techniques have been updated and refined. Production engineers must be familiar with the wide range of processes that are now available and select those that will provide quality and profitability.

Engineering processes and process techniques

The most common engineering production processes may be classified as:

- material removal
- material shaping
- joining and assembly
- heat treatment
- chemical treatment
- surface finishing.

Within these processes, there are a number of different processing methods or techniques. Some of the techniques are appropriate to small-scale production, and others can only be justified if large quantities are required. It may be possible to produce simple products and components using a single process. The more complex electromechanical products assembled from a number of parts are, however, likely to involve all the above range of processes in their production.

Material removal

Engineering components very often begin life as metal forgings, castings and barstock. Surplus material is then removed to give them their specified dimensions and surface finish. Three of the most common material removal processes are:

- drilling
- turning
- milling.

Drilling

In the drilling process, material is removed from the stationary workpiece by a rotating cutter. The most common types of drilling machine are the sensitive bench-mounted drill, the pillar drill and the radial arm drill, as shown in Figure 3.1.

The sensitive drill is used for small-diameter holes, usually less than 10 mm in relatively small workpieces. The pillar and radial arm drills can accommodate larger workpieces and produce larger diameter holes. They are made in different sizes, but they all have the same general appearance. With the sensitive drill, the spindle is driven through a belt drive from an electric motor. To change its speed, the position of the V-shaped drive belt must be changed on stepped pulleys. Pillar and radial arm drills are driven through a gearbox equipped with speed selection levers.

With each type of drill, the spindle is usually fed downwards through a rack and pinion mechanism, operated by a lever or hand wheel. The larger sizes of pillar and radial arm drill are equipped with a range of automatic feeds that can be selected from the gearbox. Sensitive and pillar drills have worktables that contain slots to which the workpiece or a machine vice can be bolted. The worktable can be raised or lowered and set over at an angle if required. With radial arm drills, the worktable is stationary but the height of the radial arm that contains the spindle can be adjusted for height.

The material removal operations that are most commonly carried out on drilling machines are drilling, reaming, countersinking and spotfacing. Holes are drilled using twist drills, which are usually made from high-speed steel. Reaming is an operation used to enlarge drilled holes to a precise diameter. Countersinking involves cutting an angled recess at the start of a hole. This is to accommodate a countersunk-headed screw or rivet that must lie flush with the surface of the component. With spotfacing, a small amount of material is removed from the material surface around a hole to provide a flat seating for a nut and washer.

The safe and efficient removal of material on drilling machines requires a knowledge of:

- workholding methods
- toolholding methods
- types of cutting tool
- selection of cutting speeds
- hole positioning and drilling.

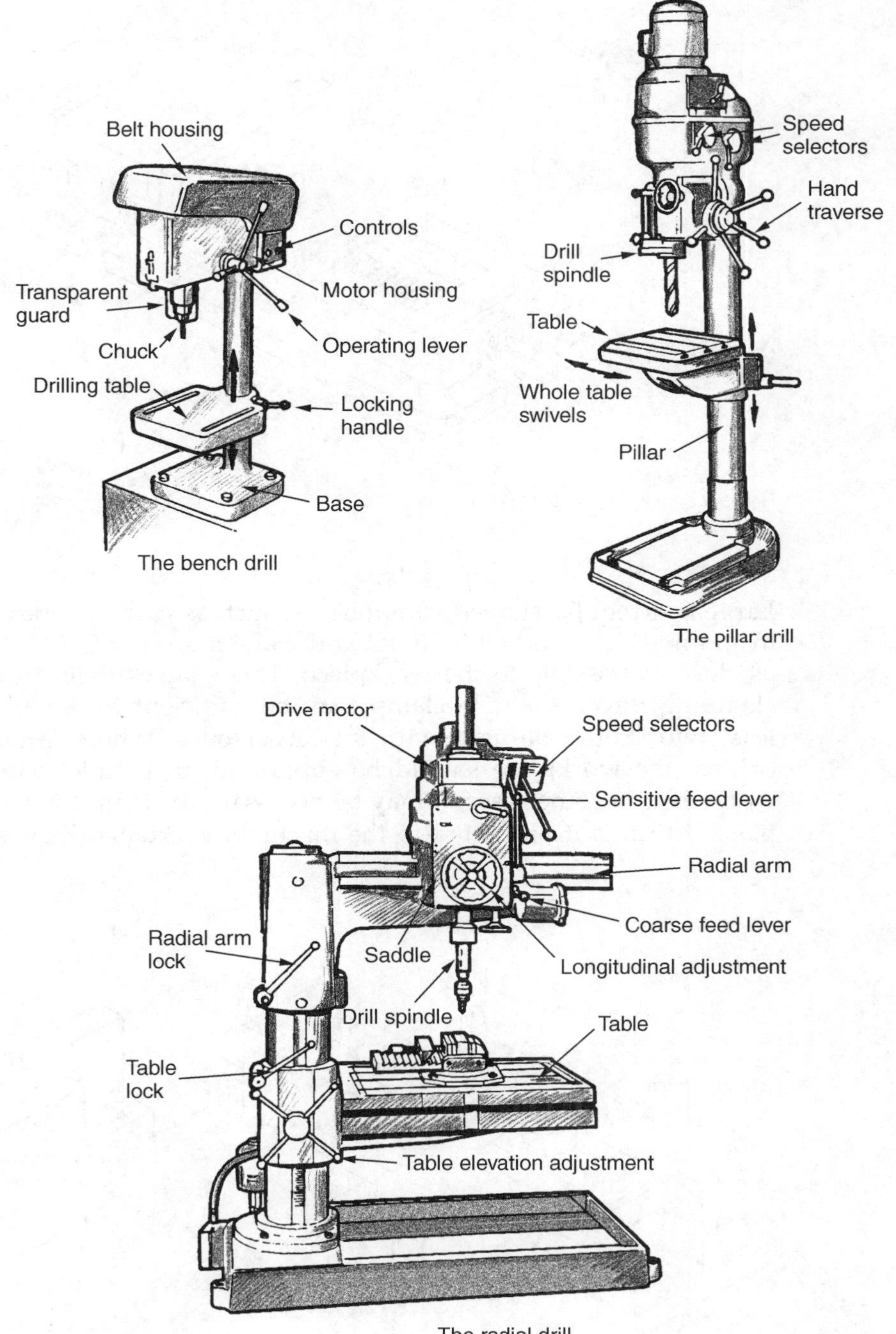

Figure 3.1 Drilling machines

Workholding methods: small components are held in a machine vice, which is bolted to the machine worktable (Figure 3.2). They are supported in the vice on parallel bars. These ensure that the drilled hole is perpendicular to the lower face of the workpiece. The parallel bars also reduce the likelihood of damage from the drill striking the vice after breaking through the material.

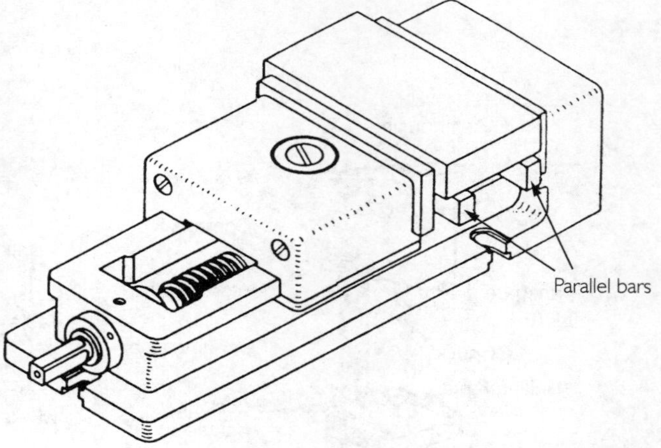

Figure 3.2 Workholding in the machine vice

Large or irregular-shaped components, such as castings, may be secured directly to the machine worktable. Bolts and clamps are used with the bolts positioned as close as possible to the workpiece. This enables them to exert the maximum clamping force. A single clamp may be sufficient for small workpieces, but at least two should be used with the larger ones. Where through holes are to be drilled, the workpiece should be supported on parallel bars to raise it off the worktable. In some cases, it may be necessary to clamp the workpiece to an angle plate, which is itself bolted to the machine worktable (Figure 3.3).

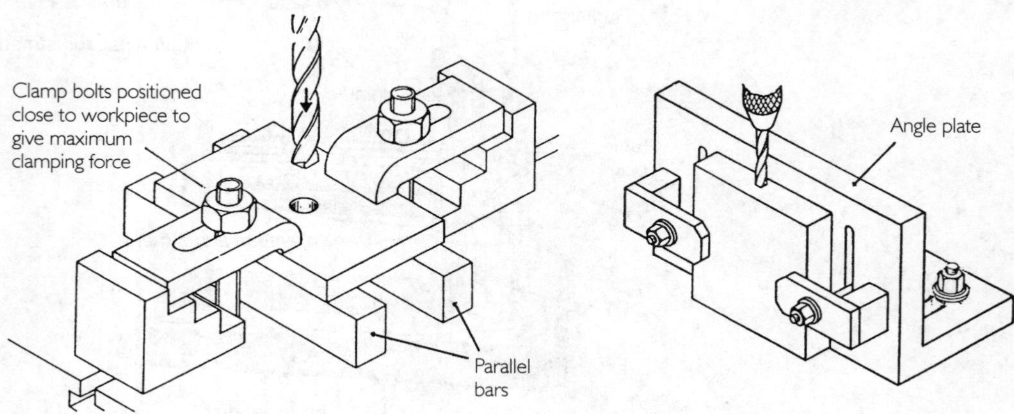

Figure 3.3 Workholding on the drilling machine

Securing cylindrical components can sometimes be a problem (Figure 3.4). Some drilling machine vices have a V-shaped groove cut in the jaws for gripping these components. Otherwise, it is good practice to clamp them on V-blocks.

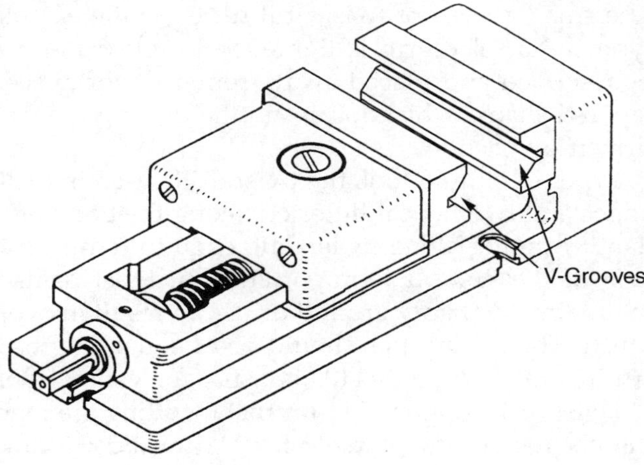

Figure 3.4 *Vice for gripping cylindrical components*

Toolholding methods: drilling machine spindles are hollow and contain an internal morse taper. The larger diameter drills have morse-tapered shanks that locate directly in the spindle or in a morse-tapered sleeve, which also locates in the spindle. They are held and driven by friction alone. The tang at the end of the tapered shank is used only when releasing the drill using a tapered drift as shown in Figure 3.5.

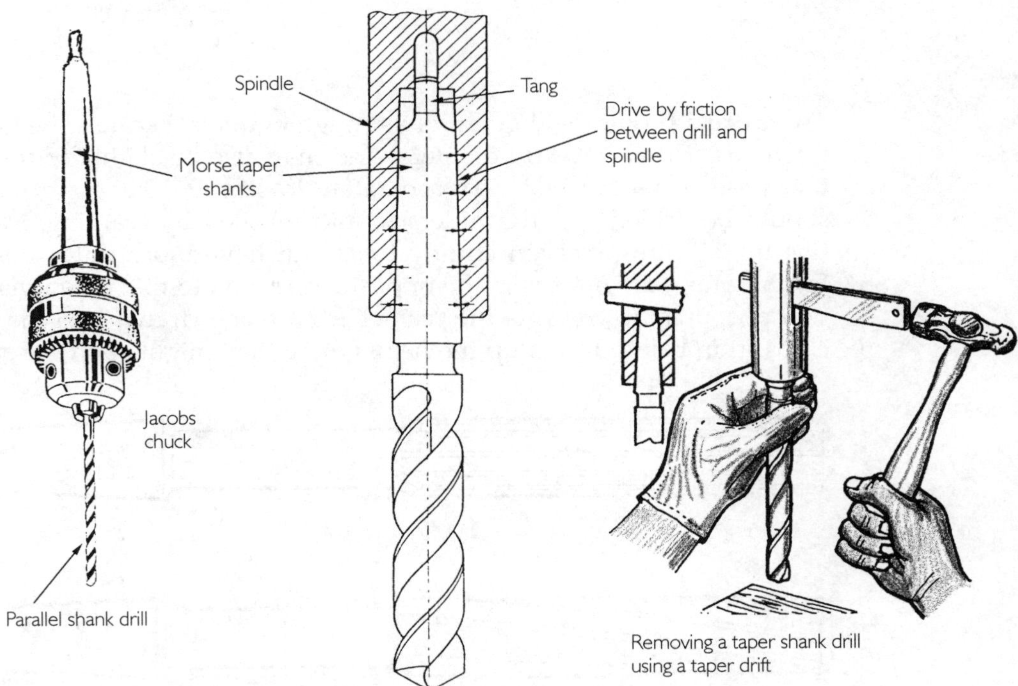

Figure 3.5 *Toolholding in the drilling machine*

The smaller sizes of twist drill have parallel shanks and are gripped in a Jacobs-type chuck. The grip is tightened by means of a chuck key, which should then be removed and placed away from the workpiece. The chuck itself has a morse-tapered shank, which locates in the drilling machine spindle and is also held and driven by friction.

Types of cutting tool: the twist drills used in drilling machines have two cutting edges and two helical flutes cut along the body of the drill. Each flute has a raised 'land' running along its leading edge to reduce friction between the drill and the side of the hole. It is important that twist drills are accurately ground. Poorly maintained or badly ground drills can result in oversized holes with a poor surface finish. The cutting lips should be of equal lengths and ground at the same angle. An included angle of 118° is usual for general-purpose drills.

Twist drills (Figure 3.6) for metal cutting are made from high-speed steel, while the cheaper types, suitable for DIY woodworking, are sometimes made from a lower grade plain-carbon steel. Twist dills are also made with different helix angles for drilling different materials. Those used for drilling metals have a faster helix than those used for drilling soft plastics, i.e. more twists.

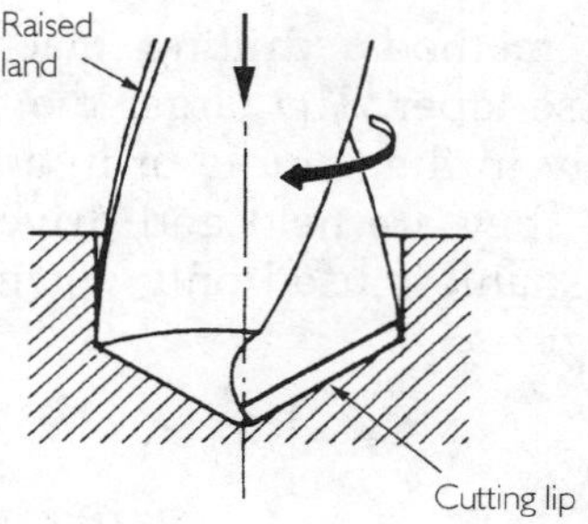

Figure 3.6 Twist drill

Twist drills should be thought of as roughing tools, because the hole produced is often larger than the drill diameter and may also be slightly out of round. Holes that need to be finished very accurately to size and have a good surface finish should be drilled slightly undersize and finished by reaming. Machine reamers (Figure 3.7) are precision cutting tools that have more flutes than a twist drill. Furthermore, the helix is in the opposite direction to that on a twist drill. This has two purposes: it prevents the reamer from being drawn into the hole and stops chips from being drawn up the helix where they might spoil the surface finish.

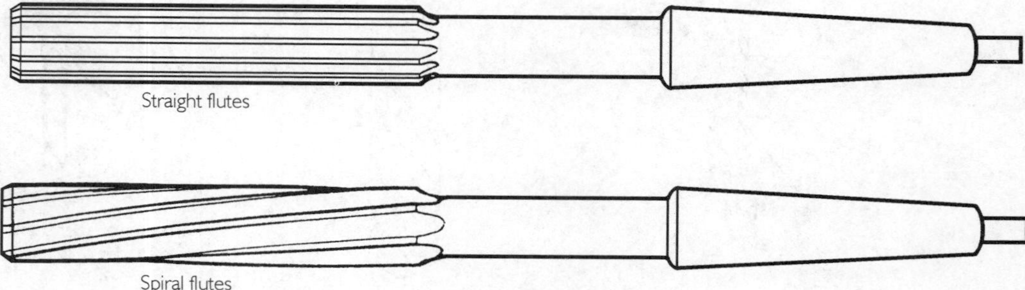

Figure 3.7 Machine reamers

There are two types of reamer. One type cuts only on the ends of the flutes and is best for use with mild steel. The other type, which is known as a rose-action reamer, cuts on both the ends and the edges of the flutes. It gives best results when used with brasses, bronzes, cast iron and plastics that have a tendency to close on, or grip, the reamer. Machine reamers should only be used with a low spindle speed.

Twist drills are not suited to drilling large-diameter holes in thin plate. When breaking through, they have a tendency to grab and tear the material, leaving a jagged hole that is oversize and out of round. A hole saw (Figure 3.8), held in the Jacobs chuck, is a more suitable cutting tool. This incorporates a twist drill to make a pilot hole and a saw blade that cuts the hole to the required diameter.

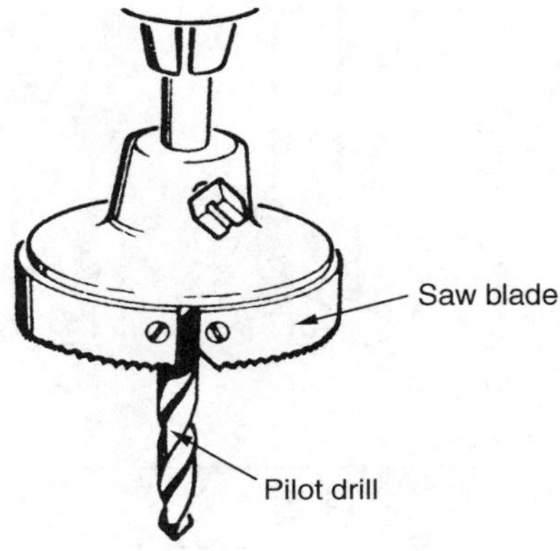

Figure 3.8 *Hole saw*

Countersinking cutters, which are also referred to as 'rose' cutters, have a conical cutting head and a parallel shank for holding in the Jacobs chuck. They are used to form countersunk recesses and also to deburr drilled holes, particularly on the side where the drill has broken through the material. Spotfacing cutters have a central pilot peg that locates in the drilled hole and acts as a pilot (Figure 3.9).

Selection of cutting speeds: the spindle speed to be selected for drilling depends on the workpiece material, the drill material and the diameter of the drill. Table 3.1 gives a general guide to the cutting speeds for different materials using high-speed steel cutting tools.

Table 3.1 Cutting speeds for different materials

Material being cut	Cutting speed (m/min)
Aluminium	70–100
Brass	70–100
Phosphor bronze	35–70
Mild steel	30–50
Grey cast iron	25–40
Thermosetting plastics	20–40

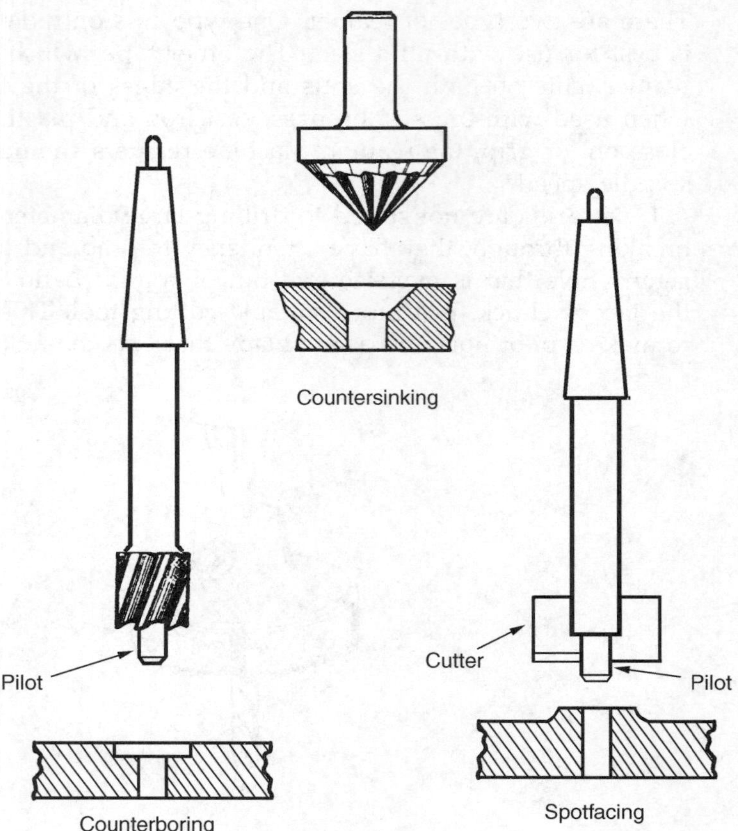

Figure 3.9 Countersinking, counterboring and spotfacing

These are cutting speeds at the drill tip from which the spindle speed can be calculated using the formula:

$$N = 1000\ S/\pi\ d$$

where N = spindle speed in r.p.m., S = cutting speed in m/min and d = the diameter of the drill in mm.

Hole positioning and drilling: after carefully marking out the centre positions of holes that are to be drilled in a component, they should be centre punched to help start the hole in the correct place. Where accurate positioning is required, or when drilling large-diameter holes, it is good practice to mark out the hole circumference using dividers, and to centre punch the circumference in four positions at 90° to each other (Figure 3.10).

Drilling should then begin with a small pilot hole. If its position is satisfactory, this can then be opened out progressively using larger drills up to the required diameter. If the hole position is seen to have shifted slightly during the intermediate drilling, it can be brought back on centre by the careful use of a found file.

When feeding the drill by hand, care should be taken not to exert too much force. Small-diameter drills are easily broken, and larger drills may overheat if they are fed too quickly. The pressure on the drill should be relieved as it is about

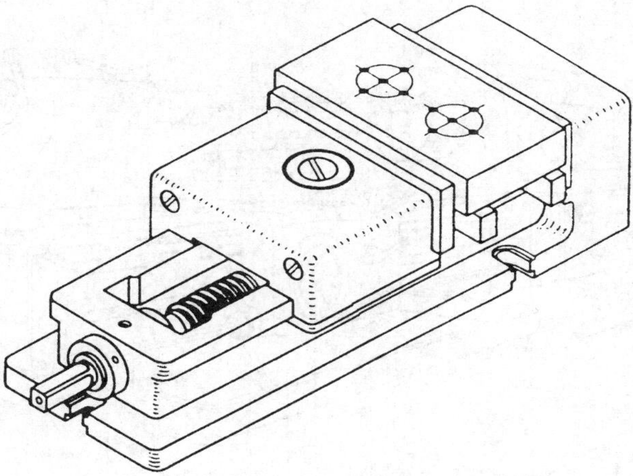

Figure 3.10 Marking out and centre punching

to break through the material. At this point, there is a tendency for the drill to 'grab'. This can cause small-diameter drills to break or the workpiece to spin dangerously with the drill if it is not securely clamped. When deep holes are being drilled, it is advisable to use a coolant and to remove the drill occasionally from the hole. This will lift out any swarf that has accumulated in the flutes of the drill.

Drilling machines are fitted with a depth gauge that can be used to measure the depth of penetration. The gauge can be set with a lock-nut to limit the travel of the drill. This is useful when drilling 'blind' holes that do not pass all the way through the workpiece. It can also be used to limit the travel of the drill after it breaks through the material. This prevents possible damage to the drill, the machine vice or the worktable.

Activity 3.1

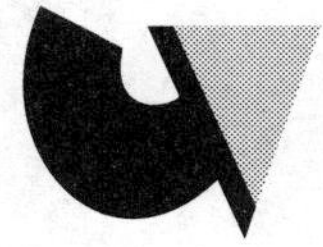

When drilling a hole, it is noticed that there is some vibration and that the continuous chips that the drill produces are of unequal thickness. Furthermore, the drilled hole is seen to have a poor surface finish and is oversize. Make a sketch of a drill cutting point showing the most likely cause of the above faults.

Turning

Turning operations are carried out in a lathe where material is removed from a rotating workpiece by a single point cutting tool or a twist drill. Material removal in the lathe enables cylindrical, tapered and flat surfaces to be produced. It also enables holes to be drilled and bored, and internal and external screw threads to be cut. The basic type of lathe is the centre lathe. Centre lathes (Figure 3.11) are not suitable for quantity production, but they are widely used for making small-quantity specialist items and for training. Many of the lathes used for quantity production are complex computer-controlled machines, but their development can be traced back to the centre lathe.

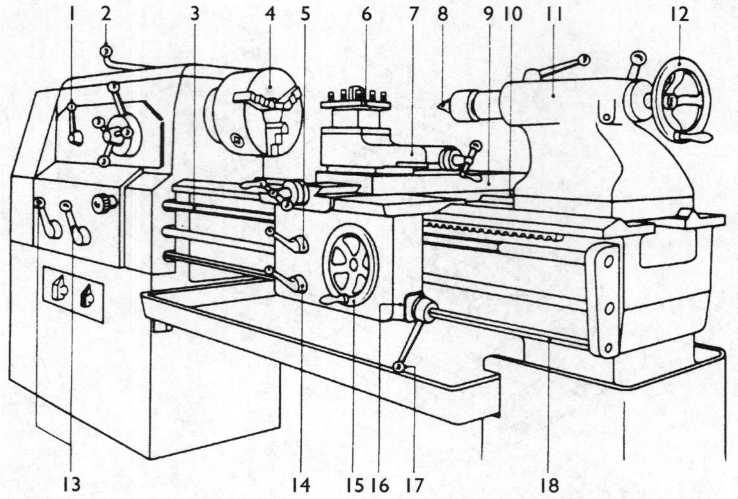

Figure 3.11 The centre lathe

The main parts of the centre lathe are the bed, the headstock, the tailstock and the saddle or carriage assembly. The lathe bed is made of cast iron with hardened V-shaped slideways along which the tailstock and saddle can be moved. The headstock contains the gearbox for selecting different cutting speeds and the spindle. The drive to the headstock is generally by V-belts from an electric motor. The spindle can carry a chuck, faceplate or a catchplate and centre for holding and rotating the workpiece.

The tailstock with its centre can be clamped in any position along the lathe bed to support the free end of a workpiece. The centre can be removed from the tailstock spindle and replaced by a taper shank drill or Jacobs-type chuck for drilling operations (Figure 3.12). A handwheel moves the tailstock spindle towards and away from the workpiece. On many centre lathes, the tailstock can be set over so that its centre is out of line with the spindle axis. This is useful for taper turning between centres, an operation that will be described later.

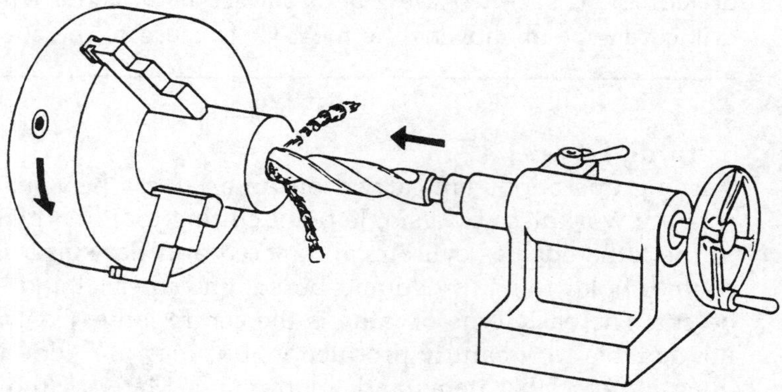

Figure 3.12 Use of tailstock for drilling

The saddle or carriage assembly rests on the slideways of the bed and contains the cross-slide, the compound slide and the toolpost. It may be traversed along the bed manually by means of a handwheel, or automatically by engaging an automatic traverse mechanism. The cross-slide moves across the bed at right angles to the spindle axis and is used for facing operations and increasing the depth of cut when cylindrical turning (Figure 3.13).

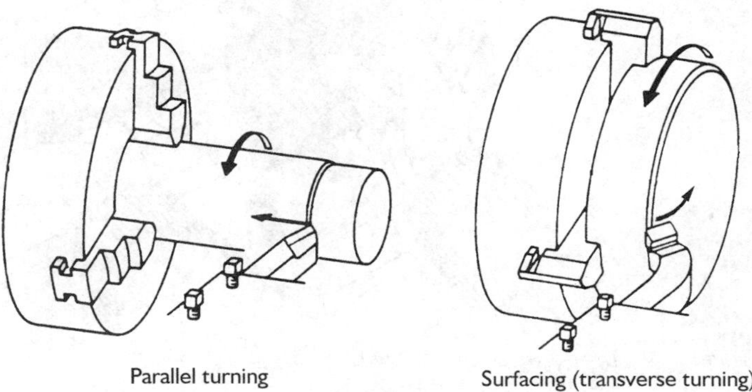

Figure 3.13 Parallel turning and facing operations

The compound slide is located on top of the cross-slide and can be set at any angle for turning tapers and chamfers. The handwheels of the cross-slide and compound slide contain scales calibrated in divisions of 0.01 mm, which enable the depth of cut to be set accurately. The toolpost rests on top of the compound slide and, depending on the type, may hold up to four cutting tools.

The front of the carriage is called the apron. It contains the manual traverse handwheels and the levers that are used to engage automatic traverse for cylindrical turning and automatic cross-traverse for facing. Automatic traverse is driven from the feedshaft that runs along the bed behind the apron. It can also be driven from the leadscrew, which runs alongside the feedshaft, but this should only be used for screw-cutting operations.

The safe and efficient removal of material in the lathe requires a knowledge of:

- workholding methods
- toolholding methods
- types of cutting tool
- selection of cutting speeds
- the use of coolants.

Workholding methods: the lathe spindle can support a chuck, a faceplate or a catchplate and centre. These are devices used for holding and rotating the workpiece. The spindle nose contains a screw thread on which they locate. This and the mating screw threads should be thoroughly cleaned when changing attachments. Lathe chucks may be of the three-jaw self-centring type (Figure 3.14) or the four-jaw independent type.

Figure 3.14 Three-jaw self-centring chuck

The three-jaw self-centring chuck is used for gripping round and hexagonal workpieces. The jaws move inwards and outwards in unison, driven by a scroll plate that is turned by a chuck key. The three jaws ensure that the centre line of the workpiece lies on the spindle axis for turning external cylindrical surfaces, end facing, drilling and boring operations. It will be noted that the outside of the chuck jaws are stepped. This enables workpieces with a large-diameter bore to be gripped internally, while the outer surface is being turned.

When turning long cylindrical workpieces, the free end needs to be supported. A conical recess is drilled in it using a specially shaped centre drill (Figure 3.15). The free end of the workpiece can then be supported on a conical centre held in the tailstock spindle. If a stationary centre is used, it must be well greased to prevent overheating. It is better to use a running centre if one is available. This rotates on ball bearings with the workpiece and generates very little heat.

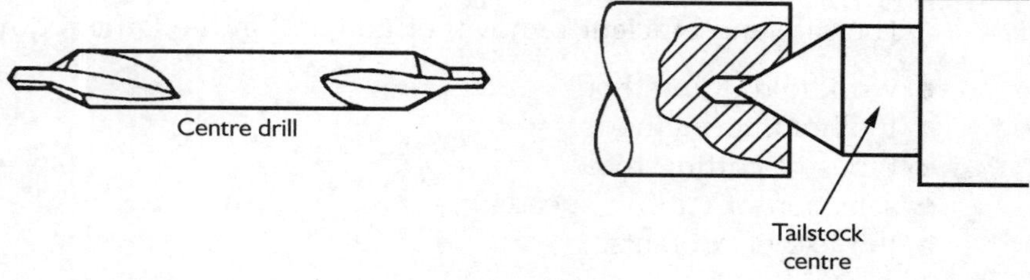

Figure 3.15 Use of centre drill

The four-jaw independent chuck (Figure 3.16) is used for holding square, rectangular and oddly shaped workpieces, such as castings and forgings. Each jaw has its own chuck key socket to move it inwards and outwards. The position of the workpiece can thus be adjusted until its centre of rotation lies on the spindle axis. The four-jaw chuck has greater gripping power than the three-jaw type but takes longer to set up correctly. Like the three-jaw chuck, it can be used to grip internally on components with a large-diameter bore.

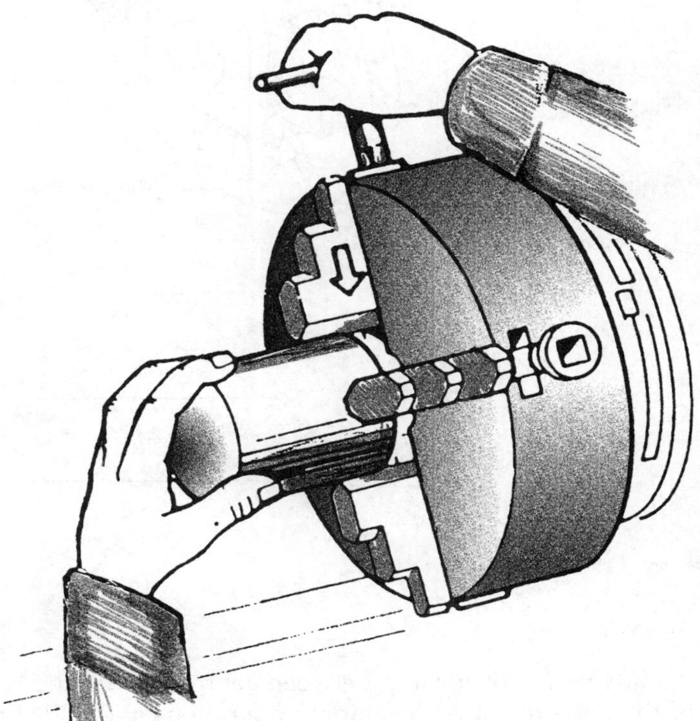

Figure 3.16 Independent four-jaw chuck

The faceplate is used to hold irregularly shaped workpieces that may be too large or otherwise unsuitable for gripping in the four-jaw chuck. The workpiece may be clamped directly to the faceplate as shown in Figure 3.17, or clamped to an angle plate. The procedure allows diameters and faces to be turned that are parallel or perpendicular to a premachined face. The premachined face locates against the faceplate, or on the angle plate. For through-boring operations, it is necessary to position parallel bars between the workpiece and the faceplate.

Irregularly shaped workpieces and the use of the angle plate can produce large out-of-balance forces. These cause vibration that can damage the machine and result in a poor surface finish. Balance can be restored by fixing weights to the faceplate. They are positioned by trial and error on the side opposite to the out-of-balance mass, until the faceplate can be placed in any angular position without rotating due to gravity.

The use of a catchplate, carrier and centres to support and rotate a workpiece is one of the oldest turning techniques. It is a very accurate method of turning long cylindrical workpieces that have been centre drilled at each end (Figure 3.18). A carrier or 'dog' is clamped to one end of the workpiece, which is then held between the live centre in the spindle nose and the tailstock centre. The carrier and workpiece are rotated by the driving peg on the catchplate.

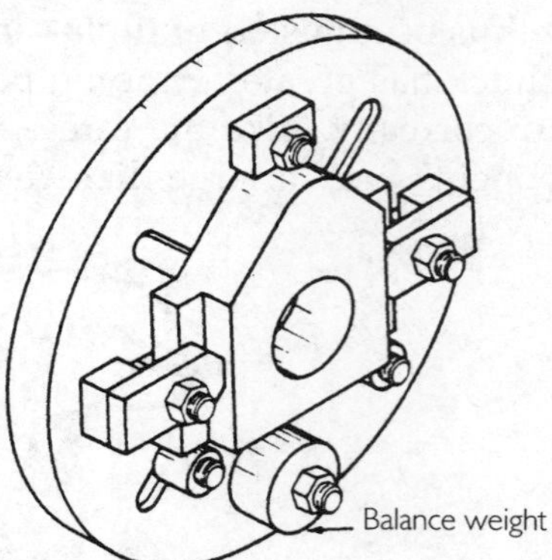

Figure 3.17 Use of faceplate

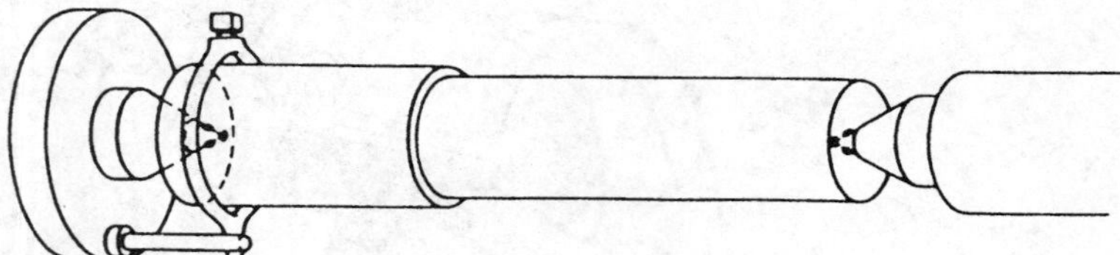

Figure 3.18 Turning between centres

An advantage of turning between centres is that the workpiece can be taken out of the lathe for other machining operations and still run true when it is replaced. It can also be turned round and replaced between the centres without any loss of concentricity. A disadvantage of the method is that boring and full-facing operations cannot be carried out.

Components that cannot easily be gripped in a chuck or that might be damaged by the chuck jaws can sometimes be turned on a tapered mandrel (Figure 3.19). This is a tapered bar held between centres and driven by the catchplate and carrier. The component, which must have a central premachined hole, is held by friction on the mandrel taper while its external surface is turned.

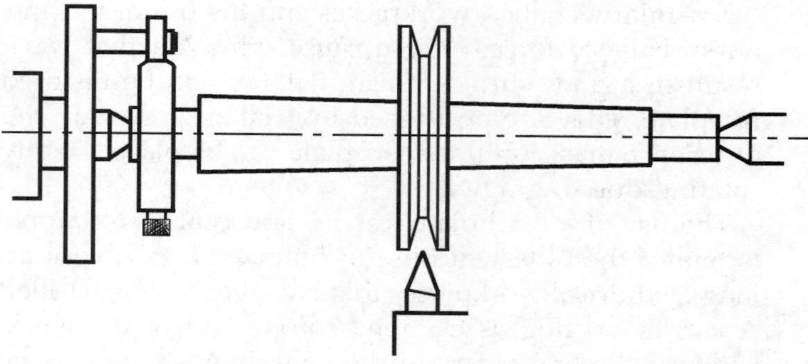

Figure 3.19 Use of tapered mandrel

Activity 3.2

Setting the tailstock over by a calculated amount while turning between centres enables a taper to be turned on the workpiece. By how much would you set the tailstock over to turn a taper of included angle 10° on a workpiece of length 450 mm between centres?

Types of cutting tool: the single point cutting tools used for cylindrical turning, facing and boring are clamped in the toolpost. A pillar-type toolpost is often used on small centre lathes, while the four-way and quick-release types are common on larger machines (Figure 3.20). The tools must be set with their cutting edges at the centre height of the lathe. A curved boat-piece enables the tool height to be adjusted with the pillar-type toolpost, and the quick-release type usually has height-adjusting screws. Packing pieces are placed beneath the cutting tools in the four-way toolpost.

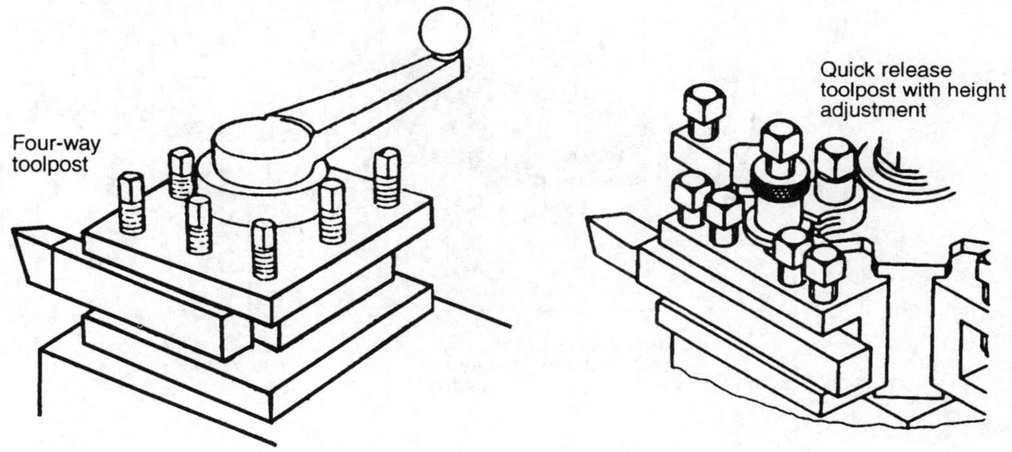

Figure 3.20 Toolposts

The quickest way of setting the cutting tools near to the machine centre height is to set the cutting edge level with the point of the tailstock centre or the centre in the spindle nose, if one is being used. Alternatively, a setting gauge or a steel rule can be used to measure the centre height above the lathe bed. A quick check on the centre height can be made by running the tool up to the stationary workpiece against an upright steel rule. The tool will be very close to centre height if the rule appears to stand vertical.

The tailstock is used in drilling operations for holding and feeding the drill. Taper-shank twist drills or a Jacobs-type chuck are held in the internal morse taper in the tailstock spindle. Before drilling, the workpiece should be centre drilled to start the hole.

The tailstock can also be used to hold taps and dies for cutting screw threads in the drilled holes. This is a hand-fed operation and should only be carried out with the machine isolated and in neutral gear. The screw-cutting taps for internal threads are held in the Jacobs chuck, while the dies for cutting external threads are held in a special die holder, which also locates in the tailstock spindle. The unclamped tailstock with tap or die is held against the workpiece, which has been oiled to aid the cutting process. The workpiece is then turned backwards and forwards in the approved manner for screw-cutting by means of the chuck key until a sufficient length of thread has been cut.

Turning tool selection: some of the more commonly used lathe-cutting tools are shown in Figure 3.21. The roughing tools are used for making initial cuts where large amounts of material are to be removed. The knife tool is used for finishing cuts and producing right-angled corners. Facing cuts are taken using the facing cranked tool, and finished components are cut off from bar using the thin parting-off tool.

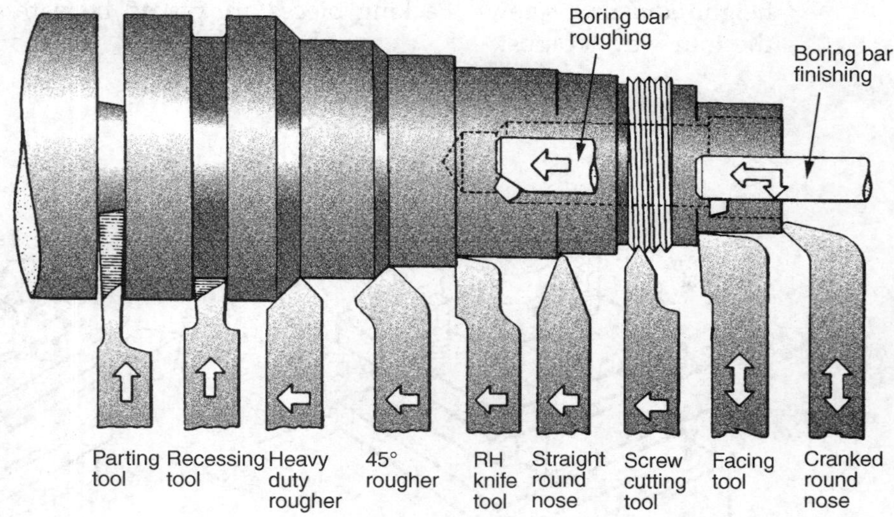

Figure 3.21 Lathe cutting tools

Selection of cutting speeds and feeds: the spindle speed to be selected for a particular turning operation depends on the material being cut, the material from which the cutting tool is made and the workpiece diameter. The cutting speeds for different materials when using a high-speed steel-cutting tool can be obtained from Table 3.1. The spindle speed is calculated using the formula:

$$N = 1000\ S/\pi\ d$$

where N = spindle speed in r.p.m., S = cutting speed in m/min and d = diameter of workpiece in mm.

Having selected the appropriate spindle speed, it is generally accepted that a deep cut and a fine feed rate is the best combination for the fast removal of material, long tool life and a good surface finish. When taking roughing cuts, the depth may be as large as the machine can handle, provided that the workpiece is securely held and the cutting tool is properly ground and set in position. Spindle speeds and automatic tool feed rates are obtained by setting the selection levers or dials on the headstock to the appropriate position.

Use of coolant: most metal-machining processes make use of a coolant. The purposes of a coolant are:

- to carry heat away from the workpiece and cutting tool
- to lubricate the chip-cutting tool interface, thus reducing tool wear
- to prevent chip particles from becoming welded to the tool face and forming a built-up edge
- to improve the surface finish and wash the swarf away from the cutting edge
- to prevent corrosion of the workpiece and the machine.

The coolant used on centre lathes is stored in a reservoir beneath the lathe bed. From here, it is delivered by a pump to a control tap and adjustable supply pipe on the carriage assembly. The most common coolant used for general machining is emulsified or soluble oil. This is a type of oil containing detergent and disinfectant that mixes with water. The mixture has a white, milky appearance and should be directed onto the cutting area in a steady stream. Cast iron can be machined without the use of a coolant. The graphite flakes in its structure make it self-lubricating, and less heat is generated than when cutting other materials

Case study

Cutting tool tips

The requirements of a single point cutting tool are that it should be wear resistant, be a good conductor of heat and have a low coefficient of friction with the workpiece material. There is also a need to keep down the costs and reduce the cycle time for material removal processes. This has been achieved in part by advances in cutting tool technology that have allowed the use of higher cutting speeds and feeds.

High-carbon steel was the only material available to the Victorian engineers, but this could only be used with low cutting speeds. High-speed steel was introduced around the time of the First World War and, a little later, the first cemented carbides made their appearance. These enabled much higher material removal rates to be achieved and are still widely used today. Up to the 1960s, hard-wearing tungsten carbide tips were brazed on to a mild steel shank but, since then, a variety of disposable-tip systems have been developed such as those shown in Figure 3.22. The tip inserts are clamped to a toolholder. When worn, they can be rotated to expose a new cutting edge and are finally discarded.

In recent years, advances in surface engineering have led to the introduction of tool coatings that improve the performance of both high-speed steel and cemented carbide cutting tools. Titanium nitride is one such coating that has found general acclaim. It is vaporised before being deposited in a thin layer on the tool material and has an attractive golden colour. It is extremely hard and wear resistant and has a low coefficient of friction. These properties have enabled higher material removal rates to be achieved while maintaining the quality of surface finish and tool life.

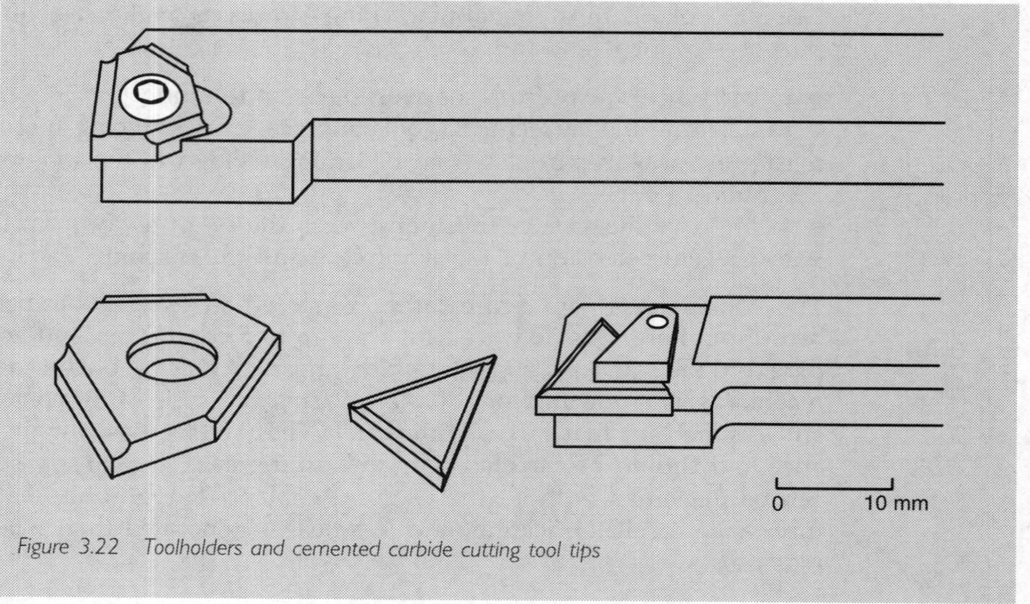

Figure 3.22 Toolholders and cemented carbide cutting tool tips

Milling

In the milling process, material is removed as the workpiece is fed past a rotating multitoothed cutter. There are two basic types of milling machine, the horizontal mill (Figure 3.23) and the vertical mill (Figure 3.24). With the horizontal mill, the cutter rotates about an axis in the horizontal plane and with the vertical mill the cutter axis is in the vertical plane. Both types of milling machine are used to machine flat surfaces, channels and grooves. The vertical type is perhaps the more versatile of the two since it can also be used to mill profiles.

The main parts of a milling machine are the body column, which is made of cast iron; the knee, which can move vertically in slides machined on the front of the column and supports the cross-slide and worktable; the cross-slide, which can be traversed horizontally from front to back on top of the knee and supports the worktable; and the worktable itself, which can be traversed horizontally from side to side on top of the cross-slide. Horizontal mills have an overarm and overarm steady, which support the arbor and cutter. Vertical mills have a spindle head which may be set over at an angle to the vertical if required.

On all except very small milling machines where the spindle may be belt driven, the drive is through a gearbox, which has speed change levers for selecting the required spindle speed and automatic worktable feed. With both types of milling machine, the worktable can be raised, lowered and traversed in two perpendicular directions in the horizontal plane by means of handwheels. As with centre lathes, milling machines are equipped with a coolant reservoir and pump.

The safe and efficient removal of material on milling machines requires a knowledge of:

- workholding methods
- cutter selection
- cutter mounting
- selection of cutting speeds
- up-cut and down-cut milling.

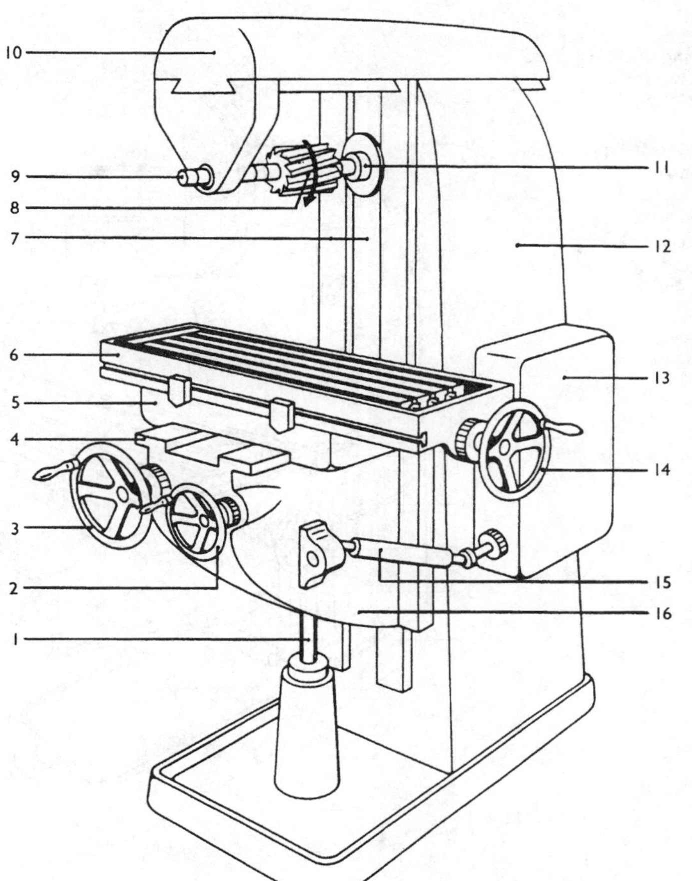

Figure 3.23 The horizontal milling machine

Workholding methods: the machine worktable on both horizontal and vertical mills is equipped with T-shaped slots to which large or awkwardly shaped workpieces may be clamped directly. Large cutting forces are present when milling and at least two clamps should be used (Figure 3.25). As when drilling, cylindrical workpieces should be supported on V-blocks.

Smaller workpieces may be held in a heavy-duty machine vice that is also bolted down to the worktable. Vices with a swivel base are often used, which can be set at any angle in the horizontal plane. Parallel bars should be used to support the workpiece in the machine vice. This ensures that the upper and lower machined surfaces of the workpiece are parallel.

Wear in the vice slideway often causes the workpiece to lift off the parallel bars. A blow from a hide hammer as the vice is being tightened usually ensures that the workpiece is sitting down tightly. If the vice is so worn that it is difficult to keep the workpiece in contact with the bars, it is then no longer suitable for precision work.

Cutter selection: four of the most commonly used cutters on the horizontal milling machine are slab cutters or slab mills, side and face cutters, slotting cutters and slitting saws (Figure 3.26).

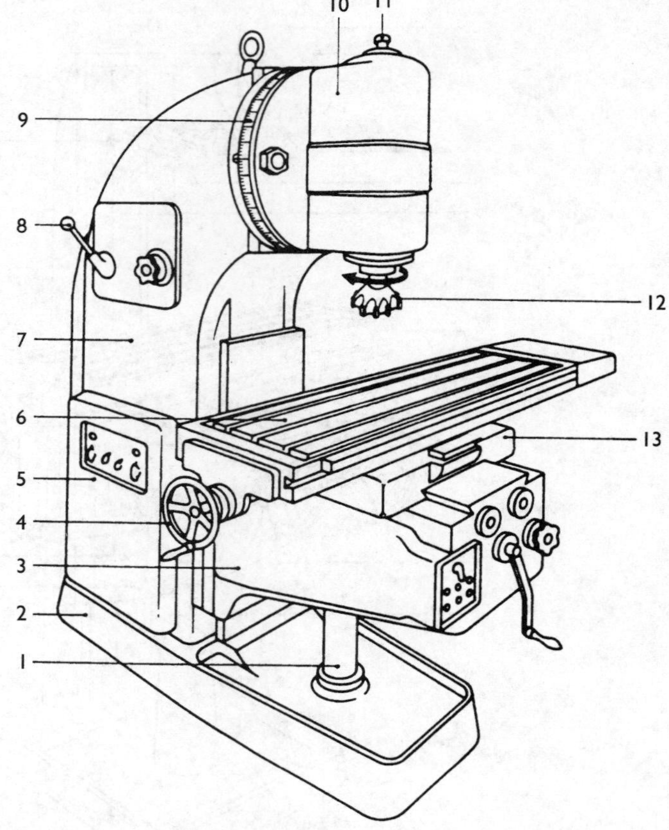

Figure 3.24 The vertical milling machine

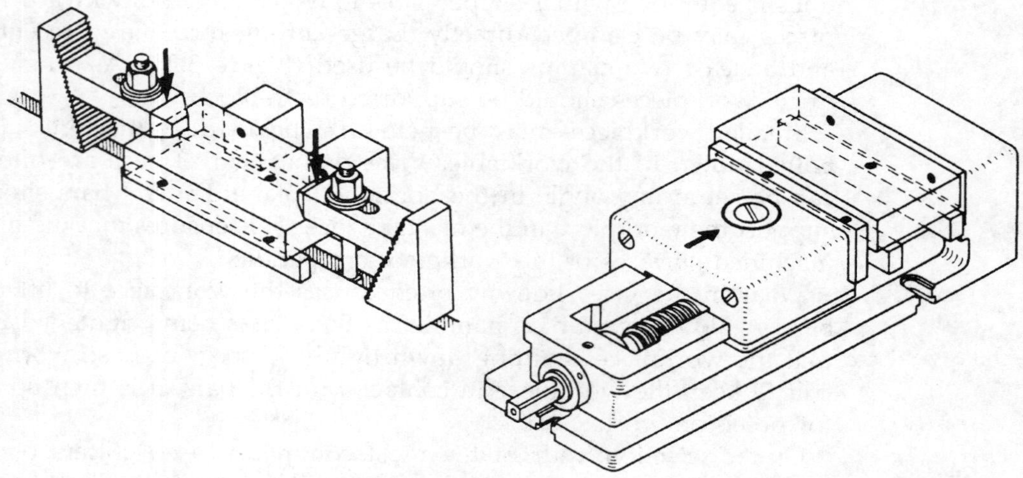

Figure 3.25 Workholding for milling

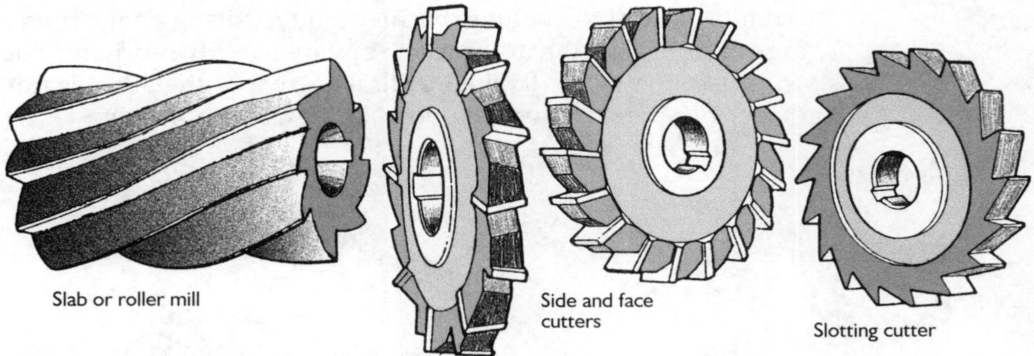

Figure 3.26 Cutters for the horizontal milling machine

Slab cutters are used to produce wide flat surfaces. Side and face cutters have cutting edges on both the periphery and the sides of the teeth and are used for light facing operations and for cutting channels, slots and steps. Slotting cutters are thinner than side and face cutters and have teeth only on the periphery. They are used for cutting narrow slots and keyways. Slitting saws are very thin cutters used for cutting narrow slots and for parting off excess material.

Four of the most commonly used cutters on the vertical milling machine are face mills, shell end mills, end mills and slot drills (Figure 3.27).

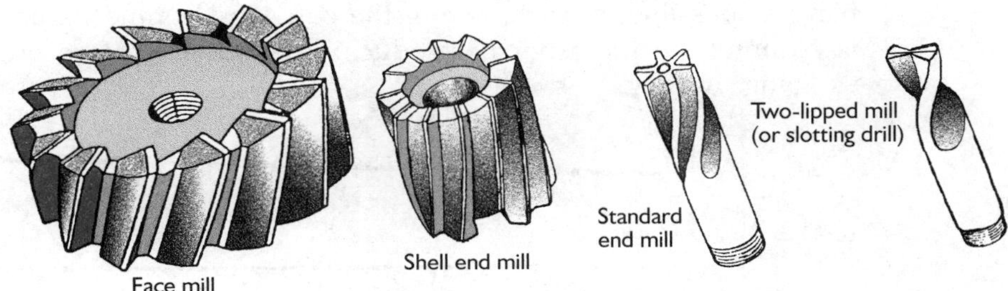

Figure 3.27 Cutters for the vertical milling machine

Face mills have cutting edges on both the end face and the sides. They are used to produce wide flat surfaces, as are slab mills on the horizontal milling machine. They do, however, produce the surface more accurately because, unlike the slab mill, every part of each tooth on the cutter face passes over the whole surface. In this way, face mills are said to 'generate' a flat surface. Shell end mills are smaller but similar to face mills. They have the same cutting action and are used for generating smaller flat surfaces.

End mills are of smaller diameter than face mills but, like them, they have cutting edges on the end face and sides. They are used for light facing, profiling, recessing and milling slots. Slot drills are similar to end mills but have only two cutting edges. They are used for accurately milling slots and keyways.

Cutter mounting: horizontal milling machine cutters are mounted on a shaft known as an 'arbor' (Figure 3.28). This has a taper at one end, which locates in the spindle nose of the milling machine, and a driving flange containing two slots. A draw bolt fastens the arbor to the spindle, and two driving dogs locate in the slots. Unlike the morse-tapered drill, the drive to the arbor is not by friction on the taper but directly from the dogs to the driving flange.

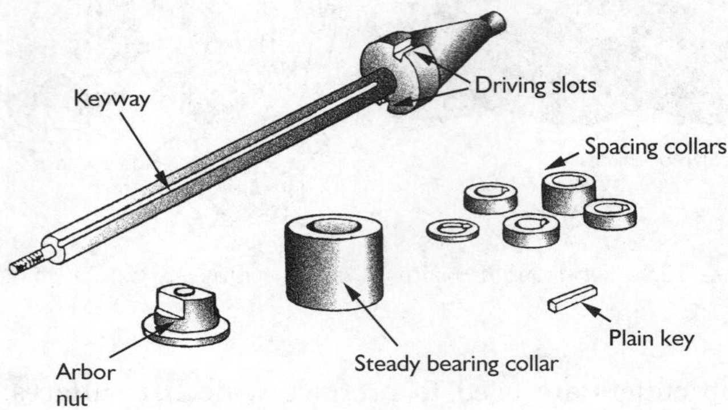

Figure 3.28 Long arbor for the horizontal mill

The arbor has a keyway running along its length. The cutter is positioned on it with a key and with spacing collars on either side. The free end of the arbor passes through the overarm steady, and a nut on the end holds the cutter and collars tightly together (Figure 3.29). There is a tendency for the arbor to flex, and the cutter and overarm steady should be positioned so that there is as little overhang as possible on each side of the cutter. Sometimes, a combination of cutters is mounted on the arbor to produce a particular surface profile. This is known as 'gang' milling.

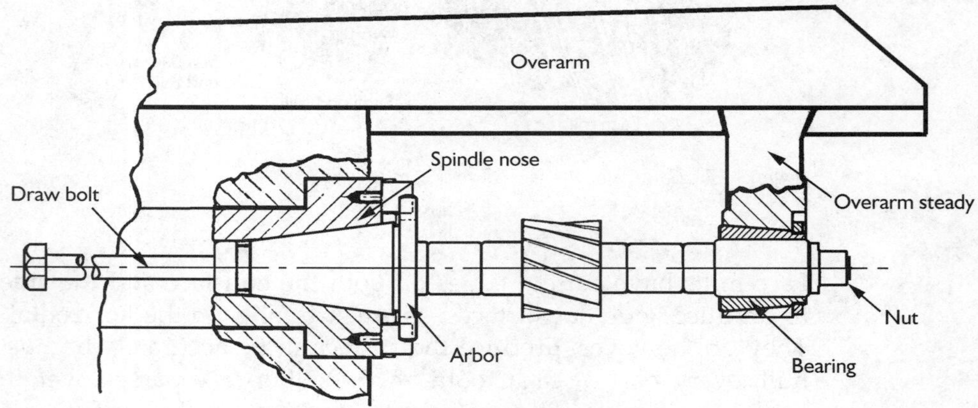

Figure 3.29 Cutter mounting for the horizontal mill

The face mills and shell end mills for vertical milling machines are mounted on a short 'stub arbor'. This has the same taper and driving flange as the long arbor used on the horizontal mill above and is fixed to the spindle nose in the same way. The cutters locate on a spigot and are held in position by a retaining screw (Figure 3.30).

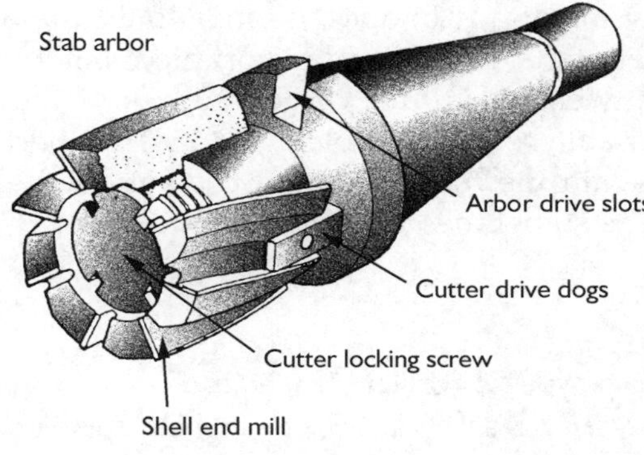

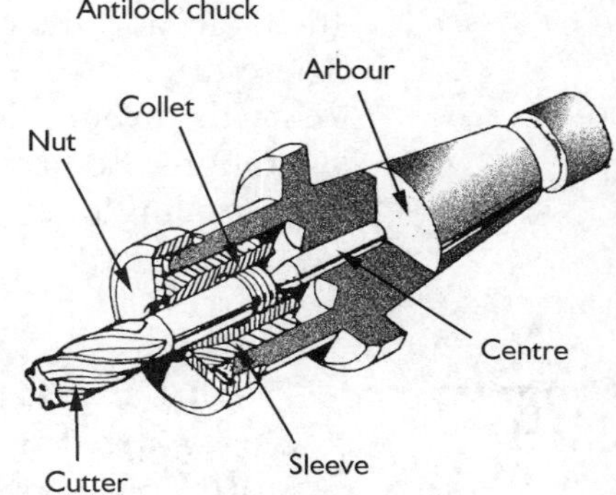

Figure 3.30 Cutter mounting for the vertical mill

End mills and slot drills for the vertical mill have parallel shanks that are threaded at the end. They are held in an 'antilock' collet chuck fixed to the spindle nose. This is more complex than the Jacobs chuck used for drills. It is designed in such a way that, as the cutting forces increase, so does the grip on the cutter.

Selection of cutting speeds: with both types of milling machine, the required spindle speed depends on the workpiece material and the diameter of the cutter and the cutter material. The cutting speed for a particular material when using a high-speed steel cutter can be obtained from Table 3.1. The spindle speed is calculated by the formula:

$$N = 1000\ S/\pi\ d$$

where N = spindle speed in r.p.m., S = cutting speed in m/min and d = the diameter of the cutter in mm.

Up-cut and down-cut milling: there are two ways in which the workpiece can be fed under the cutter when using the horizontal milling machine. They are known as 'up-cut' milling and 'down-cut' milling (Figure 3.31).

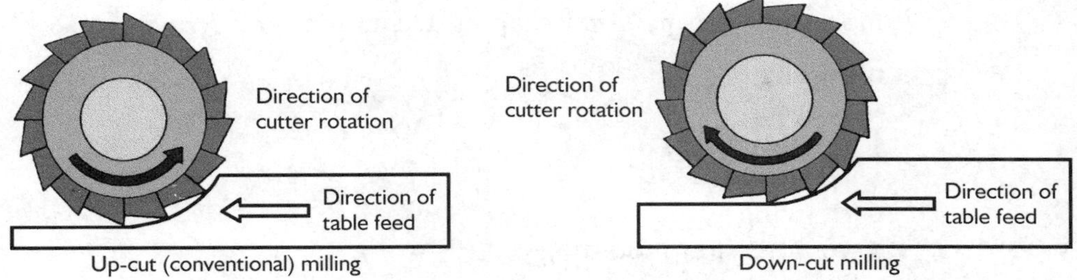

Figure 3.31 Up-cut and down-cut milling

Up-cut milling should always be used unless the machine is specially designed for the other method. With down-cut milling, there is a tendency for the cutter to climb onto the workpiece or for the workpiece to be dragged under the cutter. This is especially the case with machines that are not new and have some backlash in the worktable feed mechanism. At the best, this can cause excessive vibration and a poor surface finish. At the worst, the cutter can climb onto the workpiece, stall the machine and bend the arbor.

Up-cut milling also has some disadvantages. It tends to lift the workpiece from the table, and the cutter teeth tend to rub on the workpiece before starting to cut. These are, however, far outweighed by the disadvantages of the other method. As with turning, it is advisable to use a coolant for all but the lightest milling operations. Soluble oil is again the most common coolant used and, for heavy-duty milling operations, it is sometimes used in undiluted form.

Progress check

1. What are the three main types of drilling machine?
2. How are taper shank drills driven by a drilling machine spindle?
3. Why are parallel bars used to support components held in a machine vice?
4. What are the main parts of a centre lathe?
5. List three different ways in which the tailstock of a centre lathe can be used during material removal operations.
6. What are the advantages and disadvantages of turning between centres?
7. What are the main parts of a horizontal milling machine?
8. What is meant by 'gang' milling?
9. Why is it generally recommended that up-cut milling should be used on horizontal milling machines?
10. Why is it that a flat surface can be produced more accurately by a vertical milling machine using a face mill than by a horizontal milling machine using a slab mill?

Material shaping

Some engineering materials are processed into a standard form, such as plate, sheet and barstock, after which they are sold on for further processing. Others are formed into a specific shape. They may be finish formed into an engineered product or they may be rough formed, leading to further processing. Some of the more common material shaping techniques are:

- casting
- forging
- pressworking
- extrusion
- drawing
- compression moulding
- injection moulding
- blow moulding
- vacuum forming.

Casting

Casting is a liquid forming process in which molten metal is poured into a mould or injected into a die and allowed to solidify. Thermosetting plastic resins are also sometimes poured into open moulds, but they are generally more viscous than molten metals and cannot be cast into such intricate shapes. The main casting processes are:

- continuous casting
- sand casting
- die casting.

Continuous casting is used to produce continuous lengths of large section bar. These are known as blooms if they are above 150 mm square, billets if they are less than 150 mm square and slabs if they have a wide rectangular cross-section. Continuous casting processes are often part of an integrated steel plant in which the blooms, billets and slabs are rolled, drawn or forged into shape while still in a heated condition.

In continuous casting, the molten metal is poured into a water-cooled mould. This is usually made from copper and has a retractable base. As the mould is filling and the metal is solidifying, the base is drawn downwards and removed. The metal then passes through a system of rolls, as shown in Figure 3.32.

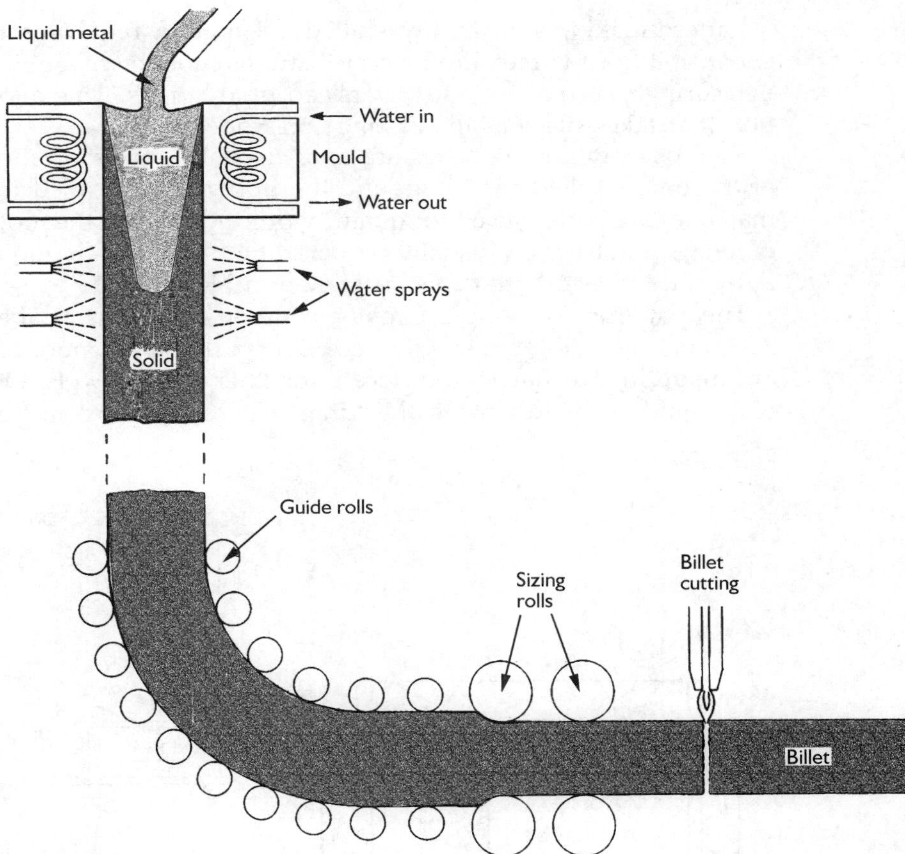

Figure 3.32 Continuous casting

The mould may be vibrated to assist the downward movement of the metal and lubricated with a graphite compound. The metal is cut into appropriate lengths before being transported on for further processing.

In *sand casting*, the materials that are formed by this technique include cast iron, steel, brass, bronze and aluminium alloys. There are two basic types of sand casting. They are known as 'green sand' casting and 'dry sand' casting. The technique can be used to produce intricately shaped components, but they do not have a very smooth surface finish or high precision. The complexity that can be achieved by casting depends on the fluidity of the molten metal. Cast iron and aluminium are very fluid, as are some of the brasses and bronzes. Steel, however, is more viscous and cannot be cast into such intricate shapes.

Green sand contains moisture, which helps it to hold its shape when moulded. Dry sand contains binding agents, which help it to hold its shape and which set when the mould is heated and dried. The first step in making a sand casting for a component is to prepare the mould. This involves packing the sand around a wooden or metal pattern in such a way that the pattern can be withdrawn to leave a cavity that has the shape of the component. Talc, which is also called 'French chalk', is often sprinkled over the pattern so that the sand does not adhere to it.

Pattern-making is a highly skilled occupation requiring a knowledge of the expansivity of the metal to be cast. Patterns must be made oversize using special measuring instruments with oversized graduations. This allows for the contraction that takes place as the casting cools down.

Two-piece moulding boxes are used to contain the sand mould. The lower half of the box is called the 'drag' and the upper half is called the 'cope'. Moulding machines are widely used for quantity production, but the moulds for single items or small quantities are usually prepared by hand. The moulding process is essentially the same in both cases, as shown in Figure 3.33.

The patterns for simple circular components, such as that shown in Figure 3.33, may be made in a single piece. The patterns for more complex components are made in two halves that locate together with dowels. One half is laid on a flat moulding board inside the 'drag' half of the box and dusted with French

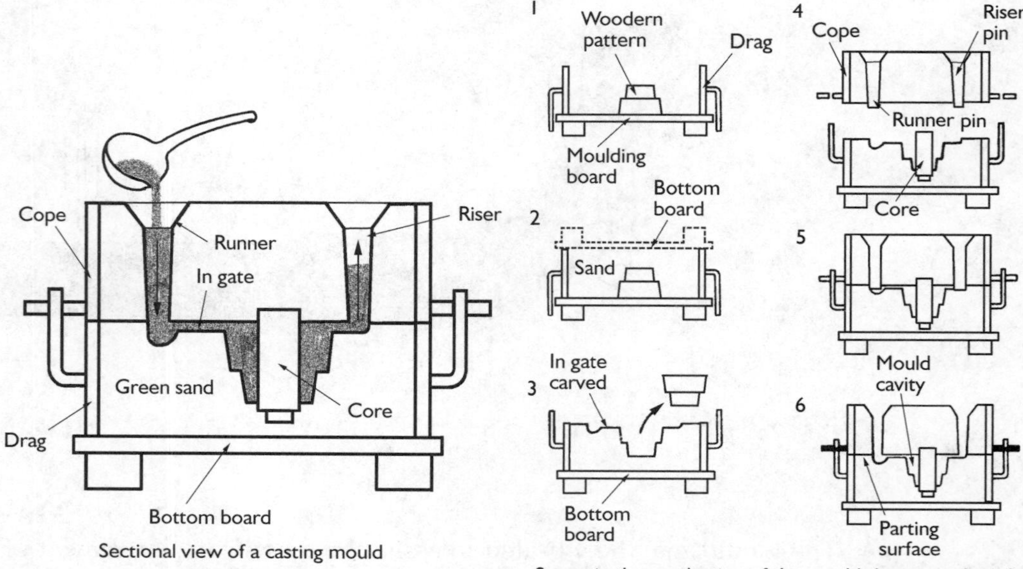

Figure 3.33 Sand moulding

chalk. Moulding sand is then riddled over the pattern and rammed down tightly until the box has been filled. The sand is then levelled off with a straight edge and the box is turned over so that the pattern is uppermost.

The other half of the pattern is now assembled in position and the 'cope' is placed on top of the drag. Locating pins hold the two boxes together. The pattern and the interface between the two halves of the mould is again dusted with French chalk, or a fine dry 'parting' sand, to enable the completed mould to be split. The tapered wooden 'runner' and 'riser' pins are then placed in and held in position while moulding sand is riddled over the exposed pattern. This is again rammed tight until the 'cope' is full and levelled off as before.

The 'runner' and 'riser' pins are now gently withdrawn and the 'cope' is lifted off the 'drag' and turned over. The two halves of the pattern are then carefully removed from the mould and the 'in gate' is cut in the sand to connect the 'runner' to the mould cavity. After dusting off any surplus sand, the 'cope' and 'drag' are again placed together and casting can proceed. The molten metal is poured into the 'runner' until it appears at the top of the 'riser'. The metal is then allowed to cool before the sand is knocked out to reveal the completed casting.

After cooling, the casting is 'dressed' or 'fettled'. This involves cutting off the runner and riser and removing any flashing where the molten metal has penetrated the joint between the two halves of the mould.

Hollow castings, such as that shown in Figure 3.33, require the additional preparation of a core that has the shape of the hollow space and around which the molten metal can flow. This is made from baked dry sand and placed in the mould cavity after the pattern has been removed. The core locates in 'core prints' within the mould and can also be supported by small metal studs known as 'chaplets'.

Die-casting processes use a permanent metal mould into which the molten metal is poured under gravity or forced under pressure. The metal moulds, or dies, are more expensive to produce than the patterns for sand casting, and large production quantities, usually several thousand, may be required to justify the cost. Compared with sand casting, however, the surface finish is much finer, and castings can be produced in finished form with greater dimensional accuracy. Die-casting is mostly carried out with aluminium, magnesium and zinc-based alloys, which have a relatively low melting point.

The process in which molten metal is poured manually or automatically into a die is known as gravity die-casting (Figure 3.34). The dies sometimes incorporate metal cores and may be made in several parts so that they can be split to remove the casting. For complex hollow components, baked sand cores are sometimes placed inside the dies.

The forming technique in which the molten metal is forced into a die under pressure is known as pressure die-casting. There are two basic types of pressure die-casting machine, the 'hot chamber' type and the 'cold chamber' type. The hot chamber process is used mainly with low melting point zinc-and tin-based alloys (Figure 3.35).

In the hot chamber process, molten metal is heated in a crucible in which there is a partially submerged goose-necked duct containing a piston. When the piston is raised, molten metal enters the duct through an intake port. When the piston is forced down, the port is covered and the molten metal in the duct is forced into the die cavity. The metal cools very quickly, the die is split and the ejector pins push out the cast component. The process is quick and efficient and can be fully automated.

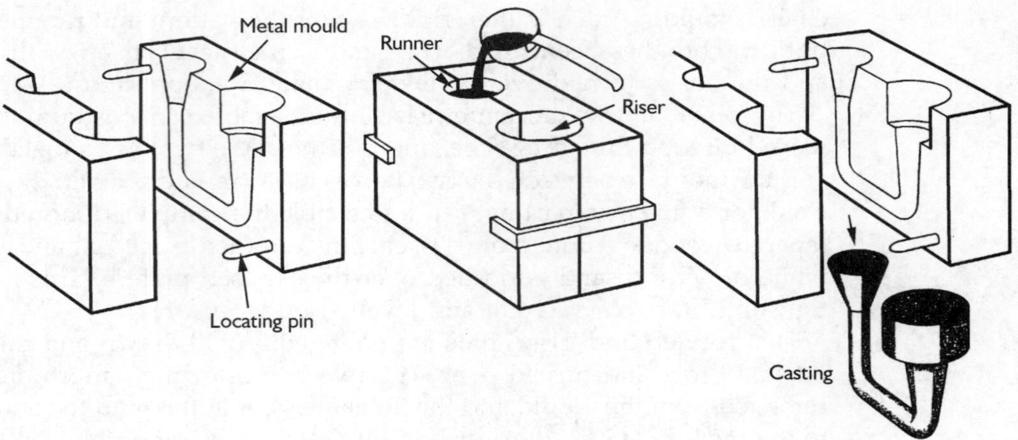

Figure 3.34 Gravity die-casting

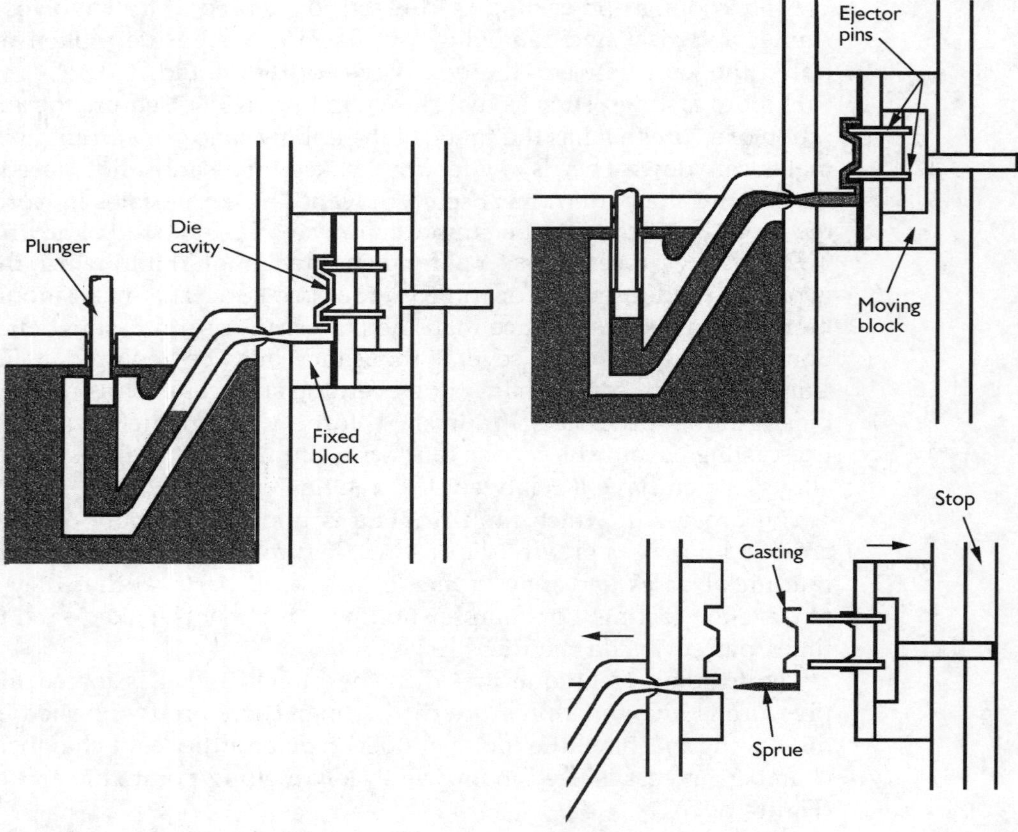

Figure 3.35 Hot chamber pressure die-casting

The cold chamber process is used with the higher melting point aluminium and magnesium alloys (Figure 3.36). The machine incorporates a 'shot' chamber, into which molten metal is ladled or automatically poured, and an injection piston.

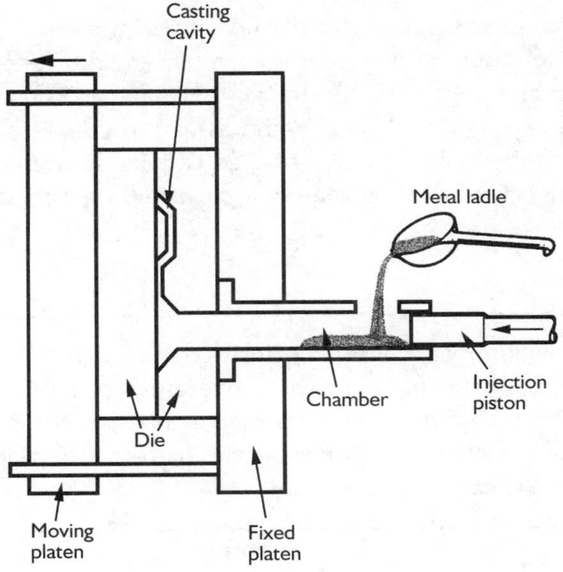

Figure 3.36 Cold chamber pressure die-casting

The piston forces the molten charge into the dies where it quickly solidifies. The die then splits and the ejector pins push out the completed casting. High pressures are involved with both the hot and the cold chamber processes, and the machines that hold the dies must be capable of applying large forces to hold the two parts of the die together. The machines are rated according to the force they can apply, ranging typically from 200 kN to 25 MN.

Case study

Centrifugal casting

The cylinders of internal combustion engines are generally fitted with cast iron liners. These are no more than a few millimetres thick with a narrow flange around the upper end. Casting such a long thin cylindrical section by means of a sand mould and core is difficult and the centrifugal casting technique is used (Figure 3.37). The molten metal is poured into a rotating cylindrical mould with its axis slightly inclined to the horizontal. The mould is lined with copper and water cooled. The molten iron is uniformly distributed around the mould by centrifugal force where it solidifies.

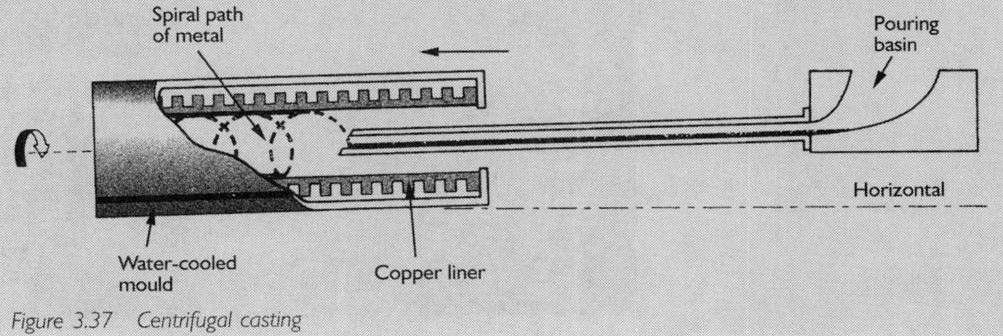

Figure 3.37 Centrifugal casting

The advantages of this method for quantity production are that there is very little mould preparation time required and the equipment is relatively cheap and uncomplicated. Furthermore, there are no runners, risers or cores that have to be removed and, after withdrawal, the casting is ready for machining. As would be expected, the outer surface is quite smooth and any non-metallic particles or slag accumulate on the inner surface because they are less dense than the iron.

Activity 3.3

Aluminium has a linear expansivity of 23×10^{-6}/K and solidifies at a temperature of 660°C. Assuming a normal temperature of 20°C, what will be the actual size of the millimetre graduations on a metre rule that has been specially made for use by a pattern maker who is making the patterns for aluminium castings? How long would such a metre rule really be?

Forging

Materials that are formed to shape by forging need to be malleable. Some may be formed to shape while cold. They are then said to have been 'cold worked'. Other materials need to be heated to increase their malleability. They are then said to have been 'hot worked'.

In the traditional forging processes practised by blacksmiths, wrought iron, steel and some brasses and bronzes are formed to shape on the anvil using hand tools. Nowadays, traditional blacksmiths are mainly involved in the shoeing of horses and producing ornamental and decorative metalwork. Large forgings for engineered products are produced using power hammers and presses, while most of the smaller forged components are produced by drop-forging and hot pressing.

An advantage of forging is that, during deformation, grain flow takes place and any impurities that are present are distributed throughout the metal. When a section through a forging is polished and etched, the grain flow is seen to follow the contours of the component (Figure 3.38). This gives the material a fibrous appearance, rather like the grain direction in timber. As a result, forged components are stronger and tougher than those that have been machined or cast to the same shape.

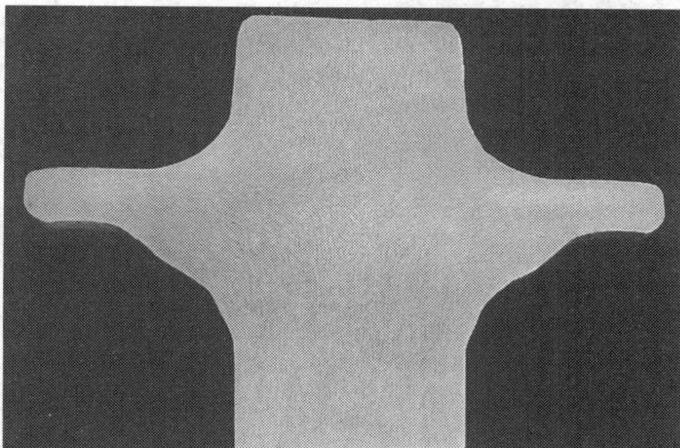

Figure 3.38 Grain flow in a forged component

With drop-forging, the material is formed between dies. One half of the die is fixed to the large anvil of a drop hammer or a power hammer. The other half is fixed to the 'tup', which is the moving part of the hammer. With the tup raised, a billet of metal is placed on the lower die ready to be formed. In the case of a drop hammer, the heavy tup is allowed to fall under its own weight to forge the component to shape between the dies. In the case of a power hammer, the tup is usually driven by compressed air.

More than one forging operation may be needed for the more complex shaped components. These require more than one die or a multistage die in which billets are moved from one impression to the next as the tup is raised (Figure 3.39).

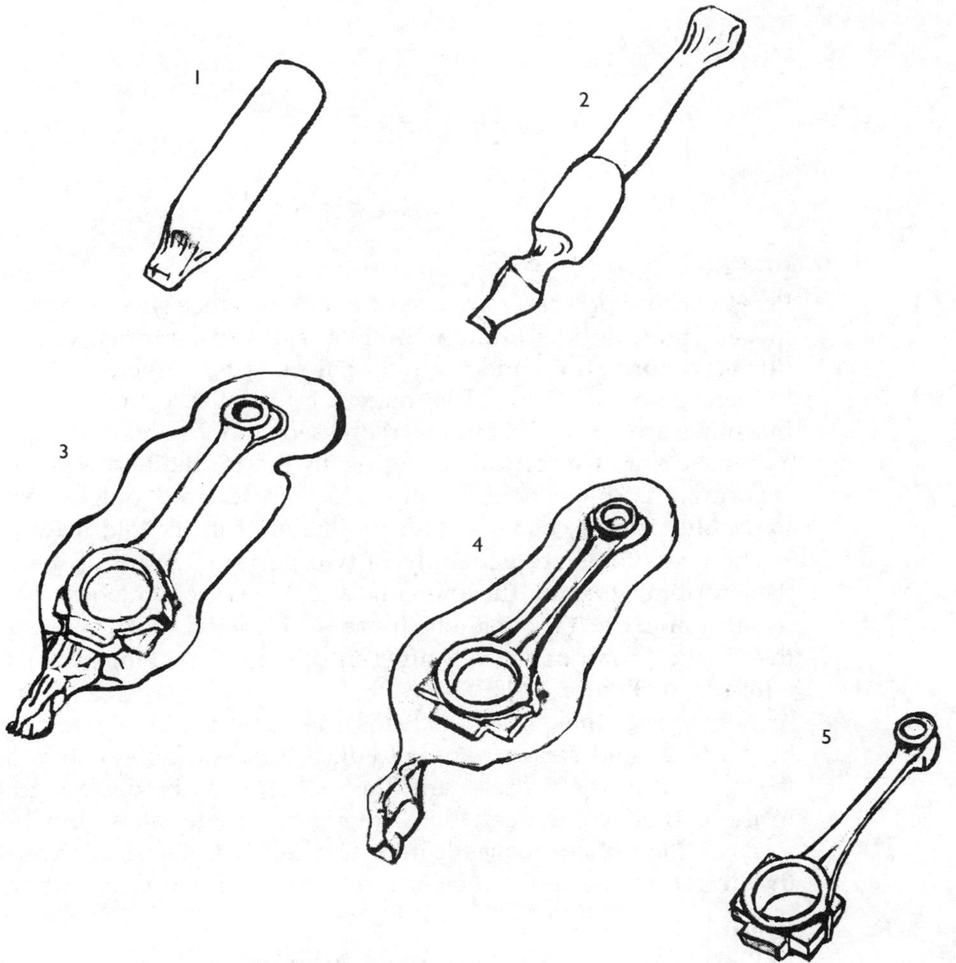

Figure 3.39 Forged component

Hot pressing is a process that has been developed from drop forging and is generally used for simple shaped components. The material is formed to shape between dies in a mechanical or hydraulic press by a steady pressure. The slower squeezing action gives a more uniform deformation of the material (Figure 3.40).

Figure 3.40 Hot-pressed component

Pressworking

Pressworking processes are used to mass produce sheet metal components, mostly in steel, brass and aluminium. Both tensile and compressive forces may be present during deformation, and the material must be sufficiently malleable and ductile to accommodate them. The panels for motor vehicles, cookers and washing machines are typical examples of pressed components and also smaller items such as washers and electrical soldering tags. The main pressworking processes are shearing, bending and drawing. They are carried out on both mechanical and hydraulic presses, usually with the material in its cold state.

The press tools are generally in two parts. The ram of the press usually holds the cropping tool or the piercing and blanking punches. The bed of the press usually holds a die over which the sheet metal strip is fed and through which the finished component falls after cropping or blanking. The process is often fully automated (Figure 3.41).

In bending, the punch deforms the material in two dimensions to produce brackets, angled strip or angled edges to sheet metal components. The material is stretched in the process, and the tools should be designed to prevent cracking in the corner of the bend. In drawing, the material is stretched in three dimensions as the punch pushes it into the die. Here again, care must be taken during the design of the tools to ensure that the material does not crack or wrinkle.

Rolling

Rolling is used to produce plate, sheet bar and other sections in both ferrous and non-ferrous metals. The material is passed through a series of rolls, which successively reduce its cross-section to the required shape and size. The deformation forces are mainly compressive, and the material must be sufficiently malleable to accommodate them. Hot rolling is used in the initial stages to produce large reductions in cross-section. Cold rolling is used only for finishing operations to produce rods, strip, sheet and foil with a smooth surface finish and dimensional accuracy.

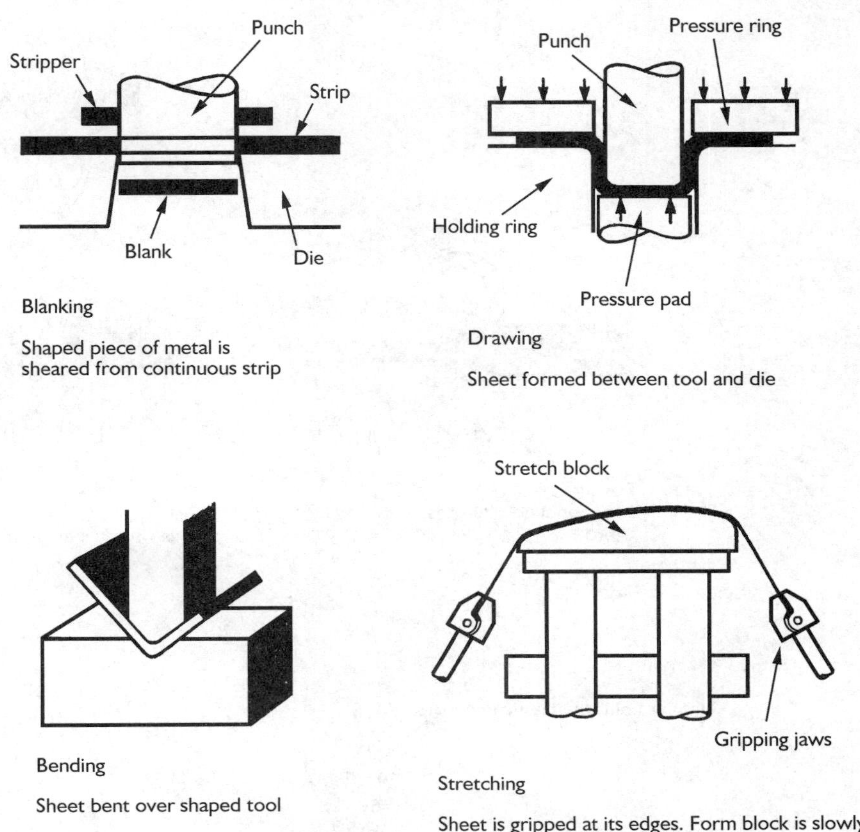

Figure 3.41 Pressworking operations

In modern integrated systems, hot rolling follows the production of blooms, billets and slabs by the continuous casting process. These are passed backwards and forwards through powerful, two-high reversing rolls, which greatly reduce their cross-section. The process continues through successive sets of rolls to produce the required size of plate, bar and structural sections such as channel, angle and I-section (Figure 3.42).

Sheet metal, metal strip and foil are finished by cold rolling. This gives a smooth, clean surface finish and closer dimensional tolerance than hot rolling. The rolls are highly polished and may be of the four-high type with backing rolls. These are required when rolling wide sheets or foil to prevent the working rolls from bending.

Some thermoplastic sheet is produced by a rolling process known as calendering (Figure 3.43). Polythene and PVC sheet is produced in this way from raw material in the form of a heated plastic dough. This passes through a series of rolls with successively narrower gaps between them. The final roll is a chilling roll, which cools the sheet before it is coiled.

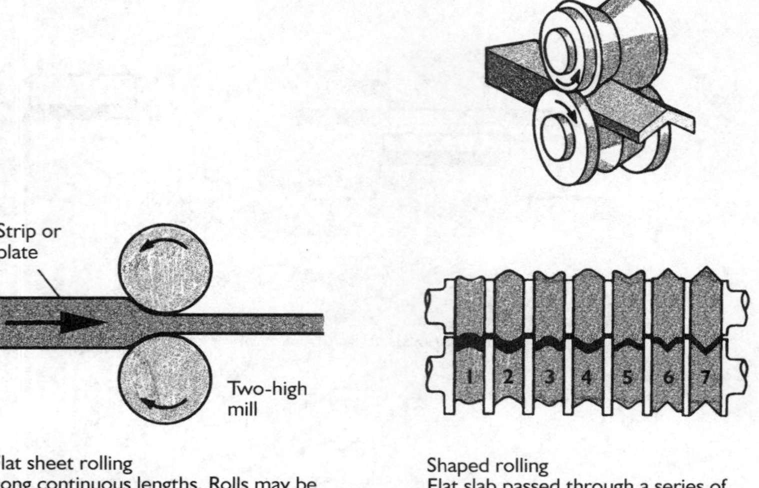

Figure 3.42 Rolling

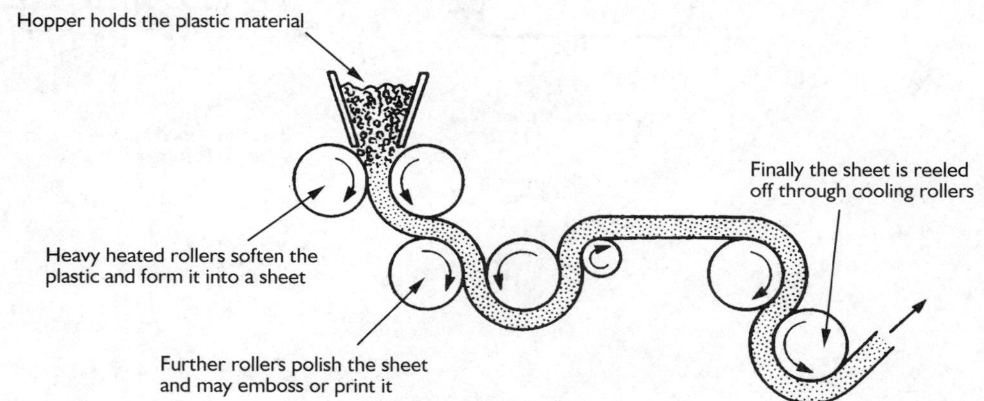

Figure 3.43 Calendering

Extrusion

Extrusion is a process in which material is forced through a die to make rod or tube of a particular cross-section. It is a similar action to squeezing toothpaste out of a tube except that the material may be a polymer or a metal and the pressure is applied by a ram. In the case of tubular sections, the die incorporates a mandrel over which the material is forced to produce the hollow section.

The forces acting on the materials are largely compressive and they must be sufficiently malleable. Extrusion is used for shaping ferrous and non-ferrous metals in both the hot and cold conditions. In particular it is used to produce a wide range of aluminium and brass sections that would not be possible by other means. It can also be used to produce hollow sections by the process of indirect or 'back' extrusion (Figure 3.44).

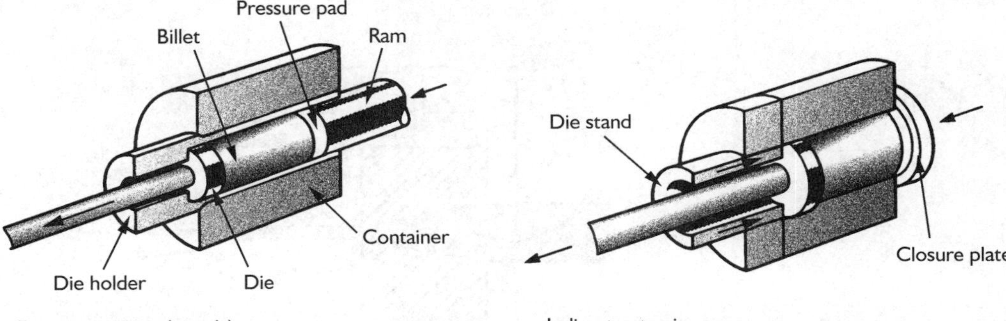

Figure 3.44 Extrusion process

With the extrusion of plastics, the raw material is in the form of granules contained in a feed hopper. The granules are carried forward from the hopper along a chamber by means of a screw mechanism. The chamber is heated electrically causing the granules to melt into a viscous liquid. This is then forced through a die to produce the required rod or tube section (Figure 3.45). On emerging from the die, the extrusion must be supported on rollers or a conveyer system and cooled, usually with an air blast.

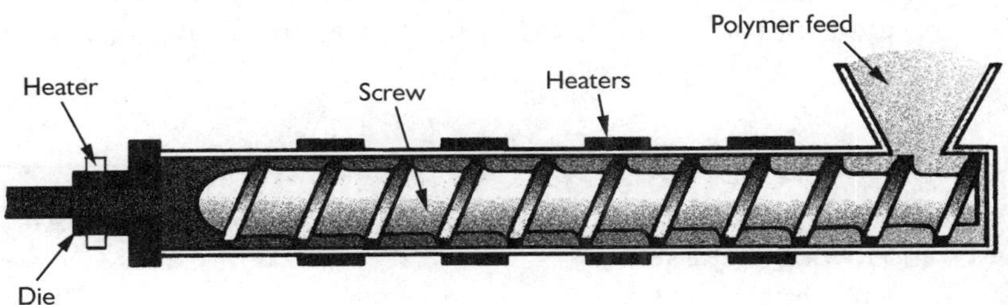

Figure 3.45 Extrusion of plastic material

Drawing

Drawing is a cold working process used to produce long lengths of wire, rod and tube from material that has been initially hot rolled or extruded. The material is pulled through a series of dies, reducing in size to the required cross-section (Figure 3.46). The forces that deform the material are mostly tensile and so it must have a high ductility plus sufficient tensile strength for the pull.

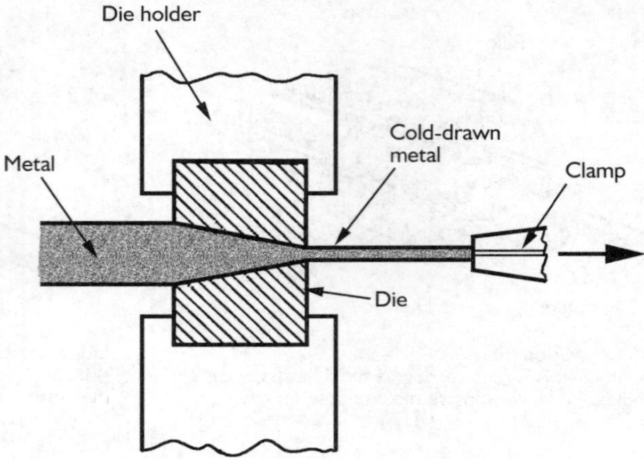

Figure 3.46 Drawing process

Before drawing, the end of the material must be reduced between rolls to allow it to pass through the die. In wire drawing, the end is then attached to a revolving drum or 'block', which pulls the material through the die. In the case of rod and tube, the end is gripped by a clamp or 'dog' and pulled along a drawbench. In both cases, the dies are lubricated with oil or a soap solution.

Drawing dies are generally made from very hard alloy steel or from tungsten carbide. Diamond is used for the dies for making very fine-gauge copper wire. Drawing produces a very smooth surface finish with close dimensional tolerances. The bright and shiny mild steel barstock used for general workshop purposes is produced in this way. It is known as 'bright drawn mild steel', which is indicated by the letters BDMS on engineering drawings.

Activity 3.4

Which method of forming do you think would be the most likely for producing the following metal items?

a the boot lid of a car
b a sheet of cooking foil
c copper pipe used for plumbing
d a domestic water tap
e the blank for a heavy duty gearwheel.

Compression moulding

Compression moulding is used for both thermoplastic and thermosetting plastic materials. It involves heating and compressing a measured charge of granular or powdered raw material between the two parts of a mould until it becomes fluid (Figure 3.47). In the case of thermosetting plastics, the heating triggers cross-linking of the polymers and the component quickly solidifies as the pressure is removed. In the case of thermoplastics, the mould must be cooled and the component allowed to solidify before being ejected.

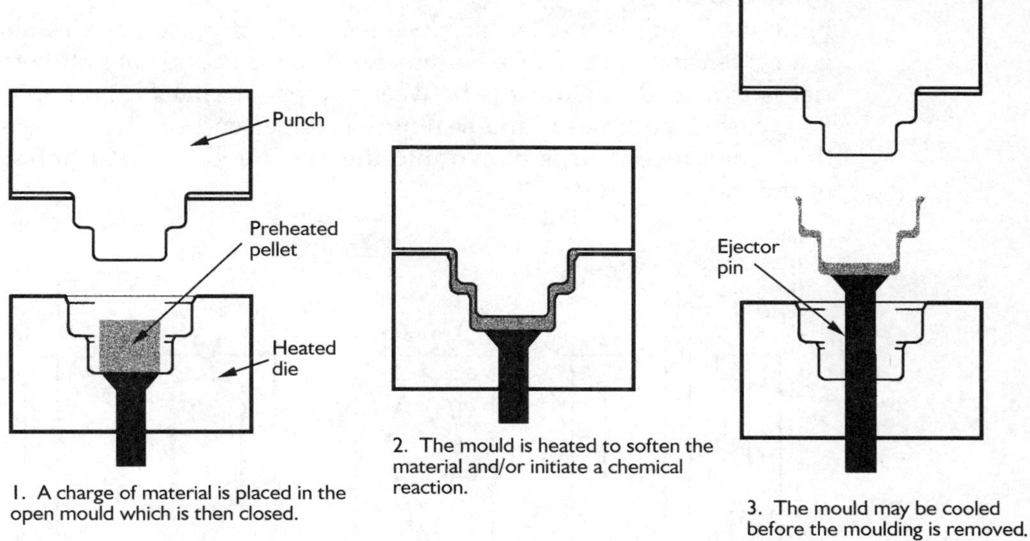

Figure 3.47 Compression moulding

Injection moulding

This process is carried out mainly using thermoplastic materials. It is very similar in principle to pressure die casting with metals and enables more complex shapes to be produced than are possible with compression moulding (Figure 3.48).

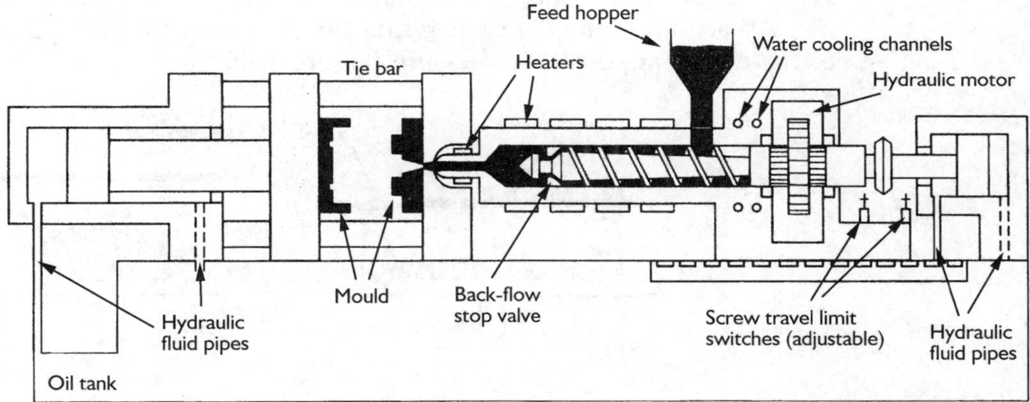

Figure 3.48 Injection moulding

The granular or powdered raw material is fed from a hopper into a chamber where it is compressed by a piston, heated until molten and then injected into the mould. The material solidifies rapidly against the cooler metal faces of the mould, after which the mould is separated and the component is ejected. As with die casting the process is very quick and can be automated.

Blow moulding

Blow moulding is used to produce a variety of plastic bottles and containers. It is an extension of the extrusion process in which a tube of hot thermoplastic material is extruded downwards between the open halves of a mould. The mould is then closed, cutting off and sealing the upper end of the extrusion. At the same time, compressed air is blown into the lower end so that it inflates to the shape of the mould (Figure 3.49).

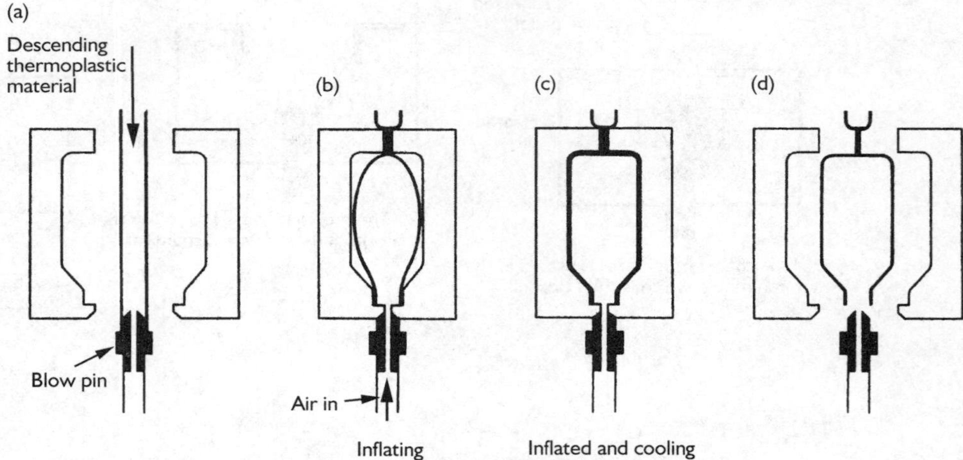

The process of blow moulding containers, showing the sequence of operations in the water-cooled mould

Figure 3.49 Blow moulding

Vacuum forming

With this process, a sheet of thermoplastic material is heated and placed over a die. A vacuum is then created inside the die cavity and the plastic is forced on to the die by atmospheric pressure (Figure 3.50).

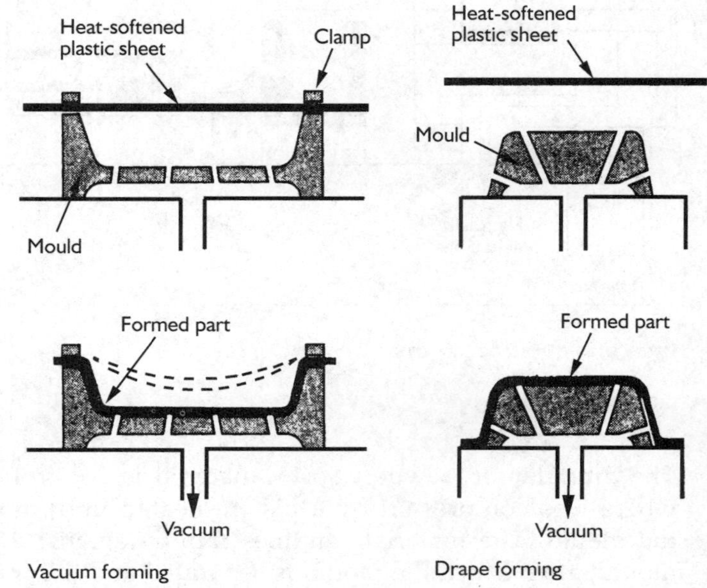

Figure 3.50 Vacuum forming

Progress check

1 What are the two basic methods of sand casting?
2 How are the grains of a die-cast component likely to differ from one which has been sand cast?
3 What effect does forging have on the grain structure of a material?
4 State three different kinds of pressworking operation.
5 How are structural steel channel, angle and I-sections formed to shape?
6 What is the difference between drawing and extrusion?
7 What is the process used to make plastic bottles?
8 What is the form of the raw thermoplastic material used for injection moulding?
9 How are polythene and PVC sheet produced?
10 What happens to thermosetting plastics during the compression moulding process?

Joining and assembly

After components have been finish machined or finish formed, the next production stage is to bring them together for joining and assembly. The most common joining and assembly techniques are:

- soldering
- welding
- adhesion
- mechanical fastening.

Soldering

Soldering is a process in which metal components are joined together by a different metal with a lower melting point. In its molten state, the joining metal combines with the joint surfaces to form an amalgam. This is a liquid solution, which becomes a solid solution as the joint cools down. The joining metal or solder must have this ability to combine with or 'wet' the joint materials. There are two basic soldering techniques:

- hard soldering
- soft soldering.

Soft soldering: here the solder is an alloy of tin and lead to which a little antimony has been added to improve its fluidity. It is used to join sheet metal components made from mild steel, copper and brass and to join the copper pipes used in plumbing. Mild steel is sometimes coated with tin and known as 'tinplate'. The tin gives it corrosion protection and also makes it easier for the soft solder to adhere. Soft solder is also widely used to make joints in electrical and electronic circuits. It is made with different proportions of tin and lead for different applications. Some of the more common soft solder compositions are shown in Table 3.2.

Table 3.2 Common soft solder compositions

British Standard type	Composition (%)			Melting temperature range (°C)
	Tin	Lead	Antimony	
A	65	34.4	0.6	183–185
K	60	39.5	0.5	183–188
F	50	49.5	0.5	183–212
G	40	59.6	0.4	183–234
J	30	69.7	0.3	183–255

Types K, F and G are used for sheet metal work and type J is used for plumbing. They are mainly supplied in stick form. Type A is used for joints in electrical and electronic circuits. It is supplied mainly in wire form with a core of resin flux. The melting temperature range gives an indication of the time that it takes for them to solidify. Type A solidifies very quickly, while type J takes the longest.

It is especially important to clean the joint surfaces before soft soldering or the solder may not form a good bond. They must wiped free of any oil or grease and metal oxides such as rust must be completely removed using emery cloth or wire wool. A flux, in the form of a liquid or paste, is then applied to the surfaces being joined.

A flux may be active or passive (Figure 3.51). A traditional active type is acidified zinc chloride known as 'Bakers fluid'. When the surfaces are heated, this has the effect of cleaning away any remaining dirt or oxide and protecting the surfaces against further oxidation. Active fluxes can be corrosive, and all remaining traces should be washed off the joint immediately after it has been made. Resin, in the form of a paste or in the core of wire solder, is a passive type of flux that does not have a chemical cleaning action. It merely protects the cleaned surfaces against oxidation and is the type used for soldered electrical connections.

Figure 3.51 Commercial fluxes

After applying a flux, the next stage of joint preparation is to 'tin' the joint surfaces (Figure 3.52). This is done using a soldering iron that has been heated in a gas flame. The iron is usually hot enough when the flame turns a greenish colour around the copper bit. The bit is then quickly cleaned with a file, dipped in flux and loaded with molten solder. This is then applied in an even coating to the joint surfaces. The solder combines with the material near the surface, which is known as the 'substrate', to form an amalgam.

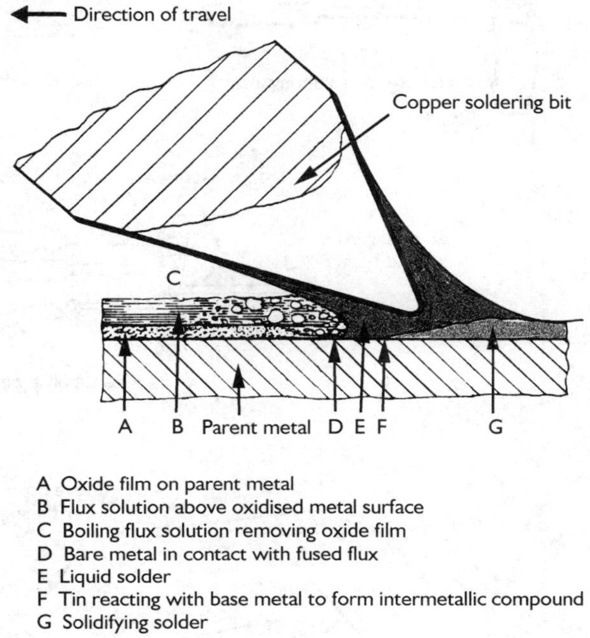

Figure 3.52 The tinning process

The tinned surfaces are then placed together and heated with a soldering iron while pressure is being applied. The soldering iron (Figure 3.54) is drawn slowly along the joint until molten solder is seen to appear around the joint edges. This process is known as 'sweating' the joint, in which the molten solder is drawn between the joint faces by capillary action. The soldering iron is then removed but pressure is still applied until the solder has solidified. Any active flux that remains can then be washed off and, if the material being joined is bare mild steel, it is a good idea to apply a rust inhibitor.

Soft soldered joints in copper pipes are generally made using standard fittings that contain an insert of solder. They include elbows, T-pieces, nipples and reducers. The pipes are cleaned with wire wool, coated with passive resin flux and the joint assembled. Heat is then applied from a blowpipe fed with natural gas or propane. As soon as solder appears at the edges of the joint, the heat is removed and the solder allowed to solidify (Figure 3.54).

For fittings that do not contain a solder insert, solder is applied to the edge of the joint when it is judged to be hot enough. The solder melts and is drawn into the joint by capillary action. The same process is carried out when soldering electrical terminations to copper cable. Here, care must be taken not to overheat the joint since excess heat may be conducted back along the conductor and damage the insulation.

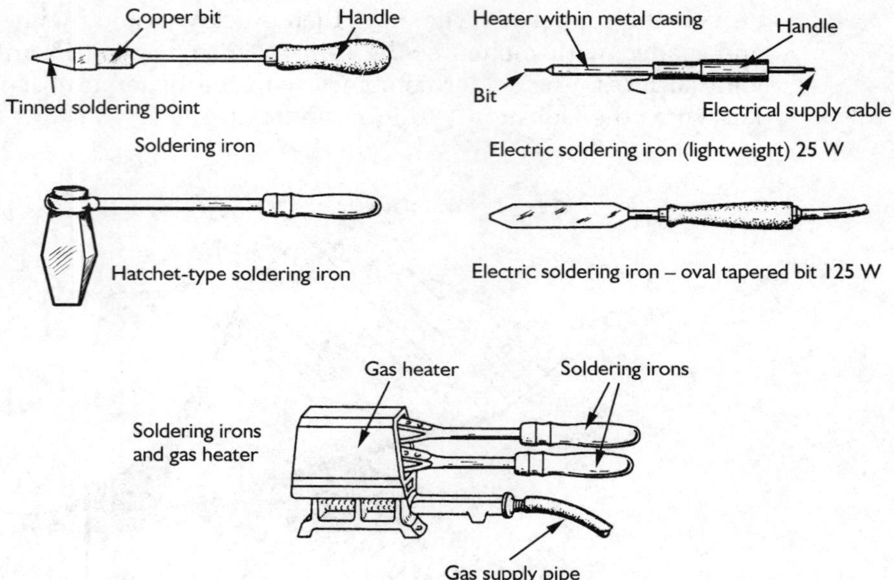

Figure 3.53 Soldering irons

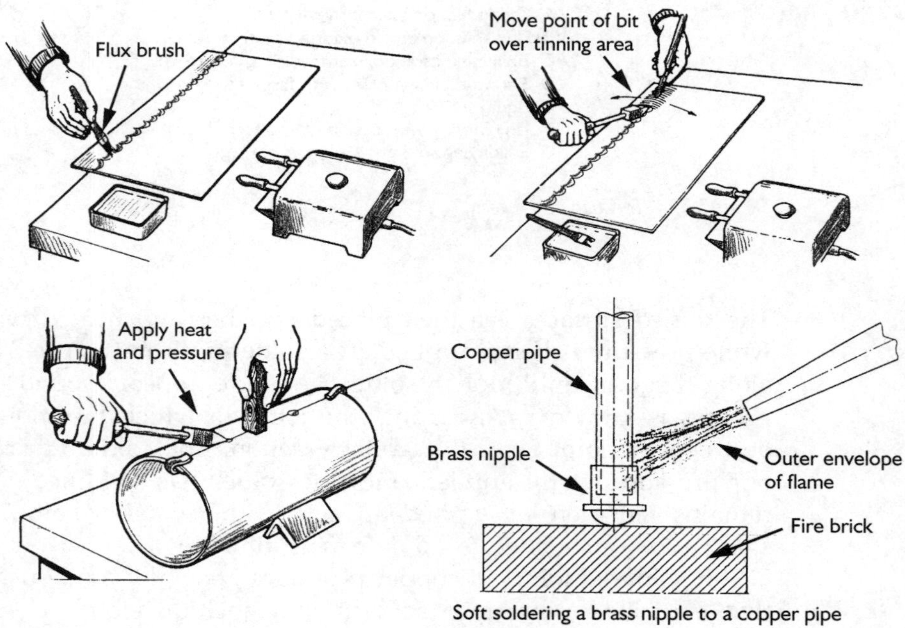

Figure 3.54 Soldering processes

The electronic components used with printed circuit boards (PCBs) usually have leads that are already coated with tin. After placing them in position and cutting the leads down to size, heat is then applied from a soldering iron bit loaded with just enough resin-cored solder to make the joint. This should be done as quickly as possible to avoid damaging sensitive components. The leads on the opposite side of the PCB to the joint can sometimes be gripped with pliers to conduct away excess heat.

Hard soldering: this includes brazing, in which brass, known as brazing 'spelter', is the bonding metal, and silver soldering, in which the solder is an alloy of silver, copper, zinc and cadmium. Hard solders can be used to join mild steel, cast iron, copper and brass. They have a higher melting point than soft solder and produce a stronger joint. Like soft solder, they combine with the material at the joint surface to form an amalgam. Silver solder has a lower melting point than brazing spelter but does not produce such a strong joint. It is used for joining brasses and bronzes where the melting point is very close to that of brazing spelter. Two typical hard solders are shown in Table 3.3.

Table 3.3 Composition of hard solders

British Standard type	Category	Composition (%)				Melting temperature range (°C)
		Silver	Copper	Zinc	Cadmium	
3	Silver solder	50	15	15	20	620–640
10	Brazing spelter	60	40	–		885–890

As with soft soldering, surface preparation is important. The joint surfaces should be cleaned free of oil or grease and surface oxide removed using a wire brush, emery cloth or wire wool. Hard soldering fluxes need to be of the active type. Their constituents vary depending on whether they are to be used for brazing or silver soldering, but mostly they contain sodium borate, known as borax. They are supplied as powder, which is mixed with water to form a stiff paste before applying it to the joint surfaces.

The fluxed components are assembled on the brazing hearth surrounded by fire bricks. This is to reflect and contain the heat, which is supplied from a blow-torch. As the temperature of the components rises, the flux will be seen to be spreading and cleaning the joint surfaces. The hard solder rod or wire can then be touched against the joint and, if the temperature is correct, it will be seen to melt and to be drawn into the joint by capillary action (Figure 3.55).

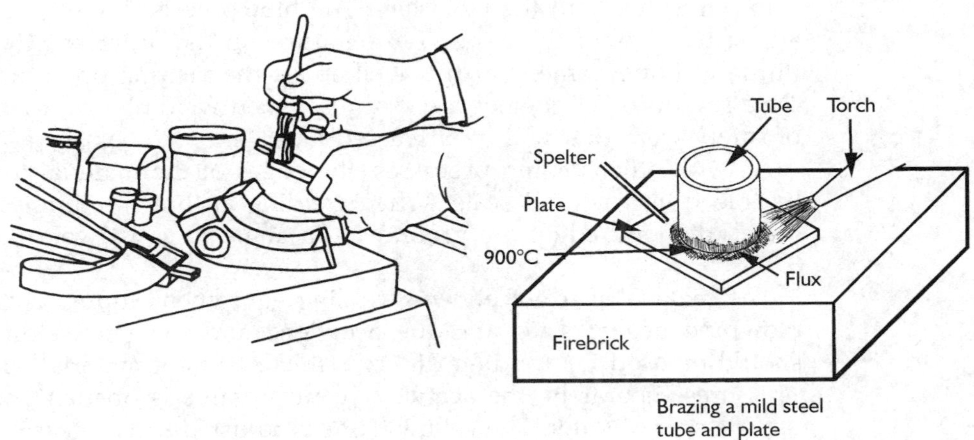

Figure 3.55 Hard soldering

After the joint has cooled, the remaining flux will be seen to have set in a glass-like film. This can usually be removed by emery cloth or wire brushing. Caustic soda and sulphuric acid solutions are also used followed by a thorough washing in water.

Activity 3.5

It has been mentioned that acidified zinc chloride or 'Bakers fluid' is an active flux for soft soldering. It is quite easy to prepare it in the workshop or chemistry laboratory. Find out its constituents, the method of preparation and the safety precautions that are necessary. If possible, and working under qualified supervision, you may then prepare a small quantity and test its effectiveness as a flux.

Welding

Fusion welding is a process in which the materials to be joined are fused together in their molten state. Additional material may be added to the joint from a filler rod that is of the same material composition. If the process is properly carried out, the strength of the joint should be very close to that of the parent materials.

Welding technology covers a wide range of procedures and techniques for joining different materials of various thicknesses. These are continually being added to as new methods of welding and improving weld quality are developed. There are two basic manual welding processes. They are:

- oxy-acetylene welding
- manual metal arc welding.

Oxy-acetylene welding: with this process, acetylene and oxygen gases are mixed together in a blowpipe and burned to supply the heat for welding. The gases are stored in steel cylinders fitted with pressure regulators and flashback arresters (Figure 3.56). Each pressure regulator has two pressure gauges. The right-hand gauge measures the pressure of the stored gas inside the cylinder, and the left-hand gauge measures the regulated welding pressure in the line to the blowpipe.

The flashback arresters prevent the gas flame from travelling back along the connecting hoses to the cylinders. The blowpipe is fitted with control valves to adjust the flow of the gases. Temperatures in the order of 3150°C are generated in the welding flame, which is well above the melting point of most engineering metals. Approved goggles are essential when welding, together with other items of safety wear that will be discussed later.

As with other joining processes, the edges of the material to be joined should be free from rust and scale. When welding all but the thinnest plate and sheet metal, the joint edges are ground to an angle to assist weld penetration (Figure 3.57).

The makers of oxy-acetylene welding equipment supply charts that show the blowpipe nozzle sizes and the acetylene and oxygen welding pressures that should be used for welding different thicknesses of material. Having set the line pressures separately, the acetylene control valve is opened and the gas ignited using a spark lighter. The yellow flame is adjusted until it ceases producing smoke and then the oxygen control valve is slowly opened. The colour of the flame now

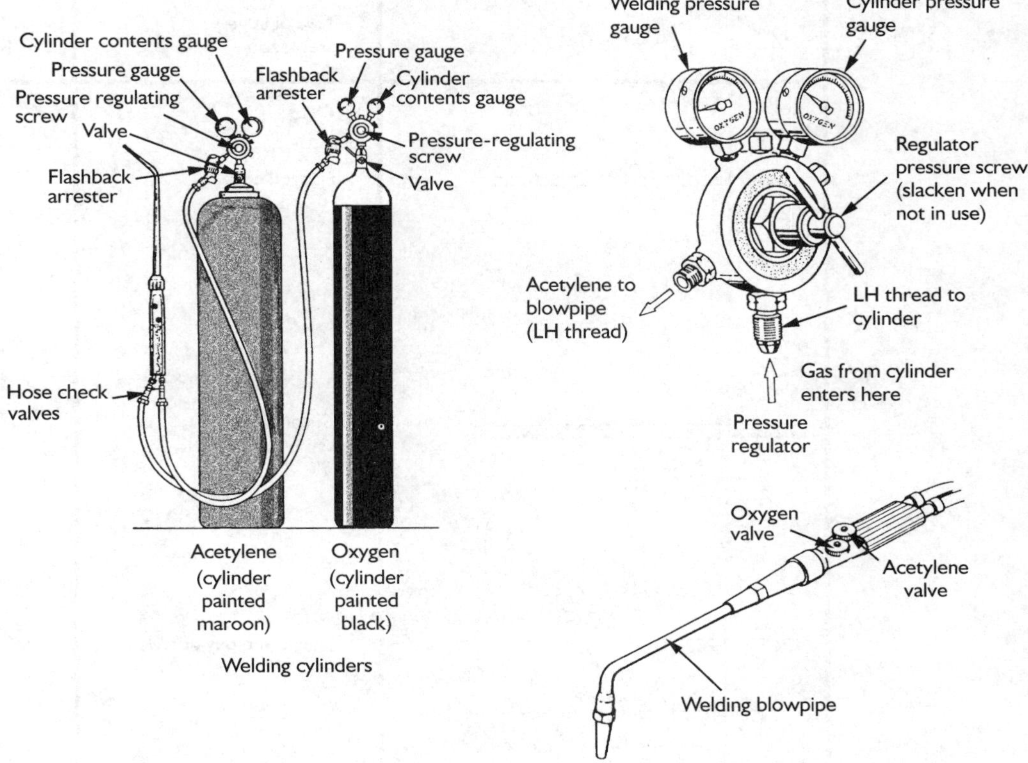

Figure 3.56 Oxy-acetylene welding equipment

changes to blue with two lighter coloured inner cones. For welding mild steel, the oxygen supply is adjusted until the outer cone just disappears to leave a sharply defined inner cone. This gives what is known as a 'neutral flame' in which acetylene and oxygen are being burned in equal volumes (Figure 3.58).

Manual welding is a highly skilled operation and a great deal of practice is required to produce welds of a reliable quality and tidy appearance. The recommended techniques for welding with oxy-acetylene equipment are shown in Figure 3.59. Leftward welding is used for plate thicknesses up to around 6 mm and the rightward method for greater thicknesses. Judgement of the correct speed and movement of the blowpipe and filler rod is critical. If it is too fast, the weld penetration will be incomplete and, if it is too slow, the flame may burn through the metal.

Activity 3.6

What do you think are the reasons why:

a the thickness of the metal around the neck of a gas cylinder is less than that of its sides and base
b oxygen and acetylene gas cylinders should always be kept in the vertical position
c there must be no trace of grease on the gas bottle and pressure regulator screw threads, particularly those of oxygen cylinders?

Joint	Type of edge preparation	Weld symbol BS 499
	Flanged edge	
Gap	Square butt	
80–85° Gap No root space	Single vee oxy-acetylene	

Figure 3.57 Edge preparation for oxy-acetylene welding. Flanged edge and square butt methods suitable for up to 3 mm plate thickness; single vee suitable for more than 3 mm plate thickness

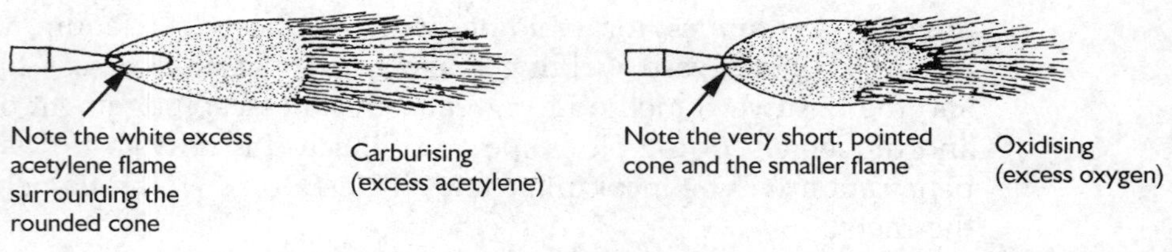

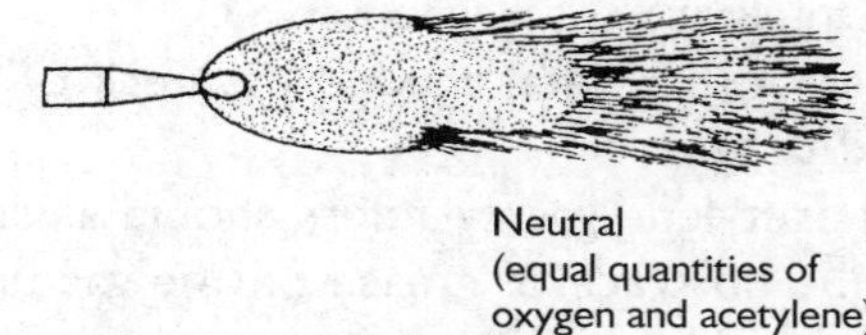

Figure 3.58 Types of welding flame

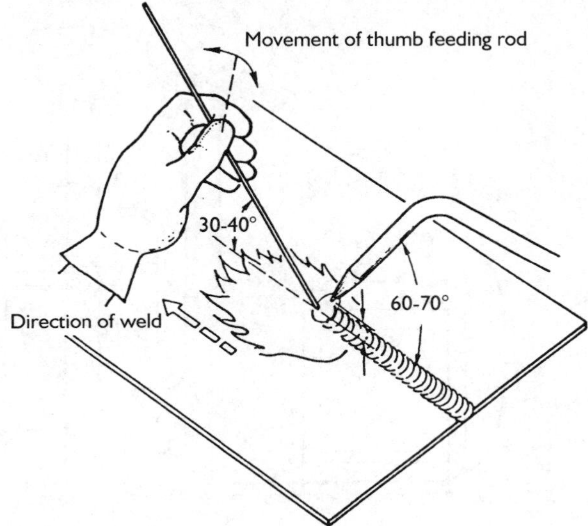

Figure 3.59 Leftward welding technique

Manual metal arc welding: with this method, an electric arc is struck between the filler rod, which acts as an electrode, and the joint material (Figure 3.60). This supplies the heat for welding, and the temperature generated in the arc can approach 6000°C. At this temperature, oxidation can affect weld quality and, to overcome this, the filler rod is coated with a flux. This burns to give a gas that shields the weld against the oxygen in the atmosphere.

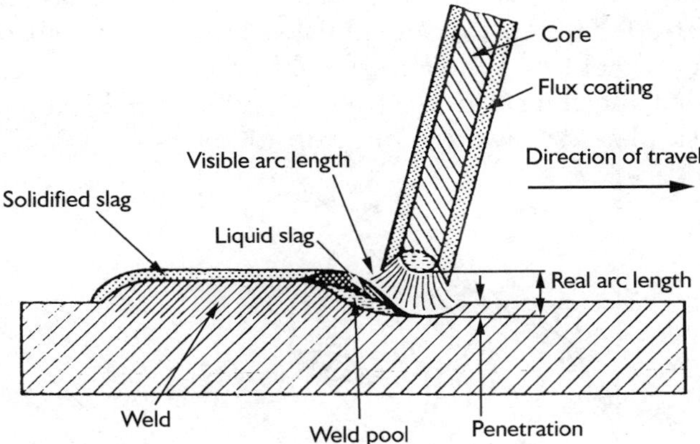

Figure 3.60 The electric arc

Arc welding machines may be powered from the AC mains supply or from a portable generator. They contain a transformer that gives an output of about 100 V before the arc is struck and falls to about 25 V during welding (Figure 3.61). The operator is able to control the current for welding, and the equipment suppliers often provide information on suitable current settings and recommended types of filler rod for different applications.

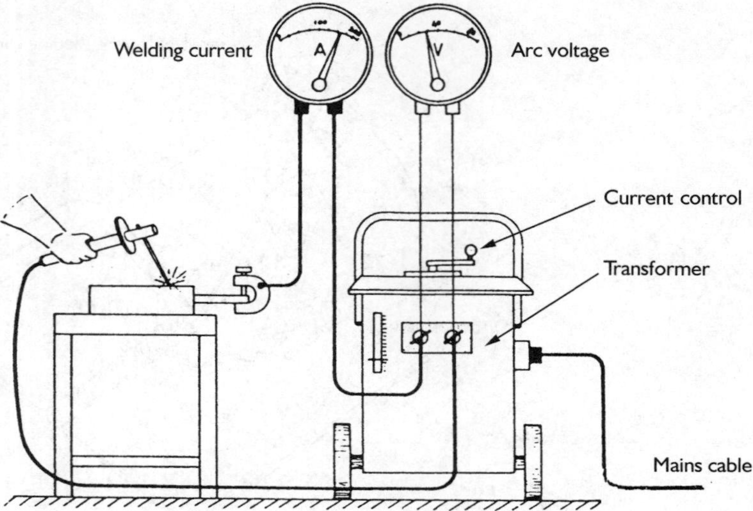

Figure 3.61 Circuit for arc welding

Ancillary equipment includes an insulated electrode holder, its lead cable, and a return lead cable and clamp, which is attached to the workpiece or the metal table on which it rests. A chipping hammer is required for removing solidified flux from the finished weld. An approved safety visor, through which the workpiece must be viewed, is essential, as are other items of safety wear, which will be discussed later.

After cleaning and preparing the joint edges, the components are assembled in position on a metal worktable. The return cable clamp is fixed to the table or directly to the workpiece, and a filler rod of the correct size is inserted in the electrode holder. Having adjusted the current control to a setting suitable for the size of weld and filler rod, welding can commence. The recommended angle at which the filler rod electrode should be held is shown in Figure 3.62 and, for thick plate, the weld is built up in layers.

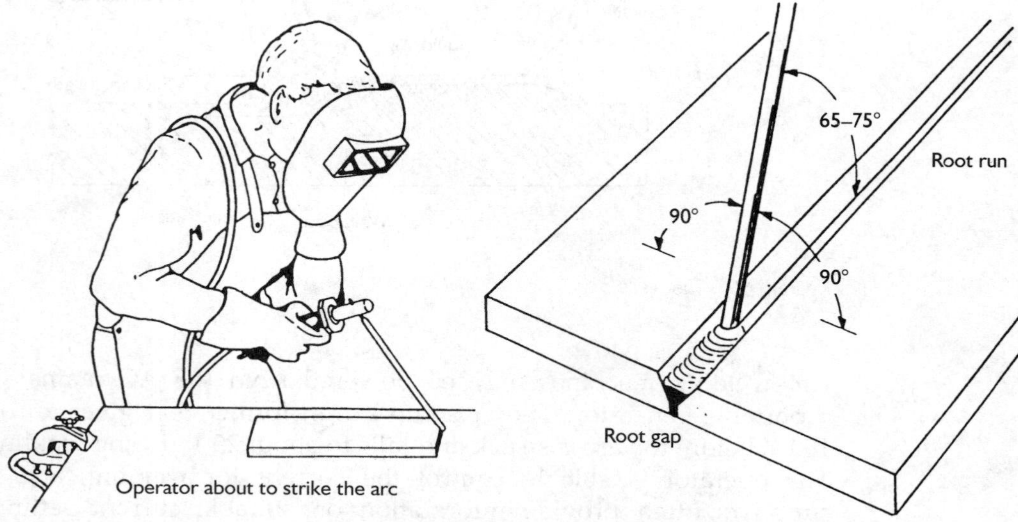

Figure 3.62 Manual metal arc welding technique edge preparation for manual metal arc welding

The view of the weld through the safety visor is more restricted than in oxy-acetylene welding. Manual metal arc welding is thus a highly skilled occupation. Striking and maintaining the arc and moving it at the correct speed takes a great deal of practice before welds of industrial quality can be made.

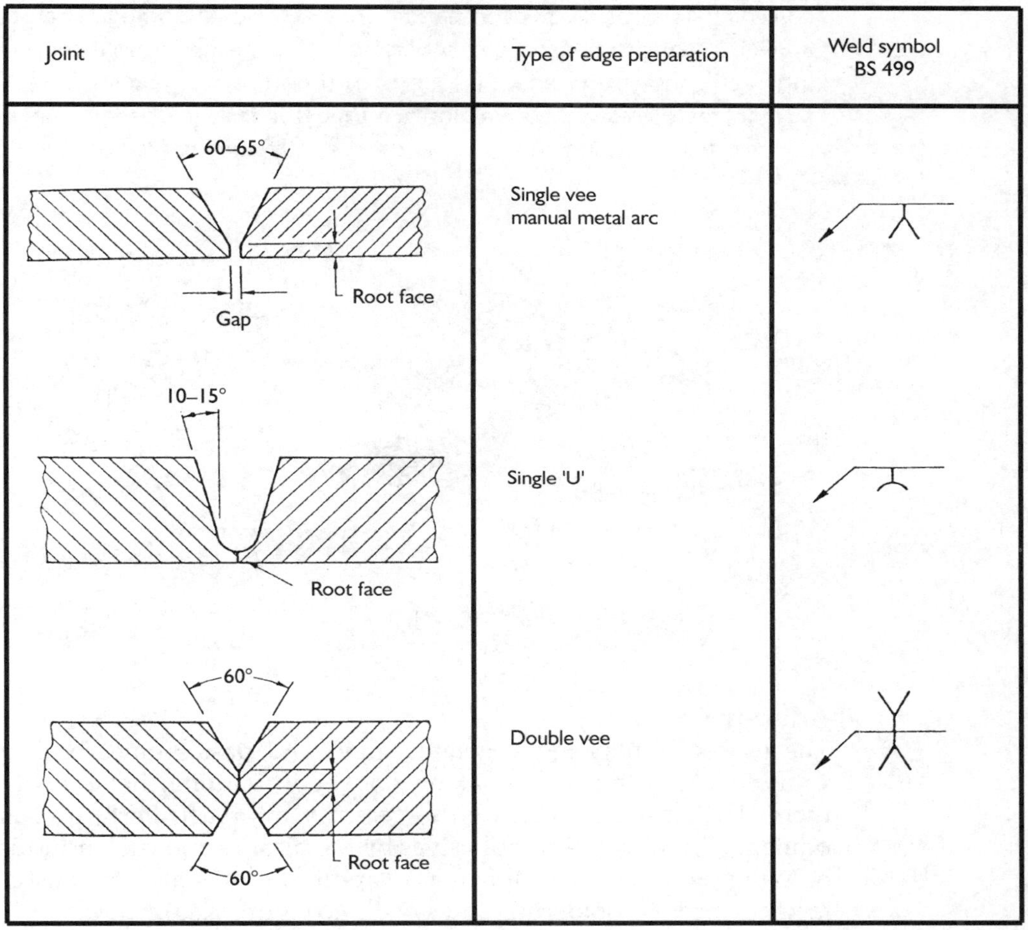

Joint	Type of edge preparation	Weld symbol BS 499
60–65°; Gap; Root face	Single vee manual metal arc	
10–15°; Root face	Single 'U'	
60°; 60°; Root face	Double vee	

Figure 3.63 Edge preparation for manual metal arc welding

Case study

Robotic resistance welding

Motor car body panels are joined together by resistance welding, the process being largely operated by robots. The principle of resistance welding is that, when an electric current flows through a metal conductor, heat is generated. Heavy currents are used in resistance welding, which are applied to the metal through copper alloy electrodes. The resistance of the assembled panels is greater at the joint interface. It is here that the temperature is highest and where fusion takes place as the melting point is reached. Depending on the thickness of the materials, the process takes little more than a second.

Resistance or 'spot' welding machines comprise a transformer, which steps the mains voltage down to between 4 V and 25 V. The output voltage is controlled by a selector switch in the transformer primary circuit. In addition to conducting current to the joint, the electrodes apply pressure and conduct heat from the outer surfaces. To assist this transfer of heat and to prevent the electrodes from softening as a result of overheating, they are water cooled. The pressure is applied through pneumatic cylinders, its effect being to hold the joint surfaces in intimate contact and effect a forging of the materials (Figure 3.64). The advantages of resistance welding are that it is fast, it does not require the use of filler rod or flux and the process can be fully automated.

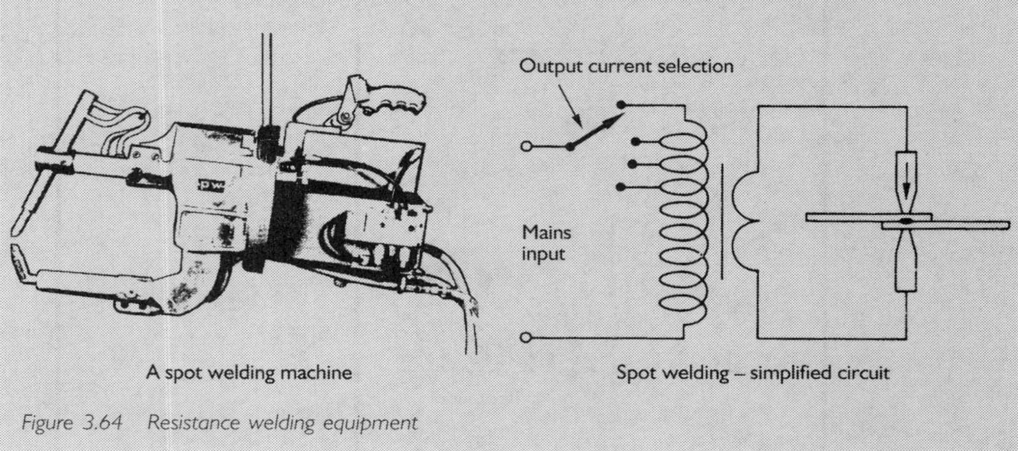

Figure 3.64 Resistance welding equipment

Adhesive bonding

The use of animal and vegetable glues and pastes dates back many centuries. Crafts, such as furniture-making and book binding, have always made use of them. Their use has increased in recent years particularly because of the development of new synthetic adhesives by the plastics and rubber industries. Adhesive bonding is now used for many applications that previously used screwed fasteners, rivets, soldering and welding (Figure 3.65).

When choosing an adhesive, consideration should be given to the loading conditions and service environment that it will encounter. Where a mechanical fixing is to be replaced by adhesive bonding, the joint may need to be redesigned. Adhesives are strong when loaded in tension and shear but tend to be weak when subjected to peeling and cleavage forces (Figure 3.66).

All surfaces to be joined by adhesives must be completely free of oil, grease and loose oxides such as rust. They should finally be cleaned with a detergent or a chemical solvent and may benefit from being roughened to provide a 'key' for the adhesive. The manufacturer's instructions for applying the adhesive should be followed carefully, particularly health and safety instructions. The safety procedures and equipment associated with adhesive bonding will be discussed later.

The most common adhesives can be categorised into:

- thermoplastic resins
- thermosetting resins

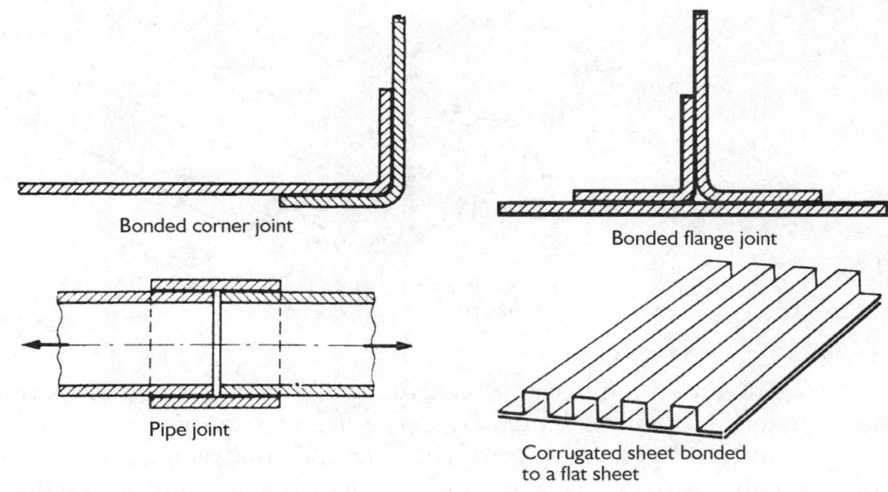

Figure 3.65 Typical adhesive joints

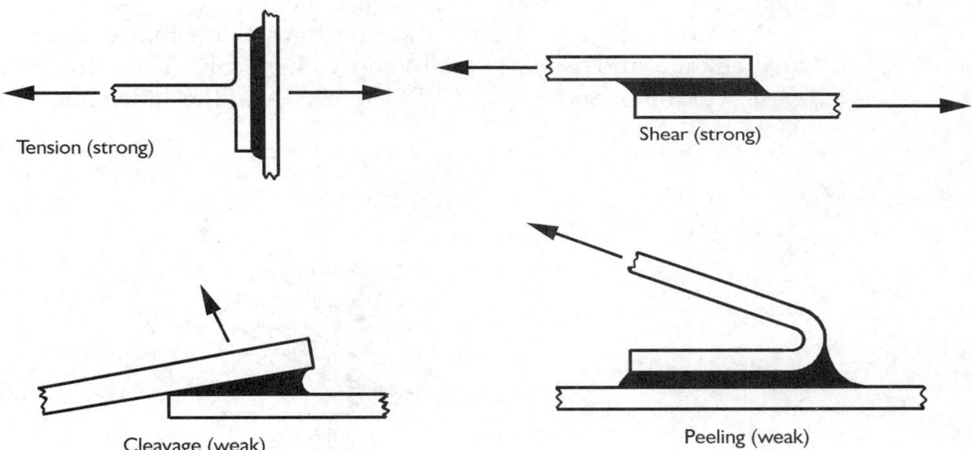

Figure 3.66 Loading of bonded joints

- elastomers
- cyanoacrylate adhesives.

Thermoplastic resins: these may be supplied in liquid or solid form. The solid form must be melted by heating it before use. It is then applied to the joint surfaces, which are held together under pressure until the adhesive has cooled and set. The liquid type is applied in the same way but sets as a result of the evaporation of a solvent.

Solvent-based adhesives should not be used on joints that have a large surface area, unless the materials are porous, since the solvent must be able to evaporate. As a general rule, thermoplastic adhesives should not be used where the joint will be subjected to heat as this may cause softening and lead to failure. Thermoplastic resins, of which 'Bostik' is a familiar brand name, are used to bond glass, Perspex and wood (Figure 3.67). Their bond strength is less than that of thermosetting adhesives, but they are less brittle and will allow a joint to flex slightly.

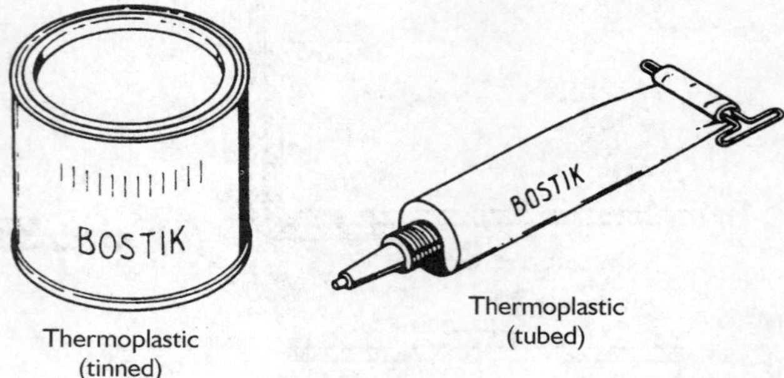

Figure 3.67 Thermoplastic resins

Thermosetting resins: these are usually two-part adhesives consisting of the resin and a chemical hardener. These are mixed together in specified amounts and applied to the joint faces. The joint is then assembled and held together while the adhesive hardens. Hardening is caused by a chemical reaction taking place in the adhesive. This generates heat, which causes cross-links to form between the polymers and the resin to solidify (Figure 3.68).

Thermosetting resins give a strong hard bond that is unaffected by temperature change and is resistant to most chemicals. They are used to join metals, glass, ceramics and plastics. Two-pack 'Araldite' adhesive is a familiar brand name.

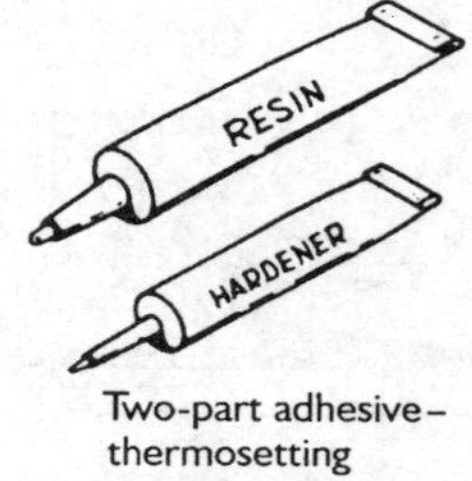

Figure 3.68 Thermosetting resins

Elastomers: these have a synthetic rubber base and contain a solvent. They are called 'contact' or 'impact' adhesives. The adhesive is applied to both joint surfaces and left for a short time until it becomes 'tacky'. The surfaces are then pressed together to exclude any air and a bond is immediately formed. It is important that they are correctly positioned since the joint cannot be adjusted once contact has been made (Figure 3.69).

Elastomers are used to join rubber plastics and laminates and, because the solvent evaporates before the joint is made, they are particularly suited to joints with large surface areas. 'Evo-stik' is a popular brand name.

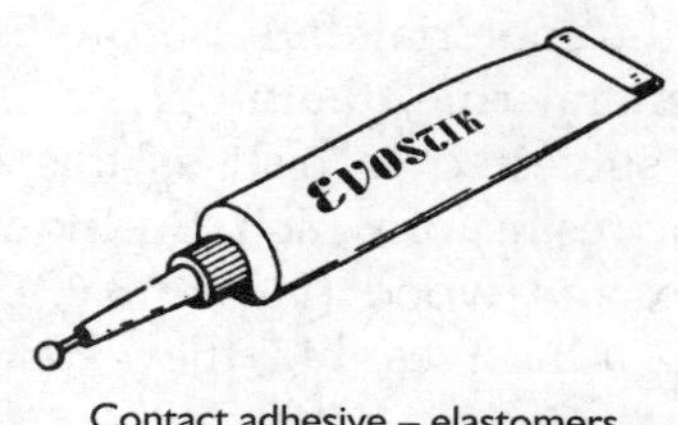

Figure 3.69 Contact adhesive

Cyanoacrylate adhesives: these are more commonly known as 'super glues'. They are supplied in thin liquid form and as a gel and cure in the presence of moisture. A thin coating is applied to one of the joint surfaces only. The surfaces are then brought into contact and light pressure is applied. A bond is formed almost immediately and, as with impact adhesives, it is important that the joint surfaces are correctly positioned when contact is made.

Cyanoacrylate adhesives will join wood, metals, ceramics and most plastics and rubbers. They are used extensively in the assembly of small electrical and electronic components.

Their use requires particular care since they will readily form a bond with human skin, which is difficult to separate.

Mechanical fastenings

A variety of mechanical methods are used in the assembly of engineering components and in the wiring of electrical and electronic circuits. They include:

- screwed fastening
- riveting
- crimping
- wire wrapping.

Screwed fastening: these include nuts and bolts, nuts and studs, setscrews and screwed electrical terminations (Figure 3.70). They have many applications, particularly for joining or securing components that may have to be removed or replaced during servicing.

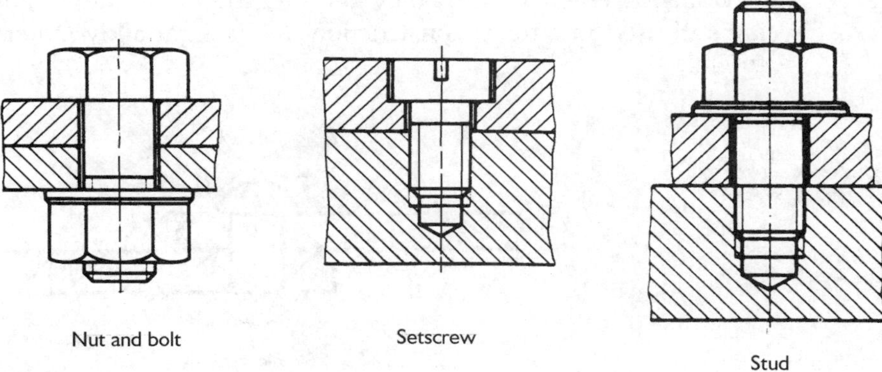

Figure 3.70 Screwed fastenings

The ISO metric thread is now used by most countries for general engineering purposes (Figure 3.71). It has replaced the older unified fine thread (UNF), British Standard fine thread (BSF) and British Standard Whitworth thread (BSW), which were used in this country before metrication. The older kinds of screwed fastening can, however, still be obtained for the servicing of plant and equipment that dates back to premetrication.

The British Association (BA) thread is used for small-diameter screwed fastenings. Although it is a British thread form dating back to long before metrication, it is still widely used internationally for small screwed fastenings and terminals in electrical and electronic equipment.

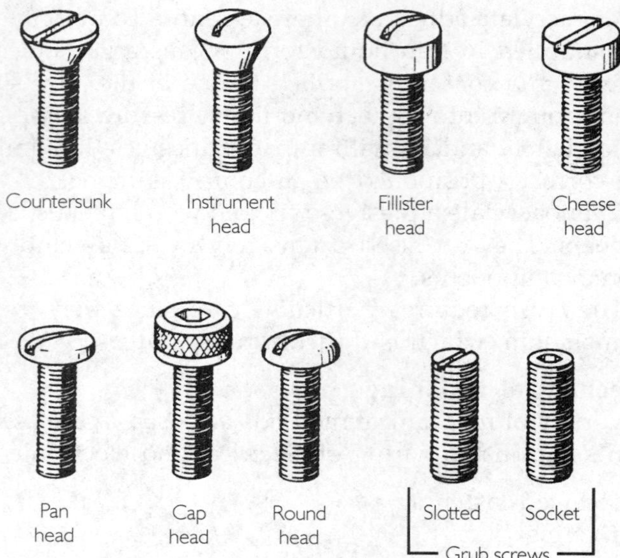

Figure 3.71 Metric screws

Riveting: riveting is used to join components together permanently (Figure 3.72). Welding has replaced many of the traditional applications of riveting, such as in ships' hulls, pressure vessels, bridges, etc. Riveted joints do, however, have some advantages, one being that they allow a structure to flex slightly under load. They are still used in aircraft construction to join light alloy materials that are difficult to weld.

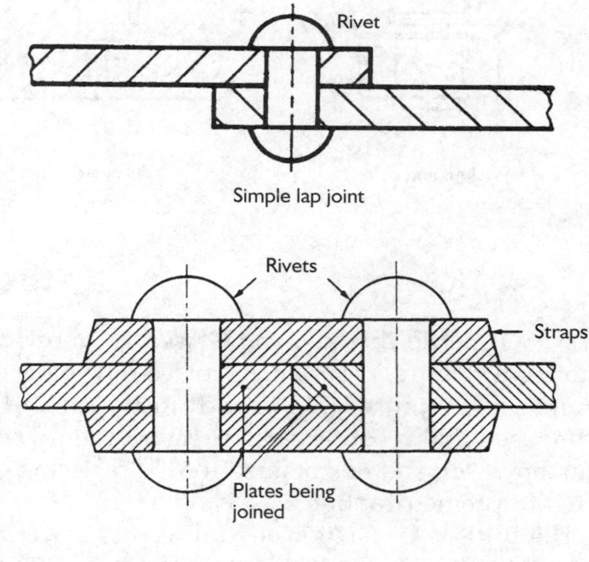

Figure 3.72 Riveted joints

Rivets are available with a variety of head shapes (Figure 3.73). The pan head and cup head types are the strongest, while the flat and mushroom types are used where the head is not required to protrude so much above the material surface. Countersunk rivets are used where a perfectly flush surface is required. Pop-rivets are useful for making joints where access is only possible from one side. They are formed by pulling the central pin through the rivet using special pliers. Eventually, when the head on the blind side has been formed, the pin breaks off.

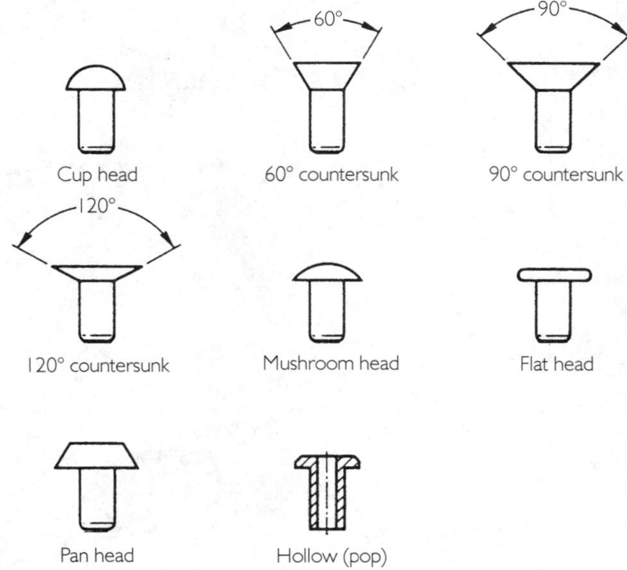

Figure 3.73 Types of rivet

Crimping: crimping is used to attach loop- or spade-type lugs to electrical cables that have screwed terminations (Figure 3.74). A short length of the insulation is removed from the cable, which is then inserted into the terminal lug. Hand-operated crimping pliers or a press is then used to compress the lug onto the cable. No heat is involved that might damage the insulation, and the process can be carried out very quickly.

Wire wrapping: wire wrapping is used to join fine wire conductors to components or terminals without the application of heat (Figure 3.75). It is an alternative to soldering that is now widely used in electronics. A special wire wrapping tool wraps the wire a specified number of times tightly around a square section terminal with sharp corners. The insulation may be stripped off the wrapped length, but sometimes it is left in place. The sharp corners of the terminal cut through the insulation and into the conductor to form a satisfactory electrical joint.

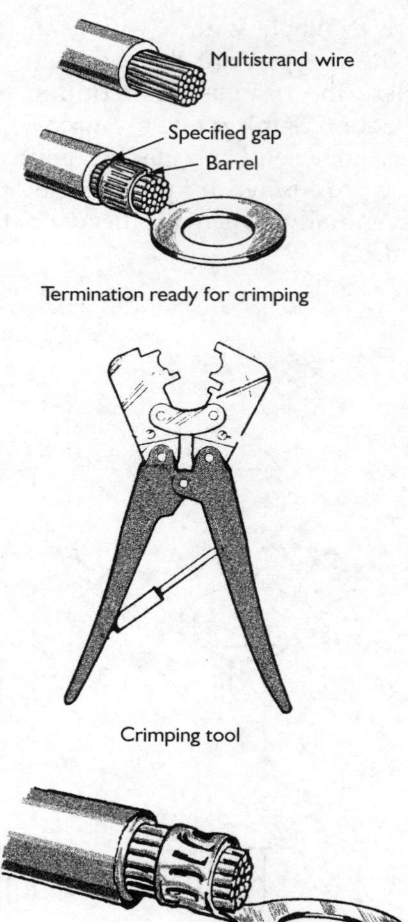

Figure 3.74 Making a crimped termination

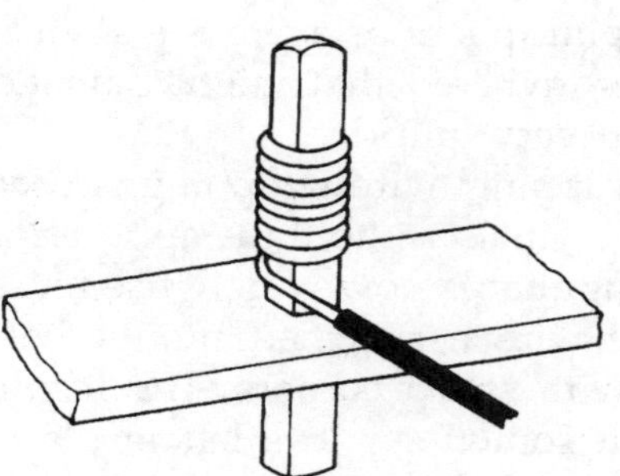

Figure 3.75 Wire wrapping

Progress check

1. Which grade of soft solder is used for making joints in electronic circuits?
2. What happens at the material substrate during the tinning process?
3. What is the action of the flux used in hard soldering?
4. What are the essential differences between soft soldering and hard soldering?
5. What purposes do the pressure gauges and flashback arresters serve on oxy-acetylene welding equipment?
6. Why are the filler rods, used in manual metallic arc welding, coated with flux while those used in oxy-acetylene welding are not?
7. What kinds of forces are adhesive-bonded joints least able to withstand?
8. What is the procedure that should be followed when joining materials by means of a contact or impact adhesive?
9. What advantage might a riveted structure have over one that has been welded?
10. Distinguish between the techniques of crimping and wire wrapping when making electrical terminations.

Heat treatment

Temperature rises affect materials in different ways. Metals oxidise on the surface and sometimes undergo a change in structure before finally melting. Thermoplastics become softer and melt before eventually starting to burn. Ceramics and thermosetting plastics are less affected by temperature rises but eventually they too will char and degrade if subjected to high enough temperatures.

Metals, alloys and some ceramics can have their properties changed by heating and cooling them in different ways. Some heat treatment processes are designed to make the material more malleable and ductile, while others seek to improve its hardness or toughness. The temperature to which the material is raised and the rate at which it is cooled down can have a critical effect on its final properties. The most common heat treatment processes are:

- annealing
- normalising
- quench hardening
- tempering
- case hardening
- precipitation hardening.

Annealing

When metals are cold worked by pressing, rolling, extruding or drawing them into shape, the crystals or 'grains' of which they are composed become distorted (Figure 3.76). This makes the material harder and it is said to have become 'work hardened'. Sometimes this is a good thing, particularly in products such as pressings, which can be formed in a single cold working operation. If it is required to deform the material further, however, it may need to be softened to avoid breaking it. The process used to restore malleability and ductility to a material is annealing.

Annealing involves heating the material in a furnace to what is known as its 'recrystallisation' temperature and holding it there for a period of time. At this temperature, new undeformed crystals or grains start to form from the points at

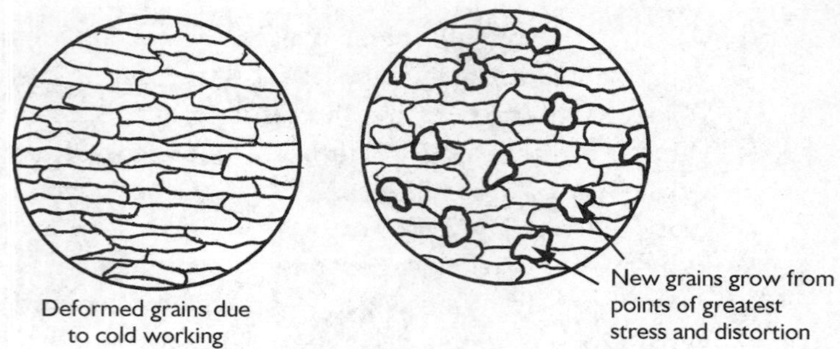

Figure 3.76 *Recrystallisation*

which the old ones were most distorted. These grow until the old structure has been completely replaced by new undeformed grains. If the period of 'soaking' is prolonged, the new grains feed off each other and grain growth is said to occur. The bigger the grains, the softer the material will be and, if the process is carried on for too long, the material may become too soft.

Materials experts are able to specify the time for which a material should be held in the annealing furnace and the way in which it should be cooled down when the new grains are of the correct size. Plain-carbon steel, aluminium, copper and brass can all be softened by annealing. For plain-carbon steels, Figure 3.77 shows the temperature band in which recrystallisation occurs for different carbon compositions.

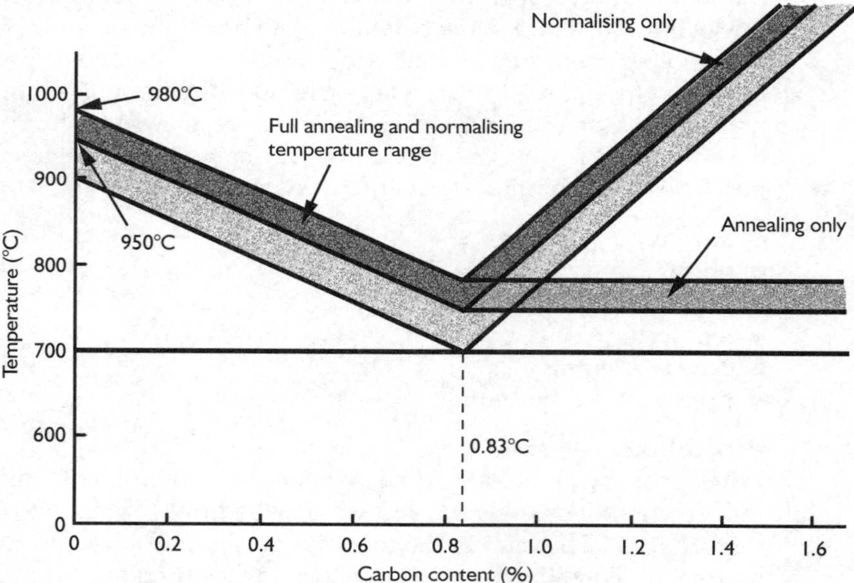

Figure 3.77 *Annealing and normalising temperatures for plain carbon steels*

When the steel has been soaked for a sufficient time, the furnace is turned off and the material is allowed to cool down slowly in the 'dying' furnace. A general guide to the recrystallisation temperatures for other materials is given in Table 3.4.

Table 3.4 Recrystallisation temperatures

Material	Temperature (°C)
Pure aluminium	500–550
Cold working brass	600–650
Copper	650–750

Unlike steels, these materials do not need to be cooled down slowly. They may be quenched after a suitable period of soaking.

Normalising

Steel components that have been formed to shape by hot working processes, such as forging and hot pressing, often contain internal stresses and grains of unequal size. This is mainly caused by uneven cooling. Normalisation is a process closely related to annealing that relieves the stresses and refines the grain structure. It involves heating the components to within the normalising temperature band shown in Table 3.4. After a suitable soaking period, during which recrystallisation takes place and the stresses are relieved, they are removed from the furnace and allowed to cool down in still air. Because the cooling rate is faster than for annealing, the normalised grains are smaller, and the material is stronger than if annealed.

It will be noted that normalising and annealing are carried out at the same temperatures for steels with a carbon content up to 0.83%. The high-carbon steels with carbon contents above this value are normalised at a higher temperature than for annealing.

Quench hardening

Steel is an alloy of iron and carbon in which the maximum amount of carbon is seldom more than about 1.5%. At normal temperatures, iron is only able to absorb a very small amount of carbon in solid solution. The majority of the carbon combines with the iron to form the intermetallic compound Fe_3C. It is called iron carbide or 'cementite'. When viewed under a microscope, steel can be seen to be made up of two kinds of grain (Figure 3.78). One of these appears plain and white, being composed of pure iron in which there is a very small amount of carbon. They are called 'ferrite' grains. The other kind of grain has a lined appearance and is made up of alternate layers of ferrite and cementite. The name given to this layered combination is 'pearlite'.

As can be seen, the higher the carbon content of steel, the greater is the amount of iron carbide or cementite that is present. At 0.8% carbon content, the structure is completely made up of pearlite grains and, at higher carbon contents, the pearlite grains are surrounded by cementite.

In the normalised condition, plain-carbon steels are relatively soft. They are easy to machine and those with a low carbon content can be cold worked by pressing and drawing. Steels with a carbon content above 0.3%, i.e. medium- and high-carbon steels, can, however, be hardened by heating them to within the same temperature band as for annealing and then quenching them in oil or water. When steel is heated to these high temperatures, a structural change takes place. The carbon is fully absorbed in solution with the iron and the material structure is then called 'austenite'.

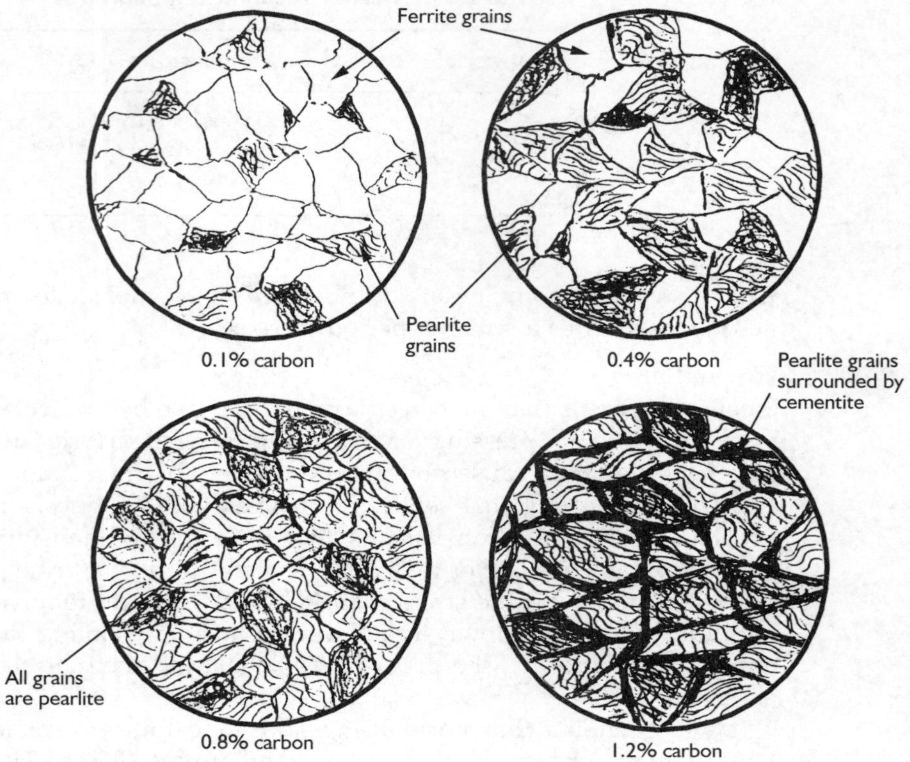

Figure 3.78 Grain structure of plain carbon steels

If the steel is allowed to cool slowly, the structure changes back again, but quenching does not allow time for this to happen. It results in the formation of needle-like crystals, known as 'martensite', and the steel becomes very hard and brittle. Water quenching gives the fastest cooling rate and maximum hardness. Oil quenching is slower and leaves the steel a little less hard and brittle. High-carbon steels with a carbon content above about 0.9% should only be quenched in oil since water quenching is too violent and can cause cracking.

Mild steels with a carbon content below 0.3% have insufficient carbon atoms in their structure for the formation of martensite and their hardness is hardly affected by quenching. They can, however, be 'case hardened', as will be described later.

Tempering

The steel components that have been quench hardened are often too hard and brittle for use. Tempering is a process that removes some of the hardness and makes the steel tougher. It involves reheating the components to between 200°C and 600°C and then quenching them again in oil or water. The tempering temperature chosen for a particular component depends on its final use.

Special furnaces, in which the temperature can be accurately controlled, are used for tempering large batches of components, but small single items can be tempered in the workshop using a gas blowtorch. Before heating, they are polished and, when placed in the flame, the coloured oxide films that spread along the surface are used as a temperature indicator. The tempering temperature for different components and the corresponding oxide colour are given in Table 3.5.

Table 3.5 Tempering temperatures for different components

Components	Tempering temperature (°C)	Oxide colour
Trimming knives and other edged tools	220	Pale straw
Turning tools	230	Medium straw
Twist drills	240	Dark straw
Screwcutting taps and dies	250	Brown
Press tools	260	Brownish-purple
Cold chisels	280	Purple
Springs	300	Blue

Other components, such as crankshafts, gears, transmission shafts, etc., which have to be both hard wearing and tough, are tempered at various temperatures up to 600°C.

Case hardening

As stated above, mild steel cannot be quench hardened. Case hardening is a method of increasing the surface hardness of the steel, while leaving a soft and tough core. This is an ideal combination of properties for components that must be both wear and impact resistant. In cases where it is not required to case harden the complete component, it is first copper plated. The surfaces to be case hardened are then machined, and it is only these that are affected by the process.

The first part of the process is carburising. This involves soaking the components in a carbon-bearing material at the temperature used for annealing. Over a period of time, the carbon soaks into the surface layer of steel and increases its carbon content. The traditional method is known as 'pack hardening' in which the components are packed in iron boxes with a carbon-bearing powder. This may be made from charcoal, bones and leather scrap and can be purchased with different trade names. After soaking and removal from the carbon, the components are reheated to refine the grain size and to harden and temper the outer case.

Precipitation hardening

Precipitation hardening is a process used to control the hardness of certain aluminium alloys that contain copper or small amounts of magnesium and silicon. The alloy 'duralumin', which contains around 4% copper, is typical of the group. When cast and allowed to cool slowly, the final structure of duralumin contains small grains of the intermetallic compound $CuAl_2$. These are sparsely distributed throughout the metal and their properties are not much different from pure aluminium.

The first stage of precipitation hardening is to heat the alloy to around 500°C. At this temperature, a structural change takes place and the copper is absorbed into a solid solution with the aluminium. The material is then quenched and the copper is held in solution giving a material that is stronger and tougher. The structure is, however, unstable and, over a period of several days, very small particles of the compound $CuAl_2$ start to appear. They are known as 'crystallites' and are evenly scattered throughout the structure.

Whereas the original $CuAl_2$ grains have little effect on the material strength, these crystallites have the effect of pinning the aluminium structure, making the metal harder and stronger. The effect, known as 'age hardening', can be accelerated by reheating the quenched alloy to between 120°C and 160°C for around 10 hours. This speeds up the precipitation of the crystallites and the final strength and hardness that it achieves are greater than for age hardening at room temperature.

Aluminium alloys containing small amounts of magnesium and silicon can also be precipitation hardened. Here, it is the intermetallic compound Mg_2Si that precipitates to improve the strength and hardness of the material.

Activity 3.7

Take an old file that is no longer fit for use, heat it to a bright cherry red for about 5 minutes and then allow it to cool slowly. Assess its hardness by making a centre punch mark and a few strokes with a hacksaw at 1-cm intervals along its length. Reheat the file to a bright cherry red and hold for about 5 minutes again. Now hold about 2 cm of it below the surface of a bath of cold water, agitating it slightly until it has all cooled down to room temperature again. Repeat the tests for hardness with the centre punch and hacksaw in approximately the same positions as before. What do your findings tell you about the heat treatability of high-carbon steel?

Chemical treatment

Materials are given chemical and electrochemical treatment to prepare them for other processes, to remove material, to add material and as a finishing process. They include:

- chemical cleaning
- pickling
- etching
- electroplating.

Chemical cleaning

Material and components are often coated with oil or grease from their forming processes. Sometimes, they are also sprayed with a film of oil or wax to protect them while being stored. It may be possible to remove the coating by washing them before use in white spirits or paraffin. In some cases, this may not leave the surfaces clean and dry enough and a more effective process is needed.

Special degreasing equipment, which uses more powerful solvents such as triclorethylene, is widely used in industry. Very often, however, the solvents give off toxic fumes and may also be flammable. Such equipment should be installed in a well-ventilated area and used only by operators who are properly trained and protected.

Pickling

Metals that have been hot formed are often covered with a layer of oxide. Steel, in particular, that has been forged or hot rolled is covered with a black flaky coat of iron oxide, known as millscale. Pickling is used to remove this layer. It is a chemical process in which the components are immersed in sulphuric acid, phosphoric acid or a mixture of the two. Phosphoric acid has an advantage when used with steel. In addition to removing the scale, it combines with the metal to form a protective layer of iron phosphate on the surface.

Etching

Chemical etching is used both for surface preparation and as a material removal process. When used for surface preparation, the etching solution removes only a small amount of surface material, mainly from around the grain boundaries. In this way, it cleans the surface and roughens it slightly ready for painting or plating with another material. Aluminium components in particular need to be etched in this way before painting.

Etching is used to remove larger amounts of material in the production of printed circuit boards and in chemical milling. It enables thin sheets or thin layers of material to be formed into very complex shapes. The areas that are to be left untouched are 'masked' off with a material not attacked by the etchant. The masking may be applied by a photographic process.

Different etching solutions are used for different materials. Ferric chloride is commonly used in printed circuit board production in which the material removed is copper. The chemical substances used in etching processes can be hazardous to health. They can cause irritation to the skin and eyes and give off toxic fumes. Their use requires the provision of good ventilation, safety wear and training. These will be discussed later.

Electroplating

This is an 'electrochemical' plating process commonly used for plating components with copper, cadmium, zinc, nickel and silver (Figure 3.79). It is mainly intended to provide corrosion protection but can also improve the wear resistance and appearance of a product.

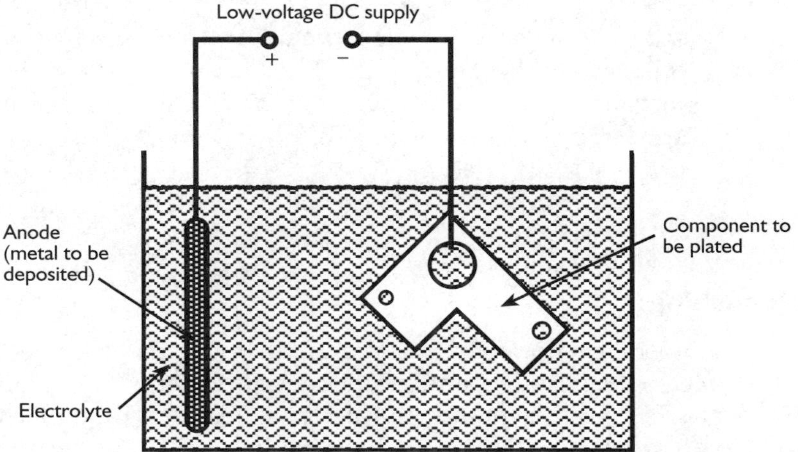

Figure 3.79 Electroplating process

Electroplating requires a DC supply and a suitable chemical solution known as an 'electrolyte'. The component to be plated is connected to the negative terminal of the supply and is known as the 'cathode'. A slab or bar of the plating metal is connected to the positive terminal and is known as the 'anode'. They are then immersed in the electrolyte and the current is switched on. Over a period of time, metal from the anode dissolves, passes through the electrolyte and is deposited on the cathode. The surface layer is evenly distributed and has a uniform thickness that can be accurately controlled.

Case study

Chromium plating

A large number of motor car and motor cycle components are chromium plated to make them hard-wearing, corrosion resistant and decorative. Chromium plating came into widespread use around 1930 before which nickel plating was used. Nickel, however, tends to become tarnished, and it was found that a very thin layer of chromium on top of the nickel gave a superior finish. Both steel and brass components can be chromium plated. The specification for a typical brass component is:

- a layer of nickel 0.03 mm in thickness
- a layer of chromium 0.003 mm in thickness.

With steel the process is a little more complex since there is a tendency for pinholes to be present in the nickel plating. To remedy this, alternate layers of nickel and copper are applied so that the copper fills the pin holes in the nickel, and vice versa, until the surface is smooth enough to receive the chromium. The specification for a typical steel component is:

- a layer of nickel 0.005 mm in thickness
- a layer of copper 0.013 mm in thickness
- a layer of nickel 0.02 mm in thickness
- a layer of chromium 0.003 mm in thickness.

The copper and nickel are applied by electrolysis in the conventional way from copper and nickel anodes. The electrolytes used are copper sulphate solution and nickel sulphate solution containing a small amount of nickel chloride from a copper or nickel anode. The chromium plating is carried out using mixtures of chromic and sulphuric acids. A chromium anode is not used; the metal is deposited from the electrolyte, which gradually becomes depleted.

Surface finishing

Finishing processes are chosen to meet the specified dimensional tolerances, service conditions and aesthetic requirements of a product. For finish machined surfaces, the tolerance, surface roughness and the machining process to be used are often specified on engineering drawings. The service conditions and aesthetic requirements might require a product to be finished by painting or plating with another material. Some of the more commonly used finishing processes are:

- grinding
- polishing
- painting
- plating
- phosphating
- oxidising.

Grinding

This is a material removal process in which components are accurately machine finished to close dimensional tolerances. The two main grinding processes are surface grinding and cylindrical grinding. In each case, the material is removed by a rotating abrasive grinding wheel as shown in Figure 3.80.

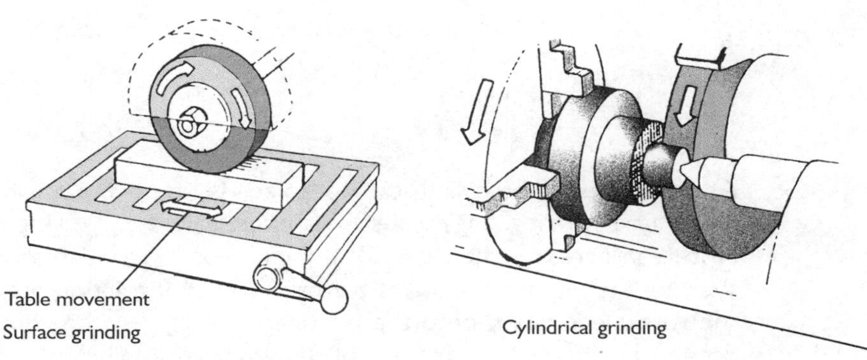

Figure 3.80 Grinding processes

Steel workpieces are held on a magnetic worktable for surface grinding. The process is rather like horizontal milling, but with the grinding wheel replacing the milling cutter. Cylindrical grinding can be used to finish external and internal cylindrical surfaces. The process is rather like turning, but with the grinding wheel replacing the cutting tool or boring bar.

Case study

Cylindrical grinding

To achieve the specified dimensional tolerance and degree of surface roughness, some cylindrical components need to be ground over their complete external surface in one operation. As a result, it is not possible to grip them in the chuck or between centres in a cylindrical grinding machine. The gudgeon pins, which connect the pistons to the connecting rods in internal combustion engines, are a case in point.

Components such as these are finished by a technique known as through-feed 'centreless grinding' (Figure 3.81). The system has three main elements, the grinding wheel, the workrest and the control wheel.

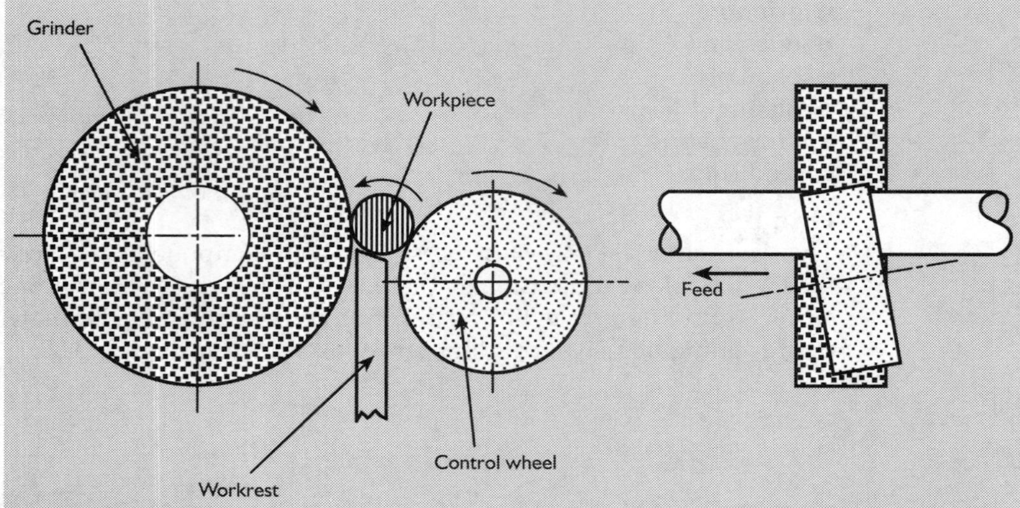

Figure 3.81 Centreless grinding

As can be seen, the cylindrical workpiece is supported on the workrest, which has a hard-wearing, angled surface and incorporates work guides. Lowering the workrest increases the depth of the cut. The larger diameter grinding wheel is the one that actually removes material, its cutting action being such as to push the workpiece down onto the workrest and against the control wheel. It will be noted that the axis of rotation of the workpiece is higher than the spindle axis of the grinding wheel.

The control wheel is made with a rubberised bond and takes no part in the grinding process. It rotates in the same direction as the grinding wheel but at a lower speed. Its purpose is to rotate the workpiece and to feed it across the periphery of the grinding wheel. The control wheel axis is below that of the grinding wheel and slightly inclined to it. It is this inclination that produces the feeding action. Some centreless grinding operations are fully automated. The machine can be loaded by gravity feed from a hopper or by a robot arm. Sensors monitor the ground diameter and the depth of cut is automatically adjusted if any drift towards the specified limits is detected.

Polishing

Polishing and the related processes, lapping and honing, use an abrasive substance in the form of a paste or a liquid containing abrasive particles. Surfaces that are to be finished by these processes should already be quite smooth since they only remove a small amount of material.

With polishing, the abrasive substance is loaded on to a pad, which is also known as a 'buff'. This rotates on the spindle of the polisher, which may be of hand-held or stationary machine type. With the stationary polishing machine, components are held against the rotating buff. With the portable hand-held type, the polisher is held against, and moved over, the workpiece. It can be used to polish the surface metal or to add shine and lustre to surface coatings such as cellulose paint used on motor vehicles.

Painting

Paint is made up of fine solid particles suspended in a liquid. The particles are called pigments and it is these that give the paint its colour. When paint is applied to a surface, the liquid dries and sets as a result of the evaporation of a solvent or by chemical action. Engineered products may be painted purely for decoration but, in the main, it is to protect them against corrosion and heavy use. They are also sometimes painted to identify them by means of a colour code.

All traces of dirt, grease, rust and scale must be removed before painting. This can be done using one or other of the chemical processes that have already been described, such as degreasing, pickling and etching. Alternatively, a mechanical process, such as wire brushing or sand blasting, may be used. Some of the more commonly used paints are:

- oil paints and varnishes
- lacquers
- stoving paints and enamels
- catalytically drying paints.

Oil paints and varnishes have their pigments suspended in natural oils, such as linseed oil, which also contains a solvent such as turpentine or its substitutes. They are generally slow drying as the solvent evaporates. Oil paints and varnishes have been greatly improved in recent years and may contain resins, synthetic rubber, bitumen and asphalt to give added protection and strength in hostile service environments. They are used both indoors and outdoors for the general protection of wood and metal.

Lacquers include enamels and cellulose paints containing chemical solvents that make them much quicker drying than oil-based paints. They may be applied by brush but are more commonly applied by a spray gun. This is much quicker and gives a more even distribution of the paint. Lacquers give a thin hard-wearing finish, which, in the case of cellulose paints, can be further enhanced by polishing with a fine abrasive.

Lacquers are widely used on furniture and motor vehicle bodies where a high-quality hard-wearing finish is required. Shellac varnish, which is a clear, quick-drying lacquer, belongs to this category of paint. It is used by the manufacturers of electric motors and transformers to insulate the coils and core laminations.

Stoving paints and enamels consist of pigments and thermosetting resins mixed with other materials called plasticisers. These are natural oils or synthetic materials, which help the paint to adhere to a surface and also make it less brittle when set. When heat is applied from an infrared source, cross-links are formed between the polymers in the resin and the paint hardens and dries in minutes. A conveyer system is often used to transport mass production components through the stages of paint spraying and drying.

Stoved paints and enamels give a very smooth and hard-wearing finish, which is resistant to temperature and solvents. They are used for domestic equipment, such as washing machines, cookers and refrigerators, and also for motor vehicle bodies and components.

Catalytically drying paints, like stoving paints, are thermosetting resin based and also contain a chemical hardener and a solvent that keeps the paint in liquid form while in its container. When the paint is applied to a surface, the solvent evaporates and a chemical reaction takes place between the resin and the hardener.

This causes the resin to harden quickly as cross-links are formed between its polymers. The paint film is hard and tough and gives better protection to wood and metal than oil-based paint.

Activity 3.8

With a little practice, it is quite easy to make enamelled items of jewellery. The enamel can be purchased in powder form in a variety of colours from craft suppliers. The enamel is plated on copper or brass that has been cut to the shape of the required ear rings, brooch or pendant. The metal must be carefully cleaned to ensure that it is completely free of grease and coated with a thin covering of wallpaper paste. The enamel powder is then sprinkled on to the paste and the item is heated in a gas flame. Eventually, the paste evaporates and the enamel liquefies and adheres to the metal surface. By using a combination of colours and trailing a wire across the enamel before it finally solidifies, attractive patterns can be created.

Plating

In addition to electroplating, which has already been described, sheet steel and steel components can be plated with other metals by:

- hot dipping
- flame spraying
- powder bonding.

Hot dipping: both tin and zinc are plated on steel using this process. Steel strip, which is plated with tin, is called 'tinplate' (Figure 3.82). The tin gives corrosion protection and is not attacked by the natural chemicals found in foodstuffs. It is for these reasons that tinplate is widely used for food canning.

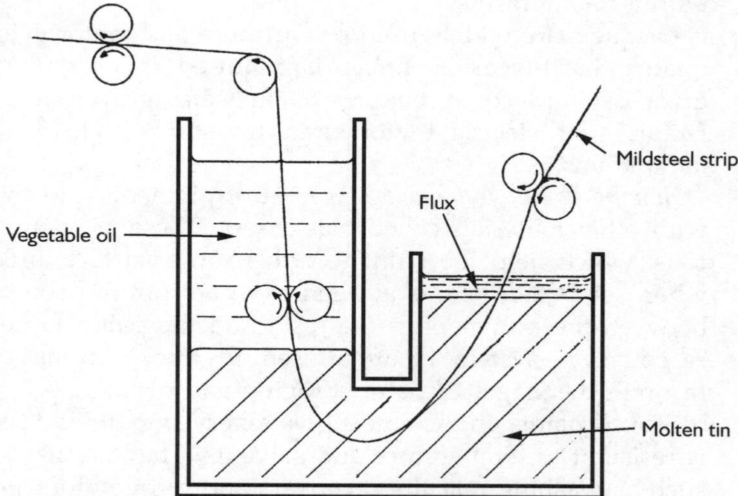

Figure 3.82 Tinplate production

When zinc is plated on steel, the process is called 'hot-dip galvanising'. Zinc gives what is called 'sacrificial' protection to the steel. That is to say, if the zinc coating is damaged and the steel is exposed, it is the zinc that will corrode in the presence of moisture and not the steel. Because zinc corrodes at a very slow rate, the protection will usually last for the working life of a steel component. Tinplate does not behave in this way, and even a thin scratch through the tin coating will allow corrosion to start.

Tinplating proceeds by passing a continuous steel strip through a flux and then through a bath of molten tin. The tin forms an amalgam with the steel surface and the strip then passes through rollers, which give the coating a uniform thickness. Hot-dip galvanising is carried out on both steel sheet and finished components. The process begins with degreasing and fluxing after which the steel is dipped in a bath of molten zinc and quenched in water.

In *flame spraying* or *'hot spraying'*, the coating metal is fed continuously into a blowpipe where it is melted and emerges as a molten spray. Aluminium, cadmium and zinc can be flame sprayed in this way. As with hot dipping, the receiving surface must be properly cleaned and fluxed. The advantage of the process is that a smooth uniform coating of metal can be applied to components with irregular shapes, rather like paint spraying.

Using *powder bonding*, zinc can be plated on steel components by a process known as 'sherardising', and aluminium can be plated by a process known as 'calorising'. The components are first cleaned by sand blasting. They are then placed in a revolving cylinder, which contains zinc and zinc oxide dust or aluminium and aluminium oxide dust. As the temperature of the components is increased, a thin film of zinc or aluminium forms on the surface.

Phosphating is a chemical finishing process in which steel components are immersed in a bath containing acid phosphates of zinc or manganese. A protective phosphate film forms on the metal surface, which, when washed and dried, gives a good base for paint. Alternatively, the components can be washed, dried and finally dipped in hot oil to seal the coating. This process is known as 'parkerising'.

In *oxidising*, steel components with a clean polished surface are heated at around 300°C, so that a blue oxide film is formed on the surface. This takes on a dark blue colour as it thickens and, after a sufficient period of time, the component is oil dipped to seal the film. The process, which is known as 'bluing', gives good corrosion protection and a decorative appearance.

Progress check

1. What is recrystallisation and how is it brought about?
2. What is the purpose of the normalising process?
3. Why is it not possible to quench harden mild steel?
4. Why is it advisable to quench harden high-carbon steels using oil rather than water?
5. Name an industrial use for trichlorethylene.
6. What is the advantage of using phosphoric acid to remove millscale from hot-worked steel?
7. What is the sequence of operations in the chromium plating of steel components?
8. Why do lacquers dry more quickly than oil-based paints?
9. Name three ways in which a steel component can be plated with zinc.
10. What is 'parkerising' ?

Evaluation and selection of processes and process techniques

In a typical engineering company, the choice of the processes that will be used to manufacture a product is ultimately the responsibility of the planning team headed by the senior production engineer. The direct production costs often form the greater part of the total cost of making a product, and it is important that they are kept as low as possible without sacrificing quality.

Before any processing decisions for a product are made, the production team must know how many are required and when, or at what rate, they must be completed. This information is usually supplied by the marketing department. For a company that has won an order to supply fifty new railway locomotives, there will be no doubt about the number required. The delivery dates will also be specified, and there will probably be a penalty clause in the contract specifying that failure to deliver on time will result in the company having to pay compensation to the customer. In such instances, it is essential that reliable processes are chosen and that raw material supplies are equally reliable.

In other instances, a firm might be planning the launch of a new consumer product such as a refrigerator. Here, the requirements are not so easy to predict, and the marketing department must carry out careful research, taking into account the competition from other manufacturers.

If too many are produced and they cannot be sold immediately, the manufacturing costs per item may be lower but the storage or carrying costs may be high. If too few are produced, the manufacturing costs may be higher and customer demand will not be satisfied. Either way, the product will be less profitable than if the requirements had been more accurately predicted.

Once the design of a product has been finalised and the quantities required have been estimated as accurately as possible, the production team must decide which components will be purchased complete and which will be produced by the firm. Generally, it is fastenings and specialist components that are bought out complete, although it must be remembered that these are also engineered products that have gone through a similar process selection procedure at their place of manufacture.

Having decided which components are to be made in-house, it is necessary to identify the transformations from the raw material that are needed to produce them in their finished form. An evaluation of the available processes can then be made in terms of the:

- material specifications
- quality specifications
- quantity and cost.

Material specifications

The specified material and the finished form of a product or component can limit the choice of manufacturing processes. Very often, the production team are consulted during the design stage and are able to suggest materials that will meet the design requirements and yet be the most economical to process.

Polymer materials are generally formed to shape in a single process. With thermoplastics, the choice is between compression moulding, injection moulding, blow moulding, vacuum forming and extrusion. The form of the raw material, whether granules, resin or sheet, and the complexity of the required shape, often dictate which process will be the most suitable.

With thermosetting plastics, the choice is rather narrower. It is limited mainly to compression moulding, injection moulding and contact moulding. The raw material form, whether resin, powder or granules, and the required shape will again dictate the process to be used. Sometimes it is not possible to finish form a polymer component in a single process. Its complexity and dimensional tolerances may prevent this, and moulding may need to be followed by a material removal process.

Bonded ceramic components are made from clays, and their process path usually begins with mixing the raw materials to the correct consistency. This is followed by moulding to shape and kiln firing for an appropriate length of time. With glass products, the raw materials are first heated to a homogeneous molten state. They are then blow moulded or pressure moulded to shape, after which they may be annealed and toughened by heat treatment.

The more simple round or cylindrical metal components are generally machined directly from barstock. This produces a good surface finish and close dimensional control. It can, however, be wasteful in material with components that have large and small stepped diameters or large internal bored holes. The more complex metal components are formed by casting, forging or pressing. Die-cast and pressed components can be finished in one operation. Very often, however, these are primary forming operations, which are followed by material removal, heat treatment and possibly a surface protection process of some kind.

The properties of a specified material can affect the choice of production processes and, likewise, the choice of process can affect the final properties of a material. Steel is perhaps the most widely used engineering material. It can be forged, pressed, machined, welded, heat treated to control its hardness and toughness and plated for surface protection. Steel, however, has poor fluidity when molten and cannot be cast into intricate shapes. Cast iron, as its name suggests, can be cast into intricate shapes, it is easy to machine and can have its toughness and strength improved by heat treatment. Its tensile strength is, however, less than that of steel and it is not easily welded.

Similar examples of suitability and unsuitability for different processes can be found among the non-ferrous metals. Some brasses, bronzes and aluminium alloys have their constituents adjusted to make them suitable for forging and are known as 'wrought' alloys. Others have constituents that improve their molten fluidity and are known as 'cast' alloys. Some can be hardened and toughened by heat treatment, while others can only be heat treated to control their grain size and to remove work hardness. Brasses and bronzes can be joined by soft and hard soldering, but aluminium alloys cannot. It is possible to weld aluminium, but the process requires special equipment and expertise.

Just as material properties affect the choice of process, similarly processing can affect the final properties of a material. Heat treatment processes are an obvious example where the different techniques are specifically intended to modify the strength, hardness and toughness of a material. Forming process can also affect these properties. Cold worked components, which have been pressed, drawn or rolled to shape, become work hardened. Sometimes this is unwanted and the effects of cold working are removed by heat treatment. In other instances, however, work hardness may be desirable as it can toughen and stiffen a component.

Hot working processes can also affect the final properties of a material. With cast components, the type of mould and the rate of cooling affects the composition and size of the grains. Regions where the cooling rate is slow can have large grains that tend to cause brittleness. This is undesirable in materials, such as cast iron, which are naturally brittle. On the other hand, forging and hot pressing can enhance the grain structure of a material. The flow of material as it is shaped or forced into the shape of the die imparts a grain direction to the material. This is not present in components that have been cast or machined from bar. The grain follows the contours of the component giving it added strength.

Quality specifications

Quality is very often defined as the fitness of a component or product for its intended purpose. The quality specifications for a product are to be found on its engineering drawings or in memoranda issued by the design team. The selected production processes must be capable of meeting these specifications, which include:

- toleranced dimensions
- surface finish
- surface hardness.

Toleranced dimensions

A tolerance states the allowable limits within which a dimension must fall. Two typical ways of showing a toleranced dimension are shown in Figure 3.83.

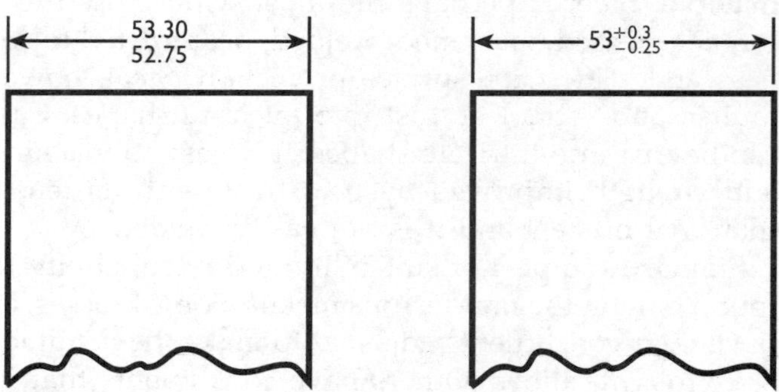

Figure 3.83 Tolerance dimension

Dimensions that are not shown with a tolerance are known as 'open dimensions', but even these can have a tolerance. Guidance as to their required degree of accuracy is very usually stated in the title block of the engineering drawing, e.g. all dimensions to be ± 0.25 mm, unless otherwise stated.

Skilled operators working on small to medium-sized manually operated machines should be able to work within the following limits.

- milling parallel surfaces ± 0.1 mm
- turning cylindrical surfaces ± 0.03 mm
- cylindrical and surface grinding ± 0.008 mm.

Where tolerances are specified on drilled holes, they should be drilled undersize and finished to tolerance by reaming. Modern computer numerical-controlled (CNC) machining centres can turn and mill components to much finer limits than those above. They have in-built quality monitoring systems that are able to compensate automatically for errors such as those that might arise from tool wear. As a result, it is now possible to achieve tolerances in milling and turning that could only previously be obtained by grinding.

Surface finish

If an engineered product is to be given some form of surface protection or decoration, this is usually specified on the engineering drawings. It might be painting, with the type of paint and the number of coats specified. Alternatively, it might be plating with another material, with the thickness of the protective layer specified. In such cases, it is not the production process but the most appropriate application technique that the production engineer has to select.

Paint may be applied by spraying, by dipping, by brush or by roller. Plating can be carried out by hot dipping, by flame spraying or by electrolysis. The production engineer must decide which technique will give the desired quality of finish at the least cost. The production quantities and the availability of equipment and skilled operators will also influence the selection.

For components whose surface is to remain in the formed or machined condition, and particularly those that will be in sliding or rolling contact with other components, the surface roughness is an important quality specification. Surface roughness is specified as the permitted amount by which the surface can deviate above or below the mean surface line. It is measured in 'microns', which are millionths of a metre (μm). The measurement is obtained using specialised equipment in which a probe is drawn along the material surface. Variations from the mean surface line are detected and the surface roughness value is displayed.

Critical machined surfaces have the required surface roughness indicated on engineering drawings by the symbol shown in Figure 3.84.

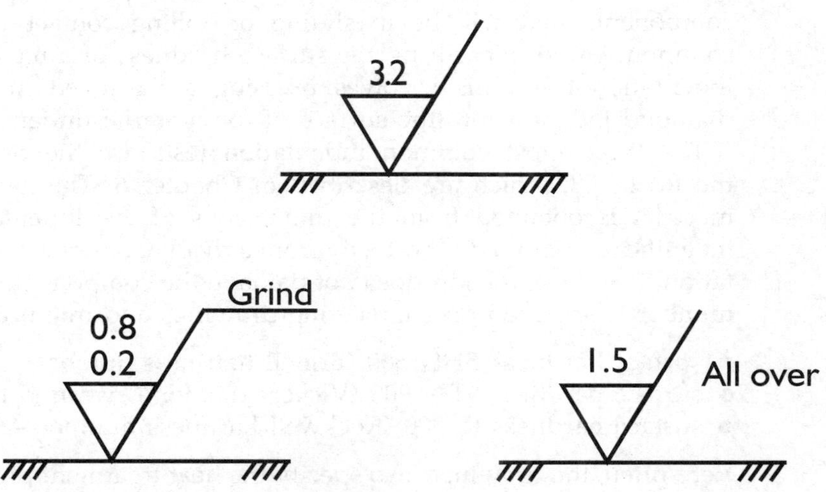

Figure 3.84 Surface roughness specifications

As can be seen from the second symbol in Figure 3.84, the design engineer may specify not only the surface roughness, but also the process technique that is to be used. In such cases, the production engineer is only required to select a grinding machine and grinding wheel type that will suit the material. Where it is the surface roughness specification alone that is given, the production engineer must select a suitable finishing process or technique. Table 3.6 is a guide to the surface roughness that can be expected from different forming and machining processes.

Table 3.6 Range of surface roughness produced by various process techniques

Process technique	Range of surface roughness (μm)
Sand casting	12.5–25
Die casting	0.8–1.6
Forging	3.2–12.5
Drilling	1.6–6.3
Turning	0.4–6.3
Milling	0.8–6.3
Grinding	0.1–1.6
Polishing	0.1–0.4

It should be noted that the condition of the equipment, the material being used and the skill of the operators are important factors that also need to be taken into account when selecting processes to give a specified surface finish.

Surface hardness

Hardness is a measure of the ability of a material to withstand wear and abrasion. As with surface roughness, it is an important quality specification for components that will be in sliding or rolling contact with others. The most common way of measuring the surface hardness of a material is to carry out an indentation test. This involves pressing a hardened metal ball or a pointed diamond indentor into the surface of the material under a controlled load.

The three most common indentation tests are the Brinell, the Vickers and the Rockwell, which are described in Chapter 6. The hardness number for the material is obtained from the dimensions of the indentation measured under magnification or read off directly from a display on the test equipment. The indentation is quite small and does not damage the component. The required hardness number is specified on engineering drawings and might appear as:

- surface hardness BHN 500 (Brinell hardness number)
- surface hardness VPN 900 (Vickers diamond pyramid hardness number)
- surface hardness R_c 300 (Rockwell hardness number – 'c' scale).

Very often, the drawings also specify the heat treatment process that is to be used. It then remains for the production engineer to select the most appropriate heating and quenching techniques for the material.

Quantity and cost

Whatever the scale of engineering production, quantities and costs are closely related. At one end of the production scale are the firms who specialise in the manufacture of single-item or small-quantity products. These might be precision instruments, forging dies, press tools or the turbines and generators for a power station. Specialist production process and highly skilled labour are needed for many of these products, and there is very often strong competition from other firms.

Production engineers would always like to select the most up-to-date processes and techniques in order to improve the quality of a product. Indeed, it is essential for a firm to have an ongoing programme of equipment replacement and retraining for it to stay competitive. It is in firms that make small quantities of precision components that flexible computer-aided manufacturing systems have made an impact. Before purchasing any new equipment, however, a specialist manufacturer must be sure that the cost can be spread over a continuing requirement for similar products.

At the opposite end of the production scale are the high-volume manufacturers who operate continuous production processes. These turn out the same products day after day using dedicated plant and equipment. Typical examples are the steel mills producing plate, bar and tube, etc. The manufacturers of consumer products, such as cars, television sets and washing machines, also operate continuous production processes.

Increasing use is being made of fully automated processes and robots for continuous processes, but the initial costs can be astronomic. With these companies, the choice of processes for a new product is a major decision involving far more people than the production engineers. Decisions can only be made after long discussions between production engineers, marketing experts and financial executives to ensure that the demand justifies the cost. Very often, the discussions also involve government officials who are able to provide grants or who are concerned with the environmental impact of the processes.

Intermediate between the two extreme scales of production are the manufacturers who operate batch production systems. A great many engineering firms fall into this category, especially those who supply components to the larger companies. When components are required intermittently, or regularly but in relatively small quantities, it is often economical to produce them in single or repeated batches that are held in stores to be issued as required. The manufacturing cost per item comes down as the size of a batch increases. This, however, is offset by the cost of storing them for longer periods until required. A compromise between the manufacturing and carrying cost per item must be reached when deciding on the size and frequency of production batches.

The processes and equipment used by firms who specialise in batch production must be adaptable. On completion of a batch, it must be possible to reset the equipment quickly for another product or component. It is thus essential to select general purpose or flexible manufacturing processes for economic batch production. The selection depends largely on the quantities in the batches. Small intermittent batches might best be produced on a flexible computer-aided manufacturing system. The initial cost is, however, high, and there must be a sufficient volume of work to justify it.

Large repeated batches of similar components, such as gears or screwed fastenings, are often produced on conventional equipment that is permanently set up for a particular kind of product. The equipment may require very little adjustment between batches and, depending on the quantity and complexity of the product, it may be automated. Such systems often make good use of the traditional types of production machinery, but engineers should always be looking to update the processes with state-of-the-art equipment when replacement becomes necessary.

Activity 3.9

Figure 3.85 shows how the manufacturing cost per item and the storage cost per item of a particular product vary with batch size. By adding the costs together, sketch a third graph to show how the combined cost per item varies with batch size. From it estimate the most economic batch size and the cost of producing this batch.

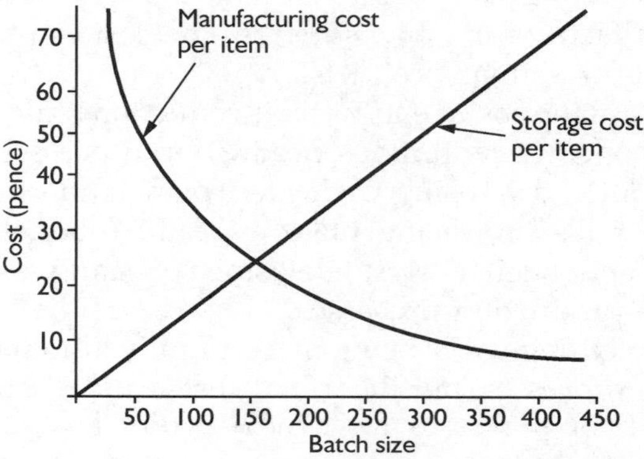

Figure 3.85

Progress check

1 What essential information from marketing personnel is needed for the evaluation and selection of production processes?
2 What kind of components are generally cheaper to purchase than to manufacture?
3 List three of the most commonly available processes for forming thermosetting plastics to shape.
4 What is the usual process path for bonded ceramic components?
5 Why are some non-ferrous metals classed as 'wrought' alloys?
6 In what way can hot pressing and forging enhance the structure of a product?
7 List three kinds of quality specification that are found on engineering drawings.
8 How do the dimensional tolerances that can be achieved using CNC machines compare with those that are possible using manually operated machines?
9 For what scale of production might it be economical to install a flexible manufacturing system?
10 What are the factors that affect the choice of batch size in repeated batch production systems?

Safety procedures and equipment

The law relating to the welfare of people at work is contained in The Health and Safety at Work etc. Act of 1974. The act provided for the creation of the Health and Safety Commission whose duties are to oversee the implementation and functioning of the act. The Commission, which is made up of members appointed from industry, the trade unions, local authorities and the general public, also oversees the working of the Health and Safety Executive.

The Health and Safety Executive is the body that is empowered to carry out workplace inspections. Their function is to ensure that all the laws and regulations are being complied with and to investigate the causes of accidents. The inspectors can prosecute employers and employees who have seriously contravened the law and safety regulations.

The Health and Safety Act lays down the employer's and the employee's responsibilities. The employer must provide a safe and healthy working environment and appropriate safety equipment. Plant and equipment must be properly maintained and guarded, and all process regulations must be followed. Installations, such as lifting gear, electrical equipment and pressure vessels, must be regularly inspected and up-to-date inspection and maintenance records kept. The employer is also required to use safe working procedures and keep a record of all accidents and dangerous incidents that occur in the workplace.

Certain types of accident that require more than first-aid treatment or cause an employee to be absent from work for more than a given number of days must be reported to the inspectors. The inspectors have a right to see the register at any time and investigate the cause of any accident or incident.

The employer is also required under the act to provide supervision, information, instruction and training, particularly for new or young and inexperienced employees. Employees must be made fully aware of hazardous substances and processes and be trained in their use. They must also know how to act in an emergency situation and be properly trained in the use of emergency equipment.

Under the Health and Safety at Work Act, the employee can also be prosecuted for breaking the law and failing to observe safety regulations. It is the legal responsibility of the employee to co-operate with the employer in fulfilling the requirements of the act. Employees must follow safe working procedures and make full use of the safety wear and safety equipment provided by the employer. They must not endanger other workers or members of the public and can be prosecuted for interfering with or misusing health and safety equipment. They thus have a responsibility to take reasonable care of their own health and act in a responsible manner towards others.

The health and safety procedures relevant to the engineering workplace may be divided into:

- general health and safety procedures
- use of health and safety equipment
- specific process safety procedures.

General health and safety procedures

The general health and safety procedures with which all engineering students and employees should be made familiar are:

- personal conduct
- reporting accidents
- responding to an alarm.

Personal conduct

Engineering students and employees should behave responsibly at all times while in the workplace. Practical jokes and horseplay can cause accidents and damage. They have no place in an industrial or training environment. Compressed air lines, pressure hoses, power tools and chemical substances can be lethal. Equipment and materials should never be used for anything other than their intended purpose.

Food and drink should not be consumed at the workstation and 'no smoking' instructions should be obeyed. On no account should anyone be under the influence of alcohol or illicit drugs whilst in the workplace.

Personal cleanliness contributes to the health of an employee or student. In the engineering workplace, it is good practice to rub a barrier cream on the hands and forearms before starting work. Barrier creams prevent dirt from entering the pores and becoming ingrained. They also have antiseptic properties that protect the skin from infection and, being water soluble, they are easily washed off. The finger nails should be kept short and any cuts or abrasions to the skin should be covered with a suitable dressing.

Chemical solvents should not be used to clean the hands since they may cause skin irritation and may be toxic. The hands should always be washed before meal breaks. Barrier cream should then be rubbed on again before restarting work. Overalls should be removed after work and washed regularly. They should not be reworn if they become soaked in oil or glazed with grease since this can penetrate the clothing and cause skin irritation. It can also be a fire risk.

Reporting accidents

All accidents and dangerous incidents should be reported to the area supervisor. In cases of personal injury, first-aid treatment should be sought. All engineering students and employees must know who and where to report to for first-aid treatment. The nature of the accident, however minor, must be recorded, together with the treatment given.

Responding to an alarm

Whenever an emergency situation is discovered, such as an outbreak of fire or the escape of toxic substances, the alarm signal should be triggered. Employees should be made aware of the location of alarm points and the way in which the alarm is sounded. It generally requires the breaking of a glass cover with a striker to release a spring-loaded button. This triggers the alarm, which is usually in the form of a siren.

On hearing the alarm, all process equipment should be turned off and isolated. The building should then be evacuated in an orderly manner without running or panic. The evacuation route should be marked with green arrows in passageways and on stairs. Employees should be made familiar with this, and the outside assembly point, during regular evacuation drills. A roll call should be taken at the assembly point to ensure that all personnel are present.

Activity 3.10

Draw a floor plan of your workshop or training block containing the following items of information:

a the location of first-aid equipment
b the emergency evacuation routes
c the positions of emergency alarms
d the positions of emergency stop buttons
e the positions of electrical isolation switches
f the position of fire doors.

Write down also the location of your assembly point outside the building, the name of the person responsible for carrying out a roll call to make sure that everyone is present and the person who you would normally report to for first-aid treatment.

Use of health and safety equipment

Under the Health and Safety at Work Act, an employer or trainer is required to make the workplace as safe as is reasonably possible. The various items of safety equipment needed to achieve this depend on the type of work being carried out and its associated hazards. Health and safety equipment can be categorised into:

- protective equipment
- safety wear
- first-aid equipment
- emergency equipment.

Protective equipment

Employers and training agents must provide guards and barriers for protection against the moving parts of machinery, hot surfaces and hazardous substances. It is an offence to sell or use machinery that is not adequately guarded. All gear trains, belt drives, chain drives and transmission shafts must be fitted with guards. Guards must also be fitted to the exposed moving parts of material removal machinery. The workpiece and cutter guards fitted to lathes, drills and milling machines will be discussed later.

It is an offence to remove or tamper with machine guards and, if they are found to be faulty or damaged, the matter should be reported to the supervisor. Before starting work on a machine, a check should be made to see that the guards are functioning correctly. They should then be used in the proper manner. Guard rails are often positioned around furnaces, chemical processes and automated equipment such as robots. They are coloured with diagonal black and yellow lines to warn of a potential hazard. Portable safety barriers, which are colour coded in the same way, are used to prevent entry to areas where maintenance work is in progress or some other temporary hazard is present. As with machine guards, it is an offence to remove or tamper with guard rails and safety barriers.

Safety wear

Safety wear is designed to give protection when working with hazardous materials and processes. It is the responsibility of an engineering student or employee to dress in a manner appropriate to the workplace and to make full use of the safety wear provided. This might include:

- protective clothing
- protective footwear
- hand and arm protection
- head and ear protection
- eye protection
- face masks and respirators.

Protective clothing: the most appropriate item of outer clothing for general engineering work is a one-piece boiler suit. This should be close-fitting but not too tight. The pockets should not be loaded down unnecessarily with tools or components, although it is reasonable to carry a small steel rule, notebook and pen or pencil. The cuffs should be fastened or safely turned back and the front should be buttoned or zipped. There must be no loose flapping edges that can become caught up in machinery (Figure 3.86).

Figure 3.86 Suitable and unsuitable dress

Long hair is as potentially dangerous as loose, flapping clothing. It should be held back in a band or, better still, contained under a cap. When working away from process areas, such as in an inspection department or stores, or when working in a seated position, it may be appropriate to wear a well-fitting warehouse coat. Certain processes require specialist safety clothing, particularly where there are sparks or splashes from chemicals, and these will be described later.

Protective footwear: protective footwear is available in a variety of boot and shoe styles. Safety boots and shoes are made with metal-reinforced, non-slip soles and metal toecaps. The soles protect the feet when treading on sharp objects, such as metal swarf or up-turned nails. The toecaps protect the feet when heavy objects are being handled, such as lathe chucks and heavy components for machining or assembly (Figure 3.87). Rubber wellington boots can also be obtained with metal-reinforced soles and toecaps.

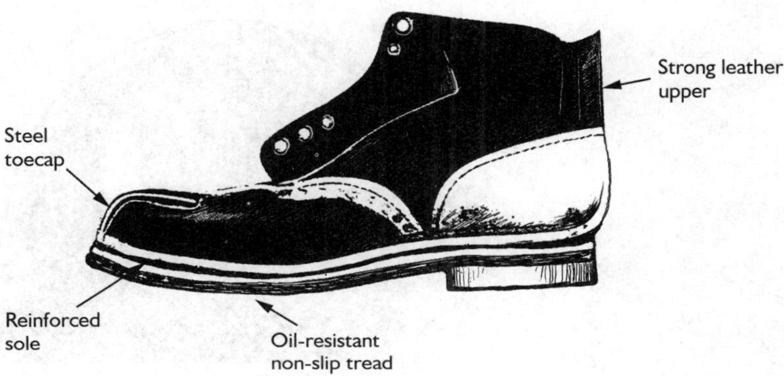

Figure 3.87 Sectioned safety boot

Hand and arm protection: a wide variety of protective gloves and gauntlets is available to protect the hands and forearms. Some are designed to give protection from sparks or when handling hot objects. Others are designed to protect the hands from cuts when handling sharp-edged materials, such as sheet metal and glass. Others still are designed to give protection from corrosive substances and irritant oils, greases and paints. Safety gloves and gauntlets are mostly made from tough leather, rubber and plastic materials (Figure 3.88).

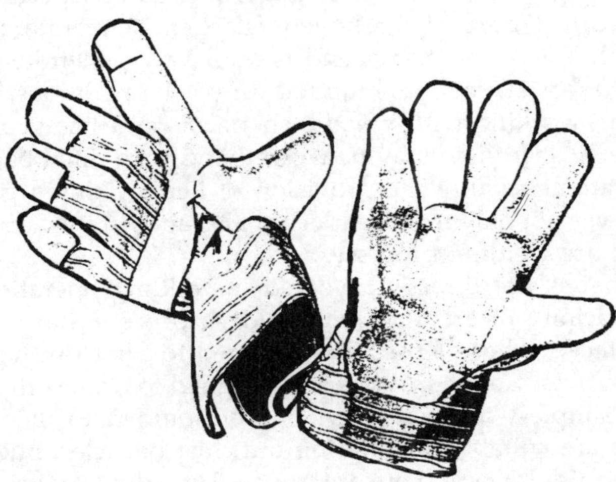

Figure 3.88 Safety gloves

Head and ear protection: this is most important when working in an area where overhead cranes are in use or where overhead maintenance or construction work is in progress. Here, a safety helmet should be worn, and very often it is forbidden to enter the area without one. Safety helmets are made from lightweight plastic materials, which are non-flammable and have a high resistance to impact (Figure 3.89). They also protect the head against electric shock and when working in confined spaces.

Figure 3.89 Safety helmet

In years gone by, deafness caused by working in a noisy environment was looked upon as an unfortunate occupational hazard and no compensation was given or expected. Nowadays, it has been realised that ear protection is necessary for noise levels above a certain intensity. Protective headsets have lightweight plastic ear protectors containing a sound-absorbing material.

A wide range of *protective eyewear*, including safety goggles, spectacles and visors, has been designed to suit different processes. Eye protection is essential for material removal processes, such as turning, drilling, milling, grinding and polishing. Here, clear safety goggles or safety spectacles with side-shields should be worn (Figure 3.90). The lenses are made from a clear toughened plastic material that will not shatter and is not easily penetrated by flying objects.

Full face protection is required for some operations, such as when quenching hot metal, working with volatile chemicals and using portable grinding equipment. Here, a clear visor may be worn that can be tilted back when not in use. Welding operations require the provision of heavily tinted goggles and visors to protect the eyes not only from sparks but also against the harmful radiation that is emitted from arc welding processes.

Persons not directly involved in welding operations but who are working in the vicinity of arc and spot welding processes are advised to wear clear safety spectacles whose lenses are designed to filter out harmful radiation.

Face masks and respirators are designed to protect the lungs from dust and fumes. The simplest disposable types are moulded to fit over the mouth and nostrils. They are suitable for filtering out dust particles, but specialised respirators may be required where there are fumes. These have cartridge-type filters and may also incorporate goggles to give full-face protection (Figure 3.91).

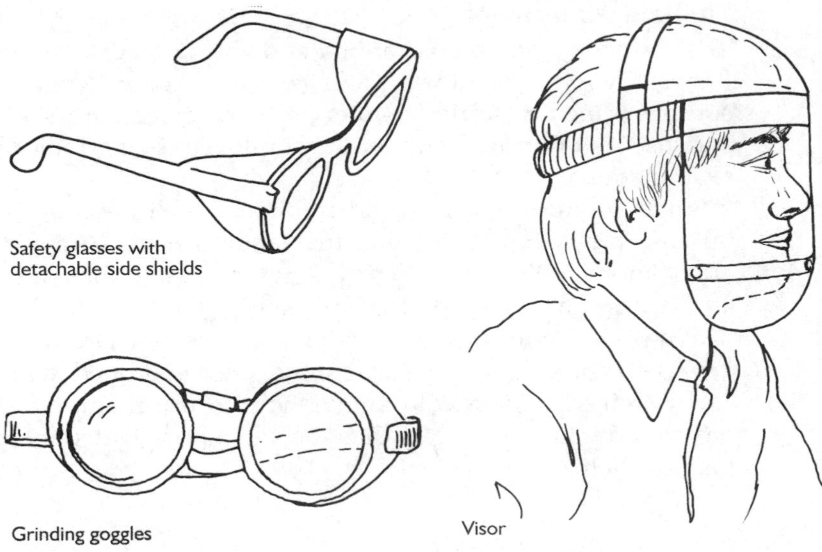

Figure 3.90 Eye and face protection

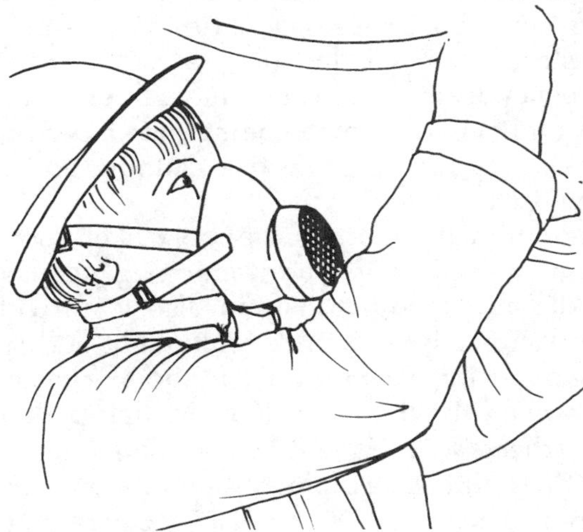

Figure 3.91 Use of a respirator

Case study

Bux v Slough Metals Ltd (1974)

In this case, a metal worker had worked for many years without wearing eye protection. The company manager then supplied goggles but, after a while, the worker went back to his old practice of working without them. One day, he was splashed with molten metal and, as a result, he lost his sight.

The Court of Appeal decided that, although the goggles would not have prevented the worker from becoming splashed, they would have reduced the seriousness of his injuries. The worker was said to have contributed to his injuries by being negligent and, as a result, the compensation payment that he received was reduced. The Court also found that the employers had been negligent because, although they had provided goggles, they had failed to instruct and persuade the worker of how important it was to wear them.

First-aid equipment

The number of first-aid stations and the number of employees trained to give first aid varies with the size of company. Large engineering firms have well-equipped medical centres supervised by trained nurses. They provide first aid and also monitor the health of the employees by periodically carrying out medical examinations.

Small engineering firms and training establishments are not so well equipped but, as an absolute minimum, they should have a trained first aider and a first-aid station equipped with dressings and solutions for the treatment of cuts and burns. They should also have slings, splints, neck braces and a stretcher for dealing with more serious accidents. In areas where toxic and caustic chemicals are used, there should be a first-aid post equipped with neutralising solutions and a trained person who knows how to use them. Affected parts of the body, particularly the eyes, can then be quickly irrigated before calling for outside medical help.

Emergency equipment

An emergency can arise because of fire or the escape of toxic fumes and high-energy radiation. It can also arise as a result of natural causes such as flooding and storm damage. In such cases, the workplace would be evacuated, but emergency teams might then have to enter the area to rectify or assess the situation and possibly to rescue stranded personnel. Emergency equipment in the form of respirators, protective clothing and monitoring equipment might then be required.

The outbreak of fire is the most likely emergency to arise in an engineering work area. Employers and training agents have a legal requirement to carry out periodic evacuation drills and also to provide appropriate fire-fighting equipment. If a fire is seen to be well alight and spreading, the best course of action is to sound the alarm, evacuate the workplace and call the emergency services. If, however, the fire is small, consisting perhaps of smouldering material caused by overheated equipment, it may be judged possible to deal with it using suitable fire-fighting equipment. Employees should be trained to recognise the different types of fire extinguisher, where they are stationed and how to operate them.

It is well known that a fire needs three things to continue burning. It needs a fuel, it needs an oxygen supply and it needs a sustained high temperature. If any one of these is removed, the fire will be extinguished. The most common items of fire-fighting equipment designed to do this are:

- fire blankets
- hose reels and pressurised water extinguishers
- foam extinguishers
- carbon dioxide extinguishers
- vaporising liquid extinguishers
- dry powder extinguishers.

Fire blankets are made of fire-retardant synthetic fabrics which put out a fire by smothering it, i.e. cutting off the oxygen supply. They are usually kept rolled up in a cylindrical container and can be released quickly by pulling on a tape.

Hose reels and pressurised water extinguishers are used to lower the temperature of a fire and, when the water has been turned into steam, have the effect of cutting off the oxygen supply. Hose reels and water extinguishers are coloured red. They should be used only for fires where the fuel is a solid material, such as wood, paper or cloth. Water should not be used on burning liquids since this tends to spread the flames, and it should not be used where there is live electrical equipment.

Figure 3.92 Fire extinguishers

Foam extinguishers contain chemicals that produce a jet of foam. They operate by cutting off the oxygen supply and are coloured cream. Foam extinguishers are the type most suitable for use on burning liquids, such as petrol, oil and chemical solvents. They can also be used effectively on solid materials.

Carbon dioxide extinguishers produce a jet of carbon dioxide gas that cuts off the oxygen supply and does not leave a residue to be cleaned up after the fire has been extinguished. They are the type that should be used for extinguishing fires in electrical equipment where there may be live conductors. They may also be used effectively with burning liquids and solid materials.

Vaporising liquid extinguishers produce a jet of vapour that cuts off the oxygen supply in the same way as does the carbon dioxide extinguisher. They are coloured green and are used for fires in electrical equipment, burning liquids and solids. The vapour is, however, toxic and they are best used outdoors.

Dry powder extinguishers cut off the oxygen supply to a fire in two ways. First, the powder has a smothering effect and, when heated, it produces carbon dioxide gas, which also cuts off the oxygen supply. They are coloured blue and are useful for fighting flammable liquid fires particularly in kitchens and food stores. The powder is non-toxic and, afterwards, it can be swept up or removed with a suction hose.

Activity 3.11

Find out and make a note of the following items of safety information:

a Where are the fire points in your workshop or classroom block?
b What types of fire-fighting appliance are they equipped with and how are they operated?
c How often are fire extinguishers checked and whose responsibility is this?
d How often should fire drills be carried out and who is responsible for conducting them?
e On hearing the alarm, what action should you take before leaving your workstation or study area?
f What is the address and telephone number of your Health and Safety Inspector?

Specific process safety procedures

The general safety procedures and items of safety equipment used in engineering work areas have been discussed above. In addition, there are certain other procedures and items that are specific to the different production processes. The process techniques that have already been described are:

- material removal
- joining and assembly
- heat treatment
- chemical treatment
- surface finishing.

Material removal

The material removal processes that have been described are turning, drilling and milling. Before using any of these process machines, it is essential that operators fully understand the controls. They should be able to identify the isolator switch, the motor start and stop switches, the clutch and the speed and feed controls. The isolator may be in the form of a key switch or a lever switch, and the stop and start switches are usually of the push-button type, coloured green and red respectively.

In the case of sensitive bench drills and very small centre lathes where there is a direct drive from the motor to the spindle, these switches also start and stop the spindle. It is thus important to know their position so that the machine can be stopped quickly in an emergency. Larger machines are fitted with a clutch mechanism operated by a lever. Here again, it is important to know exactly where the lever is positioned so that the machine can be stopped quickly.

In addition to being able to stop an individual machine in an emergency, it is important to know the position of the emergency cut-out switch for the work area. This is in case one sees a fellow worker in difficulties who might be some distance away. Emergency switches are usually of the push-button type and are situated at strategic positions around the work area.

Mention has already been made of the chucks that are used on both lathes and drilling machines. Chuck keys should not be left in the chucks. Injury to the operator or damage to the machine can result from starting the spindle with the chuck key still in position. Chuck keys and other hand tools should not be allowed to accumulate and clutter up the machine or its worktable. They should be kept in the machine tool cabinet and returned there after use.

The danger of wearing loosely hanging clothes and long hair when operating machines cannot be stressed too greatly. Some of the most horrific industrial accidents have resulted from operators becoming caught up in machines. The importance of wearing protective spectacles and goggles has also been mentioned (Figure 3.93). This applies especially where very small chips are thrown out by the cutting process.

Figure 3.93 Protective glasses and cap

Lathes and drills are fitted with different designs of chuck guard, and milling machines are fitted with cutter guards (Figure 3.94). It is essential that the guards are properly used and in position before starting the machine; also, that they are left in position until the spindle is stationary again.

Care should be taken when handling swarf. This can be both hot and sharp and should only be handled with protective gloves. Furthermore, swarf should not be allowed to accumulate on a machine, and it should be removed only when the machine is stopped. It is also advisable to wear protective gloves when handling milling cutters.

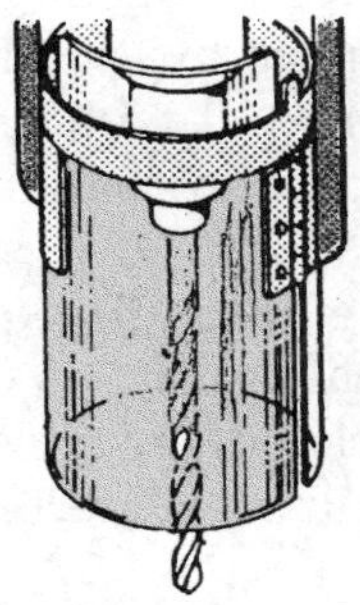

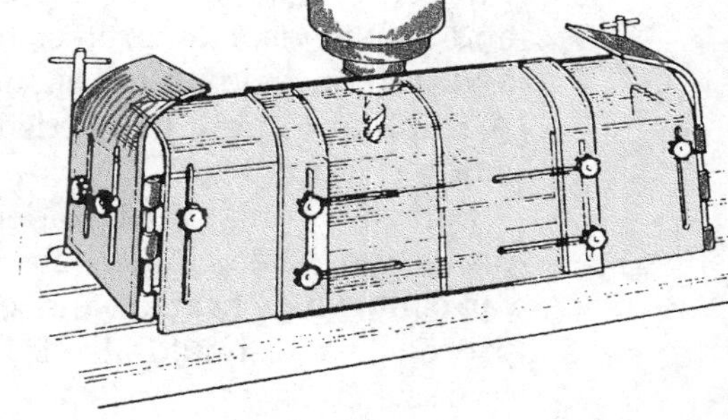

Figure 3.94 Examples of machine guards

Joining and assembly

Safe working practices and the correct use of safety equipment are essential when joining materials by soft soldering, hard soldering and welding. Heat is involved in each of these processes, and heat-resistant gloves should be made available for handling the hot materials. Welding requires the provision of additional items of safety wear because of the hot sparks and splashes of molten metal that are produced. These include leather gaiters to protect the legs and feet, a leather apron or jacket to protect the front of the body and gauntlets to protect the arms (Figure 3.95).

Depending on the type of welding, the above items need not all be worn, but it is advisable to wear gloves and a protective apron for all welding operations.

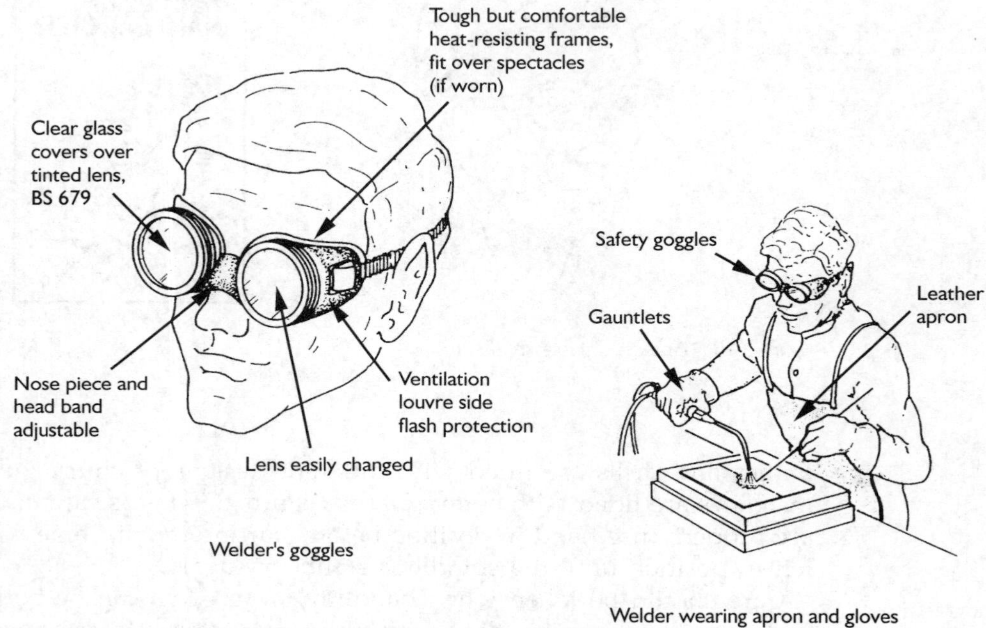

Figure 3.95 Safety equipment for oxy-acetylene welding

Heavy arc welding processes, particularly overhead welding, require full body protection. Here, all of these items are worn together with a combined helmet and safety visor to protect the head. Fumes are produced during arc welding, and indoor welding areas should be equipped with fume extraction equipment. A respirator might be needed when welding in confined spaces with poor ventilation (Figure 3.96).

Soft soldering, hard soldering and welding all require the provision of eye protection. The acidified zinc chloride that is used as a flux when soft soldering can spit when heated, and clear spectacles or goggles should be worn for this process. They should also be worn when hard soldering to protect the eyes from sparks.

As has been stated, welding requires the wearing of tinted goggles and visors. These not only protect the eyes from heat and sparks but also from the harmful

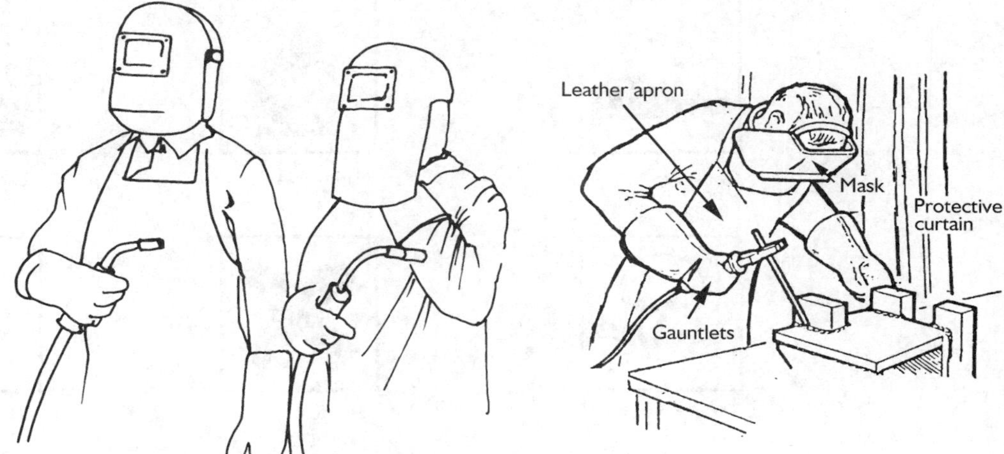

Figure 3.96 Safety equipment for arc welding

radiation that is produced when arc welding. Arc welding operations should be carried out in a screened booth or behind a portable screen to protect fellow workers from the intense flash and radiation. Where this is not possible, workers in the vicinity should be warned to turn away as the arc is struck. In areas where unscreened arc welding is an ongoing process, protective spectacles that filter out harmful radiation, should be provided for all the employees.

Heat treatment

Particular care is required in heat treatment areas when furnaces are being loaded or unloaded. This sometimes requires the expert use of lifting and positioning equipment in the presence of intense heat. Heat treatment personnel engaged in these operations should be fully equipped with heat-resistant clothing and visors. Other employees should be kept clear of the area.

Care is also required during quenching operations where hot oil or hot water and steam is likely to splash or spit out from the quenching tank. Here again, protective clothing and eye protection is required. Components that have been normalised are required to cool down in still air. The area reserved for this should be fenced off and have warning signs to indicate the presence of hot material. All heat treatment areas should be well ventilated.

Chemical processes

The chemical substances used in processes, such as pickling, etching, electroplating and phosphating, must be properly stored and labelled. In addition to the trade name, the labels should give details of the chemical composition of the substance together with one of the signs shown in Figure 3.97.

These indicate, at a glance, the hazardous nature of a substance. The Care of Substances Hazardous to Health (COSHH) regulations specify the ways in which hazardous substances should be transported and stored.

As has already been stated, the first-aid posts in chemical process areas should have neutralising solutions for the chemicals that are in use. Process operators

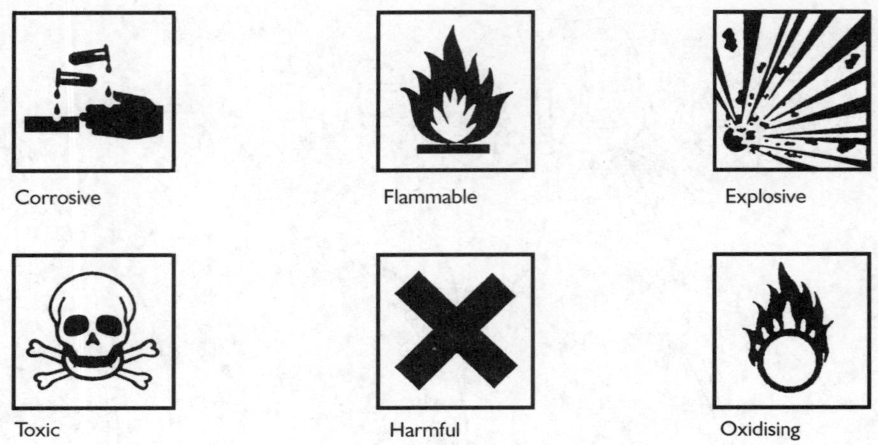

Figure 3.97 Hazardous substances

should be provided with appropriate safety wear and eye protection and there should be good ventilation. Personal cleanliness is most important where corrosive or toxic chemicals are in use. The hands should be thoroughly washed before meal breaks and food should not be consumed in the process area.

Finishing processes

Of the finishing processes described previously, grinding and polishing are essentially material removal processes. The safe working procedures described above for turning, drilling and milling generally apply, particularly the use of guards and the wearing of eye protection. These processes often produce sparks and dust. Stationary machines, such as surface and cylindrical grinders, should be fitted with dust extraction equipment. When using portable grinders and polishers, a face mask with a throw-away filter should be worn.

Grinding wheels are supplied to suit different materials and only the approved type should be used. Grinding should always by carried out using the periphery of the wheel. Using the side face can cause the wheel to shatter. The remounting and truing of grinding wheels is a skilled operation and should only be carried out by a qualified person.

Paint spraying by hand requires the provision of a face mask, eye protection and protective overalls. These often incorporate a hood to keep the paint mist out of the hair. Cotton gloves are also worn to protect the hands from the paint and solvent. The process should always be carried out in a well-ventilated area.

Hot dipping, hot spraying, powder bonding and oxidising are finishing processes that make use of molten metal or the application of heat. They have much in common with heat treatment and welding processes and share the same general safety procedures and items of safety wear.

Note that more detailed information on the items of safety equipment and safety procedures described in this chapter can be obtained from British Standard Specifications BS 3456, BS 4163, BS 5304, BS 5378 and PD 7304.

Progress check

1. What is the name of the body that is empowered to carry out workplace inspections?
2. What kind of accidents must be reported to the Inspectorate?
3. What are the legal responsibilities of an employee under health and safety legislation?
4. What course of action should a machine or process operator take on hearing the alarm?
5. What are the most appropriate items of outer clothing for the engineering workshop or stores area?
6. With what essential items should the first-aid post in a small engineering firm be equipped?
7. What is the best course of action on finding a fire that is well alight?
8. What would be the likely effect of using a hose reel or pressurised water extinguisher on a burning liquid?
9. How does a carbon dioxide fire extinguisher produce its extinguishing effect?
10. What are the regulations that govern the handling, storage and transportation of hazardous chemical substances?

Assignment 3

This assignment provides evidence for:
Element 3.1: Select processes to make electromechanical engineered producs
and the following key skills:
Communication: 3.1, 3.2, 3.3, 3.4
Information Technology: 3.1, 3.2, 3.3

Your tasks

Having obtained a knowledge of some of the more common engineering processes and process techniques, you should now be able to provide examples of products or components in which they have been used. You should also be able to identify the processes and techniques that have been used to manufacture given electromechanical products. To demonstrate this, complete the following tasks, which should take in the range of processes and techniques, and submit a report of your findings and observations.

- Identify two engineered products or components that have been formed by material removal together with the specific techniques that have been used.
- Identify two engineered products or components that have been formed by shaping together with the specific techniques that have been used.

- Identify two engineered products or components that have been fabricated or assembled and the specific techniques that have been used.
- Identify two engineered products or components that have undergone heat treatment together with the specific techniques that have been used.
- Identify two engineered products or components that have undergone chemical treatment together with the specific techniques that have been used.
- Identify two engineered products or components that have undergone a finishing process together with the specific techniques that have been used.
- Identify the processes and techniques that have been used to manufacture an electrical plug top.
- Identify the processes and techniques that have been used to manufacture a sparking plug for an internal combustion engine.
- Identify the processes and techniques that have been used to manufacture a metal filament lamp.

Chapter 4 Making electromechanical products to specification

This chapter covers:
Element 3.2: Make an electromechanical product to specification.

. . . and is divided into the following sections:
- Devising a sequence of processes to make a given product
- Selecting techniques, tools and equipment for production processes
- Maintaining tools, equipment and the work area in good order.

The production of an engineered product must be carefully planned to ensure that the selected processes are carried out efficiently and safely. Having decided on the most suitable processes, the production engineer must place them in a logical sequence. Each process must then be considered in detail, and the most appropriate processing techniques selected. Several operations might be required to complete a process, and the engineer must also decide on the best sequence for these.

The next step in production planning is to identify the tools and equipment that will be required for each processing operation. Any special tools, such as patterns, dies, templates, jigs and fixtures, will take time to produce and should be ordered well in advance of the scheduled production start date. Certain processes might also require the provision of safety equipment. It is essential that this is also in place before production begins and that the process operators have been instructed in its use.

Devising a sequence of processes to make a given product

Production planning very often takes place in parallel with product design. The design must be fit for its purpose, and this is the responsibility of the design engineer. The design must also be practicable from a manufacturing point of view. It is the responsibility of the production engineer to advise on these matters. As the design progresses, he or she must start to plan how it will be made and suggest possible modifications that will reduce production costs.

Wherever possible, a designer will specify the use of standard components, such as bearings, screwed fasteners and electrical components, which can be purchased more cheaply than they can be made in-house. A designer will also make use of existing components made by the firm, if they can be used with little or no modification. Having decided which components will be bought-out complete and which existing components can be used, the production engineer must then draw up detailed production plans.

The more complex the product, the more complex the production plans will be. They must cover all aspects of production from the manufacture of components to the assembly and finishing of the product. Having decided on the processes that will be used to manufacture a particular component, they must next be placed in logical sequence. When doing this, production engineers very often make use of block diagram flow charts. They are often referred to as process layouts. The process layout for a simple component might be as shown in Figure 4.1.

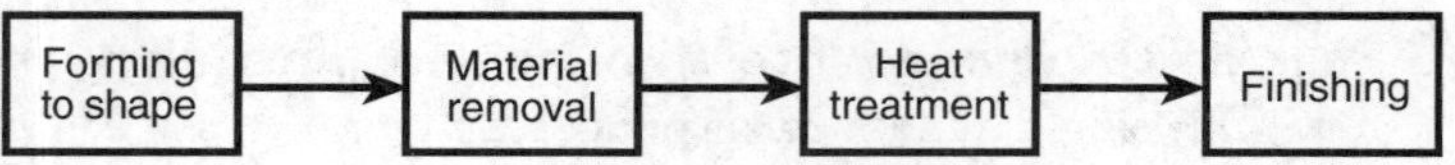

Figure 4.1 Component process layout

An estimate of the time required for each process can be included in each block. These can then be totalled together with an allowance for the time between processes to give an estimate of the overall production lead time for the component. Flow charts can also be used to show the overall sequence of processes from component manufacture to assembly and finishing. The process layout for a product might be as shown in Figure 4.2.

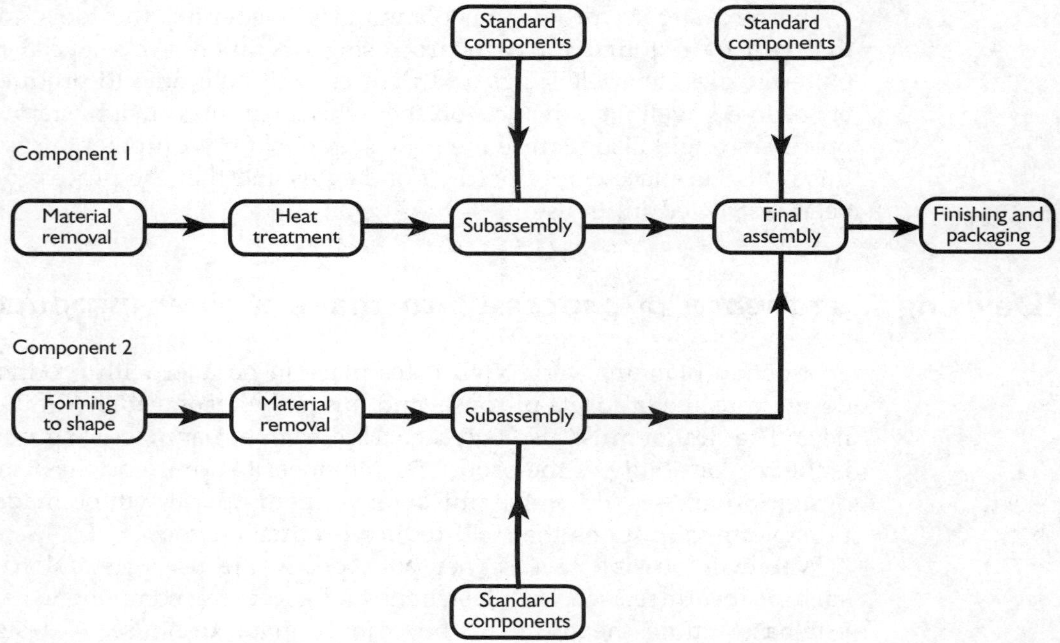

Figure 4.2 Process layout for a product

Having selected the processes that will be used to make a product, there is very often only one possible sequence in which they can be carried out. Simple components may be made by a single forming process, such as injection moulding or pressing from sheet. With the more complex components, production begins with a forming process, such as casting, forging or moulding, which might be followed by material removal. Heat treatment may be applied at some intermediate stage to soften, harden or toughen the material, and the sequence may end with a finishing process, such as plating or painting. Finished components go on to be assembled and, finally, the engineered product might also go through a finishing process for protection or decoration.

Activity 4.1

The screwdriver in Figure 4.3 has a hardened steel blade that has been forged to shape. The handle is made from hardwood which is lacquered and fitted with a metal ferrule at its tapered end. Draw up a possible process layout in block diagram format for production of the screwdriver.

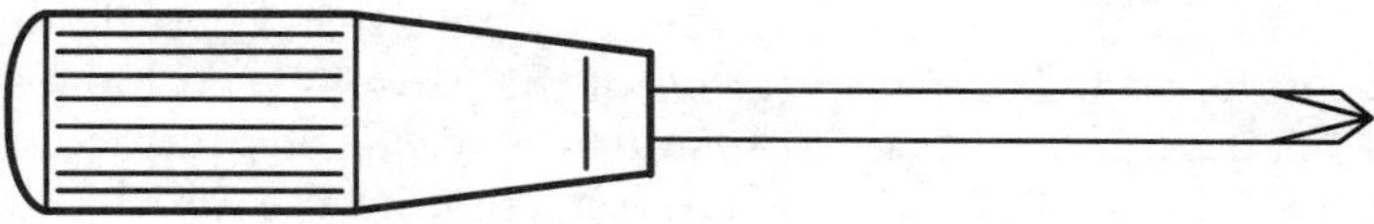

Figure 4.3 Screwdriver

Selecting techniques, tools and equipment for production processes

Having selected the processes that will be used to manufacture a product and the sequence in which they will be carried out, the production engineer must then draw up detailed process plans. Within each process, he or she must select the most appropriate techniques bearing in mind the quality specifications for the product, the available manufacturing resources and the need to keep production costs to a minimum.

Components formed by processes, such as injection moulding and die casting, can be completed in a single operation. Here, the production engineer must ensure that the dies and other specialist equipment are available for the start of production and select process plant that can deliver the required quality and quantity of a product.

Other components might only be completed after carrying out several process operations involving a number of different processing techniques. These need to be planned in a logical sequence, each with the required equipment and tools clearly specified. Detailed process plans are drawn up on a document known as a process planning or process specification sheet. Generally, each manufactured component has its own process plan, and separate ones are drawn up to cover the assembly and finishing stages of the complete product. A typical format is shown in Figure 4.4.

Part no. ____________	Description ____________________				Material ____________________	
Operation number	Description of operation	Machine/ equipment	Tools/ gauges	Spindle speed	Estimated time	Remarks
10						
20						
30						
40						

Figure 4.4 Process plan format

The format varies from firm to firm, but it is usual to number the process operations 10, 20, 30, etc. If any additional operations are required, they can be inserted as 15 or 25, etc. Process plans are generally kept on a computer database where they can easily be updated and from which hard copies can be obtained for the production departments as required. When operations are added or deleted as a result of design changes or to improve quality, it is essential to have a good line of communication with the process area. There is often a specified introduction date, and all existing copies of the process plan must be duly replaced or updated. Failure to do this can result in a waste of production time and materials.

The second column should contain the precise description of the operation such as 'Turn 25 mm diameter' or 'Normalise at 900°C'. The machine/equipment column should contain a description of the item, such as 'Vert Mill' or 'Plating Bath', with perhaps also a factory code number for the item.

The 'tools/gauges' column should list the hand tools that are needed and the gauges required for checking critical dimensions. It should also contain the details or factory code numbers of any special jigs or fixtures that are required. A fixture is a device designed to hold awkwardly shaped components. A jig is a device that is designed not only to hold a component but to position it precisely for processing and possibly also to provide a guide for a cutting tool.

The 'estimated time' column might give both the allowed setting up time for the equipment and the processing time per piece. Very often these times have been agreed between the process operators and the firm's representatives as being a reasonable and safe processing rate. With processes such as heat treatment, it is the batch processing time rather than the time per piece that is recorded. The 'remarks' column might contain quality indicators, such as a check on a critical dimension, surface finish requirements or a reading required from a test instrument.

Case study

Production plan for a bush produced on a capstan lathe

The following is the production plan for a bush that is to be produced on a capstan lathe (Figure 4.5). Capstan lathes are designed for the batch production of simple turned components. They are similar to a centre lathe except that the tailstock is replaced by a hexagonal turret. Each face of the turret can contain a different device, such as a bar stop, centre drill, twist drill, etc., and it can be indexed to bring each one into play in the correct sequence as required. As you will realise, this saves the time that it takes to change the tools in the tailstock of a centre lathe.

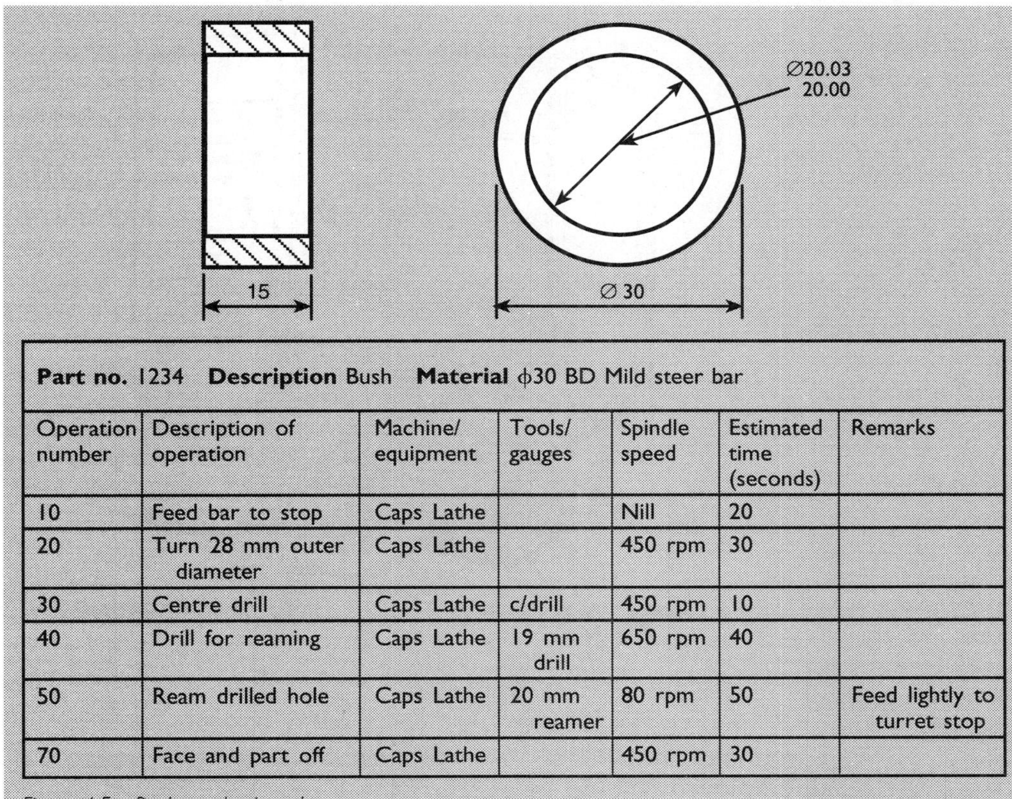

Part no. 1234 **Description** Bush **Material** ϕ30 BD Mild steer bar						
Operation number	Description of operation	Machine/ equipment	Tools/ gauges	Spindle speed	Estimated time (seconds)	Remarks
10	Feed bar to stop	Caps Lathe		Nill	20	
20	Turn 28 mm outer diameter	Caps Lathe		450 rpm	30	
30	Centre drill	Caps Lathe	c/drill	450 rpm	10	
40	Drill for reaming	Caps Lathe	19 mm drill	650 rpm	40	
50	Ream drilled hole	Caps Lathe	20 mm reamer	80 rpm	50	Feed lightly to turret stop
70	Face and part off	Caps Lathe		450 rpm	30	

Figure 4.5 Bush production plan

Progress check

1 Why is it desirable for design and production engineers to work in close co-operation?
2 What is the purpose of a process layout diagram?
3 What are the main factors that affect the choice of techniques for carrying out a particular process?
4 Why are process operations usually numbered 10, 20, 30, etc.?
5 What are jigs and fixtures?
6 What precautions are necessary when changes are made to the production plan?

Maintaining tools, equipment and the work area in good order

Basic maintenance activities are an essential part of the workplace routine. Poor maintenance of tools, equipment and the work area can cause accidents, unscheduled stoppages, waste of material and poor product quality. It is the duty of the employer to provide a safe working environment, and it is the duty of the employee to co-operate and help to maintain the workplace, tools and equipment in a safe condition.

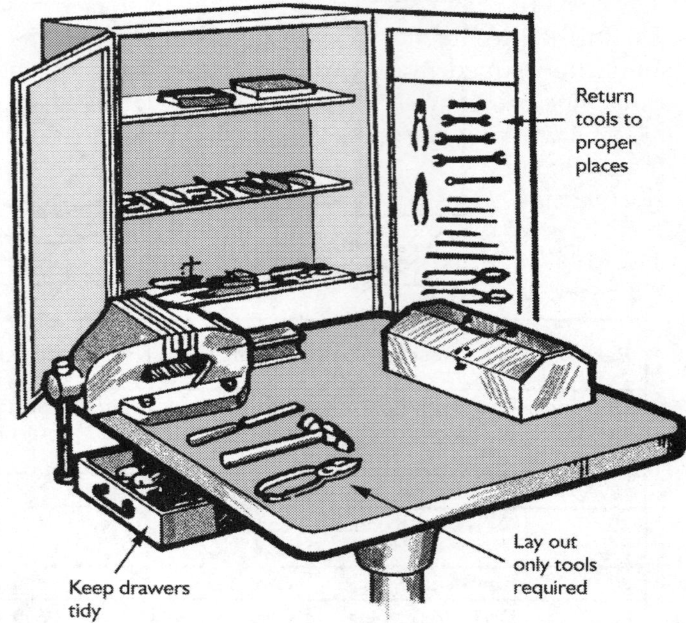

Figure 4.6 A tidy workstation

Routine maintenance a is commonsense part of safe working practice in which all employees should be instructed and trained. It should begin on entering the workplace and continue as an ongoing activity throughout the working day. If the work area has been left in an untidy state, it should be cleared of unwanted tools, unused material and components or waste material before starting work. The floor area should be clear of any obstructions, as should adjacent passageways and gangways.

The work area should be kept in a safe and tidy condition throughout the working day. Only the tools that are required should be laid out for use (Figure 4.6). Material for processing and completed work should be safely stacked in the containers or racks provided and waste material should not be allowed to accumulate. Tools and materials should be returned to their proper place when not in use, waste material placed in the containers provided and any spillages should be cleaned up immediately. The following are general guidelines for the routine maintenance of:

- machine tools
- power tools
- hand tools.

Machine tools

Lathes, drilling machines, milling machines, grinders and polishers should be given a thorough visual inspection before use. In particular, the guard should be in safe working order. If they are fitted, the machine lighting, the coolant supply and dust extraction equipment should also be in good working order. There should be no worn or damaged electrical connections and any faults that are apparent should be reported and rectified immediately.

The level in the reservoir of the coolant delivery system should be checked, as should the level of the lubricating oil in the machine gearbox. If necessary, they should be topped up. Swarf and tools should not be allowed to accumulate on a machine, particularly on the slideways or over the coolant return filter. At the end of a working day, all swarf should be removed and the slideways wiped free of any remaining coolant. From time to time, lubricating oil should be applied to the machine slideways, leadscrews and bearings, and it might be appropriate to do this at the end of a shift. Any faults that occur during the working day should be reported immediately.

Power tools

As with stationary machine tools, portable power tools, such as drills, grinders and polishers, should be inspected visually before use. The condition of guards and electrical connections should be checked, together with that of the outer casing and grips. While being held in a safe position, the tool should be switched on and off to check that it is running smoothly and that the automatic cut-off switch is working. The safety poster shown in Figure 4.7 lists other safety and maintenance hints.

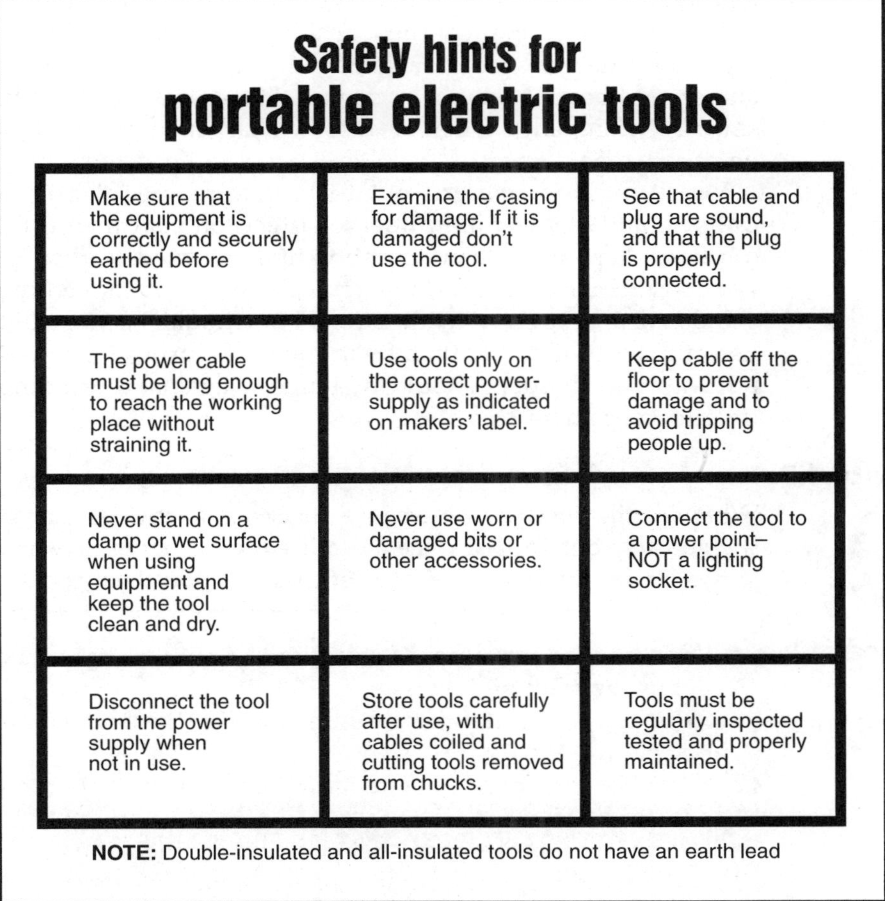

Figure 4.7 Safety poster

During the working day, power tools should be kept in a clean condition and stored in a dry place when not in use. Any signs of damage or malfunction should be reported immediately.

Hand tools

Hand cutting tools, such as files, hacksaws, punches and cold chisels, should be inspected before use. Damaged or badly fitting handles should be replaced, as should hacksaw blades if they are cracked or have broken teeth. Hammer heads should be checked frequently to ensure that they are securely attached, and also the copper, plastic and hide plugs in faced hammers and mallets.

File teeth often become clogged or 'pinned' with waste material, and they should be cleaned periodically with a wire brush or file card. File teeth are hard and brittle. They are easily damaged and can also damage other tools with which they come into contact. To prevent this, they should be stored separately, preferably on a rack.

The heads and cutting edges of punches and chisels should be checked regularly. The heads may show signs of mushrooming, and this should be ground off. The cutting edges should be inspected for sharpness and also reground if required. Chisels and punches that have been allowed to get into a very bad state can usually be reconditioned. First, they are annealed to remove any hardness. They can then be filed or machined to the correct shape and then rehardened and tempered. Finally, the cutting edge can be ground to the required sharpness.

As with files, the handles of screwdrivers and hand scrapers should be replaced if they are loose or damaged. Screwdriver blades become worn with use and should occasionally be reground. Hand scrapers must have a very sharp and smooth cutting edge. They need to be sharpened quite often during use, but only on an oilstone.

Measuring and marking-out tools require special care. Micrometers, Vernier callipers and dial test indicators are precision instruments that should be handled carefully, kept in a clean condition and returned to their protective cases when not in use. Engineers rules, squares, callipers, scribers, centre punches and dividers should also be treated with care. They should be kept lightly oiled and never used for anything other than their intended purpose. The points of scribers, centre punches and dividers should be touched against a grinding wheel from time to time to keep them sharp.

Activity 4.2

What are the routine maintenance checks and activities that every car owner should carry out at least once a week either at home or when filling up at the local garage or service station? Which of these do you think are essential safety checks?

Progress check

1 What routine maintenance activities and checks might it be necessary to carry out at the start of a working day?
2 What are the typical routine maintenance activities that need to be carried out on hand tools?
3 What are the typical routine maintenance activities that need to be carried out on machine tools?
4 How might a badly worn cold chisel be reconditioned?
5 What are the special precautions that should be taken to maintain measuring and marking out tools in good condition?

Assignment 4

This assignment provides evidence for:
Element 3.2: Make an electromechanical product to specification
and the following key skills:
Communication: 3.1, 3.2, 3.3, 3.4
Information Technology: 3.2, 3.2, 3.3
Applicaton of Number: 3.1, 3.2, 3.3

Your tasks

You should now be able to devise a sequence of processes and process techniques to make a given electromechanical product. You should also be able to select the tools and equipment, including safety equipment, which will be required. Having planned your activities, you can then proceed to realise the product. Safe working practices and procedures must be followed at all times.

A suggested product is a direct current (DC) multimeter (Figures 4.8 and 4.9) for measuring current, voltage and resistance over the following ranges:

- current: 0–100 mA
- voltage: 0–2.5 V, 0–10 V and 0–25 V
- resistance: 0 to infinity.

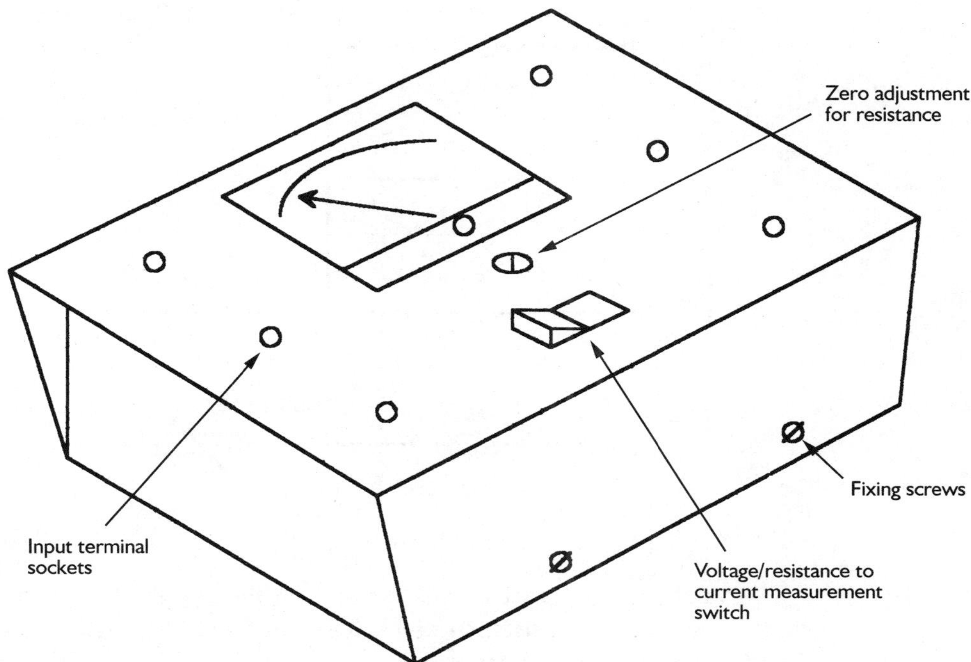

Figure 4.8 Multimeter

A possible circuit diagram is shown in Figure 4.10. Tag board or strip board might be used in its construction. A suitable battery holder and input terminal sockets will also need to be selected. The dimensions of the enclosure should be chosen to suit the selected components.

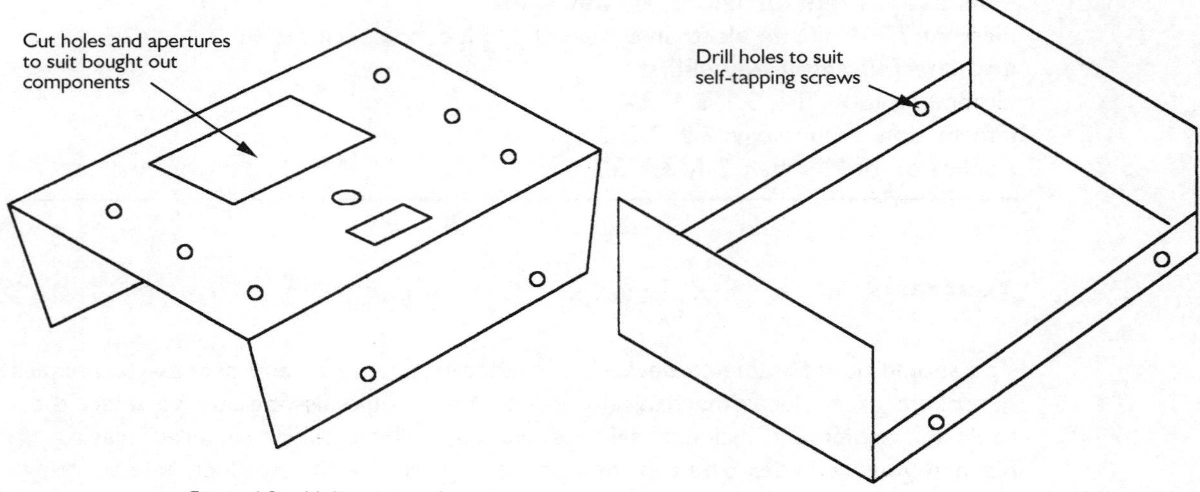

Figure 4.9 Multimeter enclosure

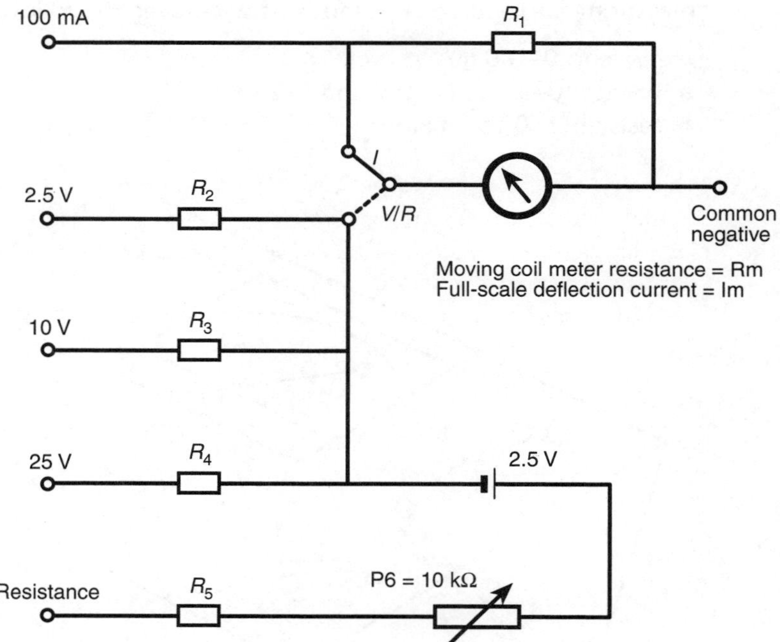

Figure 4.10 Multimeter circuit

The values of the resistors will depend on the sensitivity of the meter. A typical meter might have a resistance of 1300 Ω and a full-scale deflection current of 50 μA. The value of the resistor R_1 for the 0–100 mA current range can be calculated using the formula:

$$R_1 = I_m R_m/(I - I_m),$$

where $I = 100 \times 10^{-3}$ A. The values of R_2, R_3 and R_4 for the voltage ranges can be calculated using the formula:

$$R_{1-4} = V/I_m - R_m$$

where $V = 2.5$ V, 10 V and 25 V. The value of the resistor R_5 for the resistance range can be calculated initially using the formula:

$$R_5 = 1.5/I_m - R_m.$$

A resistor about 5 kΩ less than this calculated value should be chosen. The 10-kΩ variable resistor will make up the value and provide the adjustment needed to zero the meter before measuring resistance.

To complete the assignment, you will need to carry out the following tasks while keeping a log of all your activities.

- Calculate the values of the circuit resistors required to suit a given moving coil meter.
- Identify and obtain all of the bought out standard parts.
- Decide on dimensions for the enclosure and the size and positions of the holes for the meter, terminals, switch, variable resistor and fixing screws.
- Obtain sufficient material for the enclosure.
- Identify the processes which will be required to realise the product and draw up a process layout
- Identify the specific techniques, tools and equipment which will be required and draw up a process plan.
- Identify the safety equipment and the safety procedures relevant to the processes.
- Carry out the processes using the identified techniques and the appropriate safety procedures and equipment.
- Maintain tools, equipment and the working area in good working order.
- Submit your completed log and product as evidence that you have completed the processes together with general notes on the use of the processes and remaining tools and equipment which have not been used but are listed in the range.

Chapter 5 Engineering services

This chapter covers:
Element 3.3: Perform engineering services to specification.

. . . and is divided into the following sections:

- Identification of procedures for performing engineering services
- Identification of safety procedures and equipment for performing engineering services
- Performing engineering services to specification.

Engineering services are activities that are related to engineering products, plant and equipment but are not direct production activities. They are essentially concerned with inspection and maintenance. Inspection procedures are required to monitor product quality and also to monitor the condition of the plant and equipment. Maintenance activities are necessary to ensure that the plant and equipment are fit for their purpose and available when required.

The nature of the services depends on the type of engineering and the product. Rigorous inspection throughout manufacture is required by law for certain engineered products that are potentially hazardous. Regular inspection and certification throughout the service life of the product may also be required, as is the case with pressure vessels, motor vehicles and aircraft. In other cases, random sampling may be sufficient for monitoring product quality, and the plant and equipment may require only periodic checks, as recommended by its manufacturers.

The purpose of maintenance activity in the workplace is to prevent unscheduled stoppages and to provide a safe working environment. A programme of planned inspection and maintenance based on manufacturer's recommendations and plant experience can go a long way to achieving this. It can also contribute to product quality by identifying equipment faults before they become critical.

Identification of procedures for performing engineering services

The procedures for performing engineering services vary with the size of firm, the scale of production and the type of product. The aims of ensuring product quality, a safe working environment and uninterrupted production are, however, common. The essential engineering services in most engineering firms are:

- quality control
- scheduled maintenance.

Quality control

Quality control activities include the inspection of raw materials and bought-out components, product inspection during and at the end of manufacture and dealing with customer complaints. A great many firms have written procedures for these activities, which are contained in a quality manual and used for training and reference purposes. The guidelines for quality systems are contained in the British Standard BS 5750 and the international equivalent ISO 9000.

Depending on the quality specifications, inspection can range from a visual check to the use of specialised test equipment. Visual checks might be carried out to ensure that a product is complete and correctly assembled. They might also be carried out to check for blemishes, surface damage or uneven surface finish. A check list of quality 'attributes' that the product must have is very often used to record the activity.

The inspection of critical dimensions requires the use of measuring and test equipment. The micrometer and the Vernier calliper are instruments commonly used for checking dimensions, but they are not well suited for quick and repetitive inspection. The process has been automated on many modern manufacturing systems where special sensors keep a check on the dimensions and make corrections to the process if any drift is detected. With the more traditional or small-scale production processes, however, manual inspection is still required.

Gap or plug gauges are very often used to check diameters and hole sizes (Figure 5.1). These are of the 'go' and 'not go' type, where the gaps or the plug diameters are precisely machined to the upper and lower limits of the dimension.

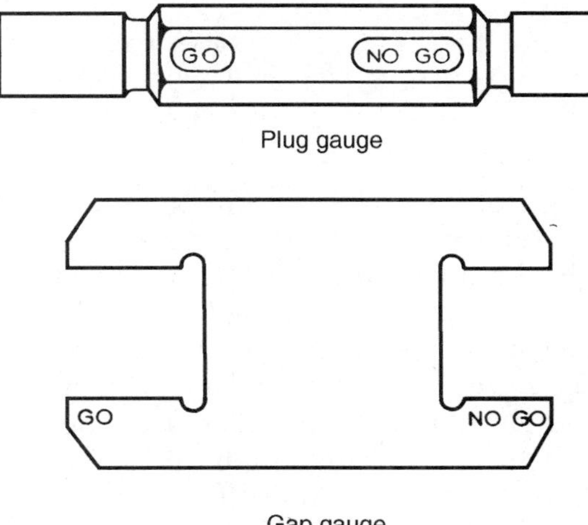

Figure 5.1 Limit gauges

An alternative method of rapid dimensional inspection is to use a comparator (Figure 5.2). The most common type is the dial test indicator (DTI), which is mounted on a rigid stand over a surface plate. A comparator does not measure a dimension directly but compares it with a known standard.

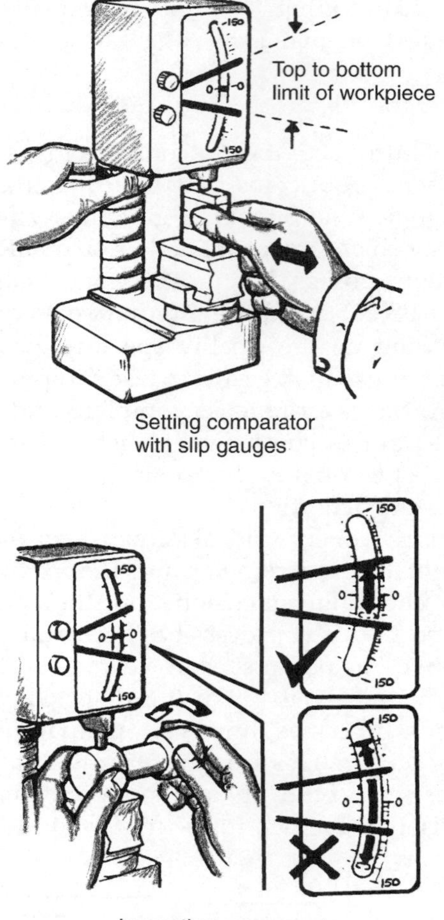

Figure 5.2 Use of a comparator

The plunger of the DTI is set to height above the surface plate using slip gauges. These are small blocks of hardened steel, precisely ground to a range of different sizes, which can be stacked or 'wrung' together to make the nominal dimension. The plunger is first set to a height slightly below the lower limit of the dimension. The slip gauges are then placed under it and its dial is set to read zero. The slip gauges are then removed and the components to be measured are passed under the plunger. The deflection of the pointer in the plus or minus directions indicates whether they are within the specified limits.

The tests that are carried out to check the mechanical properties of a material or a component may be divided into destructive tests and non-destructive tests. In destructive tests, the sample component or material specimen is loaded to destruction. The tensile and impact tests described in Chapter 6 are typical of this type. Others have been devised to simulate the conditions that a component will encounter in service. They include fatigue tests in which components are subjected to vibration or repeated loading and unloading until failure occurs.

Non-destructive tests are widely used to check for defects in castings, mouldings and welded joints. They include X-ray, ultrasonic, magnetic resonance and die-penetrant testing. The indentation tests described in Chapter 6 for measuring surface hardness are also non-destructive. In addition to these, there are a great many inspection procedures, particularly with electrical and electronic products, which require readings to be taken on specialist test equipment. The proof testing of pressure vessels to specified pressures, which are far above the level that they will experience in service, also fall into the non-destructive category.

Scheduled maintenance

Scheduled inspection and maintenance activities are programmed to be carried out at specified intervals. The activities are planned so as not to interfere with production, and are usually carried out at weekends or during night shifts. In the case of continuous round-the-clock processes, production is scheduled to stop periodically for maintenance to be carried out. Scheduled inspection and maintenance may be the duty of a maintenance team or be carried out by the production operators and technicians themselves. The maintenance and calibration of specialist measuring and control equipment is very often carried out by the suppliers or by specialist outside contractors.

Scheduled inspection and maintenance activities include:

- condition monitoring
- calibration
- fault diagnosis
- maintenance work
- recording and evaluation.

Condition monitoring

The condition of plant and equipment should be monitored as part of the daily routine by the process operators and maintenance personnel. A person who is familiar with the running characteristics of a particular process will very soon be aware of anything that is abnormal. This might be the sight of something worn, broken or detached. It might be an unusual noise or an unusual smell, which indicates that all is not well. It might also be that some part of the equipment is unusually hot to the touch. Routine observation by operators and maintenance personnel involves the critical use of all the senses, with the possible exception of taste.

Techniques have been developed in recent years that enable the condition of critical items of plant and equipment to be monitored automatically. The monitoring may be continuous or periodic. Temperature, noise and vibration monitoring using sensors placed at strategic positions can be continuous and designed to trigger alarms if a critical level is exceeded. Periodic monitoring might include the analysis of samples of lubricating oil. The nature and size of metal particles in the oil can indicate abnormal wear of a particular component, and the presence of process fluids can indicate the failure of a gland or seal.

Case study

Compressed air systems

The compressed air that is used in industry for pneumatic control systems, paint spraying and powering pneumatic tools is stored in large upright cylindrical pressure vessels with domed ends, known as receivers. The Health and Safety regulations relating to these vessels are quite stringent. They state that: a written scheme of examination for the periodic examination of prescribed components of an installed or mobile pressure system must be drawn up and implemented, or certified as being suitable, by a competent person before the system can be operated. The components requiring the written scheme are:

- protective devices
- pressure vessels/pipelines where a defect may give rise to danger
- parts of pipework where a defect may give rise to danger.

The written scheme is in fact a programme of planned inspection and maintenance. It must be approved by the firm's insurers who will themselves carry out an annual inspection and certification exercise. The protective devices that must be covered in the plan include the pressure safety valve, which will open when the air in a receiver reaches the prescribed pressure. A receiver contains inspection covers which, when removed, enable the interior to be examined for signs of excessive corrosion or other forms of defect. Receivers are also equipped with a pressure gauge, and this must also be subjected to a periodic calibration check. The parts of pipework that may give rise to danger are joints, valves and other ancillary equipment, which must be checked for damage, leaks or corrosion.

Calibration

The performance of measuring and recording instruments tends to deteriorate with time. Periodically, they need to be inspected, checked for accuracy and, where necessary, adjusted or replaced. Calibration is the process of checking the performance of an instrument against a known standard. In some cases, it is possible to make adjustments and restore the instrument to service. Others are designed to be discarded and replaced when their performance has deteriorated. The activity should be scheduled to take place at regular intervals, as specified by the equipment suppliers or as has been found to be necessary from plant experience.

Micrometers and Vernier callipers are calibrated using slip gauges. Pressure gauges are calibrated using portable pneumatic testers or bench-mounted dead weight testers. Both types generate a range of known pressures against which the gauge reading can be compared. The outputs from temperature-measuring devices, such as thermocouples, thermistors and resistance thermometers, are checked against known temperature sources. These devices cannot be adjusted and they are replaced when found to be faulty. Recording instruments, which convert an electrical input signal into an output reading on a display or chart, are injected with precise input signals against which the display output is compared. It is usually possible to make adjustments to these devices.

Some devices can only be calibrated in position and the process may require production to be halted. With others, it is possible to replace them with a calibrated instrument without interrupting the production process. The original can then be recalibrated in the instrumentation workshop and stored for reuse. Firms involved in precision engineering very often operate a scheduled recall and calibration system for all measuring tools and gauges used in production.

Fault diagnosis

The causes of unscheduled breakdowns and the faults that are found during scheduled maintenance need to be diagnosed as quickly as possible. In a great many cases, the cause is readily apparent and work can begin immediately to restore the situation. In other cases, however, particularly with intermittent faults or faults in equipment that has had a trouble-free history, the cause may not be so obvious. This very often places the maintenance technician under pressure and it is essential to keep a cool head and to work logically.

Fault diagnosis or 'troubleshooting' is a skill that, like other skills, has to be learned. The maintenance technician must first of all have a good working knowledge of the plant and equipment. This does not have to be a knowledge of every last detail but, when the occasion arises, the maintenance technician should know where to find detailed knowledge and not be slow to ask colleagues who have more experience. A systematic approach is then called for. There may be a number of causes for a fault and a logical approach is required to eliminate all but the actual one by the shortest possible route.

The correct solution can often be arrived at by common sense but, where the problem is complex, a more systematic approach may be called for. The first step is clearly to define the nature of ihe fault and to carry out a preliminary inspection for clues such as oil leakage, overheating, excessive noise or vibration, etc. Information from process operators and data from condition-monitoring equipment should be collected if available. Checks should also be made to ascertain whether a similar fault has occurred before or whether there have been any recent changes or modifications to the equipment.

The next step is to analyse the problem carefully making use, where appropriate, of process or equipment drawings, maintenance records and the manufacturer's information, which might even include a troubleshooting chart. The equipment should then be divided up into functional zones, such as the power transmission system, lubrication system, cooling system, etc. The input and output of each zone can then be examined and any zone that is malfunctioning identified.

Once the fault has been traced to a particular zone, it can be subjected to a thorough check. Checks should be made on the input and output of each element or component using test equipment where appropriate. Failing this, it may be possible to replace certain components with new ones or to bypass components to see if this eliminates the fault. Having located the cause of the fault, a decision can then be made whether to replace or repair the element or component in question. The cause is usually normal wear and tear, misuse or an inherent design weakness where a component is not suitable for the service conditions. Recognising the cause often determines the choice of corrective action.

Activity 5.1

Your electric kettle appears not to be working. Describe how you would proceed to diagnose the fault. You are equipped with a multimeter suitable for measuring the voltage of the mains supply and also for measuring electrical resistance.

Maintenance work

A considerable amount of maintenance work can be generated from the routine observations of process operators and from condition monitoring. This is unscheduled maintenance resulting from reported breakdowns or critical operating conditions. Unscheduled maintenance very often results in a stoppage or slowing down of production and is to be avoided. A programme of planned maintenance based on plant experience and the data from condition monitoring may not prevent unscheduled stoppages, but it can make them much less frequent.

Planned maintenance work includes the cleaning, lubricating and adjusting of machinery, replacing worn components and carrying out modifications to plant and machinery. It also includes the replenishing of lubricating oils and carrying out safety checks on guards and other items of emergency equipment. Finally, it might include the running and testing of equipment in readiness for production to restart.

The manufacturers of equipment very often supply a maintenance manual. Some of these are very detailed with maintenance instructions, diagrams, the frequency at which the maintenance operations should be carried out and fault diagnosis charts. The service manuals that can be obtained for motor vehicles are a good example of these. Other manufacturers supply only the essential information such as safe working instructions and recommended lubricants. In such cases, it is often necessary to draw up a detailed maintenance plan. For large items of equipment or sections of plant, this may take the form of a manual that is available for reference along with the manufacturers' information.

For smaller items of equipment, the maintenance plans may be written on a maintenance planning sheet of some kind. As with process plans, these may be filed for reference as hard copies or stored on a computer database. A typical format might be as shown in Figure 5.3.

Process ______		**Location** ______			
Maintenance operation	Description of operation	Equipment	Tools	Materials	Frequency of operation
10					
20					
30					

Figure 5.3 Maintenance plan format

In Figure 5.3, the frequency column states how often an activity should be carried out. It is usual to list the operations in descending order of frequency, i.e. that which is to be done the most often as operation 10.

Recording and evaluation

In a well-organised maintenance department, all of the planned maintenance activities that are due to be carried out over a particular period of time are issued as a maintenance schedule. This can be used by supervision to allocate work. It can also be used to track the activities that are in progress and those that have been completed. A separate record is usually kept, which must be signed and dated by the maintenance technicians as evidence that the maintenance activities have been carried out.

Recording systems vary but it is also essential that a record is kept for individual items of plant and equipment. These should contain details of all the maintenance work, scheduled and unscheduled, which has been carried out since their installation. The record should also contain details of replacement parts, design modifications and changes of purpose or function. It may be that there is a product change involving heavier usage and an increase in the number of reported faults. In this way, a historical record can be built up, which is useful for reference when faults occur and in evaluating the maintenance plan. The record might indicate a need for more frequent maintenance or additional maintenance operations.

The evaluation of maintenance activities is an essential part of the planning process. When a fault is detected or a breakdown occurs, it is important to find the real cause. If a patch-up job is effected, the fault will most probably reoccur. This is particularly the case if the fault is seen to lie in the design of the equipment. Instead of merely replacing a faulty part with a similar one, consideration should be given to ways of modifying the design. The solution may lie in improved lubrication or the use of a tougher material to suit the local operating conditions.

Activity 5.2

Draw up an inspection and maintenance plan for an upright vacuum cleaner. This should take into account the condition of the flexible mains connection the brush and drive belt. It should also include the removal and replenishment of the disposable dust bag and the removal, cleaning and reassembly of the brush. The frequency of your operations will obviously depend on use.

Progress check

1. What constitutes an engineering service?
2. Where would you expect to find a firm's quality control procedures documented?
3. How can the dimensions of toleranced diameters and holes be checked quickly?
4. What is the principle of 'comparative' measurement?
5. What kind of service conditions do fatigue tests seek to simulate?
6. What is the purpose of condition monitoring?
7. How are measuring instruments calibrated?
8. How might the maintenance record of a machine be useful in the event of a fault being reported?
9. What are the most common causes of the faults that occur in plant and equipment?
10. What is the difference between a maintenance manual and a maintenance schedule?

Identification of safety procedures and equipment for performing engineering services

Most of the comments that were made in Chapter 3 in relation to safety procedures and equipment for production processes also apply to maintenance work. This is certainly true of the recommendations for personal conduct, dress, maintaining tools and the work area in good order and the use of emergency equipment. There are, however, certain additional procedures that maintenance technicians must follow to ensure their own safety and also that of others. These include procedures relating to:

- equipment isolation
- use of danger tags
- permits to work.

Equipment isolation

Very often the guards need to be removed from production machines so that essential maintenance work can be carried out. It may also be necessary to remove the covers from items of process plant to expose the moving parts or high-voltage electrical equipment. Before any maintenance work of this kind is carried out, the machine or process must be stopped or closed down and isolated from its power supply. The key should then be removed from the electrical isolating switch to ensure that the machine cannot be started accidentally. If there is no key provided, the fuses or circuit breakers should be removed.

Use of danger tags

Where it is impractical to isolate equipment for maintenance, technicians can be protected from it being inadvertently started by attaching danger tags to the control levers, valves, electrical switches, etc. The tags contain the name of the technician and the date and time at which it was attached. When it is required to close off a section of a pipeline that contains a hazardous liquid or gas, it is common practice to insert spade blanks. These are thick metal discs, which prevent any possible flow while they are in position. They too should be fitted with danger tags.

No one must use the equipment or interfere with the tags while they are in place, and only the technician whose name is on a tag may remove it. It is of course essential that process operators are trained to recognise danger tags and act accordingly. When more than one technician is working on a section of plant or item of equipment, they must each attach their own tags even to the same controls. As each one finishes his or her activities, they remove their tags but not those of their fellow technicians.

Permits to work

Permits to work are usually issued by the supervision in charge of the area where the maintenance work is due to take place. They give maintenance technicians the authority to enter the area and proceed with their activities. They are also intended to assure technicians that all the necessary safety procedures have been followed in readiness for the maintenance work. This includes an assurance that the plant has been properly closed down and cannot be restarted until the work has been completed satisfactorily.

Special permits are sometimes required for work that is particularly specialised or hazardous. This includes entry to areas where there is high-energy radiation, high-voltage electricity or entry to storage tanks and pressure vessels. Wherever a permit system is in operation, maintenance technicians must observe it at all times. The receipt of a permit assures the technician that those responsible for the work area are aware of their presence and concerned for their safety.

Progress check

1 What are the safety precautions that should be taken before removing machine guards for maintenance?
2 How can a machine be rendered completely safe to work on if the isolator switch is not fitted with a key?
3 In what circumstances are danger tags used?
4 When working on the same section of plant, why does each technician need to attach danger tags to the controls?
5 What authority and assurance does a permit to work give the maintenance technician?

Performing engineering services to specification

You will now be aware that inspection and maintenance activities are the most common form of engineering service. You will also be aware of the basic procedures, including safety procedures, which are used to perform them. You should now be able to carry out a series of given inspection and maintenance operations using the appropriate tools and equipment and safe working practices.

Assignment 5

This assignment provides evidence for:
Element 3.3: Perform engineering services to specification
and the following key skills:
Communication: 3.1, 3.2, 3.3, 3.4
Information Technology: 3.1, 3.2, 3.3
Application of Number: 3.1, 3.2, 3.3

Your tasks

You are required to carry out an inspection of a centre lathe fitted with a three-jaw self-centring chuck and perform routine maintenance operations. Reference should be made to the manufacturer's service manual, if it is available. This will give details of how to carry out routine lubrication and replenishment operations. If you are unsure how to complete any of the following tasks or have doubts about the safety of any part of the machine, you should seek expert assistance immediately. The items of specialist equipment, machine attachments and materials that you will require are:

- a dial test indicator calibrated in 0.002-mm divisions with a magnetic base
- a catchplate and carrier
- spindle and tailstock centres
- suitable lubricants and coolant
- a 300-mm length of 20-mm-diameter bar, faced and centre drilled at both ends.

To complete the assignment, you will need to carry out the following tasks while keeping a log of all your activities:

- Inspect the condition and operation of the machine guards, chuck keys and other dedicated tools that are part of the machine's equipment.
- Inspect the electrical connections to the machine and check the functioning of the isolator switch and the on–off motor switch.
- Check that the machine light fitting is in good working order and properly adjustable.
- Check the level and the condition of the coolant in its reservoir. Replenish it if necessary and, if it is heavily contaminated, drain it off and replace with new coolant. Check also the flow of coolant and that the position of the delivery pipe is properly adjustable.
- Check the operation of the clutch, the speed selection controls, the automatic traverse and the emergency stop.
- Check the level of the lubricating oil in the headstock gearbox. If the machine is not fitted with a sight glass, a dip stick may be required. Replenish with the correct grade of lubricant if necessary.
- Having taken the appropriate safety precautions, remove the chuck from the spindle as per instructions in the manufacturer's manual or with expert assistance. Note the condition of the screw threads on the chuck and spindle nose.
- Remove the jaws from the chuck and note the condition of the scroll plate that moves them. Lightly lubricate the scroll plate and jaws and reassemble them in the correct positions.
- Remove the toolpost from the compound slide and clean away any remaining swarf particles. Dismantle the toolpost and note the condition of the components. Lightly lubricate and reassemble.
- Remove the tailstock from the bed slideways and clean away any remaining swarf particles from its underside. Dismantle and note the general condition of the tailstock components, replace in position and lubricate the spindle.
- Note the condition of the slideways on the lathe bed and on the cross-slide and compound slide. Note also the condition of the feed shaft and leadscrew. Clean away any remaining swarf particles and lightly lubricate.
- Set up the machine for turning between centres and place the 20-mm-diameter bar in position for turning. Switch on the machine and take a cut of approximate depth 1.5 mm. Take out the bar and replace it between centres in the reverse position.
- Mount the dial test indicator on the saddle with its plunger in contact with the side of the bar at the tailstock end. Zero the DTI scale and note any deflection of the pointer at 50-mm intervals as it is traversed along the turned length. This will indicate any misalignment of the centres in the horizontal plane.
- Submit your completed log as evidence that you have completed the service, together with general notes on the condition of the centre lathe and any calculations and corrective action that you think is necessary. Your log should also contain details of the safety procedures that you followed and any safety equipment that you used.

Unit test 2: Engineering processes

Processes for engineering products

1 Which process is used initially to produce a safety helmet?
   A material removal
   B heat treatment
   C material shaping
   D chemical treatment
2 Casting is the first process in the production of
   A vehicle body panels
   B plastic containers
   C lathe beds
   D grinding wheels
3 Injection moulding is a forming process used with
   A thermoplastics
   B high-carbon steels
   C semiconductors
   D wood composites
4 Which of these is a metal-shaping technique?
   A horizontal milling
   B tempering
   C die-casting
   D electroplating
5 Which joining technique is used in the automated assembly of motor car bodies?
   A soft soldering
   B oxy-acetylene welding
   C hard soldering
   D spot welding
6 Which of these is a surface-finishing technique?
   A extrusion
   B galvanising
   C annealing
   D crimping
7 Which forming technique is used to produce uPVC window frame sections?
   A drawing
   B die-casting
   C pressworking
   D extrusion
8 Normalising is a
   A chemical treatment technique
   B surface-finishing technique
   C heat treatment technique
   D material removal technique

9 The termination of an electrical cable is often attached by
   A adhesive
   B electroplating
   C etching
   D crimping

Make to specification

10 The mould for a sand casting is made using a wooden or metal
   A die
   B pattern
   C jig
   D template

11 Dies must be made before a component can be formed by
   A vertical milling
   B blow moulding
   C anodising
   D hot rolling

12 Machine guards should be inspected regularly to
   A ensure product quality
   B reduce material wastage
   C prolong the life of the machine
   D comply with Health and Safety regulations

13 Equipment breakdowns can be reduced by a programme of
   A quality control
   B scheduled maintenance
   C first-aid training
   D material testing

14 Medium- and high-carbon steels can be hardened by heating them to a specified temperature and
   A quenching in oil
   B cooling in still air
   C carburising
   D cooling in the dying furnace

15 Cleaning the surfaces, tinning, sweating and removing surplus flux is the sequence of operations when making a joint by
   A hard soldering
   B arc welding
   C soft soldering
   D adhesive bonding

Engineering services

16 Which of these is an engineering service?
   A quality inspection
   B surface finishing
   C first-aid treatment
   D material recycling

17 The size of a component is sometimes checked using a
   A comparator
   B multimeter
   C reamer
   D indentor

18 When repairing mains electrical equipment it is essential to
   A use a machine guard
   B wear eye protection
   C isolate the equipment
   D evacuate the area

19 The Brinell test is used to check
   A dimensional accuracy
   B thermal conductivity
   C electrical resistance
   D surface hardness

20 Scheduled maintenance is carried out
   A during slack periods
   B when breakdowns occur
   C at specified times
   D whenever time permits

Safety

21 The screw threads on oxygen cylinders and their pressure regulators must be free from grease to prevent
   A them becoming loose
   B leakage of the gas
   C combustion occurring
   D electrolytic corrosion

22 Before handling engineering materials and components it is advisable to
   A apply a barrier cream
   B switch off electrical equipment
   C wash the hands and face
   D check first-aid equipment

23 Machines that cannot be used unless the guards are in position have been fitted with
   A a quality monitoring system
   B danger tags
   C a safety interlock system
   D condition monitoring equipment

24 Which document authorises a technician to carry out maintenance operations that are potentially hazardous?
   A medical certificate
   B permit to work
   C maintenance schedule
   D first-aid qualification

25 Personnel working in the vicinity of unshielded arc welding equipment should be supplied with
- A leather aprons
- B radiation detectors
- C protective footwear
- D eye protection

26 Danger tags on the controls of process equipment indicate that
- A it must be used with care
- B toxic substances are being used
- C maintenance work is in progress
- D there is a fire risk

PART THREE: ENGINEERING MATERIALS

Engineered products must be fit for their intended purpose and, to this end, the choice of the materials from which they are made is of prime importance. Design engineers must be aware of the range of materials at their disposal and their different properties. They must also be aware of the structure of materials and the way in which this affects their properties. The properties of materials are determined by testing them using a range of standard tests and test equipment. Engineers must be familiar with these and be able to interpret the results.

Very often materials need to be formed to shape or processed in some other way to make the component parts of engineered products. Their properties can affect the choice of process and, in some cases, the choice of process can affect the final properties of a material. Engineers need to take these factors into account when designing a product and planning its production.

Choosing the most appropriate material will not necessarily guarantee the success of a product since there are many other factors involved in the design process, but it will make a major contribution to its fitness for purpose.

Chapter 6 Engineering materials and their properties

This chapter covers:
Element 4.1: Characterise materials in terms of their properties.

. . . and is divided into the following sections:
- The properties of materials
- The structure of materials
- Material testing and sources of reference data
- The relationship between material structures and properties.

Engineered products contain a wide range of materials. Metals, such as copper, tin, lead and iron, have been used for making tools, weapons and as structural materials since ancient times. Ceramic materials, such as glass, baked clays and cements, also date back to the dawn of civilisation. Other common materials, such as plastics and rubbers, and metals, such as aluminium and titanium, have been introduced much more recently. At the beginning of the twentieth century, there were very few plastic materials and it was not until the First World War that aluminium came into widespread use.

The second half of the twentieth century has seen a vast increase in the range of available materials. A better understanding of structure and composition has enabled new materials with specific properties to be developed for use in aerospace and electronics. New constructional materials have also been developed and the trend continues. Properties are inherent qualities that describe the characteristics and behaviour of materials. It will be useful to define exactly what the different properties are and how some of them are measured. Having defined the different material properties, it will then be possible to identify and characterise a range of the more commonly used engineering materials.

The properties of materials

The properties of engineering materials may be broadly classed as:

- mechanical properties
- electrical properties
- thermal properties
- chemical stability.

and the materials used in engineering may be subdivided into:

- metals
- polymers

- ceramics
- semiconductors
- composites.

Mechanical properties

Mechanical properties describe how a material behaves when forces act on it. They include:

- tensile strength
- elasticity
- ductility
- malleability
- hardness
- toughness.

Tensile strength: This gives a measure of the ability of a material to withstand tensile forces. It is the maximum load in newtons that each unit of cross-sectional area can carry before the material is likely to fracture. The formula for calculating the ultimate tensile strength (UTS) of a material is:

Ultimate tensile strength = maximum load carried/original cross-sectional area.

The tensile strength of a material is obtained by loading a standard-sized test-piece up to the point of fracture and noting the maximum load carried (Figure 6.1). The cross-sectional area may be measured in square metres or square millimetres. The units in which UTS is quoted are usually MN/m^2 or N/mm^2. It is useful to remember that these are numerically the same, i.e. if a material has a UTS of 500 MN/m^2, this is also 500 N/mm^2.

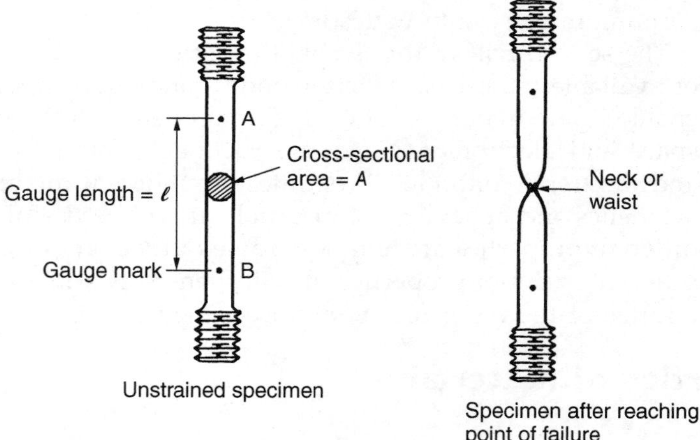

Figure 6.1 Tensile test specimen

Elasticity: An elastic material is one that obeys Hooke's law. This states that, when subjected to loading, the change in shape of an elastic material is proportional to the load applied. Furthermore, an elastic material returns to its original shape when the load is removed. The elasticity of engineering components, such as springs, is measured as their stiffness, whose symbol is 'k'. The formula for calculating spring stiffness is:

Stiffness = load applied/change in length

$$k = F/x \text{ (N/m)}.$$

This is quite satisfactory for springs, etc., but some method is also required for measuring the elasticity of a material itself, rather than a component. It is found that, when the stress in a given elastic material is divided by the strain that it produces, the value is always the same. This constant is called the modulus of elasticity of the material, or Young's modulus, after the naval architect who first had the idea about 200 years ago. It is given the symbol E, and gives a means of comparing the elasticity of one material with that of another. Modulus of elasticity is calculated using the formula:

Modulus of elasticity = stress in material/strain produced

$$E = \sigma/\epsilon.$$

Now, the stress, σ is the load carried per unit area and the strain, ϵ, is the ratio of the change in length to the original length, that is:

$$\sigma = F/A \text{ and } \epsilon = x/l.$$

The modulus of elasticity can thus be written as:

$$E = F/A \div x/l \text{ or}$$

$$E = F/A \times l/x.$$

This, in turn, can be rearranged as:

$$E = F/x \times l/A.$$

But $F/x = k$, the stiffness of the specimen, and the original length, l, and the cross-sectional area, A, are constants for a standard-sized testpiece. The modulus of elasticity of a material specimen can thus be expressed as:

Modulus of elasticity = constant × specimen stiffness.

Ductility: Ductility is a measure of the amount by which a material can be drawn out in tension before fracturing. Drawing processes are used to produce wire and tube, but they can only be carried out using materials that are highly ductile. There are different ways of measuring ductility.

Two common methods can be incorporated into a tensile test in which a material specimen is loaded until it fractures. Its elongated length and the reduced diameter at the point of fracture are then measured and the ductility is calculated as:

- percentage elongation of the test specimen; or
- percentage reduction in cross-sectional area of the test specimen.

It should be noted that the two methods do not give the same value, and it is essential to state which one the percentage value refers to.

Malleability: Malleability is sometimes confused with ductility, but they are not the same. Malleability is a measure of the ability of a material to be deformed in different directions at the same time by compressive forces. This kind of deformation occurs during forging and rolling processes, which can only be carried out on materials with a high degree of malleability. Materials that are neither ductile nor malleable are said to be brittle.

Hardness: This is a measure of the ability of a material to withstand wear and abrasion. The most common means of measuring the surface hardness of a material is by carrying out an indentation test (Figure 6.2). For softer materials, this involves pressing a hard steel ball into the material surface under a controlled load. A hardness number is then calculated from the dimensions of the resulting indentation. For harder materials, the steel ball is replaced by a conical or pyramid-shaped diamond indentor. Components, such as ball and roller bearings and cutting tools, must have a high degree of hardness and wear resistance.

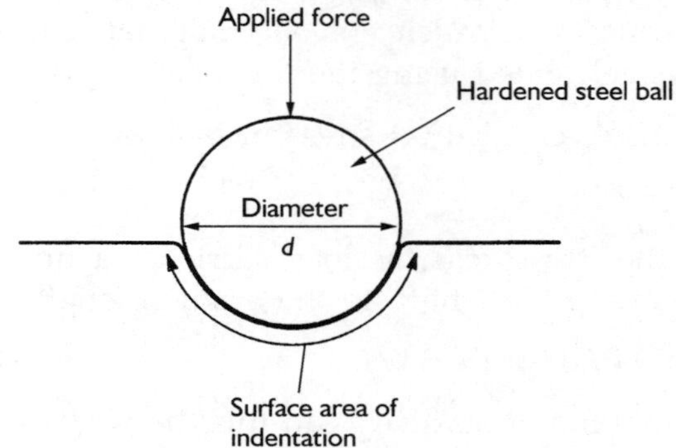

Figure 6.2 Indentation test for hardness

Toughness: Toughness is a measure of the ability of a material to withstand impact and absorb shock loads. The materials used for press tools, car suspension units and aircraft undercarriages must be tough. A common method of measuring toughness is to apply a sudden force to a specimen by means of a swinging pendulum or a spring-loaded punch. The amount of energy, measured in joules, which the specimen can absorb before fracturing, gives a measure of the toughness of the material.

Activity 6.1

The steel tie bar shown in Figure 6.3 is 2 m long and has a cross-sectional area of 200 mm^2. When it carries a tensile load of 25 kN, it undergoes an elastic extension of 2.5 mm. What is the stiffness of the bar, measured in MN/m, and what is the modulus of elasticity of the steel, measured in GN/m^2?

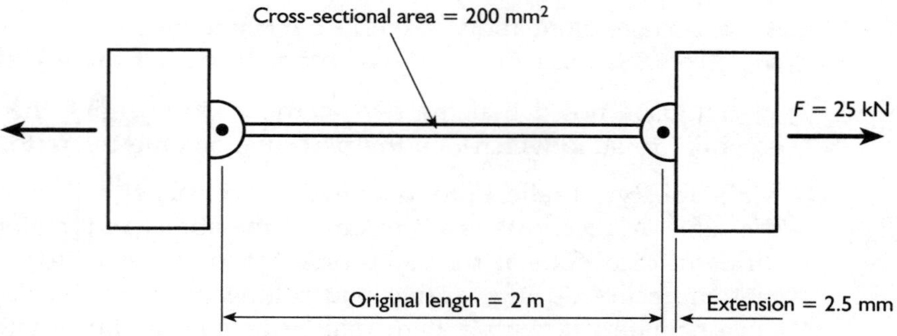

Figure 6.3 Tie bar

Suppose that the steel bar were replaced by one of the same dimensions but made out of aluminium alloy with a modulus of elasticity of 100 GN m^{-2}. Would its elastic extension be more or less than the steel bar when carrying the same load, and by how much?

Electrical properties

The electrical properties of materials describe their behaviour in the presence of electromagnetic fields and when subjected to electrical potential difference. They include:

- conductivity
- permeability
- permittivity.

Conductivity: This is the ability of a material to allow the passage of electric current. The electrical resistance, R, of a material specimen is found to be proportional to its length, l, and inversely proportional to its cross-sectional area, A. The property is usually measured as the 'resistivity' of the material, which can be thought of as the reciprocal of conductivity, i.e. the lower the resistivity value, the higher is the conductivity of a material. Resistivity is defined as the resistance of a 1-m length of the material of cross-sectional area 1 m^2. The units of resistivity are ohm metres, and it is given the Greek letter ρ (rho) as a symbol. If the resistivity of a material is known, the resistance of a sample length can be calculated using the formula:

$$R = \rho l / A.$$

Alternatively, the resistivity of a specimen can be calculated using:

$$\rho = R\,A/l.$$

Materials whose resistance is not much affected by temperature change are said to be 'ohmic' materials. Carbon is a common example of this group, and its stability is the reason for its wide use as a material for making resistors. The resistance of most metals increases uniformly with temperature, and they are said to be 'non-ohmic'. The effect is measured as the temperature coefficient of resistance. Its symbol is the Greek letter α (alpha), and its units are $°C^{-1}$. It is defined as the ratio of the increase in resistance per °C rise in temperature to the resistance at 0°C. If the resistance of a metal conductor is R_0 at 0°C, its resistance at a temperature of t°C can be found using the formula:

$$R = R_0\,(1 + \alpha t)$$

Alternatively, the temperature coefficient of resistance of a material can be calculated using:

$$\alpha = (R - R_0)/R_0\,t$$

Certain semiconductor materials, such as silicon and germanium, behave differently from metals. Their resistance falls with temperature rise and they are said to have negative temperature coefficients of resistance. However, the fall in resistance with temperature rise is not uniform. This means that α is not a constant, as it is with metals. As a result, the above formulae cannot be used for semiconductor materials.

Permeability: This is a property of a material that relates to its use as the core of a current-carrying coil. The current produces a magnetic field in and around the coil. Examples are to be found in electromagnets, electric motors and transformers. The material from which the core is made can greatly affect the intensity of the magnetic field. Certain metals, known as ferromagnetic materials, have the greatest effect. They include iron, nickel and cobalt and a variety of alloys that contain them. Their effect is to greatly increase the intensity of the magnetic field.

A comparison is made with the magnetic field produced in a vacuum, i.e. with no material whatsoever present in the coil, to give a measure of the relative permeability, μ_r, of a core material. Materials other than those listed above have little effect on a magnetic field and thus have a relative permeability of $\mu_r = 1$. With ferromagnetic materials, the value varies with the amount of current flowing in a coil and can be as high as $\mu_r = 10^5$.

Reference is sometimes made to 'hard' and 'soft' magnetic materials. Hard magnetic materials are hard alloy steels from which permanent magnets are made. Although they have a low permeability, they retain their magnetism for long periods once they have been magnetised. Soft magnetic materials are low-carbon steels from which the cores of transformers and electromagnets are made. They have a high permeability and are easily magnetised but quickly loose their magnetism.

Permittivity: This is a property of a material that relates to its use as the insulating material, or 'dielectric', between the plates of a capacitor. When a potential difference is applied across the plates, an electric field is set up between them. The intensity of the field, and hence also the amount of energy that a capacitor can store, is dependent on the dielectric material. As with electromagnetism, a comparison is made with the intensity of the electric field produced between the plates in a vacuum to give a measure of the relative permittivity, ϵ_r, of a material.

Typical values for common dielectric materials are $\epsilon_r = 1.0006$ for air, $\epsilon_r = 2\text{–}2.5$ for paper, and $\epsilon_r = 3\text{–}7$ for mica.

Thermal properties

Two of the most important thermal properties of materials describe their ability to conduct heat energy and how their dimensions are affected by temperature change. They are:

- thermal conductivity
- thermal expansion.

Thermal conductivity: As with electrical conductivity, metals are generally good conductors of heat, whereas plastics and ceramics are poor conductors. The thermal conductivity of a material is defined as the amount of heat energy per second that will pass through a specimen of length 1 m and cross-sectional area 1 m^2 resulting from a temperature difference of 1°C. It is given the symbol, k, and its units are W/m /°C.

The formula used to calculate the heat energy per second, Q, which is transmitted through a material is:

$$Q = [k\,A\,(t_1 - t_2)]/l.$$

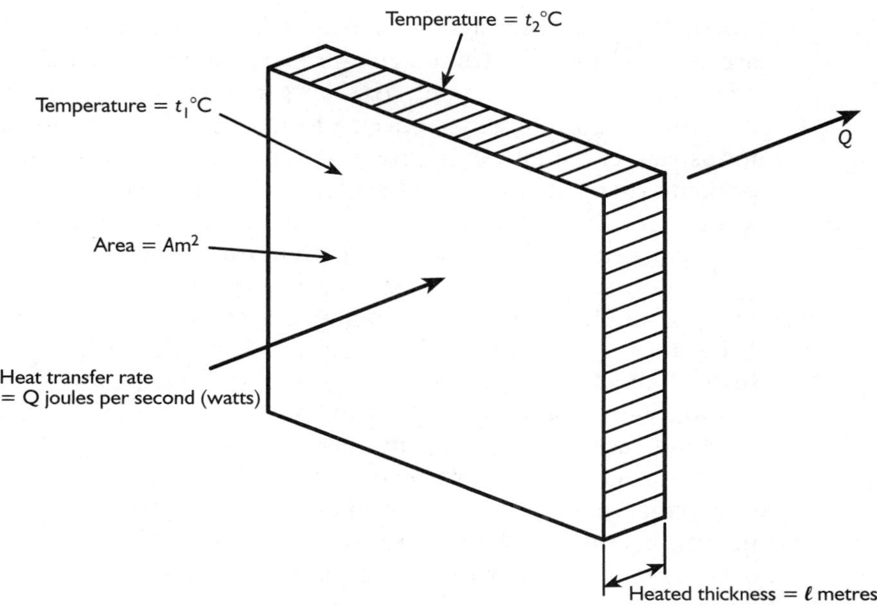

Figure 6.4 Heat transfer by conduction

Alternatively, the thermal conductivity of a material can be calculated using:

$$k = Q\,l/A\ (t_1 - t_2)$$

where A is the cross-sectional area, l is the length or thickness and $(t_1 - t_2)$ is the temperature difference between the end faces of the material.

Thermal expansion: This is a measure of the effect that a change in temperature has on the dimensions of a material. The effect is measured as the thermal expansivity of the material. It is defined as the change in length per unit of original length per degree of temperature rise. The symbol for thermal expansivity is α (not to be confused with electrical resistivity, which has the same symbol), and its units are $°C^{-1}$. The change in length, x, of a material is given by the formula:

$$x = l\,\alpha\,(t_2 - t_1).$$

Alternatively, the thermal expansivity of a material can be calculated using:

$$\alpha = x/[l\,(t_2 - t_1)]$$

where l is the original length and $(t_2 - t_1)$ is the temperature change. The units of the expansion are the same as those used to measure the original length.

Chemical stability

Over a period of time, there is a tendency for engineering materials to deteriorate owing to the effects of their service environment. The ability to resist deterioration may take the form of:

- corrosion resistance
- solvent resistance
- radiation resistance.

Corrosion resistance Corrosion is a form of deterioration to which all metals are subject. It results from a chemical reaction between the metal and some other element in its environment. This is usually oxygen from the atmosphere and in moisture, although other special forms of corrosion do occur in the chemical industries and marine engineering. Ferrous metals, i.e. iron and its alloys, are particularly prone to atmospheric corrosion. The rust that forms at normal temperatures and the millscale that forms at high temperatures are loose and porous, allowing continued attack from the environment.

Non-ferrous metals that contain no iron, or in which the iron content is very low, generally have a much higher resistance to corrosion. This is because the oxide film that forms on the surface is more dense and gives protection against further attack. The corrosion resistance of metals can be increased by painting, coating with a polymer material or plating with a more corrosion-resistant metal. In some cases, it may also be increased by artificially thickening the protective oxide surface film by processes such as anodising.

Polymer materials and ceramics do not combine with oxygen as readily as metals and thus have a high resistance to this kind of corrosion. They do, however, deteriorate in other ways. Some plastics are susceptible to solvent attack and can become brittle if exposed to the ultraviolet radiation that is present in sunlight. Ceramics can deteriorate in a high-temperature environment and are also attacked by certain chemical substances.

Solvent resistance Polymer materials are attacked by chemical substances called solvents. They can cause a material to dissolve or bring about a deterioration in its mechanical properties. Some polymer materials, in particular thermosetting plastics, have a high resistance to solvents, while some thermoplastics and rubbers have a low resistance. Care should be taken when selecting polymer materials that will come into contact with chemicals. Solvents can include petrol, fuel oils and lubricating oils. Metals and ceramics generally have a high resistance to the solvents that attack polymers, but they can be attacked by other kinds of chemical substance, particularly acids.

Radiation resistance Some thermoplastics and rubbers can be affected by the ultraviolet radiation that is present in sunlight. They become progressively more brittle with exposure, and discolouring can also occur. The effect can be reduced by adding colouring agents, black being the most effective.

Case study

Mountain bike design

Considerable advances have been made in the design of bicycles in recent years. This is particularly true in the case of mountain bikes and those used in velodrome racing.

Mountain bikes are strengthened for use on rough terrain and, in addition to the wide range of gears, some of them are also equipped with shock absorbers in the front forks. Steel is the traditional material used for the frame and forks, but high-strength, low-weight aluminium and titanium alloys are used in the more expensive models (Figure 6.5).

Figure 6.5 Mountain bike

Radical changes in the design of speed racing cycles have also occurred in recent years. Here, there has been a move towards the use of lightweight composite materials, such as carbon fibre-reinforced plastics, and injection moulding forming techniques. Disc wheels have been introduced and the combined effect of the changes has been to reduce the weight and improve the aerodynamics of the machine. With both types of bicycle, the identification and adoption of materials with superior properties has contributed to the evolution and success of the designs.

Metals

Metals may be subdivided into ferrous metals, in which iron is a major constituent, and non-ferrous metals, which contain no iron or in which iron is only present in small amounts. Some metals are used in an almost pure form, whereas others are used as the constituents of alloys. An alloy is a mixture of metals or a mixture of a metal and another substance. Some of the more common ferrous metals and alloys used in engineering are:

- plain-carbon steels
- cast iron
- alloy steels.

Plain-carbon steels

Iron in its pure form is not used as an engineering material. It is a soft metal that is not easily machined to a good surface finish. Furthermore, when molten, it tends to be pasty and difficult to cast. The addition of small amounts of carbon, however, greatly improves its properties to give a range of alloys called plain-carbon steels. The different grades are known as dead mild steel, mild steel, medium-carbon steel and high-carbon steel (Table 6.1).

Dead mild steel has a carbon content of between 0.1% and 0.15%. Like pure iron, it is ductile and malleable, but the carbon gives it added strength. It is used to make wire, rods and tubes. It is also used for rivets and for pressings.

Mild steel has a carbon content of between 0.15% and 0.3% and is the most common structural metal in use. It machines well and, when heated, it is easily

Table 6.1 Ferrous metals

Name	Percentage carbon and other elements	Tensile strength	Uses
Dead mild steel	0.1–0.15 carbon	400 N/mm	Wire, drawn tube, nails, rivets sheet steel for pressings
Mild steel	0.15–0.3 carbon	500 N/mm	Girders, boiler plate, nuts and bolts, bar for general use
Medium-carbon steel	0.3–0.8 carbon	750 N/mm	Crankshafts, axles, couplings, gears, wire ropes, hammer heads, cold chisels
High-carbon steel	0.8–1.4 carbon	900 N/mm	Springs, knives, scewcutting taps and dies, saw blades
Grey cast iron	3.2–3.5 carbon	200 N/mm	Machine beds, water pipes, columns, ornamental castings
High-speed steel	0.7 carbon 18 tungsten 4 chromium 1 vanadium	–	Twist drills, milling cutters, turning tools
Stainless steel	0.04 carbon 0.45 manganese 14 chromium	600 N/mm	Kitchenware, food containers, pressings

formed to shape by forging, rolling and drawing. It has good tensile strength and may have its surface hardness increased by the heat treatment process known as case hardening. Mild steel is supplied in the form of bars, tubes, sheets, plates and girders. It is widely used in engineering as a general workshop material and by the construction industry as building material.

Medium-carbon steel has a carbon content of between 0.3% and 0.8%. The increased carbon content makes it stronger and tougher than mild steel, and its hardness and toughness can be further increased by the heat treatment processes known as hardening and tempering. Medium-carbon steel is widely used for forged components, such as those used in vehicle suspension and trailer coupling units, which need to have good resistance to impact and shock loading.

High-carbon steel has a carbon content of between 0.8% and 1.4%. This has the highest tensile strength of the range of plain-carbon steels and, like medium-carbon steel, its hardness and toughness may be further improved by heat treatment. It is widely used to make springs and for cutting tools, such as files, screwcutting taps and dies, wood chisels and trimming knives.

Cast iron

When carbon is added to molten iron in quantities above about 1.7%, it is found that, when the metal has cooled down, flakes of carbon in the form of graphite are present. The alloy is then called grey cast iron, which usually has a carbon content of between 3.2% and 3.5%. The high carbon content has several effects on the properties of the material. It increases its fluidity when molten, which enables it to be cast into intricate shapes. It is easily machined without the use of a cutting fluid owing to the self-lubricating effect of the graphite flakes. The graphite enables cast iron to absorb the energy from vibrations. This makes it an ideal material for machine beds.

Although grey cast iron is strong in compression, the high carbon content reduces its tensile strength and makes the material brittle. Nevertheless, cast iron is an ideal material for a wide range of engineering components that have an intricate shape and are not subjected to high tensile loads. It has recently made a comeback as a decorative structural material for lamp posts, columns and garden furniture.

The properties of cast iron can be improved by adding certain other alloying elements and by heat treatment. When small amounts of magnesium are added, the effect is to produce small spheres of graphite instead of flakes. The material is then known as spheroidal graphite cast iron (SG iron), which has a tensile strength similar to plain-carbon steels. The structure of castings made from grey cast iron may also be modified by heat treatment to produce what are known as malleable cast irons. As the name suggests, they are less brittle and much stronger than the original grey cast iron.

Alloy steels

These are steels that contain alloying elements other than carbon to give them special properties. Two alloy steels that are widely used in engineering are high-speed steel and stainless steel. High-speed steel is used for cutting tools, such as twist drills and milling cutters. In addition to carbon, it contains tungsten, chromium and vanadium. These enable the tools to retain their hardness and sharp cutting edges when operating at high speeds.

Stainless steel is used for a variety of products that must have a high resistance to corrosion. In addition to carbon, it contains chromium, manganese and nickel. Stainless steel is corrosion resistant because the thin oxide film that forms on its surface is very dense and protects it from further attack.

Some of the more common non-ferrous metals and alloys used in engineering are:

- copper
- zinc
- tin
- aluminium
- brasses
- tin-bronzes
- cupronickels
- aluminium bronzes
- aluminium alloys.

Copper Copper was one of the earliest metals to be used by man. It does not have a high tensile strength, but it is malleable, ductile, corrosion resistant and an excellent conductor of both heat and electricity. It is used in its pure form for electrical conductors and for the piping used in domestic plumbing. Copper is also widely used as a constituent of brasses, tin-bronzes and cupronickel alloys.

Zinc Zinc tends to be rather brittle and does not have a high tensile strength. However, like copper and most other non-ferrous metals, it is highly corrosion resistant. It is used to give a protective coating to mild steel by a process known as galvanising, and it is also used for alloying with copper to make brasses.

Tin Tin has a very low tensile strength and also a low melting point. It is soft, very malleable and highly corrosion resistant. Tin is used as a protective coating for the mild steel used in food canning. It is also widely used for alloying with copper to make tin-bronzes and with lead to make soft solders.

Aluminium Aluminium in its pure form is very malleable and ductile and an excellent conductor of heat and electricity. It is the lightest non-ferrous metal in common use with good corrosion resistance but low tensile strength. Aluminium is widely used in its pure form and also for alloying with copper and other elements to make a range of aluminium alloys and aluminium bronzes (see Table 6.2 for a summary of the properties of pure non-ferrous metals).

Table 6.2 Pure non-ferrous metals

Name	Tensile strength	Uses
Copper	220 N/mm^2	Electrical wire and cable, water pipes, heat exchangers, alloying to make brasses and bronzes
Zinc	110 N/mm^2	Protective coatings, alloying to make brasses and die-casting alloys
Tin	15 N/mm^2	Protective coatings, alloying to make tin-bronzes and soft solders
Aluminium	60 N/mm^2	A variety of engineering and domestic products, alloying to make aluminium alloys and bronzes

Brasses Brasses are an alloy of copper and zinc to which small amounts of other elements such as tin and lead are sometimes added (Table 6.3). Brasses have good corrosion resistance and they are stronger and tougher than either of their main constituents. They are produced with different proportions of copper and zinc for different applications. Brasses with a high copper content are used for cold forming operations such as drawing and pressing. Brasses with a high zinc content are used for hot forming operations such as forging, hot stamping and extrusion.

Tin-bronzes The main constituents of tin-bronzes are copper and tin (Table 6.4). Small amounts of phosphorous or zinc are also added to prevent the tin from oxidising when the metal is molten. Sometimes lead is added to improve the machinability of the metal. Tin-bronzes containing phosphorous are called phosphor bronzes. They are very malleable and ductile and are widely used as bearing materials. They also have good elasticity when cold worked, which makes them suitable for use as springs. Tin-bronzes containing zinc are called gunmetals. They have good fluidity and are used for casting. After solidifying, they are tough and, like all tin-bronzes, they are corrosion resistant.

Table 6.3 Common brasses

Name	Percentage composition	Properties	Uses
Cartridge brass	70 copper 30 zinc	Very ductile	Cold formed deep-drawn components, e.g. cartridge cases, condenser tubes
Admiralty brass	70 copper 29 zinc 1 tin	Very ductile and corrosion resistant	Cold formed deep-drawn components for marine and other uses
Standard brass	65 copper 35 zinc	Ductile and tough	Cold pressings
Muntz metal	60 copper 40 zinc	Strong and tough	Hot forming, e.g. hot rolled plate, forgings and castings
Naval brass	62 copper 37 zinc 1 tin	Strong, tough and corrosion resistant	Hot formed components for marine and other uses

Table 6.4 Common tin-bronzes

Name	Percentage composition	Properties	Uses
Low-tin bronze	96 copper 3.9 tin 0.1 phosphorus	Very malleable and ductile when annealed. Elastic when cold worked	Springs, electrical contacts, instrument parts
Cast phosphor bronze	90 copper 9.5 tin 0.5 phosphorus	Tough with good anti-friction properties	Bearings and worm gears
Admiralty gunmetal	88 copper 10 tin 2 zinc	Tough with good fluidity and corrosion resistance	Miscellaneous castings, e.g. valve and pump components
Bell metal	78 copper 22 tin	Sonorous and tough with good fluidity	Bells and other castings

Cupronickel alloys These contain copper and nickel with smaller quantities of manganese and sometimes iron (Table 6.5). They are very tough and strong with high corrosion resistance. Cupronickels are widely used for 'silver' coinage and also in chemical and marine installations where high corrosion resistance is required.

Aluminium bronzes These contain copper and aluminium, with copper being the main constituent. They can also contain smaller quantities of nickel manganese

Table 6.5 Common cupronickel alloys

Name	Percentage composition	Properties	Uses
Coinage cupronickel	74.75 copper 25 nickel 0.25 manganese	Tough, strong and corrosion resistant	'Silver' coinage
Monel metal	29.5 copper 68 nickel 1.25 iron 1.25 manganese	Tough, strong, very corrosion resistant	Chemical plant and marine components

and tin (Table 6.6). Aluminium bronzes are ductile, malleable and tough with good corrosion resistance. They are made with different proportions of the above ingredients for both cold working and casting.

Table 6.6 Common aluminium bronzes

Name	Percentage composition	Properties	Uses
Wrought aluminium bronze	91 copper 5 aluminium 4 nickel and manganese	Ductile and malleable with good corrosion resistance	Boiler and condenser tubes, chemical plant components
Cast aluminium bronze	86 copper 9.5 aluminium 1 nickel 1 manganese	Tough, good fluidity and good corrosion resistance	Sand and die-cast, e.g. valve and pump parts, gears, propellers.

Activity 6.2

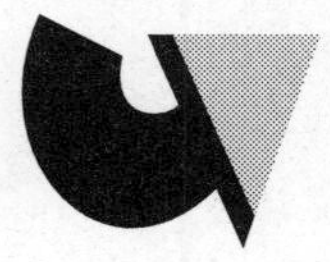

Have a look at a penny, a 10p piece and a £1 coin. They are made from three different alloys. One of them is listed in Table 6.6, but what are the other two, and what are the properties that make them suitable for use as coinage metals?

Aluminium alloys These have aluminium as the main constituent but may also contain copper, silicon, silver, nickel, manganese, magnesium and tin (Table 6.7). The effect is to increase the tensile strength of the aluminium, while retaining low weight and good corrosion-resistant properties (Figure 6.6). Some are produced for cold working and others for casting. It is also possible to increase the hardness of some types of aluminium alloy by heat treatment.

Table 6.7 Common aluminium alloys

Name	Percentage composition	Properties	Uses
Casting alloy BS 1490/LM4M	92 aluminium 5 silicon 3 copper	Good fluidity and moderate strength	Sand and die castings for light-duty applications
Casting alloy BS 1490/LM6M	88 aluminium 12 silicon	Very good fluidity and strength	Sand and die castings for motor vehicle and marine components
'Y' alloy	92 aluminium 2 nickel 1.5 manganese	Good fluidity and can have its hardness modified by heat treatment	Motor vehicle engine parts, e.g. pistons and cylinder heads
Duralumin	94 aluminium 4 copper 0.8 magnesium 0.7 manganese 0.5 silicon	Ductile and malleable in soft condition, can be heat treated	Structural uses, e.g. motor vehicle and aircraft panels
Wrought alloy BS 1470/5:H30	97.3 aluminium 1 magnesium 1 silicon 0.7 manganese	Ductile, with good strength can be heat treated	Structural uses, e.g. ladders, scaffold tubes. Overhead power lines

Figure 6.6 The airbus, which is largely made from aluminium alloys

Progress check

1 What property would you expect the spokes of a bicycle wheel to have?
2 Material is sometimes formed by pressing it to shape between specially made dies. What kind of forces do you think act on the material during this process and what property must the material have?
3 What essential property must a lathe-cutting tool or a twist drill have?
4 Why do you think that overhead transmission lines used to supply electricity have a core of steel wire surrounded by an outer layer of aluminium wire?
5 A metal filament lamp and a carbon filament lamp have the same resistance before they are switched on. What will happen to their resistances when they are working?
6 Describe two methods used to protect bicycle components against corrosion.
7 What effect does the carbon content of plain-carbon steel have on:
   a its tensile strength
   b its ductility?
8 Describe the ways in which the high carbon content of grey cast iron affects its mechanical properties.
9 Although it is highly resistant to corrosion, stainless steel does not seem to have been used very much to make motor car bodies. Can you think of two reasons why this may be so?
10 What are the properties that make copper and aluminium ideal materials for making saucepans and other cooking utensils?

Polymers

Polymers include the different kinds of plastic and rubber. They have been developed gradually over the last 100 years, giving rise to many new industrial and consumer products. Plastics are synthetic materials that fall into two main categories. These are thermoplastics and thermosetting plastics. Thermoplastics may be softened and remoulded by heating, whereas thermosetting plastics cannot be softened once they have been formed to shape. Rubbers are polymer materials known as 'elastomers'. While they are not truly elastic as defined by Hooke's law, they do have the property of returning to their original shape after undergoing large amounts of deformation. Figure 6.7 shows examples of domestic products and containers made from plastics.

Some of the main thermoplastic materials used in engineering are:

- polythene
- polypropene
- PVC
- polystyrene
- Perspex
- PTFE
- nylon
- Terylene.

Figure 6.7 Domestic products and containers made from plastics

Low-density polythene This is tough and durable and resistant to water and chemical solvents. It is used for plastic bags and wrappings, waterproof covers and cable insulation.

High-density polythene This has longer polymers than low-density polythene, which are packed closer together. It is more useful as an engineering material, being harder, stiffer and stronger than low-density polythene and with a higher melting temperature of around 135°C. It is moulded into bottles, pipes, tubs, crates, tanks and other containers.

Polypropene This has similar properties to high-density polythene but is stronger, harder, and has a higher melting point of around 150°C. It has similar uses to high-density polythene and can also be produced as a fibre for use in ropes and nets. Its higher melting point makes it suitable for hospital equipment, such as bowls and trays, which frequently need to be sterilised.

PVC This is produced from the polychloroethene polymer, which was previously called polyvinyl chloride or PVC. In its flexible form, it is known as pPVC, which is tough and has good solvent resistance. It is used for the insulation on electric wiring and cables, for wellington boots and for items of clothing. In its rigid form, it is known as uPVC, which is tough and rigid. It is used for plastic window frames and doors and also for safety helmets.

Polystyrene Polystyrene is produced from the polyphenylethene polymer. It can be made tough and dense for household and industrial use, or in expanded cellular form for use as a light packaging material. The linings for refrigerators, ball-point pens, disposable cups and the cartons for margarine are all familiar items made from polystyrene. A disadvantage is that, while flexible up to a point, it tends to be brittle and is easily broken.

Perspex Perspex is produced from the polymer, methyl 2-methylpropenoate, which was formerly called methyl methacrylate. It is strong, rigid and transparent and can be softened and remoulded using boiling water. Unfortunately, it is easily scratched and is attacked by petrol.

PTFE This name is derived from the polymer polytetrafluoroethylene. It is a tough, flexible, heat-resistant thermoplastic, which is not attacked by any solvent. It also has a very low coefficient of friction. PTFE is widely used as a bearing and gasket material in industry and as a non-stick coating for domestic utensils under the trade name of Teflon.

Nylon Nylons are produced from polymers known as polyamides. They are strong, tough and flexible and have good solvent resistance. They are used in fibre form for fishing line, ropes and nets. Nylons are also hard wearing, which makes them suitable for use as a material for bearings and gears. A disadvantage that they share with some other thermoplastics is that they deteriorate after long exposure to sunlight.

Terylene Terylene and Dacron are trade names for materials produced from polymers known as polyesters (Figure 6.8). In fibre form, they are strong and stable with good solvent resistance, but they can be attacked by strong acids and alkalis. Polyester materials are also used for recording tapes and for electrical insulation.

Table 6.8 lists the properties and uses of some common thermoplastic materials.

Thermosetting plastic materials are generally used with a filler material to improve their mechanical properties. These include wood flour, shredded textiles, paper and various other fibres and powdered materials. Some of the main types of thermosetting plastic material used in engineering are:

Figure 6.8 Parachute made from nylon and polyester materials

Table 6.8 Common thermoplastic materials

Name	Polymer	Properties	Uses
Low-density polythene	Polyethene	Tough, flexible, easily moulded, solvent resistant, gradual deterioration if exposed to light	Packaging, piping, squeeze containers, cable and wire insulation
High-density polythene	Polyethene	Similar to above but harder, stiffer and with good tensile strength	Pipes, mouldings, tubs, crates, kitchen bowls, food containers, medical equipment
Polypropene	Polypropene.	High strength, hard, high melting temperature, can be produced as a fibre	Tubes, pipes, fibres, ropes electronic components, kitchen utensils, medical equipment
PVC (polyvinyl chloride)	Polychloroethene	Can be made tough and hard or soft and flexible, solvent resistant, soft form tends to harden with time	When hard, window frames piping and guttering. When soft, cable and wire insulation, upholstery
Polystyrene	Polyphenylethene	Tough, hard, rigid but somewhat brittle, can be made into a light cellular foam, can be attacked by petrol-based solvents	Mouldings for refrigerators and other appliances, moulded foam used for packaging
Perspex	Methyl-2-methyl-propenoate	Strong, rigid, transparent but easily scratched, easily softened and moulded, can be attacked by petrol-based solvents	Lenses, protective shields, aircraft windows, light fittings, corrugated sheets for roof lights
PTFE	Polytetrafluoro-ethylene	Tough, flexible, heat-resistant, highly solvent resistant, has a waxy low-friction surface	Bearings, seals, gaskets, non-stick coatings, tape
Nylon	Polyamide	Tough, flexible and very strong, good solvent resistance but does absorb water and deteriorates with outdoor exposure	Bearings, gears, cams, bristles for brushes, textiles
Terylene	Thermoplastic polyester	Strong, flexible and solvent resistant, can be made as a fibre, tape or sheet	Textile fibres, recording tape, electrical insulation tape

- bakelite
- formica
- Melamine
- epoxy resins
- polyester resins.

Bakelite Bakelite was one of the earliest thermosetting plastics to be developed, dating from 1872 when it was patented in America by Belgian-born Leo Baekeland. It is produced from phenol–methanol resins and was formerly known as phenol formaldehyde.

Bakelite is an excellent electrical and thermal insulator and is resistant to most common solvents. It is hard and scratch resistant but tends to be brittle if used alone. It is usually moulded with a filler material, such as wood flour, to improve its strength. Its colours are limited to black and brown and, although it is still used for laminates and moulded insulating parts in electrical equipment, many of its former uses have been taken over by urea–methanol resins.

Formica Formica is the trade name for the thermosetting plastic material produced from urea–methanol resins. It has many of the properties of bakelite and was previously called urea formaldehyde. Formica is moulded with a range of filler materials to improve its properties and, although naturally colourless, it is available in a variety of colours. It is widely used in laminates and for trays, toilet seats, kitchen ware and the moulded parts of electrical equipment.

Melamine This trade name is derived from the methanol–melamine resins from which it is produced. They have similar properties to phenol–methanol and urea–methanol resins but give a product that is harder and a better heat insulator. Melamine can be produced in a variety of colours with a very smooth surface finish. It is widely used in plastic laminates, cups, saucers and plates, control knobs, handles and the moulded parts of electrical equipment.

Epoxy resins Whereas the above thermosetting plastics are moulded in the presence of heat, these are mixed with a hardener and usually reinforced with glass fibre, carbon fibre or cloth to improve their strength. They adhere to metals and have good thermal insulation and solvent-resistant properties. They are used for laminates, motor panels and as a protective internal coating for metal containers.

Polyester resins These are also moulded with a hardener and with glass fibre and carbon fibre reinforcement.

They have a hard and tough surface and are widely used for boat hulls (Figure 6.9). Polyester resins are also used in paints and enamels to improve their toughness.

Table 6.9 lists the properties and uses of some common thermosetting plastic materials.

Some of the main types of rubber used in engineering are:

- natural rubber
- styrene rubber
- neoprene
- butyl rubber
- silicone rubber.

Figure 6.9 Glass-reinforced polyester boat hull

Table 6.9 Common thermosetting plastic materials

Name	Polymer	Properties	Uses
Bakelite	Phenolic resins	Hard, resistant to heat and solvents, good electrical insulator and machinable, colours limited to brown and black	Electrical components, vehicle distributor caps, saucepan handles, glues, laminates
Formica	Urea–methanol resins	As above, but naturally transparent and can be produced in a variety of colours	Electrical fittings, toilet seats, kitchenware, trays, laminates
Melamine	Methanol–melamine resins	As above, but harder and with better resistance to heat. Very smooth surface finish	Electrical equipment, tableware, control knobs, handles, laminates
Epoxy resins	Epoxy resins	Strong, tough, good chemical and thermal stability, good electrical insulator, good adhesive	Container linings, flooring material, laminates, adhesives
Polyester resins	Polyester resins	Strong, tough, good wear resistance and resistance to heat and water	Boat hulls, motor panels, aircraft parts, fishing rods, skis, laminates

Natural rubber Natural rubber from the rubber tree is readily attacked by solvents that cause it to perish. It is not used in engineering except when mixed with synthetic rubbers.

Styrene rubber This was developed in America during the Second World War to overcome the shortage of natural rubber. It is also known as GR-S rubber. It is resistant to oils and petrol and is sometimes blended with natural rubber for use in vehicle tyres and footwear.

Neoprene This synthetic rubber has a close resemblance to natural rubber, but it is resistant to mineral and vegetable oils and can withstand high temperatures. It is widely used for gaskets, oil seals and hoses in engineering applications.

Butyl rubber Butyl rubber has good resistance to heat and chemicals. It is also impermeable to gases and is widely used for moulded diaphragms, inner tubes, air bags and tank linings.

Silicone rubber This synthetic rubber retains its properties over a wider temperature range than the other types. It is flexible in temperatures down to –80°C and can withstand temperatures up to 300°C. Silicone rubber also has good solvent resistance. It is widely used for gaskets and seals in chemical plant and in aircraft where a wide variation in operating temperature is likely.

Activity 6.3

Although rubber seems to be an elastic material, it is not perfectly elastic. To prove this, take a good-sized rubber band and measure its unstretched length. Then hang it from a hook or some other suitable support such as a retort stand. Hang weights on the lower end and measure the extension that they produce. You will probably find that the rubber takes time to stretch so allow a few moments for it to settle before taking measurements. You will be surprised how much weight the rubber will take but be careful not to load it to breaking point.

After taking sufficient readings, remove the load, measure the length of the band and compare this with the original unstretched length. Plot a graph of load against extension and examine its shape. Try to think of some explanation for the behaviour of the rubber.

Ceramics

Ceramics are a wide-ranging group of inorganic materials whose chief ingredients are clays, sand and feldspar. Clays contain silicon, aluminium and other elements in chemical combinations known as silicates. Potassium, magnesium and calcium compounds may also be present, depending on the type of rock that the clay was formed from by the action of rain. Sand contains silica, which is a compound of silicon and oxygen. It also contains feldspar, or aluminium potassium silicate, which is a white or pinkish mineral found naturally in some rocks.

Bricks, tiles, sewer pipes, grinding wheels, glassware and porcelain are all examples of ceramic products. The range also includes the new types of ceramic that have been developed in recent years for the space programme, cutting tools, surgical implants and engine components (Figure 6.10).

Figure 6.10 Space shuttle protected with ceramic tiles

The main properties of ceramics are that they are hard, wear resistant, strong in compression and good electrical insulators. Many of them can also withstand very high temperatures. This makes them suitable for use as refractory materials that are used to line furnaces and kilns. The main disadvantages of ceramics are that they tend to be brittle, weak in tension and not very tough. Some of the main types of ceramics are:

- amorphous ceramics
- crystalline ceramics
- bonded ceramics
- cements.

Amorphous ceramics These include the different types of glass used for windows, containers, lenses and the glass fibres used to reinforce plastic materials. The main ingredient of glass is the silicon oxide or silica (SiO_2), contained in sand. Other ingredients include sodium carbonate and borate, feldspar and oxides of calcium, magnesium and boron. When the mix is cooled very slowly from the molten state, it solidifies with a crystal structure, as do metals. When it is cooled more rapidly, however, crystallisation does not take place and the solidified structure consists of disordered and linked chains of silicon and oxygen atoms.

The term 'amorphous' as applied to materials means that their atoms are not arranged in a regular geometric pattern. The fact that glass is amorphous accounts for its transparency, since it is the crystal or grain boundaries and regular planes of atoms in a material that reflect light to make it opaque. Three of the most common forms of glass used in engineering are:

- soda-lime glass
- lead glass
- borosilicate glass.

Soda-lime glass accounts for about 95% of all the glass that is manufactured. It is the glass used for bottles and jars, etc. and also for the plate glass used in windows (Figure 6.11). It can be toughened by heat treatment for use in vehicle windows and plate glass doors, and is then known as 'tempered glass'. It is also produced in laminated form with a layer of clear plastic bonded between two layers of glass for use in vehicle windscreens. The plastic holds the glass fragments when breakage occurs.

Figure 6.11 Plate glass formed by the float glass process in which the glass solidifies on the surface of the molten tin

Lead glass contains lead oxide, which has the effect of increasing its refractive index. This is a measure of the amount by which light rays are bent or refracted as they pass through it. Lead glass is used for electric light bulbs, neon signs, lenses and for cut glass vases and ornaments.

Borosilicate glass contains boron oxide and is better known under its trade name of Pyrex. It has a low thermal expansivity, is tough, chemically stable and has a high electrical resistance. Borosilicate glass is widely used for laboratory equipment, electrical insulators and kitchen glassware.

Crystalline ceramics These are generally good electrical insulators and capable of withstanding high temperatures. Magnesium oxide, whose crystals are used as an electrical insulator in copper-sheathed mineral insulated cables, is one of this group. The group includes crystals that are very hard and used as abrasives. Aluminium oxide crystals, or 'emery', beryllium oxide, silicon carbide and boron carbide crystals are the ones most widely used.

They are bonded to cloth and paper, suspended in oil as grinding paste and bonded together to make grinding wheels. The group also includes some of the hard carbides and nitrides that are bonded together in a metal matrix for use as cutting tool tips. Tungsten carbide is perhaps the best known of these, others being zirconium carbide, titanium carbide and silicon nitride. Some of these combinations are called 'cermets'. They will be described later in the section dealing with composite materials.

Bonded clay ceramics These are products made from natural clays or mixtures of clays and added crystalline ceramics. They are moulded and heated, or 'fired', in ovens and kilns. The ingredients undergo a process of 'vitrification' in which crystals become bonded together in a glass-like matrix. The temperature at which they are 'fired' varies with the ingredients. Bonded clay ceramics include:

- whiteware
- structural clay products
- refractory ceramics.

Whiteware includes earthenware and china pottery, porcelain, stoneware, sanitary ware and decorative tiles. It is made from selected and refined clays to which a suitable fluxing agent, such as 'borax' (sodium borate) or 'cryolite' (sodium aluminium fluoride), has been added. These determine the degree of vitrification and also lower the vitrification temperature. They are usually fired at temperatures between 1200°C and 1500°C. Whiteware products, such as pottery, are often glazed with a thin layer of molten glass.

Structural clay products include building bricks, drain pipes, terracotta and roofing tiles. They are moulded from common clays and fired at a higher temperature than whiteware. Natural impurities in the clays act as a flux, and products such as drain or sewer pipes are given a glazed finish.

Refractory ceramics are used to make firebricks for lining furnaces, kilns and ladles for carrying molten metal. They are able to withstand sustained high temperatures, chemical attack and physical contact. Refractory ceramics are made from fireclays that have a high silicon oxide or a high aluminium oxide content.

Cements These are made from a variety of calcium, silicon, magnesium and other metallic oxides. When mixed with water, a chemical reaction occurs, which causes the cement to set into a hard crystalline structure. Cements have moderate compressive strength and hardness but are particularly weak in tension. The type most commonly used in building is Portland cement to which stone aggregates may be added to make concrete. Other types that have been developed for special applications include high-alumina cements that are resistant to sea water and silicate cements that are resistant to acid attack.

Semiconductors

Semiconductor materials are used in the manufacture of transistors and integrated circuits. Without them, the advances that have taken place in telecommunications, information technology and control systems would have been impossible. Improved semiconductor materials are continually being developed to increase the reliability of the above devices and the speed at which they operate.

Semiconductor materials include silicon, germanium and compounds such as cadmium sulphide and gallium arsenide. In their pure form, silicon and germanium are known as intrinsic semiconductors. At low temperatures, they are poor conductors of electricity but their conductivity improves with temperature rises. Their electrical properties are modified by the addition of impurity elements. The materials are then known as extrinsic semiconductors. The process is known as 'doping' and gives what are known as:

- ' n-type' semiconductors
- ' p-type' semiconductors.

n-type materials These have been doped with small amounts of phosphorous, antimony or arsenic, which have one more electron in their outermost atomic shells than the intrinsic semiconductors. The effect is to increase the number of electrons available to become charge carriers.

p-type materials These have been doped with small amounts of aluminium, indium or gallium, which have one less electron in their outermost atomic shells than the intrinsic semiconductors. The effect is to produce vacancies or 'holes' in the atomic structure of the material, which, under certain circumstances, can themselves behave as charge carriers.

The diodes, transistors and integrated circuits used in electronic equipment consist of n-type and p-type semiconductors arranged in different ways to give the components their particular characteristics. The silicon chips in integrated circuits have alternate layers of n-type and p-type material from which the circuit components are formed by an etching process.

Composites

Composites are made up of two or more materials, which may be bonded together as laminae or combined so that one material acts as a matrix surrounding fibres or particles of another. Plywood and concrete reinforced by steel are examples that have been in use for many years in the construction industry. Composites may be categorised as:

- particulate
- fibrous
- laminated.

Particulate Concrete in which aggregate and sand are bonded in a matrix of Portland cement is perhaps the oldest of this type of composite. Other types include particle-reinforced polymers in which silica, glass beads and rubber particles are used to strengthen and toughen the polymers. High-impact polystyrene is reinforced with rubber particles in this way.

Certain metals also have their properties improved by the addition of particles. The tensile strength of aluminium is improved by the inclusion of aluminium oxide particles, particularly its strength at high temperatures. Bearing metals, such as phosphor bronze, sometimes incorporate particles of graphite and PTFE to reduce frictional resistance and improve their bearing qualities.

The range of materials known as 'cermets' may also be included in this category of composite materials. Cermets are formed by compressing and heating a mixture of powdered ceramic and metal particles. Some cermets are extremely hard and wear resistant and retain their properties at high temperatures. The cemented carbides used for the tips of cutting tools and wire drawing dies are perhaps the best known of these materials. Tungsten, titanium, molybdenum and silicon carbides are some of the ceramic materials used, and the metal that forms the matrix around them is generally cobalt. In other cermets, the ceramic forms the matrix around oxidised metal particles. This type has excellent thermal insulation properties and has been used for heat shields in aerospace applications.

Fibrous The most common fibrous composites are fibre-reinforced thermosetting plastics, which use epoxy and polyester resins as the matrix material. The role of the fibres is to carry the greater part of the load and provide stiffness. Glass fibres and carbon fibres are the most common reinforcing materials, although 'kevlar', an extremely strong polymer, and ceramic fibres are also used.

Carbon fibre-reinforced plastics are very strong, stiff and light in weight. They are used in aircraft and for fishing rods. Glass fibres have more tensile strength but tend to be heavier. They have been in use for many years for small boats, motor panels, surf boards and many other applications.

Kevlar is a polymer that was developed in the 1960s and is used to make bullet-proof body armour. This, and ceramic fibres made from silicon carbide and aluminium oxide, are used to make reinforced plastic composites that are extremely tough and able to absorb shock loads. They are used for the frames of tennis racquets. Kevlar is also used to reinforce the rubber in tyres.

Laminates Laminates are layers of material that are bonded together to give improved strength and rigidity. They may be divided into:

- wood laminates
- polymer laminates
- polymer–metal combinations.

Wood laminates include plywood, blockboard and laminated chipboard. Plywood is made up of thin layers of wood arranged with their grain directions running alternately perpendicular to each other (Figure 6.12). Blockboard is a composite made up of strips of wood bonded together and sandwiched between two thin sheets. Laminated chipboard consists of wood shavings and sawdust bonded together and sandwiched between two thin outer sheets. Thermosetting resins are used as the bonding agent. The resulting composites are flexible, resistant to warping and do not have any weaknesses caused by grain direction.

Polymer laminates are made from thermosetting plastic resins and filler materials such as paper and cloth. The paper or cloth passes through a tank containing the resin and becomes impregnated. It then passes through a dyeing process where it is partly 'cured', i.e. cross-linking has started to take place. The material is then cut into lengths, which are stacked in a number of layers and placed in a hot press. The heat and pressure cause the partly cured resin to fill all the cavities between the layers and complete the cross-linking process. The resulting laminates are tough and hard wearing and widely used for trays and worktops.

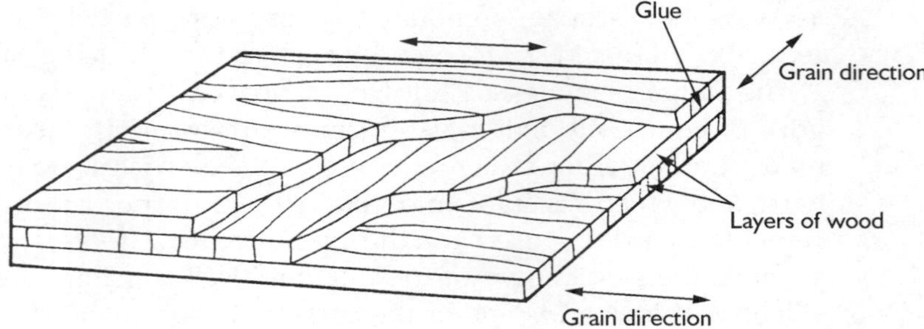

Figure 6.12 Plywood

Polymer–metal combinations are increasingly used in aircraft because of their lightness and rigidity. They consist of a low-density lightweight core contained between two high-strength skins. A common arrangement is shown in Figure 6.13.

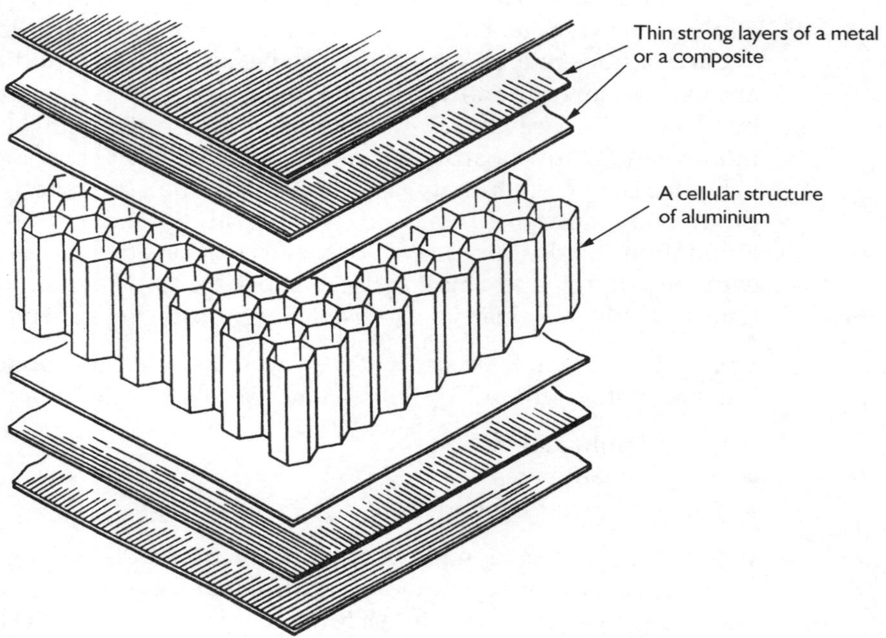

Figure 6.13 Polymer-metal composite

The core is an aluminium honeycomb structure sandwiched between sheets of aluminium and carbon fibre-reinforced thermoset. In other arrangements, the cellular core may be made from titanium, steel or a plastic foam.

Case study

Domestic central heating system

It is now taken for granted that a new house will be equipped with some form of central heating system. The Romans had a form of underfloor heating 2000 years ago, but it is not so long since the vast majority of houses in the United Kingdom were heated by open fires. Domestic central heating systems at an affordable price started to become available round about the 1960s and, since then, they have become increasingly popular with improvements to their reliability and efficiency. Natural gas is the most popular and convenient fuel, but systems are also available that use oil and solid fuels.

A central heating and hot water system can be divided into a number of interconnected subsystems. There is the heat exchanger in which water is heated by burning the fuel. There is a pump that circulates the hot water around the system. There are pipes and radiators that supply heat to the various rooms, and there must also be a control system of some kind. This is required to switch the central heating and hot water system on and off at the times of day required and also to maintain the rooms at a comfortable temperature when the system is in operation.

Central heating systems incorporate a variety of materials. Take a close look at the system in your home or ask a person in authority to explain the system in your school or college.

Progress check

1. What is the metal used for the pipes connecting the radiators of a domestic central heating system, and what properties make it suitable for this purpose?
2. The inlet and outlet valves on central heating radiators are generally made from an alloy that is plated with another metal. What are these metals and what are the properties that make them suitable for this use?
3. Central heating units are required to have a suitable metal flue that carries the waste gases away through an outside wall or up a chimney. What is the metal most often used for this purpose and why?
4. Name two things that can cause thermoplastics to deteriorate.
5. What are the differences in the properties of uPVC and pPVC?
6. Electrical plug tops and wall sockets are commonly made from a white thermosetting plastic. What properties must it have, and which material do you think it is most likely to be?
7. The plates, bowls, cups and saucers used in caravans, boats and for camping are sometimes made from a thermoplastic material. Which one is it most likely to be and what are its properties that make it suitable for this use?
8. What are the chief raw materials from which ceramic products are made?
9. What is 'emery' and in what kinds of engineering product is it used?
10. Where would you expect to find semiconductor materials used in a central heating system?

The structure of materials

There are 92 naturally occurring chemical elements. They range from hydrogen, whose atoms are small and light, to uranium whose atoms are large and heavy. In addition, there are a number of artificial elements whose atoms are more massive than those of uranium. These have been created in the laboratory or as a by-product of nuclear engineering processes. They are called 'transuranic' elements, of which plutonium is perhaps the best known.

Atoms are made up of particles. The major ones are protons, neutrons and electrons. Protons and neutrons are roughly the same size. Together, they form the central core or nucleus of an atom, where most of its mass is concentrated. Electrons have only about two-thousandths of the mass of protons and neutrons. They are in orbit around the nucleus and carry a small negative charge of electricity. The protons carry a small positive charge and are equal in number to the electrons so that the atom is electrically neutral. The forces that bind the atoms of a material together arise from these electrical charges.

The atoms of the chemical elements have different numbers of protons, neutrons and electrons. The electrons are to be found orbiting in distinct shells, of which there are seven naturally occurring. They are lettered K, L, M, N, O, P and Q. The lighter elements only have electrons in the inner shells. The heavier the element, the more shells there are that contain electrons. The elements found in common engineering materials generally have electrons in up to the fifth or O shell. The maximum number of electrons that these first five shells may contain are 2, 8, 18, 32 and 9 respectively.

Hydrogen is the simplest element with one proton in the nucleus and one orbiting electron in the K shell. Iron has 28 protons and 20 neutrons in the nucleus and 28 orbiting electrons. Their distribution is two in the K shell, eight in the L shell, 14 in the M shell and two in the N shell.

The outermost electrons influence the structure and properties of an element and its readiness to combine chemically with other elements to form chemical compounds. An atom with one or two electrons in the outer shell will readily donate them to an atom whose outer shell is almost complete. This results in the donor becoming positively charged and the receiver becoming negatively charged. As a result, they become joined together by electrostatic forces known as ionic or electrovalent bonds (Figure 6.14).

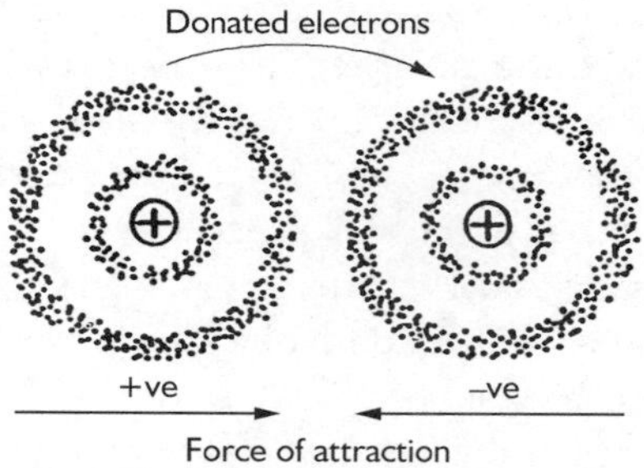

Figure 6.14 Ionic bonding

Atoms also combine together by sharing electrons in order to produce a common outer shell containing a stable number of electrons. The shared electrons have the effect of holding the atoms together with what are known as covalent bonds (Figure 6.15). With both kinds of bonding, the combined atoms are known as molecules.

Metals have a small number of electrons in the outer shell and it is thought that they are shared between all the atoms in the metal. The strong bonds that are formed between the atoms as a result are called metallic bonds (Figure 6.16). They are similar to covalent bonds but more complicated.

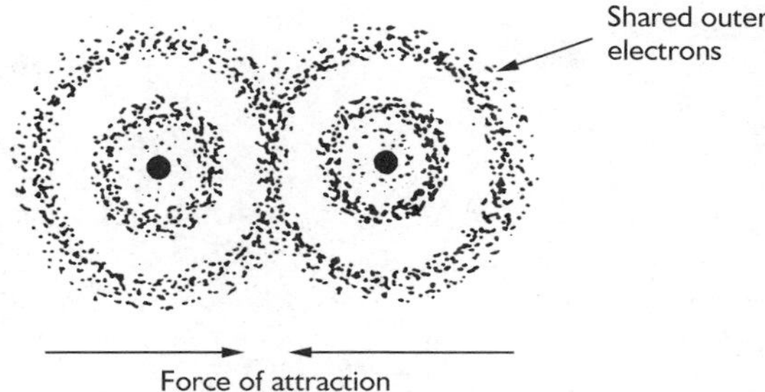

Figure 6.15 Covalent bonding

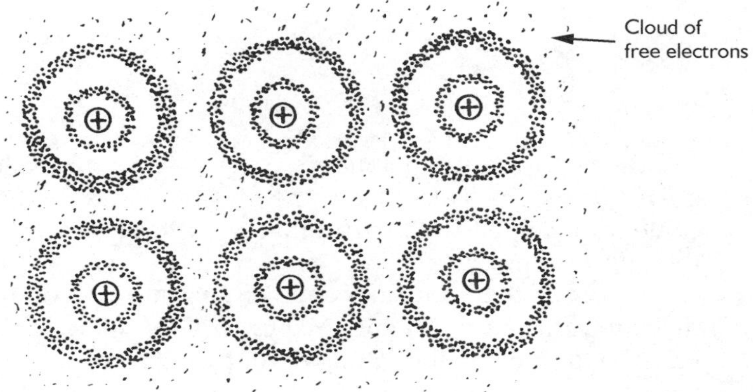

Figure 6.16 Metallic bonding

The metallic and non-metallic materials used in engineering are thus made up of atoms or molecules that are held together by strong bonds. Closer examination of their structure reveals that the atoms or molecules are arranged in particular ways. The structure of a material may be:

- amorphous
- crystalline
- polymer.

Amorphous materials

In an amorphous solid, the atoms or molecules are not arranged in any definite geometrical pattern. They are disordered, rather like they are in liquids. There are not many engineering materials that are truly amorphous. Glass and pitch are perhaps the most common examples.

Amorphous solids do not have a clearly defined melting point. As they are heated, they soften, behaving like a very viscous liquid. In fact, glass at normal temperatures will deform, or 'creep', under its own weight over a long period of time.

Crystalline materials

All metals, and a great many ceramics, solidify into a crystalline structure. Solidification starts at different points within the liquid phase as heat energy is being given off in the form of latent heat. Here, the atoms pack themselves

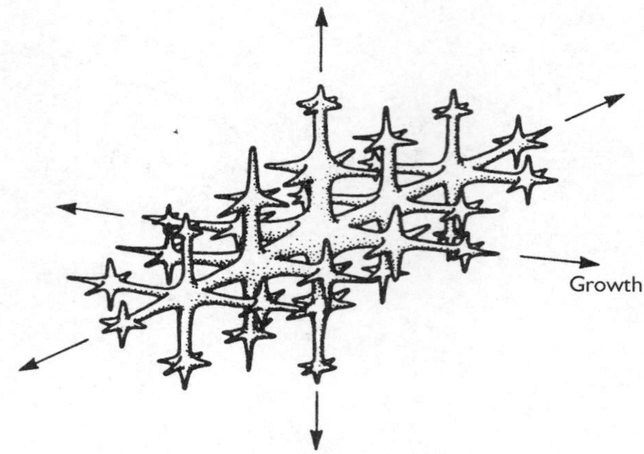

Figure 6.17 Dendritic growth

together in a regular geometric pattern. They are called 'dendrites', which grow outwards as more atoms join them (Figure 6.17). Eventually, when the dendrites come into contact with each other, solidification is complete. The dendrites have then grown into crystals or 'grains'.

The size of the grains depends on the rate of cooling and, generally, the slower the rate of cooling, the larger they will be. Often, they are clearly visible to the naked eye, such as the grains of zinc on galvanised steel. In other cases, they are very small and must be viewed through a microscope after polishing the material surface and etching it with chemicals to show up the grain boundaries. Grain size affects the properties of a material. Large grains tend to increase the brittleness and reduce the material strength.

As the dendrites form, their orientation to each other is different and so they come into contact at irregular angles. For this reason, some grains appear darker than others, even in a pure metal. In the more important engineering metals, there are three different crystal lattice structures into which the atoms arrange themselves. They are:

- body-centred cubic (BCC)
- face-centred cubic (FCC)
- close-packed hexagonal (CPH).

With the body-centred cubic structure, the atoms are arranged at the eight corners of a cube touching another atom at the centre of the cube (Figure 6.18). This is known as a unit cell.

It must be remembered that the structure is continuous within a grain and that each cube shares its faces with adjacent ones. The body-centred cubic structure is relatively open packed with more free space than the other two forms of packing.

When forces are applied to a crystal, deformation always takes place along planes of atoms that are most tightly packed. It follows that, because of its open structure, a body-centred cubic material will be more difficult to deform than the others. This turns out to be the case and examples of BCC metals are chromium, tungsten, molybdenum, vanadium and iron at normal temperatures, which are not very ductile or malleable.

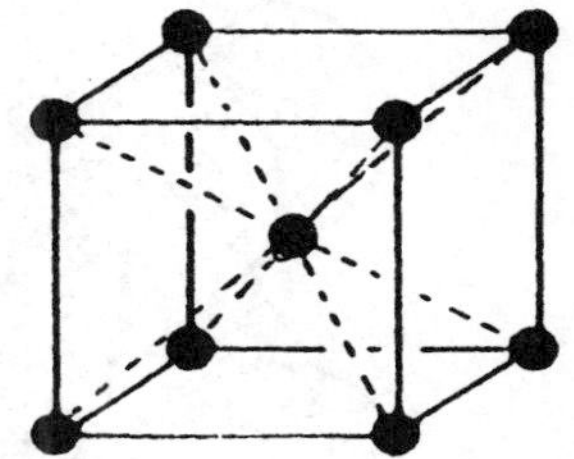

Figure 6.18 The body-centred cubic structure

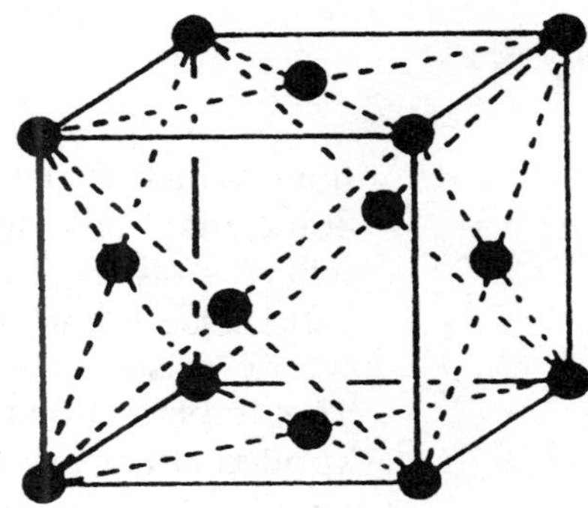

Figure 6.19 The face-centred cubic structure

With the face-centred cubic structure, the atoms are arranged at the eight corners of a cube, which also has atoms at the centre of each face (Figure 6.19).

Here, the atoms are more tightly packed especially in the planes that run in diagonal directions. It is along these planes that slip occurs when the forces act on the grains. As might be expected, FCC materials tend to be the most malleable and ductile. They include aluminium, copper, lead, nickel and iron at high temperatures.

The close-packed hexagonal structure is also made up of tightly packed layers of atoms. A unit cell extends through three of the layers. It consists of the atoms at the corners of a hexagonal prism with single atoms at the centres of the hexagonal faces (Figure 6.20). A further three atoms from the central plane are sandwiched between them.

Materials with a CPH structure do not have as many closely packed planes along which slipping can take place when forces act on the grains. They are more easily deformed than BCC materials but tend not to be as ductile or malleable as FCC materials. They include zinc, cadmium, beryllium and magnesium, all of which tend to be rather brittle.

The atoms at the grain boundaries of a crystalline material are the last to solidify and tend to be disordered. As a result, there is a thin amorphous region surrounding the grains, and this has an effect on the mechanical properties of a material. A small-grained structure has a greater proportion of amorphous material and is generally stronger than one with coarse grains. Small-grained structures are, however, more prone to high-temperature creep. Creep is a form of viscous

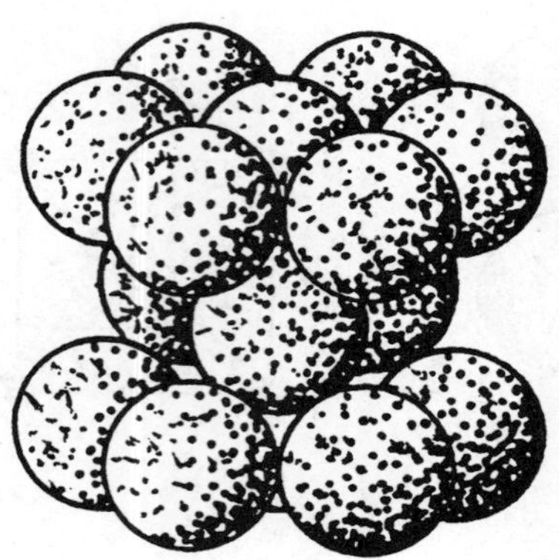

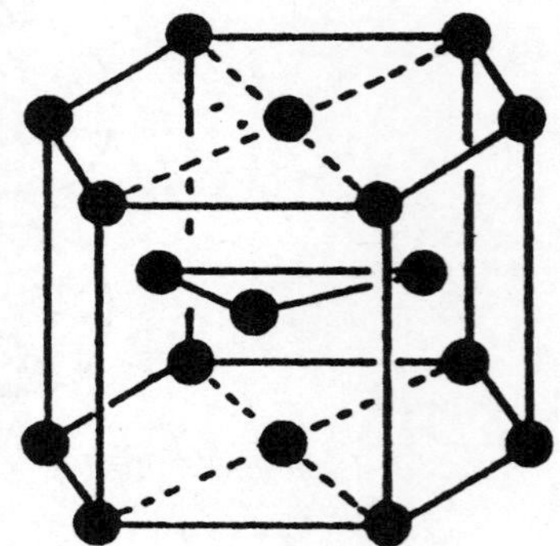

Figure 6.20 The close-packed hexagonal structure

flow, which is thought to occur in the amorphous regions of a loaded material. Some materials, such as lead, will creep when loaded at normal temperatures, whereas others, such as steels, will only creep at elevated temperatures.

It is also found that there are imperfections in the lattice structure of the grains, whose existence allows the planes of atoms to slip over each other more easily. The imperfections are known as 'dislocations' and, when a sufficient load is applied, they are thought to move through the grains rather like ripples. The result is that the layers of atoms move over each other with a rippling motion to produce plastic deformation. This occurs at loads much less than those required to make the planes of atoms slide bodily over each other.

Quite a lot of elements can exist in more than one crystalline form. They are said to be 'allotropic' or 'polymorphic'. It has already been stated that iron can exist in body-centred cubic (BCC) form at normal temperatures and in face-centred cubic (FCC) form at high temperatures. Iron is thus polymorphic. Tin is also polymorphic. It exists in the ductile FCC form at normal temperatures, which is called 'white tin', and collapses into a powdery form, known as 'grey tin', which has a tetragonal crystal structure, at subzero temperatures.

Carbon is another common example of polymorphism. It can exist as graphite and diamond. Graphite has layers of atoms arranged in a hexagonal pattern that can easily slide over each other. This makes a good lubricant. Diamond has atoms arranged in a rigid tetrahedral structure and is the hardest naturally occurring material.

When metals, and also sometimes metals and non-metals, are mixed together in the molten state, the material that results after cooling may be:

- an interstitial alloy
- a substitutional alloy
- an intermetallic compound.

An interstitial alloy results when the atoms of one material are very small compared with the atoms of another (Figure 6.21). The smaller atoms are able to occupy the spaces between the larger atoms. This is what happens with some of the carbon atoms in steel. The effect is to strengthen and toughen the material. The small atoms make it less easy for the layers of the larger atoms to slip over each other. They are said to 'pin' the structure.

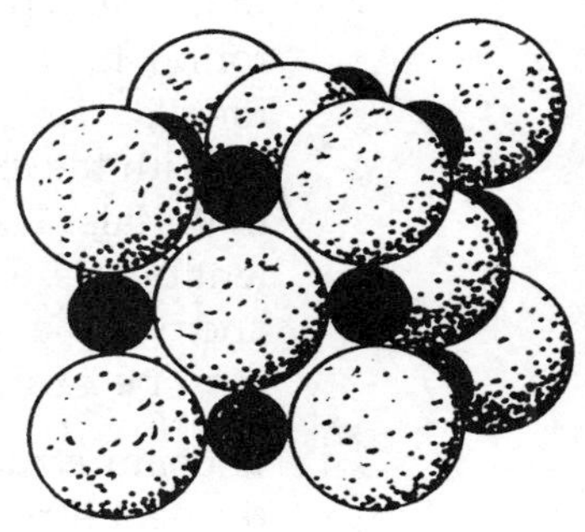

Figure 6.21 An interstitial alloy

A substitutional alloy results when two or more materials with atoms of roughly the same size are mixed together. The atoms of one material replace the atoms of the other in the crystal lattice structure (Figure 6.22). Differences in the sizes of the atoms tend to distort the lattice, making it more difficult for the layers of atoms to slip over each other. As with interstitial alloying, this can produce a material that is stronger and tougher than its main constituent.

Brasses, tin-bronzes, aluminium bronzes and cupronickels are all examples of substitutional alloys.

An intermetallic compound is the result of a chemical reaction occurring between the elements in an alloy. The compound is usually completely different in its colour and properties to each of the parent materials. Intermetallic compounds are often excessively brittle and of little use as an engineering material by themselves. In small quantities however, in a substitutional or interstitial alloy, they can enhance the properties of a material.

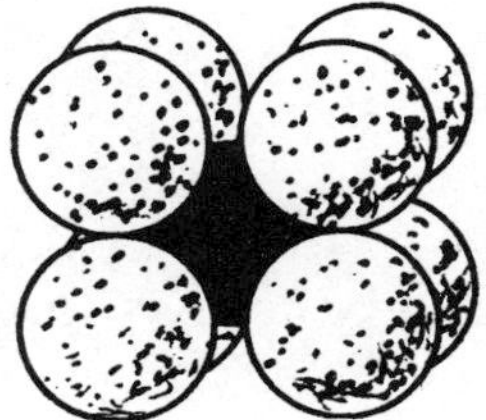

Figure 6.22 A substitutional alloy

The intermetallic compound Fe_3C, known as iron carbide or cementite, is present in steels and cast iron. It makes a contribution to the toughness of the material. The compound $Cu_{31}Sn_8$ is present in some tin-bronze materials. It is found to improve their load-carrying properties when they are used as bearing metals.

With the exception of glass, regular arrangement of atoms also occurs in ceramic materials. Here, however, the atoms are held together by strong ionic and covalent bonds. These produce a structure that is more rigid than in metals and plastics and with a high melting point. The atoms in crystalline and bonded ceramics may be ordered in:

- chain structures
- sheet structures
- three-dimensional structures.

Silica (SiO_2), whose chemical name is silicon(IV) oxide, is a constituent of a great many ceramics. It has a tetrahedral structure as shown in Figure 6.23.

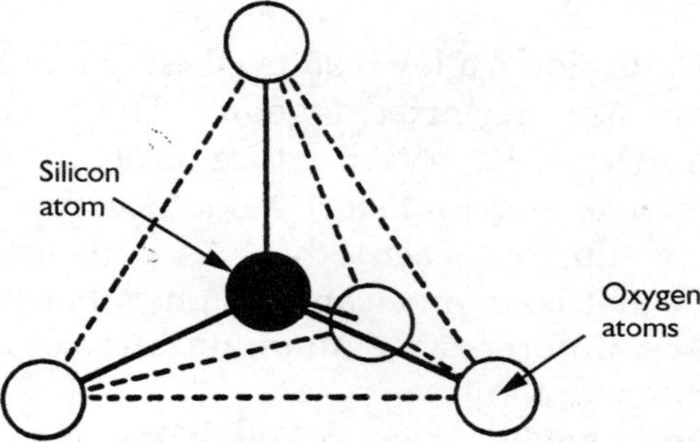

Figure 6.23 Tetrahedral unit of silica

In materials, such as asbestos, these tetrahedra become linked together in double-stranded chain structures as in Figure 6.24.

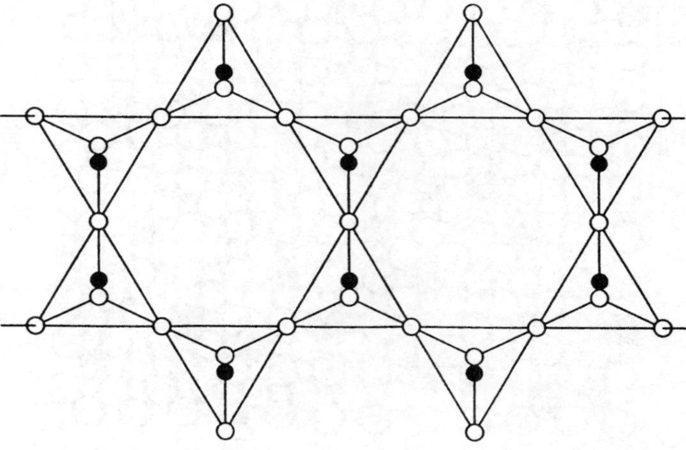

Figure 6.24 Double-stranded chain structure

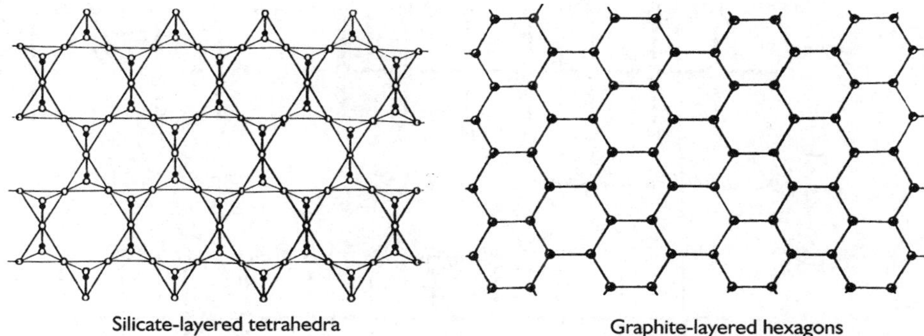

Figure 6.25 Layered structure of silicate and graphite

In other ceramics, such as mica, the tetrahedra combine to form silicates. These have a sheet structure in which the sheets are loosely held together, and this accounts for the flaky texture of the material. The carbon atoms in graphite are also layered in a hexagonal pattern (Figure 6.25). Weak bonds hold the layers together, but these are easily overcome when shearing forces are applied. The layers then slide easily over each other, giving graphite its properties as a lubricant.

The clays used to form bonded ceramics also have a sheet structure, but the presence of aluminium atoms in place of some of the silicon atoms results in strong covalent bonds being formed between the sheets. The material is more rigid but has quite large interstitial spaces. This accounts for the porosity of many unglazed clay products and their ability to absorb large amounts of water.

In quartz and other crystalline forms of silicon(IV) oxide, the silicon and oxygen tetrahedra are arranged in a three-dimensional structure. Ceramics with this kind of structure include diamond in which carbon atoms are arranged in this way. They are extremely hard, since the bonding forces are the same in all directions. There are no planes along which cleavage can take place and, when fracture occurs, it tends to be irregular.

Activity 6.4

What kind of crystal lattice structure would you expect silver and gold to have? Do a little research in the metallurgy or chemistry section of your library to see if you have made the correct choice.

A new allotrope of carbon was discovered a short time ago with its atoms arranged rather like the pattern of the panels in a soccer football. Can you find out the name of this material and whether it has any potential engineering use?

Polymer materials

The atoms in plastic and rubber materials are arranged in long molecular chains known as polymers. Each polymer can contain several thousand atoms, which are bonded together. Their lengths vary but, if fully extended, this might be anything up to a millimetre or possibly more. The chains are largely made up of carbon and hydrogen atoms held together by covalent bonds to which nitrogen, chlorine, fluorine and silicon atoms may also be attached. The raw materials used

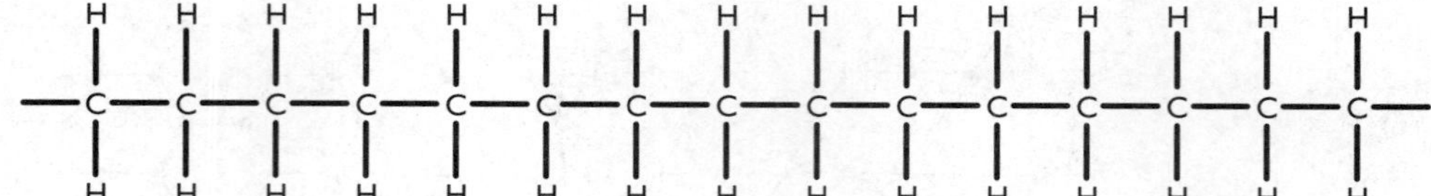

Figure 6.26 Polythene polymer

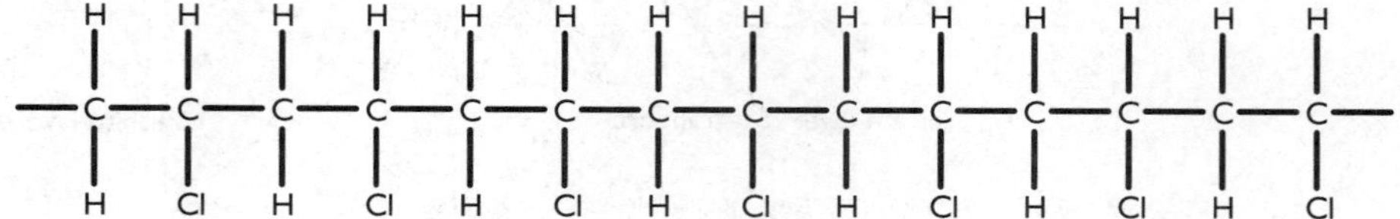

Figure 6.27 PVC polymer

to make plastics and rubbers come from animal and vegetable products, coal and crude oil. These are organic materials, or organic in their origin, as is the case with coal and oil.

Two everyday plastic materials, polyethylene (polythene), and polychlorethene (PVC), have basic polymers as shown in Figures 6.26 and 6.27. Simple polymers, which can only extend by the addition of atoms to their ends, are said to be 'bifunctional'. The more complex polymers become branched and interconnected to form three-dimensional networks and are said to be 'trifunctional'.

Many plastics are found to be part amorphous and part crystalline. In some regions, the polymers are arranged in an ordered geometrical pattern. These are known as 'crystallites'. In other regions, the polymers are randomly entangled in an amorphous mix, rather like spaghetti. The relative proportions of the crystalline and amorphous regions varies with the different materials and the way in which they have been processed (Figure 6.28).

There are forces of attraction set up between adjacent polymers as they form. They are known as van der Waals forces and cause the material to become more and more viscous as polymerisation proceeds. Eventually, when the process is complete, it is the van der Waals forces that hold the polymers, such as those of polythene and PVC, together in a solid mass.

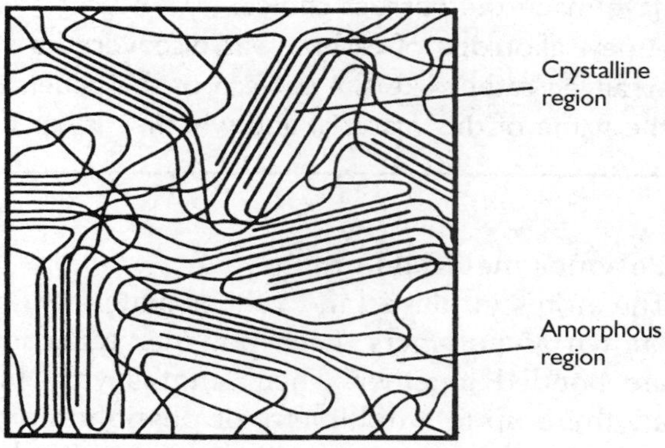

Figure 6.28 Crystalline and amorphous polymers

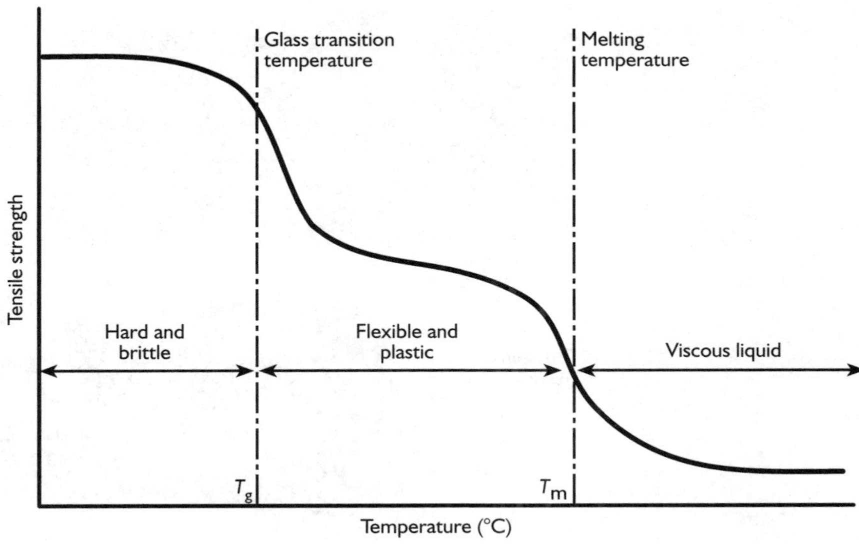

Figure 6.29 Graph of strength against temperature for a thermoplastic

At normal temperatures, some thermoplastics are soft and flexible while others are hard and brittle. If they are heated, hard thermoplastics will eventually reach a temperature where they soften and become flexible. This is called the 'glass transition temperature', which has the symbol T_g. If the heating continues, the material eventually reaches its melting temperature, T_m. Those thermoplastics that are flexible at normal temperatures are already above their glass transition temperature (Figure 6.29). Polythene, pPVC and rubber are typical examples but, if cooled, they eventually become hard and brittle as their glass transition temperatures are reached.

Activity 6.5

Find out and list the glass transition temperatures for polythene, polypropylene, polystyrene, PVC, Perspex, PTFE, nylon, Terylene and natural rubber.

In thermosetting plastics, a chemical change occurs during the moulding process, and strong chemical bonds are formed between the polymers. These are much greater than the van der Waals forces and are known as 'cross-links'. Once formed, the cross-links cannot be broken. As a result, thermosetting plastics tend to be more rigid than thermoplastics and cannot be softened by reheating. The cross-links may consist of a small chain of atoms or a single atom (Figure 6.30). Single-atom cross-links occur in rubbers in which atoms of sulphur form the bonds.

The polymers in rubber are longer and more complex than some others. They are known as 'elastomers' and may contain over 40 000 atoms (Figure 6.31). They are coiled and folded over each other and, when loads are applied to the material, they behave like coiled springs, giving the material its elastic properties. In the raw state, however, natural and synthetic rubbers can be permanently deformed if the loading is excessive. The forces of attraction between the polymers may not be sufficient to prevent them from sliding over each other and, in this condition, the rubber can be said to be both elastomeric and thermoplastic.

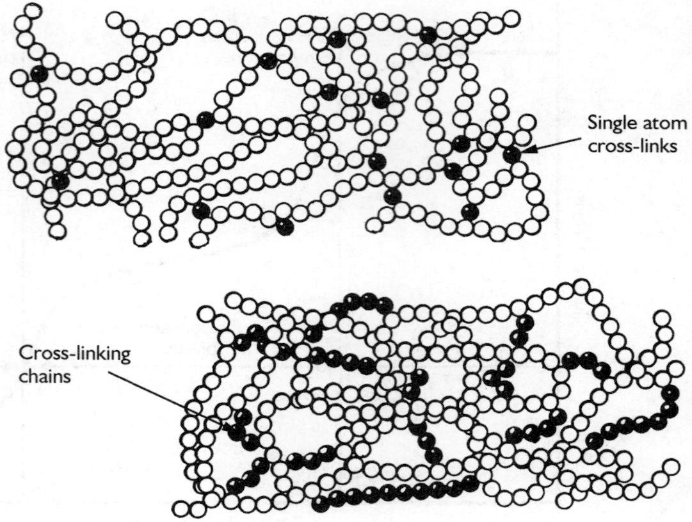

Figure 6.30 Cross-linking in thermosetting plastics and rubbers

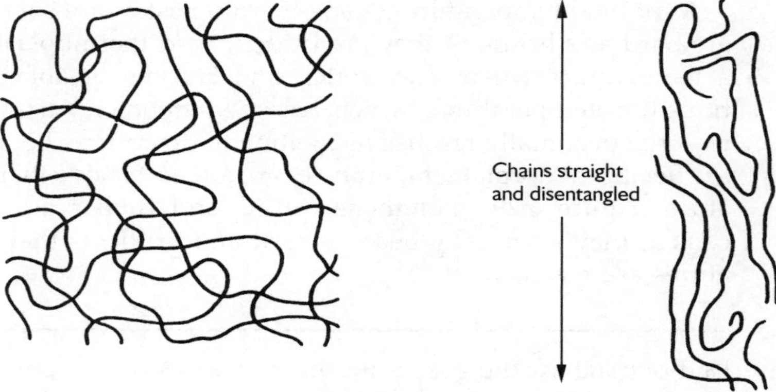

Figure 6.31 Unstretched and stretched elastomers

The addition of sulphur, and in some cases oxygen in the form of metal oxide, produces cross-linking between the elastomers similar to that which occurs in thermosetting plastics. The process is known as 'vulcanising'. It enables the rubber to be elastic while permanently retaining its shape. If more sulphur is used, the rubber starts to lose its elasticity. It can be made hard and brittle if sufficient is added, and its properties are then similar to those of the other thermosetting plastic materials.

Progress check

1 What effect does the rate of cooling of a molten metal have on the final grain size?
2 When the surface of a pure metal, such as copper or aluminium, is polished and etched with an acid to show up the grains, they appear with different shades of colour. Why is this?

3 What is the difference between an interstitial and a substitutional alloy?
4 How do the bonds between the atoms of a ceramic material differ from those that hold the atoms of a metal together?
5 What is meant by 'cross-linking' in thermosetting plastics and what effect does it have on their material properties?
6 What are the main chemical elements in plastic materials and what are the raw material sources from which they are obtained?
7 How do rubber elastomers differ from the polymers in other plastic materials?

Material testing and sources of reference data

When new materials are being developed, they are subjected to exhaustive tests to evaluate their properties. Routine material tests are also carried out by the firms that manufacture and supply the materials used in engineering. This is to make sure that the properties of the materials are within specified limits before they are delivered to the customers. Samples from the delivered batches are again tested by engineering companies to check that the correct material has been supplied, and further tests are sometimes carried out during manufacture.

It follows that material testing is an important quality control activity. British Standards specifications have been drawn up for testing the different properties of materials, and much of the testing carried out by industry follows standard procedures and uses standard equipment. Some of the more common material tests are:

- tensile tests
- hardness tests
- toughness tests
- thermal conductivity tests
- electrical conductivity tests
- chemical stability tests.

Tensile tests

There are different designs of tensile testing machine with different capacities (Figure 6.32). Two of the most common are the hydraulically operated Avery–Denison testing machine with a capacity up to 1 MN, and the portable Hounsfield tensometer with a capacity up to 20 kN. The Avery–Denison type is more widely used by industry for the tensile testing of metals. The Hounsfield type is mechanically operated and excellent for educational purposes. Machines of the Hounsfield type are also used by industry for carrying out tests on plastic materials. Industrial tests are carried out according to the British Standards specifications BSEN 10002 Tensile Testing (Metals) and BS 278 Plastics Testing.

The most common reason for carrying out a tensile test in industry is to measure or check the ultimate tensile strength (UTS) of a material. It involves loading a standard specimen of the material up to the point of fracture and recording the maximum load that it can carry (Figure 6.33).

The UTS is then calculated using the formula:

UTS = maximum load carried by specimen/original cross-sectional area of specimen.

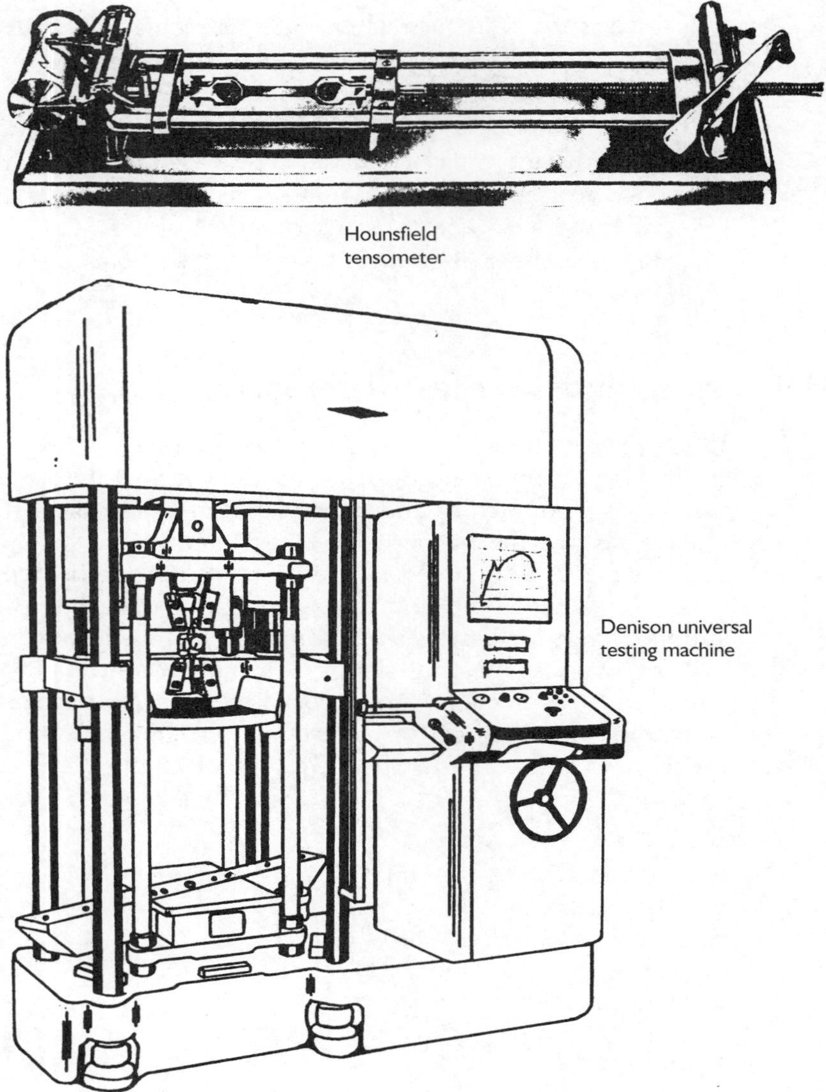

Figure 6.32 Tensile testing machines

It is usually convenient to measure the cross-sectional area in mm^2. The units of UTS will then be N/mm^2. It is useful to remember that a value in N/mm^2 is numerically the same if measured in MN/m^2. It is also found that a value in kN/mm^2 is numerically the same if measured in GN/m^2.

It might be wondered why the original cross-sectional area is used and not the final one, especially since there is quite a considerable reduction in area with some of the more ductile materials. The reason is that such materials form a waist or neck at the point of fracture, and it is often difficult to measure the final diameter accurately enough. This is why the original cross-sectional area is used in the calculation.

The ductility of a material can also be found from a tensile test. It can be measured as the 'percentage elongation' or the 'percentage reduction in area'.

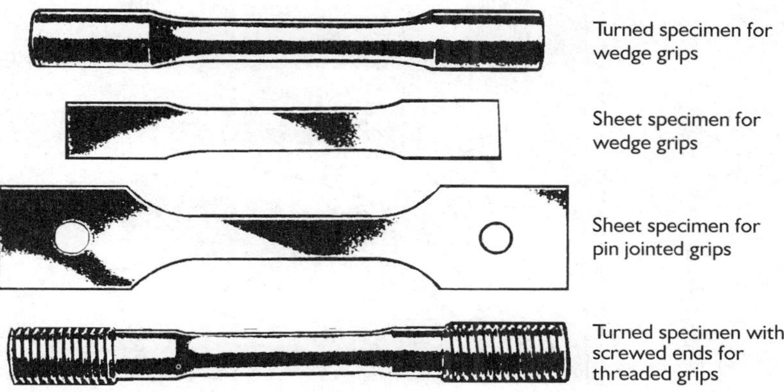

Figure 6.33 Tensile test specimens

With the larger standard test specimens, the percentage elongation is generally measured over a 50-mm gauge length, which is marked by a centre punch on the unstrained specimen. The final distance between the centre punch marks is measured using dividers with the two parts of the fractured specimen held together.

$$\text{Percentage elongation} = \frac{\text{increase in gauge length}}{\text{original gauge length}} \times 100$$

$$\text{Percentage reduction in area} = \frac{\text{reduction in cross-sectional area}}{\text{original cross-sectional area}} \times 100$$

The percentage reduction in area is not so easy to measure accurately for the reasons given above, particularly with the smaller test specimens used with the Hounsfield machines. To overcome this, gauges are available that can be set to the particular specimen size and give a direct reading of ductility. As has been stated previously, the two methods of measuring ductility do not give the same percentage value and the method used should always be quoted.

A tensile test can also be used to determine the modulus of elasticity of a material. Here, the load is applied in steady increments up to the elastic limit of the material, and the extension produced at each increment is recorded. Typical load versus extension graphs obtained from a plotter or plotted manually for different materials are shown in Figure 6.34.

It will be noted that the graph for steel is distinctive. All of the graphs begin with a linear portion in which the material displays elastic behaviour. With steels, however, the graph peaks soon after the end of the elastic stage. This is known as the yield point, and it is often an objective of a tensile test to find the stress at which the yield point occurs.

Steel contains grains in which carbon atoms occupy some of the spaces between the iron atoms. They are known as 'ferrite' grains, and the carbon atoms are said to 'pin' the layers of iron atoms, i.e. they stop the layers from sliding over each other when force is applied. Eventually, however, as the load is increased, they are no longer able to prevent this movement, and the material suddenly yields. Plastic deformation then occurs but soon the material becomes work hardened as the crystal lattice structure becomes disordered.

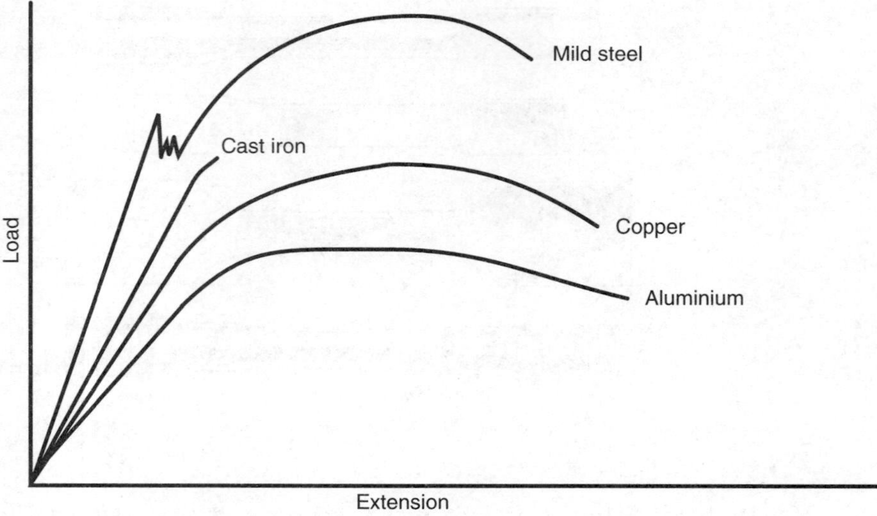

Figure 6.34 Load versus extension graphs

The load must then be increased to produce further deformation and the load versus extension graph follows a curve up to the point of fracture as with the other materials. If a specimen is unloaded during the plastic stage, some permanent deformation will remain.

To determine the modulus of elasticity of a material, it is usual to obtain or manually plot the load versus extension graph and calculate the slope of its elastic region. It has been stated previously that the modulus of elasticity is given by the formula:

$$E = \frac{F}{x} \times \frac{l}{A}.$$

But F/x is the slope of the elastic part of the graph and so:

$$E = \frac{\text{slope of load versus extension graph} \times \text{specimen gauge length}}{\text{Cross-sectional area of specimen}}$$

It is, of course, important to use the same units of length for measuring gauge length, extension and cross-section. The value of E is usually expressed in GN/m^2 for most metals, but it is useful to remember that it is numerically the same in kN/mm^2.

Case study

Extensometers

The extension of a tensile test specimen is measured using a device called an extensometer (Figure 6.35). There are several different kinds of which the Lindley, the Monsanto and the linear variable differential transformer (LVDT) types are perhaps the most common.

The Lindley extensometer measures the extension on a dial test indicator. The Monsanto extensometer incorporates a micrometer, a lamp and electrical contacts. For each increment of extension, the micrometer is adjusted until the contacts close and the lamp lights. A reading of extension can then be recorded. They are both basically mechanical devices that must be securely attached to the

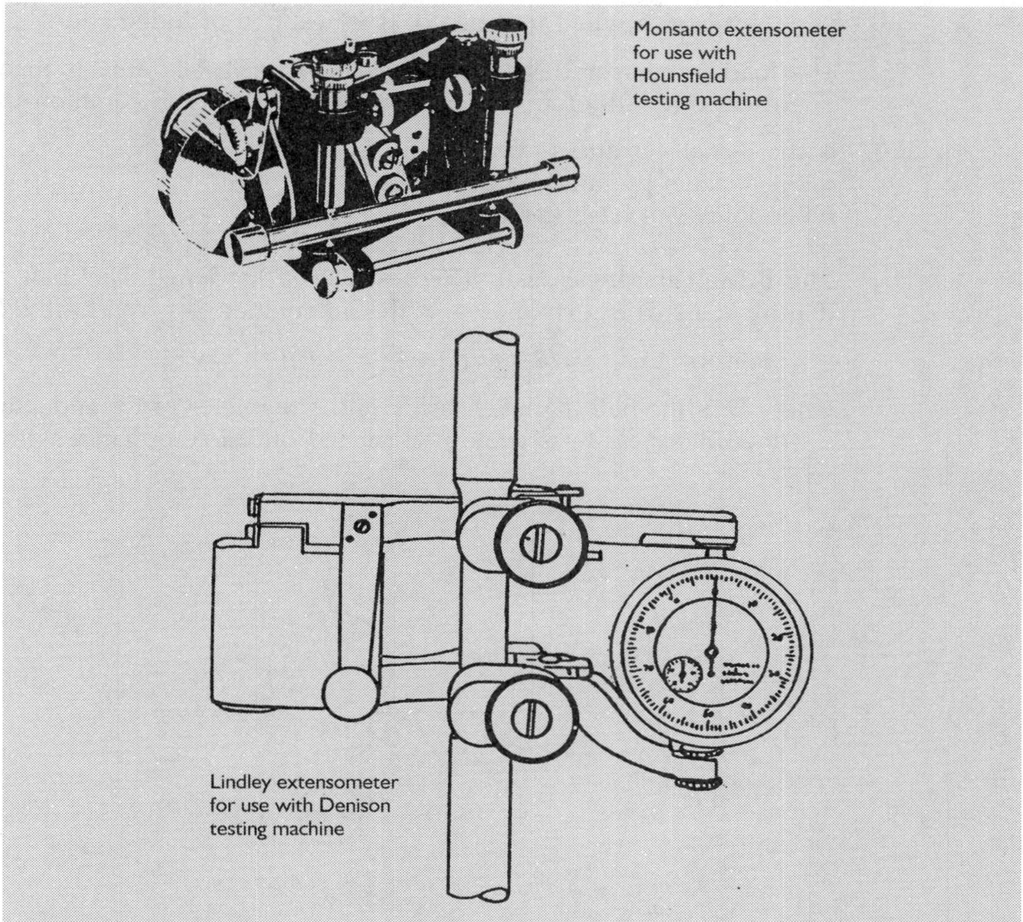

Figure 6.35 Extensometers

specimen and read with care as the load is increased. The Lindley extensometer is more sensitive than the Monsanto type. Sensitivity is the smallest increment of extension that can be recorded reliably. With the Lindley extensometer, this is generally 0.01 mm and, with the Monsanto type, 0.002 mm.

The LVDT extensometer is an electrical device. It produces a voltage signal that is proportional to the extension. This, together with a signal from a load cell that is proportional to the applied load, can be used to produce a graph of load versus extension on an *x–y* plotter. With each of these types of extensometer, it is usual to measure extension over a 50-mm gauge length on the specimen.

Hardness tests

The most common method of measuring the surface hardness of a material is to carry out an indentation test. This involves pressing an indentor in the form of a hardened steel ball or a pointed diamond into the surface of the material. The test is non-destructive, and the dimensions of the resulting indentation are used to determine a hardness number for the material. The formula used to calculate the hardness number is:

H = load applied to indentor/surface area of indentation.

The load is measured in kilograms and the surface area is measured in mm^2. There are three different indentation hardness tests in common use. They are:

- the Brinell hardness test
- the Vickers pyramid hardness test
- the Rockwell hardness test.

The Brinell hardness test This test uses a hardened steel ball as the indentor (Figure 3.36). The surface area of the indentation is given by the formula:

$$\text{Surface area} = \pi/2 \times D\ [D - \sqrt{(D^2 - d^2)}],$$

where D is the ball diameter and d is the diameter of the indentation.

Brinell hardness tests should be carried out in compliance with BS 240.

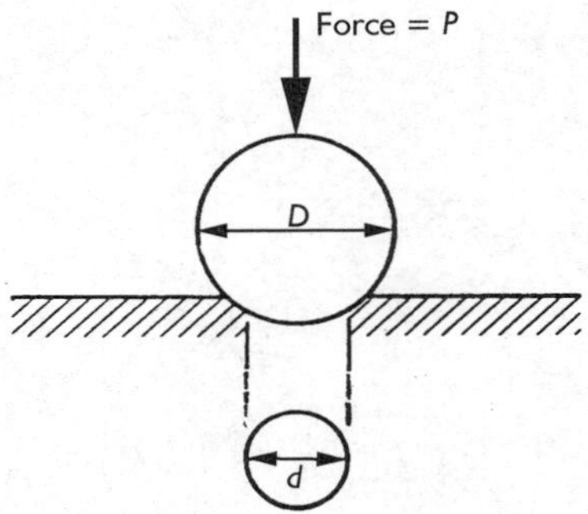

Figure 6.36 Brinell hardness test

The balls are available in 10 mm, 5 mm and 1 mm diameter sizes. The diameter of ball that should be used depends on the thickness of the material to be tested. If too large a diameter is used with a thin material, it is possible that the worktable of the tester will be helping to support the load.

It is also important not to use too light or too heavy a load on the ball. Too light a load gives a shallow indentation, the diameter of which is difficult to measure accurately, whereas too heavy a load might cause the ball to sink into the material up to its full diameter. The Brinell test cannot be used on very hard materials since they cause the ball to deform. It is recommended that the material thickness should be more than eight times the depth of the indentation, and that the indentation should be more than three times the ball diameter from the edge of the material.

For different materials, the recommended ratio of applied load (P) to ball diameter squared (D^2) is as shown in Table 6.10.

Brinell tests can be performed using the Hounsfield tensometer fitted with a hardness-testing attachment. The diameter of the indentation is measured using a calibrated microscope. In industry, however, the test is usually carried out on an Avery–Denison hardness tester in which an image of the indentation is projected onto a screen for measurement against a calibrated scale. The Brinell

Table 6.10 Recommended ratios for the Brinell hardness test

Material	P/D^2
Steel	30
Copper and its alloys	10
Aluminium and its alloys	5
Lead and tin alloys	1

hardness number (BHN), can then be obtained from tables or calculated using the formula:

$$\text{BHN} = \frac{P}{\frac{\pi}{2}D(D - \sqrt{D_2 - d_2})}$$

Vickers pyramid hardness test This is generally carried out on the same Avery–Denison machine as the Brinell test. Instead of a hardened steel ball, a square-based diamond pyramid indentor is used, whose sloping faces have an included angle of 136°. The load is automatically applied for a period of 15 seconds, and the surface area of the indentation is given by the formula:

$$\text{Surface area} = \frac{d^2}{2\sin\frac{136^\circ}{2}}$$

where d is the mean length of the diagonals of the indentation measured in millimetres.

The indentation is again projected onto a screen and the mean length of its diagonals is measured (Figure 6.37).

The Vickers Pyramid hardness number (VPN) can be obtained from tables or calculated using the formula:

$$\text{VPN} = \frac{2P\sin\frac{136^\circ}{2}}{d_2}$$

where P is the applied load measured in kilograms.

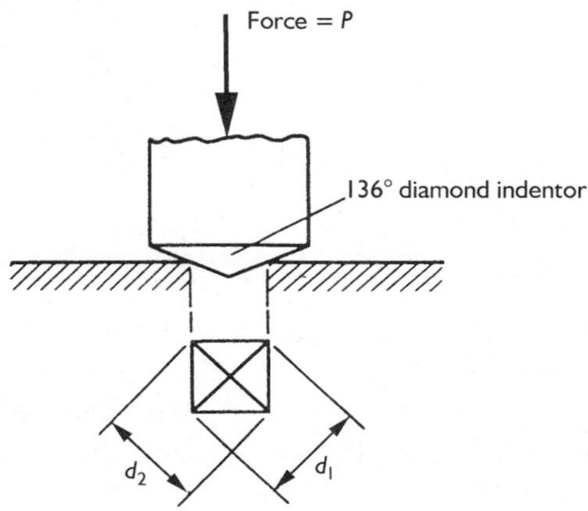

Figure 6.37 Vickers pyramid hardness test

Vickers pyramid hardness tests should be carried out in compliance with BS 427: Vickers hardness test. Unlike the Brinell test, where too great a load can press the ball into the material to its full diameter, the accuracy of the Vickers hardness number does not vary with the depth of the impression. As a result, there is no need to worry about observing the P/D^2 ratio.

Care should still be taken, however, when measuring the hardness of thin materials, and it is specified that the material thickness should be at least 1.5 times the diagonal length of the indentation. Because the indentation is very small, it is also important that the material surface is smooth and free from dirt, rust or oxide flakes, since this can affect the accuracy of the result. Before carrying out a Vickers hardness test, the material surface should be cleaned using fine grit emery paper.

The Rockwell hardness test This has its origins in America and uses both ball and conical diamond indentors (Figure 6.38). The Rockwell tester is particularly easy to operate and is widely used for routine hardness testing. No calculation or measurement of the indentation is required since the hardness number is displayed directly on a dial. The tests should be carried out in compliance with BS 4175: Rockwell hardness test.

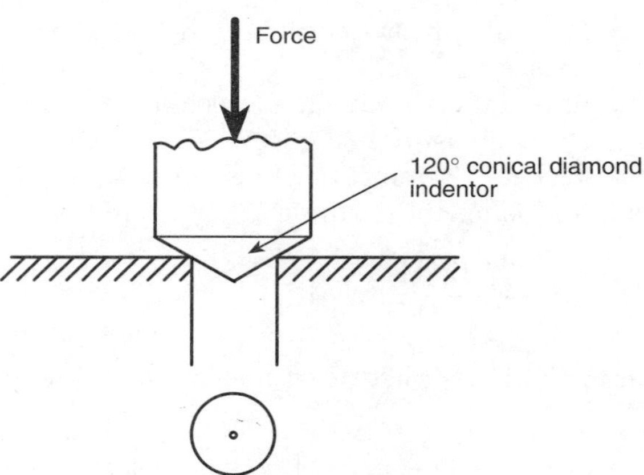

Figure 6.38 Rockwell hardness test

There are three scales on the dial lettered A, B and C. Scale A is used with a 60-kg load and the 120° conical diamond indentor. Scale B is used with a 100-kg load and a 1/16″ diameter-hardened steel ball indentor. This scale is the most useful for measuring the hardness of softer materials such as unhardened steels, copper, brasses, bronzes, etc. Scale C is used with a 150-kg load and the 120° conical diamond indentor. This scale is the most useful for hardened steels and other hard alloys.

Before carrying out a Rockwell hardness test, the surface of the material should be cleaned with emery to remove dirt, rust or oxide scale. The material is then placed on the worktable of the tester beneath the indentor. First, a light load is applied to take up any backlash in the operating mechanism and the scale is set to zero. The full load is then applied and released after a specified time, leaving the light load in operation. The machine senses the depth of penetration and displays the Rockwell hardness number on the appropriate scale.

It is important to realise that the different tests will give a different hardness number for the same material. For this reason, the type of test should always be stated together with the hardness number when specifying surface hardness, e.g. VPN 250 for Vickers hardness.

Toughness tests

Toughness is a measure of the ability of a material to withstand mechanical shock. It is usually measured by carrying out an impact test on a specimen of the material. Impact tests should be carried out in compliance with BS 131: Impact tests. The two main types of impact test are:

- the Izod impact test
- the Charpy impact test.

The two tests are very similar and can be carried out on the same testing machine. This comprises a heavy pendulum that swings from a preset position to strike a notched specimen of the material (Figure 6.39). The difference between the two tests lies in the preset height of the pendulum, the dimensions of the specimen and the way in which it is mounted in the machine.

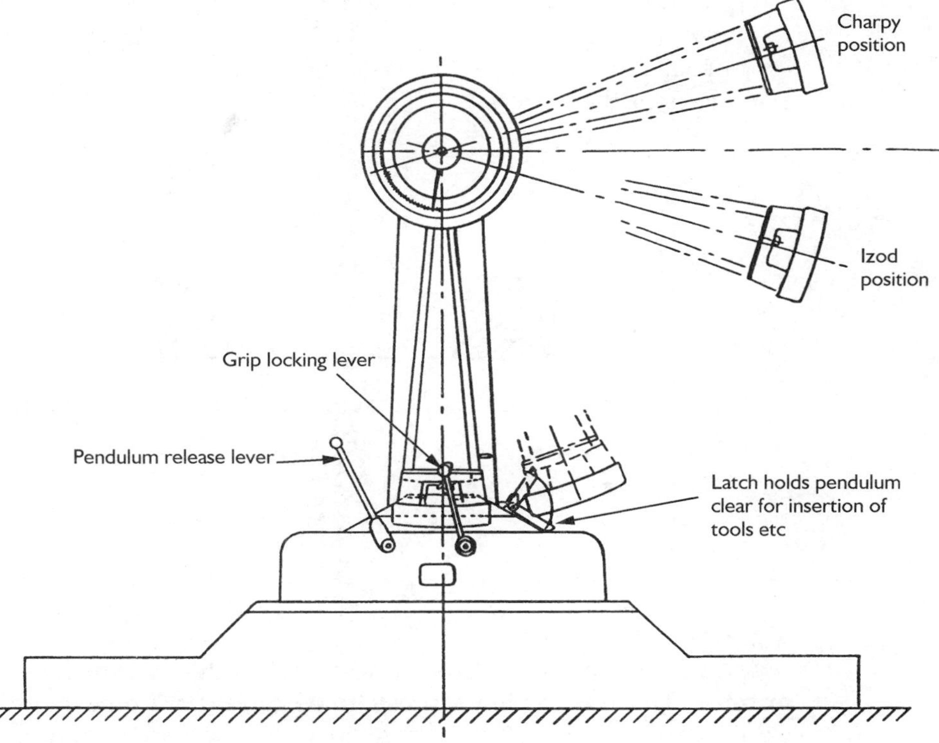

Figure 6.39 Impact testing machine

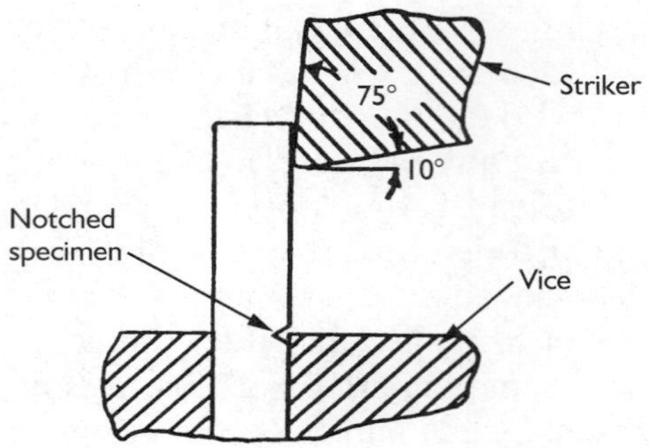

Figure 6.40 Izod specimen

The Izod test With the Izod impact test, the specimen is as shown in Figure 6.40. It is held vertically in a vice with the notch facing the pendulum striker.

In its raised position, the pendulum has a potential energy of 170 joules. This is changed to kinetic energy when the pendulum is released and picks up speed. Depending on the toughness of the specimen, some of this energy is absorbed by the material on impact. As a result, the pendulum does not swing through to its original height, and a pointer indicates on a scale the energy absorbed by the specimen. This is recorded as a measure of the toughness of the material.

The Charpy test The Charpy impact test uses a notched specimen that is supported horizontally as a simply supported beam, as shown in Figure 6.41.

The pendulum is raised to a higher initial position and its mass can be altered to give a striking energy of 150 or 300 J. As with the Izod test, the energy absorbed on impact is recorded on the scale of the machine and gives a measure of the toughness of the material.

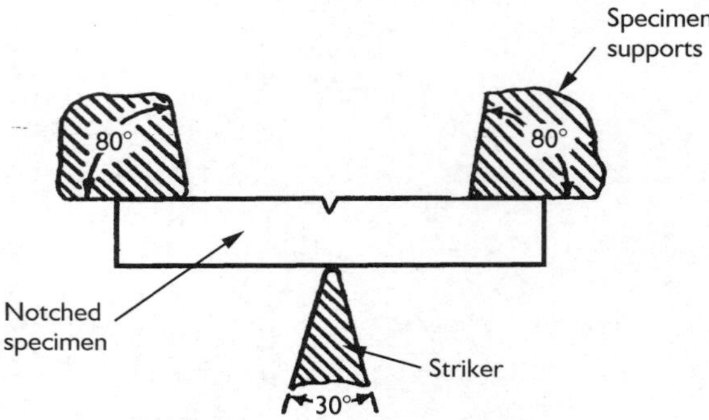

Figure 6.41 Charpy specimen (plan view)

An advantage of the Charpy test is that the specimen is supported but not gripped. This enables impact tests to be carried out quickly on very hot and very cold materials with little change in the temperature of the specimen. Further details of Charpy tests can be found in BSEN 10045: Charpy impact tests.

Activity 6.6

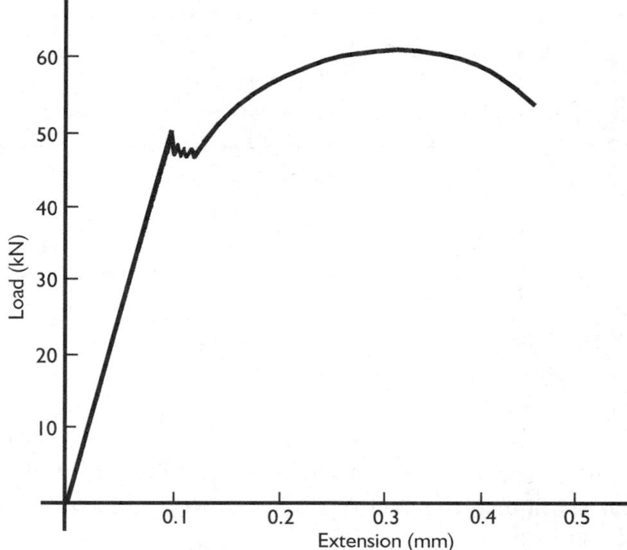

Figure 6.42 Load versus extension graph

The load versus extension graph in Figure 6.42 is for a steel specimen of original diameter 13.8 mm and gauge length 50 mm. After fracture, the gauge length was seen to have increased to 65.2 mm and the diameter at the point of fracture was reduced to 10.15 mm. Determine:

a the UTS of the material
b the yield point stress
c the modulus of elasticity
d the percentage elongation
e the percentage reduction in area.

Note

The mechanical properties of materials can be greatly affected by small variations in the amounts of their constituents and their processing. For this reason, the mechanical tests described above need to be carried out frequently to ensure that the material used in engineered products is of the specified quality. With the exception of semiconductor materials, the processing of which is highly specialised, the electrical and thermal properties of materials and their chemical stability are not so variable and do not generally require such frequent testing. The following laboratory tests for thermal conductivity, electrical conductivity and chemical stability are more likely to be carried out by material manufacturers rather than engineering firms which use the materials.

Thermal conductivity tests

The traditional laboratory method of measuring the thermal conductivity of a good conductor, such as a metal, is by the use of Searle's apparatus. Details of this can be found in textbooks on advanced level physics, but it does not lend itself directly to industrial use. Practical test equipment uses a similar principle

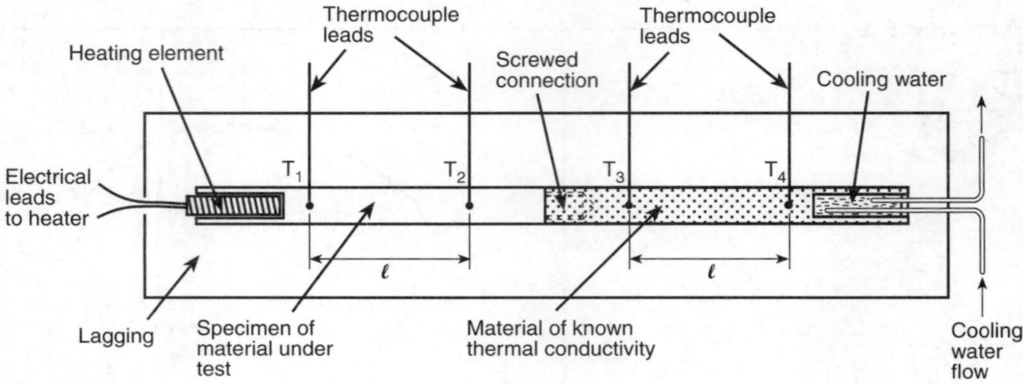

Figure 6.43 Thermal conductivity test for a good conductor

and one version is shown in Figure 6.43. It measures the thermal conductivity k_1 of a material by comparing it with a material whose thermal conductivity value k_2 is known.

Sample bars of the two materials are screwed together end-on and placed in a heavily lagged container. The joined specimens are heated at one end by an electrical heater while the other end is maintained at a lower temperature by means of a coil through which cooling water is flowing. The temperatures at points equal distances apart along the two materials are recorded using thermocouples bonded to their surfaces. The steady flow of heat energy per second, Q, is given by the formula:

$$Q = [k_1 A\ (T_1 - T_2)]/l = [k_2 A\ (T_3 - T_4)]/l$$

i.e.

$$k_1\ (T_1 - T_2) = k_2\ (T_3 - T_4)$$

$$k_1 = [k_2\ (T_3 - T_4)]/(T_1 - T_2).$$

The traditional laboratory method of finding the thermal conductivity of poor conducting materials, such as plastics and ceramics, is by the use of Lees' disc apparatus. Details of this can also be found in advanced level physics textbooks but, like Searle's apparatus, it does not lend itself directly to industrial use. Practical test equipment that operates on the same principle has, however, been developed from it, one version of which is shown in Figure 6.44.

Two identical specimens of the material under test are placed on each side of an electrically heated plate and sandwiched between two water-cooled plates. The whole assembly is surrounded by lagging to prevent heat loss from the edges of the plates. Thermocouples measure the temperatures, T_1 and T_2, on the hot and cool sides of the specimens, and the electrical power, Q, supplied to the heater is recorded. Half of the energy supplied per second passes through each specimen whose area is A and thickness l. Its value is given by:

$$Q/2 = [kA\ (T_1 - T_2)]/l$$

from which:

$$k = Ql/[2A\ (T_1 - T_2)].$$

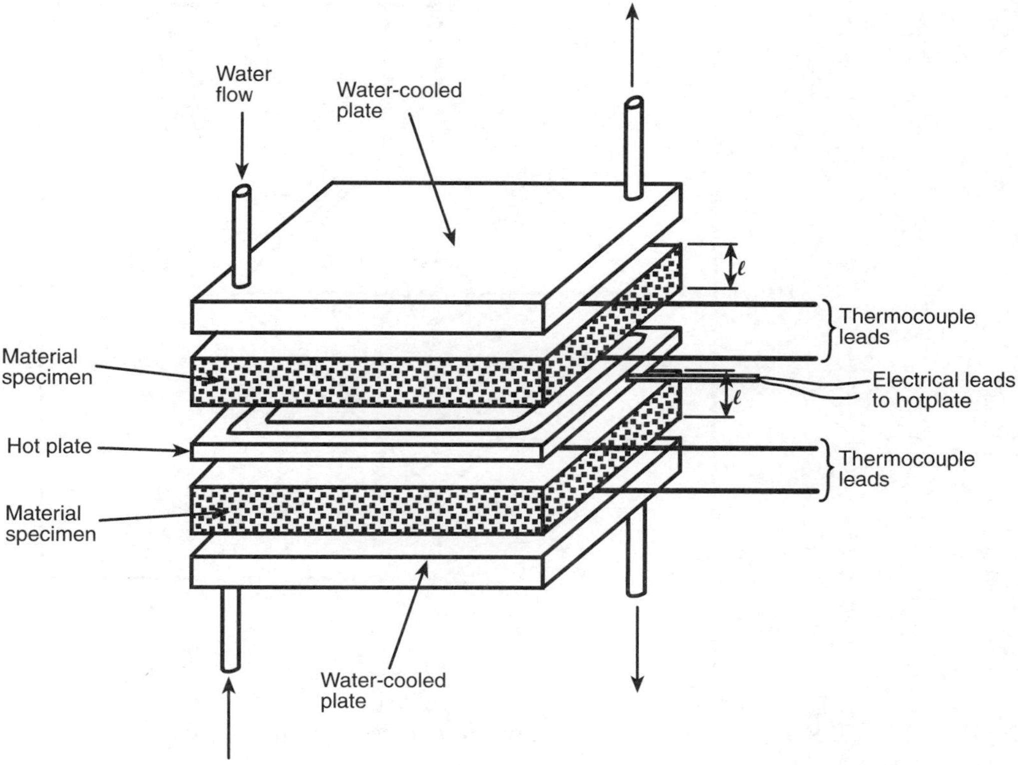

Figure 6.44 Thermal conductivity test for a poor conductor

Activity 6.7

You are supplied with rods of glass, steel, aluminium and copper, which are all of the same size and coated in paraffin wax. Devise a simple laboratory test that can be used to compare their thermal conductivity values.

Electrical conductivity tests

As has been stated, the ability of a material to conduct electric current is usually measured as its resistivity given in ohm metres, Ωm, i.e. the resistance in ohms of a cube of the material with sides of length 1 m. To determine this, it is necessary to measure accurately the resistance of a sample of known length and cross-sectional area. For good conductors, this can be done using a Wheatstone bridge circuit in which the sample forms one of the arms (Figure 6.45).

The most common practical forms of the Wheatstone bridge consist of a resistance box containing the resistances R_1, R_2 and R_3 with internal connections. Their values can be set using dials, and terminals are provided to which the material sample, battery and centre-zero galvanometer are connected. It is usual to set R_1 and R_2 to the same value. The battery and galvanometer switches are then closed, and the value of R_3 is adjusted until the galvanometer reads zero. In this position, the bridge circuit is said to be 'balanced' and the resistance, R_4, of the sample is given by the formula:

$$R_4 = (R_{2.}/R_1)\ R_3.$$

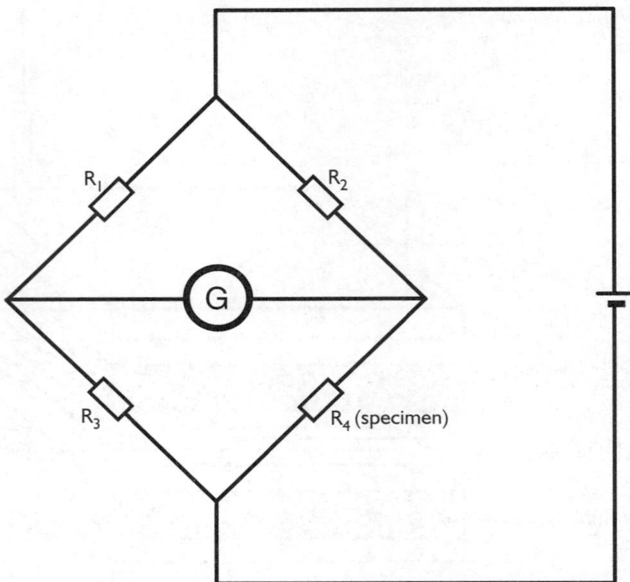

Figure 6.45 Wheatstone bridge circuit

If R_1 and R_2 have been set to the same value, this becomes:

$$R_4 = R_3.$$

The resistivity of the material can then be found using the formula:

$$\rho = R_3\, A/l$$

where A is the cross-sectional area of the material and l is its length.

The resistance of insulating materials is most commonly checked using a 'megger' tester. This is a combination of an ohm-meter and a hand-driven generator. The specifications for the insulation resistance in electrical installations are laid down in IEE regulations. The regulations also specify the methods of connecting the leads from the megger to measure the resistance of the insulation between conductors, between conductors and the metal parts of equipment and between conductors and earth. When properly connected, the generator handle is cranked and the insulation resistance is shown on the meter scale.

Chemical stability tests

The number of material tests that have been devised to determine corrosion rates and likely degradation resulting from solvents and radiation are many and varied. Some are designed to test the stability of the material itself. Others seek to test the effectiveness of the various kinds of protective coating that are applied to materials.

In all cases, the objective is to simulate severe service conditions. This might involve placing the material in its proposed service environment and observing its condition periodically over a long period of time. Generally, however, data is required urgently so that the characteristics of a new material, or an existing material in a new environment, can be specified with confidence. To do this, accelerated tests are devised, which subject the material to extreme conditions. These might include rapid heating and cooling, exposure to steam, exposure to concentrated corrosive liquids, gases and solvents, exposure to intense ultraviolet rays or a combination of these conditions.

Activity 6.8

Take six identical mild steel nails and clean them free of oil or grease using methylated spirits. Place two in an uncorked dry test tube and another two in an uncorked test tube containing a little water. Place the final two nails in a test tube that also contains water and boil the water for a few minutes until the steam has displaced the air in the test tube and then securely cork it. Let the test tubes stand for a week and then comment on the different degrees of corrosion that have occurred.

Progress check

1 What causes steel to yield suddenly at the end of the elastic range?
2 What are the indentors used in:
  a the Brinell hardness test
  b the Vickers hardness test
  c the Rockwell hardness test?
3 What are the limitations of the Brinell hardness test?
4 What advantages does the Rockwell hardness test have?
5 In what different ways are the test specimens supported for the Izod and Charpy impact tests?
6 Why is the Charpy test more suitable than the Izod for measuring the toughness of heated material specimens?

The relationship between material structures and properties

A number of references have been made in the previous sections of this chapter as to how the structure of an engineering material affects its properties. These relationships can now be summarised. The main structural groups are:

- crystalline
- polymer
- amorphous.

Crystalline materials

These include metals and ceramics. Their properties are affected by their:

- atomic structure
- crystal lattice structure
- grain size and shape.

Atomic structure Metals have a small number of electrons in the outer shell of their atoms. These are easily detached to become randomly circulating free electrons. It is almost as if they form a cloud around the metal atoms, although the true picture is probably much more complicated. It is this cloud of free electrons that accounts for some of the peculiar properties of metals.

The free electrons are thought to be responsible for the metallic bonds that bind the metal atoms together. When an electrical potential difference is applied between the ends of a metal conductor, it is thought that the free electrons drift slowly towards the positive potential, so forming an electric current. The presence of free electrons thus accounts for the fact that metals are generally good conductors of electricity.

When heat energy is supplied to one end of a metal rod, it is thought to have two effects. The atoms of the metal are supposed to be in a constant state of vibration about their fixed positions, and this increases as heat energy is received. The supply of heat energy also causes the random movement of the free electrons in that area to be speeded up. This increased activity spreads throughout the metal as more heat energy is received. It is thought that the free electrons carry the greater part of the heat energy, which accounts for the fact that metals are good conductors of heat.

The atoms of crystalline ceramics do not have free electrons and, as a result, they are generally poor conductors of both heat energy and electricity. Semiconductor materials, which are also crystalline, lie between these extremes. At low temperatures, they have few free electrons and are poor conductors. As the temperature rises, however, more electrons break free of the atoms and their conductivity increases greatly.

Activity 6.9

Some electric cookers are produced with ceramic hobs. Find out exactly what the material is, what kind of a structure it has and the properties that make it suitable for this use.

Crystal lattice structure Within a crystal lattice structure, the atoms are held together by strong metallic, covalent or ionic bonds, depending on the type of material. As has been stated, engineering metals can have a body-centred cubic (BCC), a face-centred cubic (FCC) or a close-packed hexagonal (CPH) lattice structure. These affect the mechanical properties of a material. The face-centred cubic metals are generally the most malleable and ductile, whereas the body-centred cubic metals tend to be the most hard and tough. The close-packed hexagonal metals come between the two extremes.

The mechanical properties of metals can be modified by alloying. As has been described, the mixing of molten metals and other materials can produce an interstitial alloy, a substitutional alloy or an intermetallic compound. With interstitial alloys, the smaller atoms of one material occupy the spaces between the larger atoms of a parent metal. Substitutional alloys have atoms of roughly the same size, and some atoms of the parent metal are displaced in the crystal lattice by atoms of that which is added.

In both cases, the alloy tends to be stronger, tougher and harder to deform than the parent metal. This is because alloying tends to distort and 'pin' the crystal lattice, making it more difficult for the planes of atoms to slip over each other in plastic deformation.

Intermetallic compounds are formed when a chemical reaction occurs between two materials. These tend to be hard and brittle, an example being the iron carbide (Fe_3C), which is present in steels. The occurrence of grains that contain an intermetallic compound can also make a material more difficult to deform and increase its strength, hardness and toughness.

The crystalline and bonded ceramic materials used in engineering have lattice structures that are different from metals. The atoms are held together by strong covalent or ionic bonds and, as has been stated, there are no free electrons. As a result, ceramics tend to be poor conductors of both heat and electricity. Ceramics have a wider variety of crystal lattice structures than do metals. Some display a cubic structure, but the crystalline and bonded ceramics used as engineering materials, particularly those containing silicates, tend to have a tetrahedral structure. As has been stated, this is held together by strong covalent or ionic bonds with no cleavage planes, i.e. planes where the atoms are more tightly packed than others, along which plastic deformation can occur. As a result, these ceramics tend to be hard and brittle.

It will be remembered that some materials can exist in more than one crystalline form, i.e. they are polymorphic or allotropic. Carbon and iron are common examples. Carbon can exist as diamond, which has a tetrahedral structure and is very hard. It can also exist as graphite, which has a layered hexagonal lattice structure. This is very soft and flaky and makes graphite a good lubricant. Iron can exist in body-centred cubic form at normal temperatures, which is relatively hard and tough. It can also exist in face-centred cubic form at high temperatures. This is relatively soft and malleable and enables the material to be hot worked to shape by forging or hot rolling, etc.

Grain size and shape In addition to the atomic and the crystal lattice structures, the mechanical properties of crystalline materials also depend on the size and shape of the crystals or grains. The grain size is largely dependent on the rate at which the material is cooled when solidifying. The slower the rate of cooling, the larger are the grains. Large or coarse grains tend to reduce the tensile strength, hardness and toughness of a material, whereas a small or fine grain structure tends to improve these properties.

Cold working processes, such as cold rolling, pressing or drawing, distort the shape of the grains in a metal. They become elongated in the direction of working, with a distorted internal crystal lattice structure. This increases the hardness and toughness of the material, and the effect is called work hardening. Sometimes this is desirable but, in other cases, it may need to be removed by heat treatment. Heat treatment processes that allow the size and shape of the grains to be adjusted and the material properties to be modified have been described in Chapter 3.

Case study

Creep failure

The presence of grain boundaries is undesirable in some metal components, particularly those that must operate under load at high temperatures. It is under these conditions that creep failure can occur. Creep is thought to be a kind of viscous flow that takes place in the amorphous regions at the grain boundaries. It was a problem that dogged the early development of gas turbines in which the sustained high temperatures and the forces acting on the turbine blades frequently led to creep failure (Figure 6.46).

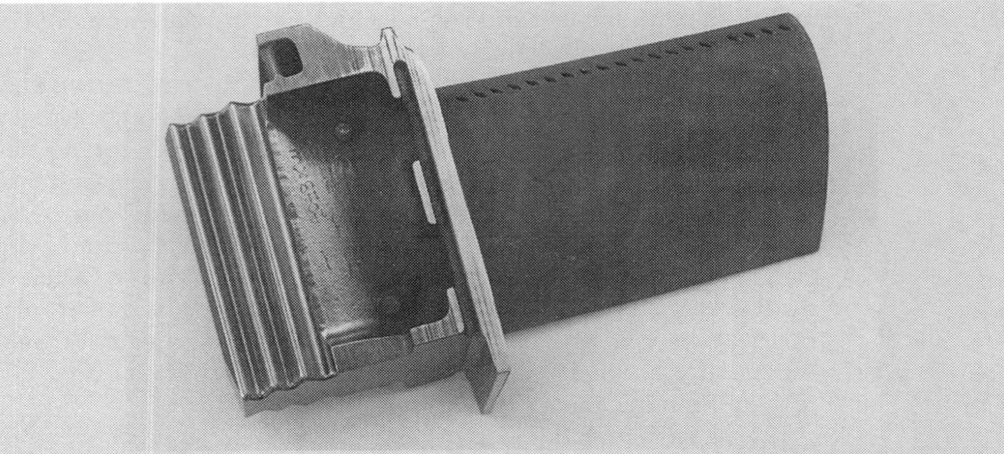

Figure 6.46 A single crystal turbine blade

In recent years, techniques have been developed that enable molten metal to solidify as a large single crystal. They involve the examination of single grains using X-rays and selecting one that has no defects. This 'seed' grain is then lowered until one of its faces touches the surface of the molten metal. It is then slowly raised and rotated so that the metal solidifies on to it as a single crystal. Components, such as the turbine blade in Figure 6.46, can now be made using single-crystal technology, making it possible to operate gas turbines at higher temperatures. This, in turn, has contributed to improvements in power output and fuel economy.

Polymer materials

Polymers are the long-chain molecules that form the structure of plastics and rubbers. They basically consist of carbon atoms linked to each other by strong covalent bonds, and they may be several thousand atoms in length. The properties of the different polymers depend on:

- polymer complexity
- polymer cross-links.

Polymer complexity Hydrogen atoms and sometimes also oxygen, chlorine and fluorine atoms are attached to the chains of carbon atoms, and it is these that give the various plastics and rubbers their different properties. Plastics tend to be part crystalline, where the polymers have arranged themselves in a regular pattern, and part amorphous, where the polymers are randomly entangled.

The mechanical properties of a plastic are related to the degree of crystallinity, the length of the polymers and the degree of branching into side chains that is present. The more crystalline a polymer material, the greater is its tensile strength. The longer the polymers and the more they are branched, the more they will be intertwined and the greater are the forces of attraction (van der Waals forces) between them. Tensile strength thus also tends to increase with chain length and with the amount of branching into side chains that takes place.

With thermoplastics, the van der Waals forces decrease with temperature increase and the material becomes less rigid. The polymers slide over each other more easily when force is applied, and the material is easily moulded into shape.

At normal temperatures, the thermoplastic materials used in engineering are generally tough and strong, but not so strong as metals. They are, however, lighter and generally less expensive.

Some thermoplastics, such as polystyrene and Perspex, are more rigid than others and some, such as nylon and Terylene, have a relatively high tensile strength. They all display some elasticity, although this decreases with temperature increase when the material will start to creep under load. All thermoplastics are poor conductors of heat and electricity owing to the absence of free electrons in their structure.

Polymer cross-links The polymers in thermosetting plastics consist of the same long chains of carbon atoms to which other atoms are attached. The difference is that, during the moulding or curing processes, strong covalent bonds called 'cross-links' are formed between the polymers. These are many times stronger than the van der Waals forces and hold the material rigid irrespective of its temperature.

The structure of thermosetting plastics generally makes them harder than thermoplastics. Some, such as Bakelite and formica, tend to be brittle, while others, such as thermosetting polyurethanes, are quite strong and tough. Thermosetting plastics do not have as high a tensile strength as some thermoplastics unless reinforced with glass or carbon fibres. Like thermoplastics, they are poor conductors of heat and electricity and for the same reasons.

The polymers in natural and synthetic rubbers are extremely long in comparison with plastics. They can contain 40 000–50 000 carbon atoms and are folded and coiled around each other. When loads are applied, the polymers act like coiled springs, which gives the material its elastic characteristics. Unless the polymers in rubbers are cross-linked, creep or 'drift' will occur under load, resulting in permanent distortion. The traditional name for the cross-linking process in rubbers is 'vulcanising'.

Cross-linking occurs during the moulding process if sulphur or certain other elements have been added. Too much cross-linking makes rubber hard and less flexible. In a particular application, it is important to have sufficient cross-links to prevent distortion, while retaining the required degree of flexibility and elasticity.

Amorphous materials

In materials that are truly amorphous, the atoms or molecules are positioned randomly with no recognisable geometric structure. Glass is the most common amorphous engineering material, although some metals can be given an amorphous structure by cooling them very rapidly during solidification so that crystals have no time to form. Even within crystalline materials, some amorphous regions may be present at the grain boundaries. These are the last places to solidify as the grains come into contact with each other, and the result is a thin region of disordered atoms.

Amorphous materials do not have a definite melting point but become progressively softer with temperature increase. They also have a tendency to creep under load. Glass will creep when loaded for long periods, and metals have a tendency to creep more when the grains are small than when they are large. This is because of the presence of more amorphous material at the grain boundaries.

Progress check

1 What aspect of the atomic structure of metals accounts for them being relatively good conductors of electricity and heat energy?
2 Give two examples of allotropic materials and describe the structure and properties of their different allotropes.
3 Explain how grain size and shape affects the mechanical properties of metals.
4 What are the main constituent elements of polymers?
5 Why do thermoplastic materials become less rigid as their temperature rises?
6 What is 'vulcanising' and what effect does it have on the properties of rubber?
7 Where would you expect to find amorphous regions in a crystalline material?
8 How do amorphous materials behave as their temperature rises?

Assignment 6

This assignment provides evidence for:
Element 4.1: Characterise materials in terms of their properties
and the following key skills:
Communication : 3.1, 3.2, 3.3, 3.4
Information Technology: 3.1, 3.2, 3.3

The modern motor car is a complex engineering system whose purpose is to convert the energy stored in the fuel into kinetic energy as efficiently, as safely and as comfortably as possible. A complex system can usually be broken down into a number of linked and interdependent subsystems, and so it is with the motor car.

The prime energy conversion system is the engine, and this itself can be further broken down into the fuel supply system, the cooling system, the lubrication system and the exhaust system. The transmission system, containing the gearbox and drive shafts, delivers the energy to the road wheels. The suspension and steering systems seek to give comfort and safety. The electrical system generates and supplies electrical power to the various parts of the car. This can be further broken down into the starter, the lighting, the instrumentation, the engine management and the in-car entertainment systems. All of these are contained in the body shell with its safety cage for the driver and passengers.

Each new model brings improvements, as designers incorporate new ideas and new materials. The materials used for some parts of motor vehicles have not changed for many years. For other parts, new and improved materials have been specially developed, and new materials developed initially for aerospace have been adopted. Research is continuing with a view to making the car more friendly to the environment, perhaps using hydrogen as the fuel, or using fuel cells to give electric power. Whatever problems arise, you may be sure that material scientists will have a major part to play in their solution.

Your tasks

Carry out the following tasks and submit a report that looks at some of the different materials used in the manufacture of motor cars.

- Identify and list a part or component of a car that is made from or contains:
  - a a ferrous metal
  - b a non-ferrous metal
  - c a thermoplastic material
  - d a thermosetting plastic material
  - e an amorphous ceramic
  - f a bonded clay ceramic
  - g two different semiconductor materials
  - h a thermosetting plastic composite material
  - i an elastomer composite material.
- Identify each material and describe its structure and the properties that make it suitable for its purpose.
- Obtain the relevant property values for each material from reference sources or from testing.
- Draw up a table that shows the structural group to which each material belongs and in which each material is ranked against the range of properties identified above.

Chapter 7 Relating material characteristics to processing methods

This chapter covers:
Element 4.2: Relate materials' characteristics to processing methods.

. . . and is divided into the following sections:
- The implications of different property values
- Appropriate processing methods
- The effects of processing on materials.

The properties of materials influence the uses to which they can be put and the ways in which they can be processed. Metals have perhaps the widest range of properties. Some have a high tensile strength, some are ductile and malleable and some are hard and brittle. Some are very fluid when molten and some can have their properties changed by heat treatment. This range of properties is reflected in the many different engineering applications for which metals are used and the many different ways in which they are processed.

Polymers also have a wide range of properties. As has been described, they may be divided broadly into thermoplastics and thermosetting plastics. It is this distinction and the form in which the raw materials are supplied that determines the most appropriate processing method and the effect that processing has on the material.

Ceramics do not have the same wide range of properties or processing methods. They tend to be hard, brittle and poor conductors of heat and electricity. The extent of these properties depends on the processing method. This generally involves moulding and heating or heating followed by moulding, depending on the type of ceramic.

The implications of different property values

The property values obtained by testing or from reference sources enable the engineer to judge a material's:

- resistance to wear
- resistance to corrosion and degradation
- shock absorption
- workability
- thermal and electrical conductivity
- magnetic properties.

Resistance to wear

The hardness of a material is by definition its resistance to wear and abrasion. It is a property that bearing metals must have. Ball and roller bearings are made from heat-treated steels with a high carbon content and might typically be expected to have Vickers pyramid hardness of around VPN 1000. They are also ground to a very smooth surface finish so that, when used with a suitable lubricant, their rolling resistance will be low.

Other bearing metals achieve their wear-resistant properties from the way they are alloyed. The 'white bearing metals' used for the main crankshaft and big-end bearings in internal combustion engines, and the phosphor bronzes used for plain bush bearings and worm gears are examples of these alloys. At first glance, they appear to be very soft compared with ball and roller bearings. Closer inspection, however, reveals that they have two kinds of crystal or grain. One that is very soft and one that is very hard. The harder material is usually an intermetallic compound whose crystals are held in a matrix of the softer material.

During the 'running-in' period, the softer material gets worn away, leaving the harder crystals standing proud. This provides channels between the hard crystals along which lubricant can flow. The combination of materials gives a bearing metal that is wear resistant and at the same time tough and ductile enough to absorb shocks. With white bearing metals, the soft matrix is tin–lead alloy and the hard crystals are compounds of antimony and tin. With phosphor bronzes, the matrix is copper–tin alloy and the hard crystals are compounds of copper and tin. Certain polymers have wear-resistant properties. Nylon and PTFE, in particular, are used as bearing materials.

Wear resistance is a property required in cutting tools, particularly those that operate at high speeds. The materials used for these are high-carbon steels, high-speed steel and cermets. As with ball and roller bearings, the hardness and wear resistance of tool steels is achieved by heat treatment. The cermets, however, have a matrix structure that is similar to some of the bearing alloys. They are composed of very hard ceramic crystals, such as tungsten carbide, held in a matrix of a tough metal, such as cobalt.

Activity 7.1

The Moh scale is an old method of assessing the relative hardness of materials, particularly minerals. Find out how many different grades of hardness it contained and how the grade was determined for a particular material.

Resistance to corrosion and degradation

The rate at which a metal corrodes depends on its affinity with oxygen, the nature of its oxide film, the surface treatment or protection it has received and its service environment. Many non-ferrous metals and alloys combine more readily with oxygen in the atmosphere than do ferrous metals. Usually, however, they have better corrosion resistance owing to the dense oxide film that forms on their surface and protects them from further attack. This can also be a disadvantage since the oxide makes the material difficult to join by welding, as is the case with aluminium and its alloys, stainless steels and titanium alloys.

Ferrous metals, with the exception of stainless steels, corrode progressively because their oxide film is loose and porous. At high temperatures, ferrous metals corrode with a black scale, known as millscale. At normal temperatures, the usual form of corrosion is rusting as a result of moisture in the atmosphere. This is known as electrolytic corrosion, which is made worse by atmospheric pollutants in the moisture.

Electrolytic corrosion occurs when two dissimilar metals are in contact and covered with moisture. An electric cell is set up with one metal becoming the anode and the other the cathode. An electric current then flows around the system and the anode starts to corrode. Unfortunately, steel contains a number of grains made up of alternate layers of iron and the intermetallic compound, iron carbide. In the presence of moisture, the iron carbide layers become cathodes and the iron layers become anodes. The iron then corrodes to form rust, which, once formed, also acts as a cathode, and the process accelerates (Figure 7.1).

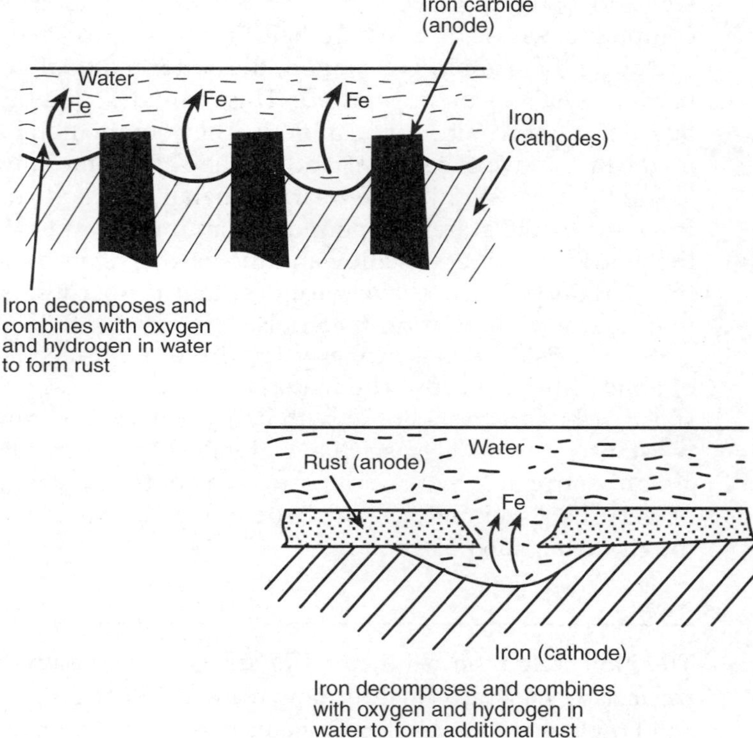

Figure 7.1 Corrosion of ferrous metals

Whenever two dissimilar metals are joined, there is always the danger of electrolytic corrosion taking place. The one that corrodes depends on their relative positions in a table known as the 'electrochemical' series.

For any two of the metals listed below in contact, it is the one that is lower in the table that will become the anode and corrode when moisture is present. It will be noted that zinc is lower down than iron. This accounts for the fact that, when steel is plated with zinc, i.e. galvanised, it does not rust even if the steel is exposed. In the presence of moisture, it is the zinc that corrodes at a very slow rate.

The electrochemical series
Gold
Platinum
Silver
Copper
Lead
Tin
Nickel
Cadmium
Iron
Chromium
Zinc
Aluminium
Magnesium

The process is known as 'sacrificial protection'. The submerged parts of ships and marine oil rigs are protected in this way. Zinc ingots are strapped at intervals to their hulls, legs and cross-members, which act as sacrificial anodes (Figure 7.2). It will be noted that cadmium, nickel and tin, which are also used to protect steel, are above iron in the electrochemical series. As a result, they do not give sacrificial protection and, if the steel is exposed, it will corrode.

Figure 7.2 Sacrificial zinc anodes on the hull of a ship

To minimise the occurrence of electrolytic corrosion, screwed fastenings, rivets and the filler material used during welding should be made from the same material as that being joined. Exposed joints between dissimilar metals should be avoided and, where this is not possible, they should be adequately protected.

Care should be taken when designing components that are to be highly stressed or formed by excessive cold working. In the presence of moisture, the stressed part can become an anode and the remainder of the component a cathode. The corrosion that results is called stress corrosion. Motor vehicle bodies and aircraft components are prone to this if the metal is not adequately protected and moisture is allowed to collect in or around load-bearing parts.

Polymer materials do not suffer from corrosion caused by oxidation, but they can degrade as a result of attack from solvents, heat and ultraviolet radiation. Thermoplastics and rubbers are generally more susceptible to degradation than thermosetting plastics. Polystyrene, Perspex and rubbers are affected by petrol and other organic solvents, which cause them to soften and dissolve. Polythene and nylon are affected by exposure to strong sunlight, which causes them to become brittle. Some protection can be given by the inclusion of coloured pigments, which slows down the process.

Activity 7.2

Up to the First World War, it was fashionable to have wrought iron railings on top of the boundary walls surrounding houses and public buildings. These were set in holes drilled in the coping stones of the walls, which were then filled with molten lead. Many of these were removed during the two world wars when iron was in short supply for munitions. Of those that do remain, many show signs of corrosion at the lead–iron interface. By referring to the electrochemical series, decide which of the two metals is the one most likely to corrode and, next time you see some old wrought iron railings, look to see if you are correct.

Shock absorption

Engineering components that are subjected to impact or suddenly applied loads need to be tough. It will be remembered that toughness is measured by the amount of energy that a material specimen can absorb before fracture during an Izod or Charpy impact test. Tough materials have a high resistance to fracture. They are able to absorb energy and dissipate it in a way that does them no damage.

Tough materials, such as heat-treated steels, are also very elastic. A shock load produces elastic deformation during which the energy they receive is changed into 'strain energy'. As soon as the load is removed, the strain energy is released as the material assumes its original shape. In the suspension springs of a motor vehicle, it is released in a controlled way through oil-filled dampers where it is dissipated as heat energy.

If the load is only applied momentarily and there is no damping device, the energy received will cause an elastic material to vibrate. In this way, the energy is dissipated in the form of heat within the material or as sound. A bell is a typical example of this and such materials are sonorous when struck. Cast iron has an in-built damping mechanism. The flakes of graphite in its structure are able to absorb and dissipate energy. This makes it an ideal material for machine beds and frames. Cast iron is, of course, also brittle and should not be subjected to tensile shock loads.

There are metals and alloys within the range of engineering materials that are extremely tough. Many polymers, such as nylon and kevlar, display toughness, as do composite materials, such as glass and carbon fibre-reinforced plastics. At the other end of the material spectrum, ceramics are generally unsuited to shock loading because of their hard and brittle nature.

Workability

The structure and property values of a material have direct implications on the processing methods that can be used to form it to shape. Metals may be cast to shape and polymers and ceramics may be moulded. In each case, the final form is determined by the contours of a mould or a die. The complexity of the shape depends on the fluidity or plasticity of the material.

Some metals are formed to shape by hot working processes, such as forging, pressing, rolling and extrusion. Metals are hot worked for a number of reasons. One of these is that they may be hard to deform when cold, but become malleable when heated. Another is that hot working often improves the structure and strength of a metal by giving it a grain direction and a refined grain structure.

Other metals are ductile and malleable enough at normal temperatures to be cold worked to shape. The most common processes are drawing, pressing, cold rolling and extrusion. Drawing into bar, wire and tube requires a high degree of ductility, whereas the other processes require the material to be malleable. Cold working causes the material to become work hardened, and this often limits the amount by which it can be deformed. It will also be remembered that excessive cold working can produce stress concentrations, which can become the source of corrosion and fatigue cracks.

Components are often finished by machining processes. Metals and plastics generally have good machinability, the limiting property being the hardness of the material. The softer metals and plastics are machined using single-point or multitoothed cutting tools. The hard metals can often only be machined by grinding or by specialist spark erosion techniques. The hardness and brittleness of ceramics makes them particularly difficult to machine, grinding being the usual method used as in the manufacture of lenses.

Thermal and electrical conductivity

Metals are generally good conductors of electricity and heat, whereas polymers and ceramics are poor conductors, i.e. good electrical and thermal insulators. Conducting and insulating materials are of equal importance for the safe and efficient working of electric circuits and high-temperature processes. Between the two extremes are semiconductor materials, which are poor conductors at low temperatures but whose conductivity increases sharply with temperature increase.

Metals behave in the opposite way. Their conductivity gradually improves as the temperature falls, becoming what are known as 'super conductors' with very low resistance as the temperature approaches absolute zero, i.e. –273°C. The electrical and thermal conductivity of polymers and ceramics is little affected by temperature change.

Although metals are good electrical conductors compared with other materials, some are much better than others. Silver, copper and aluminium are very good conductors. They are widely used where low resistance is required to minimise energy loss in the form of heat. For applications, such as lamp filaments and heater elements, however, metals with a much higher resistance are required so that they will glow red and white hot. Tungsten is used for lamp filaments, and alloys containing chromium and nickel are used for heater elements.

The ability of metals to conduct electricity is essential to joining processes such as arc and resistance welding. It is also essential to electroplating processes and for metal removal by spark erosion. The ability of metals to conduct heat is essential for efficient and rapid heat transfer processes. Boilers, heat exchangers and radiators need to have a high thermal conductivity, as do soldering irons, smoothing irons and cooking utensils. Materials that are good thermal insulators are also important to thermal processes. They are used as lagging to prevent heat loss and for operating handles and control knobs.

Case study

Body scanner

Most large hospitals have a body scanner, which is able to detect cancers and other irregularities by a technique known as magnetic resonance (Figure 7.3). The scanner produces a very strong magnetic field, which causes the atoms of the body to emit tiny radio signals. These are picked up and fed into a computer, which is able to build up a cross-sectional picture of the organs being scanned. This can then be viewed on a screen with no discomfort or harm to the patient, as might occur with prolonged exposure to X-rays.

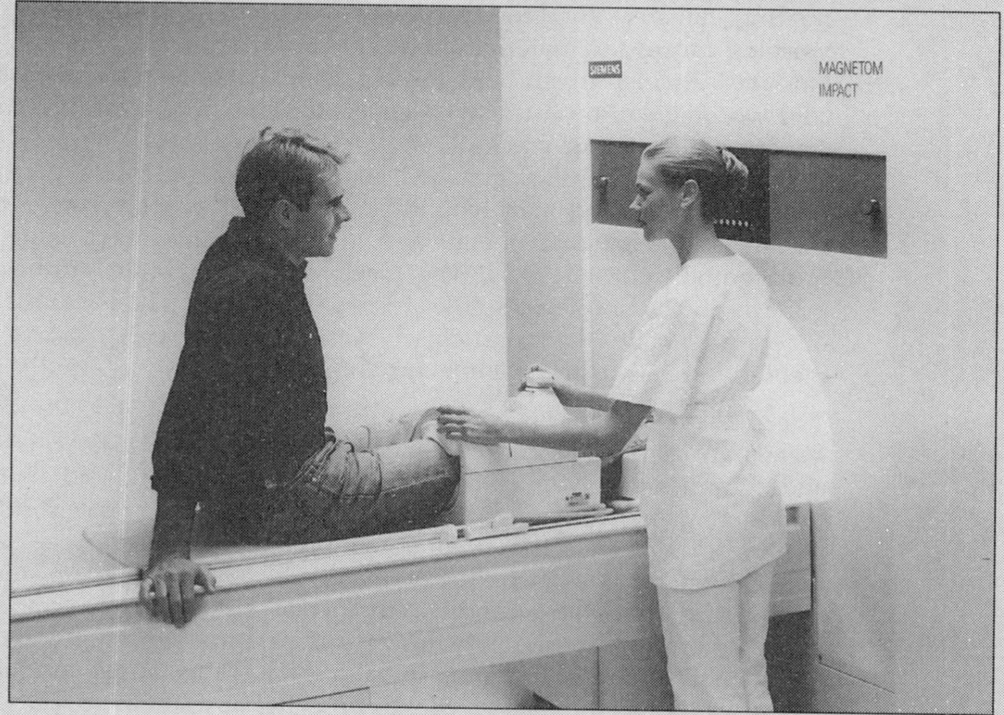

Figure 7.3 A body scanner

The strong magnetic field is produced by current-carrying coils. The currents required are very high and would normally be impossible to achieve because of the heat that they would generate. They can only be obtained by the use of a superconducting material that has a very low resistance if maintained at a low temperature. Early designs had metal conductors that had to be cooled with liquid helium to a temperature of 4.2 K (–269°C). This was very expensive and limited the use of the technique.

In the 1980s, a breakthrough occurred when ceramics with a superconducting temperature of 93 K (–180°C) were developed by the Swiss scientists Muller and Bednortz. This enabled liquid nitrogen to be used for cooling, which is much less difficult and less expensive to use. Research continues and no doubt superconductors will be developed in the near future that can operate at higher temperatures and, perhaps eventually, even at normal temperatures.

Magnetic properties

Materials are classified as ferromagnetic, diamagnetic or paramagnetic. Of these, it is only the ferromagnetic materials that exhibit strong magnetic properties. They are iron, nickel, cobalt and gadolinium. The diamagnetic group includes a few metals and all non-metals. A rod of diamagnetic material will react to a strong magnetic field by settling at right angles to it, but the force it experiences is extremely weak. The paramagnetic group includes the majority of metals. A rod made from these will align itself with a magnetic field but, as with paramagnetic materials, the force experienced is very weak.

Ferromagnetic materials and alloys containing them have many engineering applications. They include the permanent magnets used in loudspeakers and magnetic chucks. These are made from hard magnetic materials, which, once magnetised, retain their magnetism. Other applications include the cores of electromagnets, solenoids, relays and transformers, which are made from soft magnetic materials. These quickly loose their magnetism when the magnetising electric current is switched off.

The composition of hard and soft magnetic materials has been described previously. A phenomenon that should be noted is that, at a certain temperature, ferromagnetic materials loose their magnetic properties. This temperature is known as the 'Curie point'. Its value for pure iron is 768°C but, for some of the alloys used in the above applications, it can be much lower. It is essential to take this into account when designing magnetic devices that are required to operate at high temperatures.

Table 7.1 gives a comparison of the properties of metals, ceramics and polymers.

Progress check

1 What are the beneficial and detrimental effects of the dense oxide film that forms on the surface of aluminium alloys?
2 What mechanical properties are desirable in engineering components, such as vehicle suspension springs, which are subjected to shock loads?
3 What are the possible effects of temperature rise on:
   a metallic conductors
   b semiconductor materials
   c ferromagnetic materials?
4 What is meant by 'sacrificial protection'?
5 Why is cast iron widely used for machine beds and frames?

Appropriate processing methods

Some engineering materials are processed into a standard form, such as plate, sheet and barstock, after which they are sold on for further processing. Others are formed to a specific shape by processes, such as forging, casting and moulding. They may be finish formed into an engineered product or component by these processes or they may be rough formed, leading to further processing. At some point in the production cycle, a component may undergo a heat treatment process to change or modify some of its properties and, if it is part of a fabrication, it will be joined to its mating components. The forming and assembly processing methods may be grouped into:

Table 7.1 Comparison of material properties

Property	Metals	Ceramics	Polymers
Tensile strength	Up to 2500 MPa	Up to 400 MPa	Up to 120 MPa
Compressive strength	Up to 2500 MPa	Up to 5000 MPa	Up to 350 MPa
Modulus of elasticity	40–400 GPa	150–450 GPa	0.7–3.5 GPa
Malleability and ductility	Low to high	Low	Thermoplastics – temperature dependent Thermosets – low
Hardness	Medium to high	High	Low
Toughness	High	Low	Medium to high
Machinability	Good	Poor	Good
Melting temperature	Wide range – 200°C to 4000°C	High – up to 4000°C	Low – up to 300°C
Thermal expansivity	Medium to high	Low to medium	High
Thermal conductivity	Medium to high	Low to medium	Low
Electrical conductivity	Good	Poor	Poor
Magnetic permeability	Very high for ferromagnetic materials, others low	Low	Low
Electrical permittivity	Not applicable	Medium to high	Medium
Resistance to corrosion	Good except for ferrous metals	Good	Good
Resistance to chemical attack	Low to medium	Good	Good

- liquid forming processes
- solid forming processes
- gaseous processes
- thermal processes
- joining processes.

Liquid forming processes

This group of processes includes the various methods of forming molten or liquid materials to the shape of a mould or die. Most of the metals used in engineering are extracted in the molten state from the minerals in which they occur naturally.

While they are molten, impurities can be removed and alloying materials added. They are then usually cast into ingots that can be forged, rolled or remelted for further casting.

Pure metals solidify at a constant temperature (Table 7.2). Alloys solidify over a range of temperature, but this is generally small. In the liquid form, the high-energy state of the atoms makes the material very mobile, although some molten metals are more fluid than others. Steel, in particular, does not have a high fluidity and can only be cast into relatively simple shapes that do not have thin sections. Its use is confined mainly to the continuous casting of slabs and billets and the sand casting of bulky items. Cast iron, brasses, bronzes and aluminium alloys are much more fluid and can be used with a variety of casting processes to make much more complex items.

Table 7.2 Melting points of metals

Metal	Melting point (°C)
Iron	1536
Nickel	1453
Copper	1083
Silver	960
Aluminium	660
Zinc	420
Lead	327
Tin	232

The melting point and expansivity of a material affects the choice of casting process. Metals with a high melting point are generally cast in sand moulds or in ceramic moulds made by the investment or lost wax technique. Gravity die casting is carried out using a range of materials, while pressure die casting is confined mainly to the lower melting point aluminium-, magnesium- and zinc-based alloys. Care must be taken when designing concave or hollow components for die casting, since the metal may grip the die and crack as it cools. Ideally, die casting metals should have a low thermal expansivity.

Fluidity and expansivity are factors that must also be taken into account when selecting moulding processes for polymer materials. Although it is possible to cast some thermosetting resin and liquid hardener combinations in simple moulds, polymers are not generally cast by pouring. The van der Waals forces between the molecular chains make plastics too viscous for casting. Heating increases the fluidity of the raw material but not sufficiently for casting in the same way as molten metals. Attempts to increase liquid mobility by excessive heating only causes the material to decompose.

Products and components made from polymers are formed to shape by a number of processes that involve heat and pressure. Thermoplastics and some thermosets and rubbers are formed by injection moulding, extrusion, blow moulding and vacuum forming. With thermoplastics, the material is heated until it becomes sufficiently fluid. Pressure is then applied, forcing it into the shape of a mould or die. The die may be water cooled to speed solidification.

When thermosetting resins are injection moulded, the die is heated so that the polymers will cross-link and cause the material to solidify. Thermosetting plastics in powdered form are formed to shape by compression moulding. This involves compressing and heating the powder between dies until cross-linking and solidification take place.

Reinforced plastic products may be hand formed by applying alternate layers of resin with hardener and reinforcement to the contours of a mould by means of brush and roller. Alternatively, a mixture of the resin, reactive hardener and chopped glass or carbon fibre reinforcement may be sprayed on to the mould by a process known as contact moulding.

Ceramics in the form of cement, concrete and plaster can be cast in a way that is not unlike the casting of metals. Here, however, solidification is brought about at normal temperatures by a chemical change rather than a physical change. Pottery, e.g. plates and saucers, are formed by slip casting. Here, the powdered raw material is mixed with a liquid to form a slurry. This is then poured into porous moulds from which the liquid is drawn out by capillary action.

Glass is also formed to shape in the liquid form as plate glass or as lamp bulbs, bottles, TV tubes, etc. Plate glass, which solidifies at around 600°C, is produced by the float glass process in which it solidifies on the surface of molten tin held at a lower temperature. Glass containers, such as those listed above, are formed to shape in dies by blow moulding, the process being the forerunner of plastic blow moulding.

Activity 7.3

Mention was made above of the 'lost wax' or 'investment' method of casting. This is a very old method, but one that is still widely used. Find out about the method and the types of component that it can be used to cast.

Solid forming processes

Engineering materials are formed to shape in their solid condition by processes that include forging, pressing, rolling, extrusion and drawing. The materials formed by these processes need to be ductile and malleable. Some materials may be formed to shape while cold. They are then said to have been 'cold worked'. Other materials need to be heated to increase their malleability and ductility before they can be formed. They are then said to have been 'hot worked'.

Cold working is generally appropriate for those metals that have a face-centred cubic (FCC) structure at normal temperatures. Copper and aluminium are typical examples. Steel contains iron with a body-centred cubic (BCC) structure and cannot be cold worked to the same extent. Most metals become more malleable and ductile at elevated temperatures at which they can be successfully hot worked. This is especially true of steel in which the body-centred cubic structure changes to the more malleable face-centred structure at temperatures above about 900°C.

At normal temperatures, the force of attraction between the polymers in plastic materials are generally too great to permit cold forming. Plastics may be flexible and elastic, but they lack the ductility and malleability required for cold working. In ceramics, the atoms are held together by strong covalent and ionic bonding forces. These, together with the tetrahedral crystal lattice structure that is often present, makes them hard, brittle and completely unsuitable for cold working. In the form in which they are moulded, it is more appropriate to treat polymers and ceramics as viscous liquids rather than as malleable solids.

Metals, ceramics and cermets, whose hardness, brittleness and high melting point make it impossible to work them in other ways, are generally formed to shape by sintering. Here, the raw materials are powdered and compressed into shape. Heat is then applied and bonding takes place through recrystallisation or melting of one of the constituents.

Gaseous processes

These are sometimes referred to as 'surface engineering' processes in which the constituents of a gas or vapour react with, or diffuse into, the surface of a material. They include chemical vapour deposition (CVD) and physical vapour deposition (PVD). With chemical vapour deposition, a reactive gas is circulated around the recipient material. With physical vapour deposition, a substance is vaporised in its presence.

Chemical vapour deposition processes are used to 'dope' semiconductor materials, such as silicon, by heating them in the presence of vapours containing dopant materials, such as boron and phosphorus. The p-type and n-type semiconductor materials in integrated circuit chips are formed by atoms from the dopant gases diffusing into the silicon and displacing the silicon atoms.

With physical vapour deposition, the coating material is heated at low pressure until it melts and then vaporises. The vapour is then deposited onto the material that is to be coated, and the process takes place at lower temperatures than those required for CVD. The metal conductors that connect the different areas of an integrated circuit are deposited on the silicon in this way. The method is also used to deposit titanium nitride on high-speed steel cutting tools (Figure 7.4).

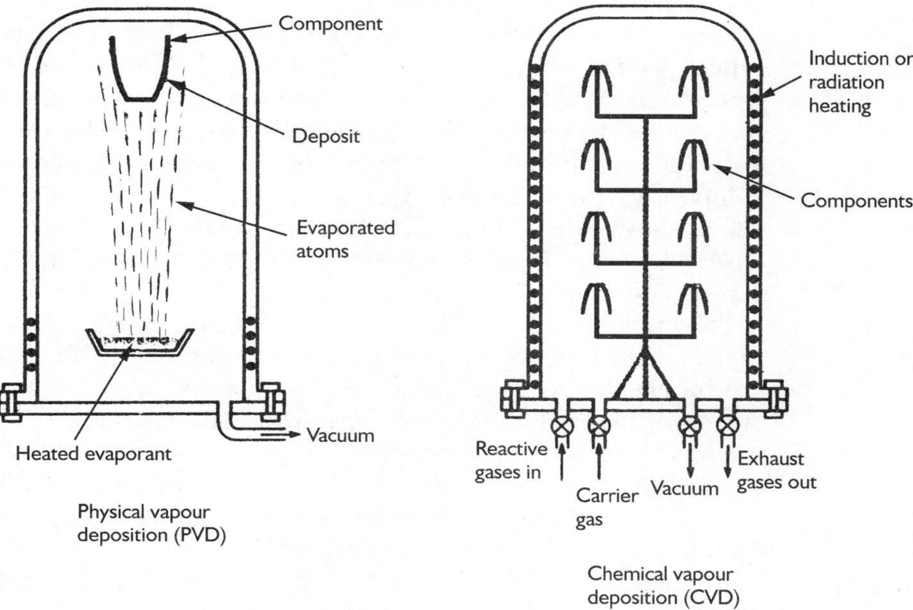

Figure 7.4 Chemical and physical vapour deposition

Thermal processes

Thermal processes are used to change, modify or fix the properties of engineering materials. The heat treatment of metals has already been described in the preceding chapter. The curing of thermosetting plastics and the firing of ceramics has also been mentioned.

Metals that have been hardened by cold working can be annealed by heating them to their recrystallisation temperature. At this temperature, which is approximately half the melting temperature in °C, new undeformed crystals grow from the points of maximum internal stress. The process continues until the material is completely recrystallised. The time of soaking can be critical since the grains will start to feed off each other, producing grain growth if the material is heated for too long a period. Large grains reduce the strength of a material and are generally undesirable. Some materials, such as copper, may be quenched after annealing but alloys, such as steel, must be allowed to cool slowly in the dying furnace.

Materials that have been hot worked or cast to shape often contain internal stresses and a non-uniform grain structure. Heat treatment processes similar to the above can be used to relieve the stresses and refine the grain structure of the metal. With steel, the process is called 'normalising'. It is similar to annealing except that the material is removed from the furnace and allowed to cool in still air.

Certain alloys can be hardened and toughened by heat treatment. Medium- and high-carbon steels can be hardened by heating them to above the recrystallisation temperature and quenching in oil or water. At these temperatures, which are round about 900°C, the carbon is fully absorbed in solid solution with the iron. Quenching traps the carbon within the grains of iron, which take on a needle-like appearance. The structure is called 'martensite', which is very hard and brittle. The steel is then tempered to toughen it. This involves reheating it to allow some of the carbon to come out of solution and precipitate as iron carbide at temperatures between 200°C and 400°C.

Heat treatment processes have been developed to change or modify the properties of other alloys. Certain aluminium alloys of the 'duralumin' type can be 'precipitation hardened'. These, which have a copper content of around 4%, are first heated to around 500°C, where the copper is absorbed in a solid solution with the aluminium. The material is then quenched and the copper is held in solution, giving a material that is both strong and tough. Reheating to around 120°C allows some of the copper to precipitate as hard particles of the compound $CuAl_2$. By controlling the amount of precipitation, the final hardness of the alloy can be adjusted.

Polymers cannot be heat treated to bring about property changes as can metals. The thermal process generally associated with them is the heating of thermoplastics to their glass transition temperature before moulding, and the curing of thermosets, i.e. heating to promote cross-linking of the polymers.

With amorphous and bonded ceramics, heating or 'firing' is an essential part of the manufacturing process. Glass products are formed at temperatures of around 700°C, whereas bonded ceramics may be fired at temperatures up to 2000°C, depending on the constituents.

After forming, glass products need to be annealed to remove internal stresses. This involves reheating the glass to around 400°C so that some plastic flow can take place in the stressed areas. Having thus relieved the stress, or reduced it to a more acceptable level, the glass is then allowed to cool down slowly. Glass can also be toughened by tempering. This is done by heating it to the annealing temperature and cooling it down by means of air jets or by quenching in oil. The safety glass used in plate glass doors and vehicle side windows is toughened in this way.

Activity 7.4

Some components are heat treated by a process called 'induction hardening'. Find out how the components are heated in this process and the kind of component that it is suitable for.

Joining processes

The components of engineered products may be joined together using screwed fastenings, rivets, soft soldering, hard soldering, welding or adhesives. The method chosen depends on the degree of permanence required, the service conditions and the properties of the materials being joined. Screwed fastenings are used where access is required for maintenance and where components may need to be removed or replaced during servicing. The other methods are used to make permanent joints, which are intended to last throughout the service life of a product.

Care must be taken when joining soft materials with nuts and bolts or setscrews, since it is possible to damage the material under the nut or bolt and screw head as they are being tightened. For this reason, washers should be used to spread the load. Where self-tapping screws are used, the type selected must suit the materials being joined. The thread-forming type is suitable for soft materials, while the thread-cutting type may need to be used with harder materials.

Wherever dissimilar materials are joined together, there is a danger of electrolytic corrosion occurring, and the joints need to be protected from moisture. Screwed fasteners and rivets should, if possible, be of the same composition as the component materials to minimise the risk of this kind of corrosion. The same applies in welding, where the filler rod should be of the same composition as the materials being joined.

Soft soldering, which uses a tin–lead alloy as the joining material, is only successful where the solder adheres to the materials being joined. Steel, copper, brasses and tin-bronzes all accept soft solder, which combines with them to form an amalgam. Hard soldering uses brass and silver solder as the joining materials. It gives a stronger joint but has a much higher melting point than soft solder. Steel, cast iron and copper readily accept hard solder, which combines with them to form an amalgam. Other materials, which do not readily accept hard or soft solder, or which have too low a melting point for hard solder, must be joined by other means.

In fusion welding, the materials are fused together in the molten state. Welding is associated with metals, but some thermoplastics are also welded. The choice of welding process depends on the material being joined and its thickness. Gas welding is generally used for thin sheet metal and can also be used to weld cast iron. Electric resistance and arc welding processes have been developed to join a range of thicknesses and material types.

When selecting a welding process, the melting point and the fusability of the material need to be considered. Thermoplastics melt at temperatures of no more than 200–300°C and can be welded using a hot air jet as the heat source. Temperatures up to 3000°C are possible with oxy-acetylene welding, while arc welding can generate temperatures many times higher than this.

Plain-carbon steels fuse readily and are easily welded using oxy-acetylene and conventional manual metal arc welding techniques. Stainless steel and alloys containing aluminium, magnesium and titanium are more difficult to weld, mainly owing to the formation of oxides that prevent the metal from fusing. Submerged arc welding techniques, where the weld area is shielded by an inert gas, are used with these materials.

The use of adhesives to join engineering components has increased in recent years. Adhesive bonding has many advantages. It enables permanent joints to be made between dissimilar materials; there are no drilled holes to weaken the materials; no high temperatures to distort them; and the joints can have a large contact area. Care is, however, required at the design stage where the choice of adhesive must take account of the service conditions, the properties of the materials being joined and the properties of the adhesive itself.

Preparation of the surfaces to be joined is very important. They should have the appropriate surface texture and be free from grease, dirt and loose flakes of material. For maximum strength, adhesives should be chosen that combine chemically with the materials being joined. Ideally, the adhesive should penetrate into the substrate of the materials, i.e. the region near the surface, forming chemical and mechanical bonds. This property of an adhesive is known as adhesion. The adhesive should be strong enough within itself to carry the loads that will be applied, and this property is known as 'cohesion'.

The main types of adhesive and their applications have been described previously, and the makers of proprietary brands, such as Bostik, Araldite, etc., are always willing to supply technical data and advice.

Case study

Solid state pressure welds

Metals can be welded in the solid state. The technique, which is known as pressure welding, was used by blacksmiths and armourers. It is still used today by craftsman blacksmiths, who specialise in decorative ironwork, and in electronics to join the gold connecting wires to silicon chips (Figure 7.5).

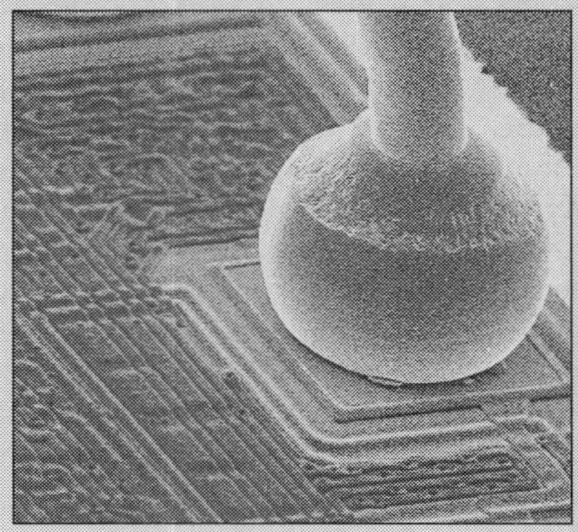

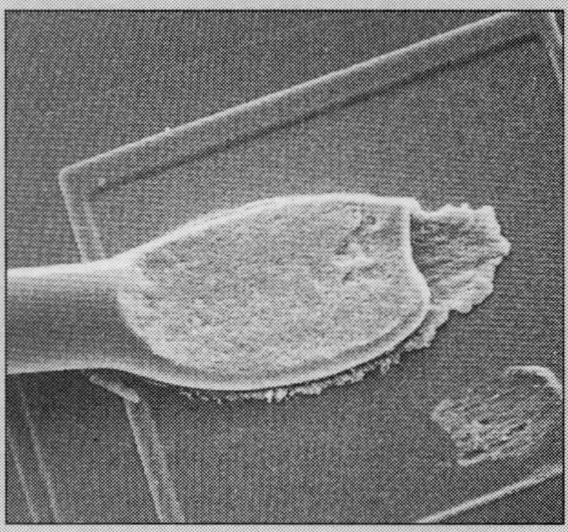

Figure 7.5 Solid-state pressure welds – ball bonded (left) and wedge bonded (right)

With blacksmiths' welds, the components are raised to a white heat, placed in position and hammered together using hand or power hammers. The pressure that is applied forces out any oxide or slag, and metallic bonds are formed between the joint surfaces. Recrystallisation also occurs in the deformed grains and this helps with the bonding.
The process is similar for making the pressure welded joints to silicon chips. The joint surfaces are cleaned chemically and the gold wires are then placed in position. Heat and pressure are then applied and pressure bonding takes place quickly at around 400°C. This is about half the melting temperature of gold. The connections may be 'ball bonded' or 'wedge bonded' as shown in Figure 7.5, depending on the shape of the tool that is used to apply the pressure.

Progress check

1 What are the material properties and physical characteristics that have to be taken into account in liquid forming processes?
2 Why can steel only be cold worked to a limited extent and why is it much easier to form it by hot working?
3 What do the initials CVD and PVD mean in relation to gaseous forming processes?
4 What danger exists where dissimilar metals are joined together?
5 What is the meaning of the terms 'adhesion' and 'cohesion' as applied to adhesives?

The effects of processing on materials

As a result of processing, some of the properties of an engineering material may change. In some cases, this is beneficial and may be a reason for choosing a particular process, but in other cases the changes are detrimental and further processing may be needed to reverse them.

As in previous discussions, engineering materials may be grouped into:

- metals
- polymers
- ceramics
- composites
- semiconductors.

Metals

Casting enables metal to be formed into intricate shapes that are impossible to achieve by other means. Sometimes, however, there are detrimental effects. The size of the grains affects the strength of the metal and depends on the cooling rate. With castings, it is common to find that the outer surface has been chilled by the mould or dies and has smaller grains than the interior metal. This is particularly the case with sand castings, which tend to have relatively low strength owing to a slow rate of cooling and large interior grains.

With some alloys, it is found that the composition of the grains is not uniform. The centres of the grains are rich in the constituent with the higher melting point and the outer part is rich in the constituent with the lower melting point. The effect is known as 'coring', which sometimes accompanies the variation in grain size described above. Both effects can be modified by heat treatment, which allows the grains to recrystallise with a uniform composition and grain size.

Hot working processes can have beneficial effects on the structure of a metal. Forging, rolling, hot pressing and extrusion produce grain flow and distribute impurities. As a result, hot worked components tend to be stronger than those of a similar shape that have been formed by casting or machining. Uneven cooling can result in variations in the grain size and the presence of internal stresses, but these detrimental effects can be removed by heat treatment.

Cold working processes, such as drawing, rolling and pressing, produce a superior surface finish to hot working. They also produce work hardness in the areas in which the grain structure is most deformed. This can be beneficial in that it stiffens the material, but the internal stresses can also produce stress corrosion, particularly in steels. The hardness and stress can be removed by heat treatment but at the expense of the bright surface finish.

Annealing and normalising are the heat treatment processes used to modify the structure and properties of cast, hot worked and cold worked metals. Annealing restores the malleability and ductility to a metal by making use of its ability to recrystallise when heated. Recrystallisation can then be followed by controlled cooling to control the size of the reformed grains. Normalising refines the grain structure and removes internal stresses from materials that have been hot worked. As with annealing, it makes use of the ability of the material to recrystallise in the solid state to produce the required structure.

Certain materials, such as medium- and high-carbon steels and some aluminium alloys, undergo structural changes at elevated temperatures, which enable their hardness and toughness to be adjusted. In steel, the iron is able to take more carbon into solution, while, in aluminium alloys, such as 'duralumin', the aluminium is able to take more copper into solution. This structure is trapped when the material is quenched and, as a result, its properties are changed.

Reheating allows some of the trapped carbon and copper to come out of solution and combine with the parent metal to form an intermetallic compound. These changes, and the accompanying changes in the strength, hardness and toughness of the metals, form the basis for the hardening and tempering of steels and the precipitation hardening of aluminium alloys that have been described in Chapter 3.

Surface treatment processes that are designed to improve the corrosion resistance, wear resistance and the aesthetic appearance of metals include plating, painting, plastic coating, phosphating and anodising. The techniques and their effects have also been described in Chapter 3.

Case study

Cast iron

As has been mentioned, lathe beds are made of cast iron. This is ideal for casting into the complex shapes required, and the graphite flakes in its structure enable it to absorb vibrations that are created during the machining process. Cast iron is, however, a relatively soft material and the slideways for the carriage and tailstock must be wear resistant.

This is overcome by case hardening the slideways after they have been milled to shape. Cast iron has a high carbon content and, if heated and quenched, it can be hardened in the same way as high-carbon steels. The heating is carried out by a gas flame, which is slowly traversed along the slideways, heating them locally to a bright red heat. This is followed by a water jet, which quenches the material, leaving it with a hard case 2–3 mm deep. Afterwards, the slideways are machined to the required surface finish by surface grinding.

Polymers

The properties of metals are dependent on the way that they have been formed to shape and by subsequent heat treatment. This is not the case with plastics and rubbers. Their properties can be varied, but this is usually achieved by making adjustments to their ingredients and to the chemical processes that are used to produce the polymers before forming.

The tensile strength and modulus of elasticity of plastics can be increased by increasing the length of the polymers and promoting side branching. This makes it more difficult for the chains to move because of the increased forces of attraction between them. Stretching the polymers so that they are aligned in a particular direction increases the amount of crystallinity, and this too can increase the density, tensile strength and modulus of elasticity of the material.

The hardness and rigidity of thermosetting plastics depends on the amount of cross-linking that can occur. This can be controlled by adjusting the ingredients during the initial mixing processes. Thermoplastics can be made more flexible by the addition of materials known as 'plasticisers'. These have the effect of filling the spaces between the chains, making it easier for them to move over each other.

The required properties can sometimes be obtained by mixing different polymers together in the liquid state and allowing them to solidify together. The resulting material can be described as a polymer alloy, although the mixture is not homogeneous, as with metal alloys. The molecular groups that join together to form a polymer chain are called 'monomers'. Sometimes, it is possible to combine different ones together to modify the structure of a polymer. The process is called 'copolymerisation'. With some polymers, it is used to adjust the melting temperature and, with others, it can give improved solvent resistance and toughness.

The properties of some polymers, such as polystyrene, polyurethane and rubber, can be greatly altered by injecting them with gases to form plastic foams. Polystyrene foam is light and an excellent thermal insulator. It is widely used for packaging and as insulation in buildings. Foam rubber and polyurethane foam are light and elastic for use in upholstery and for vibration damping.

Activity 7.5

Glass is recyclable, and bottle banks are a familiar sight in municipal and supermarket car parks. Discuss the reasons why waste plastic materials and containers are not collected and recycled in the same way.

Ceramics

The structure and properties of amorphous and bonded ceramics is determined by their ingredients, the temperature at which they are fired and the time for which they are fired. The final structure consists of two parts, a glassy or 'vitreous' matrix and crystals. Fluxes, such as sodium borate (borax), sodium carbonate, potassium carbonate, calcium fluoride and sodium aluminium fluoride (cryolite), are added to the clay and sand ingredients to lower the temperature at which vitrification occurs.

The firing temperature can vary from around 800°C to 2000°C, depending on the product. Table 7.3 gives a comparison of the proportion of fluxes, the temperature of firing and the amount of vitrification that takes place in different ceramics.

Table 7.3 Firing temperatures

Ceramic material	Proportion of fluxes	Temperature of firing (°C)	Extent of vitrification
Glasses	Moderate	1500–2000	Complete
Refractory materials	Small	1500–2000	Little
Whitewares	Varying	1200–1500	Little
Structural clay products	Large	800–1300	Little

The annealing of glass products to reduce internal stresses and the toughening of plate glass by tempering have already been described. If glass is held at its annealing temperature and slowly cooled, it can be made to crystallise. The material is then called 'glass ceramic'. This process of controlled devitrification produces crystals that are much smaller and more uniform than in other bonded ceramics. Depending on their composition, glass ceramics can be made with better thermal properties and mechanical properties than amorphous glass. The 'pyroceramics' used for ceramic cooker hobs and for aerospace applications belong to this class of materials.

Activity 7.6

Extracted teeth can now be replaced by an implant, which is inserted into the vacant socket. After a while, this becomes bonded to the bone and a crown can be built on top of it. Find out from your dentist what are the materials used for the implant and the crown, and how they are bonded together.

Composites

In Chapter 3, composite materials were divided into particulate, fibrous and laminated types. The final properties of the particulate type depend on the nature and size of the particles and on the nature of the matrix material that binds them together. This may be a metal, a thermosetting polymer, vitreous material or a ceramic cement. A description of the different processing methods is beyond the scope of this book but, in general terms, the final properties of a particulate composite depend on the successful formation of the matrix and the degree to which it bonds with the particles.

With fibrous composites, such as glass and carbon fibre-reinforced plastics, the final properties depend on the alignment of the fibres, the proportion of fibres to matrix material and the choice of matrix material. A composite in which the fibres are highly aligned will tend to be stiff in the direction of alignment. A composite in which the fibres are chopped and oriented in different directions will be less stiff but will have a very low expansivity. In fact, with careful design, composites can be made that have zero expansivity.

After processing, the properties of laminates are dependent on the nature and orientation of the different layers and on the bonding material. If the layers have a grain direction, as is the case with plywood, they are generally set at right angles to each other. This ensures that, after bonding, the composite will be isotropic, i.e. have the same properties in all directions. Use of a cellular material sandwiched between layers of dense polymer or metal produces a composite that is rigid but of low density.

Activity 7.7

Tufnol is an inexpensive composite material, which can be obtained in rod and sheet form for use in the workshop. It is easily machined and is used for making control knobs, handles and lightly loaded gear wheels. Find out exactly what Tufnol is, and what are its mechanical and electrical properties?

Semiconductors

The structure of n-type and p-type semiconductors was described in Chapter 3. The intrinsic semiconductor materials, silicon and germanium, are poor conductors at low temperatures. This is because very few of the outer electrons are able to break free of the atoms and become current carriers. During processing, silicon and germanium are doped with impurity atoms that have one more or one less electron in the outermost electron shell. The dopant used and the extent of the doping determine the final conductive properties of the material.

The dopants used to produce n-type material include phosphorous, antimony and arsenic, which have five electrons in the outer shell as opposed to four of silicon and germanium. The outer electrons of the dopant form covalent bonds with the silicon or germanium atoms, but they have one electron too many. This is released and is able to move through the crystal lattice as a free electron. The electrical conductivity of the material is thus increased, since there are now more free electrons to act as current carriers when a potential difference is applied across the material.

The dopants used to produce p-type material include aluminium, indium and gallium, which have three electrons in the outer shell. These form covalent bonds with the silicon and germanium atoms but have one electron less. There is thus a vacancy or 'hole' in the atomic lattice. This acts as an 'acceptor', which is able to capture an electron leaving another hole elsewhere. This results in increased electrical conductivity since, when a potential difference is applied, electrons are able to hop from hole to hole producing an electric current. In fact, it seems as if it is the holes that are moving towards the negative terminal, behaving as if they were positively charged current carriers.

Progress check

1. What are the beneficial effects that forging can have on the structure of a material?
2. What are the undesirable effects that casting can have on the structure of a material?
3. Compare the effects of quenching high-carbon steel from a temperature of 920°C and duralumin from a temperature of 520°C.
4. What are the dopants used to produce p-type semiconductor materials?
5. What are glass ceramics?

Assignment 7

This assignment provides evidence for:
Element 4.2: Relate materials' characteristics to processing methods
and the following key skills:
Communication: 3.1, 3.2, 3.3, 3.4
Information Technoloyg: 3.1, 3.2, 3.3

You should now be able to relate the structure and properties of engineering materials to processing methods. Demonstrate this by drawing up a table in an appropriate format that identifies the implications of property values for ten different materials. In doing so, you will need to complete the following tasks:

Your tasks

- Select one example each from the following material types:
  a a ferrous alloy
  b a non-ferrous alloy
  c a thermoplastic
  d a thermosetting plastic
  e an amorphous ceramic
  f a bonded ceramic
  g a p-type semiconductor
  h an n-type semiconductor
  i a reinforced plastic composite
  j a cermet.
- Identify appropriate processing methods for each material and briefly describe the effect of the properties and structure of the material on the use of each method.
- Identify and describe the effects of the processing methods on the final properties and structure of each material.

Chapter 8 Selecting materials for engineered products

This chapter covers:
Element 4.3: Select materials for engineered products.

. . . and is divided into the following sections:
- Criteria for selecting materials.

The materials selected for engineered products must meet the design specifications. The product must be fit for its intended purpose and, however good the design, it will fail if the materials chosen are not suitable for their service conditions. The product must also be safe to use, and materials must be chosen that comply with the relevant safety standards. The processing that the selected materials will require must also be considered and, where possible, they should be chosen to make the best use of existing plant and equipment.

Criteria for selecting materials

The main criteria for the selection of materials to meet given design specifications for engineered products are thus:

- material properties
- costs
- availability
- processability.

Material properties

The properties of the most common engineering materials have been described in some detail. Other textbooks, which are written specifically on engineering materials, contain much more information. The titles often contain the words 'Metallurgy', 'Polymer Science' or 'Science of Materials', etc., and they are to be found in libraries under the Dewey code 620 or 621. Articles on the use and properties of materials are also to be found in trade journals and the journals published by the engineering institutions. These are particularly useful for keeping abreast of developments in materials technology.

Manufacturers' and stockholders' catalogues contain a wealth of information on the composition and properties of materials. This is particularly so in the case of standard forms of supply, such as barstock, sheet, and rolled and extruded sections. Trade directories containing the names and addresses of suppliers and stockholders are distributed to engineering firms, and they are often held in the reference section of college and public libraries.

In recent years, the selection of materials has been made easier in many areas by the use of computer databases. These have been compiled by manufacturers and research bodies and are available on disk and network. Access costs are, of course, involved but the better versions quickly lead the user to one or more suitable materials through a structured set of questions on the material specifications and service requirements.

Reference often needs to be made to British Standards (BS) specifications or International Standards Organisation (ISO) specifications when selecting materials for a particular product or for use in a particular sector of industry. This is particularly true of the electrical and constructional industries, where safety is of the utmost importance. For instance, the British Standards specification for domestic products is BSEN 60335, which lays down the durability and safety specifications for materials. Where a particular grade of material is called for, this may also be covered by a British Standard specification. Examples are BS 970, which covers plain-carbon and alloy steels, and BS 2870 for brasses.

Costs

For a firm to survive, its products must be competitive in terms of cost and quality. The history of engineering is littered with examples of products that have been uneconomic. Often this has been caused by 'over-engineering', i.e. over-complex design and extravagant use of materials. Material costs often form a considerable part of the overall manufacturing costs, and they should be kept as low as possible. The cost of a particular material depends on:

- how scarce is its natural resource
- how much processing has been required to convert it into the form in which it is purchased
- the quantity and regularity of its purchase.

When selecting materials, it often happens that the one that best meets the requirements appears to be too expensive. A cheaper alternative must then be found, which is perhaps of a lower quality. Here, however, some care is required since the cheaper material may cost more to process and the economy may be false.

Wherever possible, materials should be selected that can be purchased in standard forms and require a minimum of processing. Some discount can often be obtained by purchasing material in large quantities or placing regular repeat orders with a supplier. Here again, however, there are pitfalls. The cost of storage until required may outweigh the discount, and there is always a danger of becoming too reliant on a single supplier.

Availability

Engineering materials are supplied in a variety of forms. These include:

- ingots
- granules
- liquids
- barstock
- pipe and tube

- rolled sections
- extruded sections
- sheet and plate
- castings and mouldings
- forgings and pressings.

Availability varies with the material type and the form in which it is supplied. The more common plain-carbon steels and aluminium alloys are readily available as bar, sheet, tube, etc. in a range of standard sizes and from a number of alternative suppliers. The same applies to the more common raw polymer materials, which are readily available in the form of granules or resins. Newly developed materials may have desirable properties, but they may be available only from a single supplier in limited quantities.

Castings, forgings, mouldings and pressings require the production of patterns and dies before material can be supplied in these forms. This is an expensive operation that can only be justified if large quantities will be required. It can also take a considerable time, which needs to be taken into account when planning the production and agreeing delivery dates for a new product. Once the patterns and dies have been made, supplies can be scheduled to arrive at regular intervals.

As has been stated, there is a danger of becoming too reliant on a single supplier, especially when operating a low stock or 'just in time' policy. An interruption in the supply of an essential material can lead to costly production stoppages and the cancellation of orders. If reliability is critical, it is therefore good practice to have a back-up supplier. When choosing materials and suppliers, care should also be taken to ensure that supplies will be available to meet possible increases in production.

Processability

This refers to the suitability of a material for processing to create an engineering product or component. It is no use selecting a material with the required strength, hardness, corrosion resistance, etc. if it will be difficult or expensive to process and is unsuitable for quantity production. The processed material must meet the design specifications of the product, i.e. be fit for its intended purpose, without incurring excessive production costs.

Materials that need to be formed to shape by casting must have sufficient fluidity. Steel has many desirable properties, but it has already been stated that it is difficult to cast into thin sections and intricate shapes. Where these are an essential part of a design, a material other than steel, or an alternative forming process, may be required. Similarly, materials that require forging, pressing, drawing and extrusion must have the necessary degree of malleability and ductility. It has also been stated that specialist equipment is required to weld certain materials, particularly aluminium. If the equipment and skills are not available, an alternative material or joining process may need to be selected.

Other factors that need to be taken into account when matching materials with processes are the required dimensional tolerances, the quality of the surface finish and the likely demand for a product. Wherever possible, the best use should be made of existing plant, equipment and skills and materials should be selected with these in mind.

Progress check

1 What are the main selection criteria that must be considered when choosing materials for an engineered product?
2 List three sources of information that might be used to obtain data on the composition and properties of materials.
3 What are the main factors that affect the cost of engineering materials?
4 What can initially delay the availability of castings and forgings for a component?
5 Why can the purchase of large quantities of material at a discount price be uneconomic in the long term?

Having obtained an understanding of some of the more common engineering materials, it should now be possible to make a reasoned selection to meet the design specification for given engineered products.

Assignment 8

This assignment provides evidence for:
Element 4.3: Select materials for engineered products
and the following key skills:
Communication: 3.1, 3.2, 3.3, 3.4
Information Technology: 3.1, 3.2, 3.3

Product 1
Figure 8.1 is a design proposal for a lightweight mechanical carpet sweeper. The materials selected must be readily available and suitable for high-volume production on conventional forming and machining equipment.

Product 2
Figure 8.2 is a design proposal for a portable electric fan heater suitable for 230 V mains operation. The materials selected must be readily available and suitable for high-volume production.

Product 3
Figure 8.3 is a design proposal for a regulated low-voltage power supply unit. The unit is intended for use with laboratory, audio or computer equipment in which a sudden fluctuation in voltage might be detrimental to stored information or circuit components. It is required to deliver a smoothed and regulated 12 V supply from the 230 V mains.

Your tasks

- For each product, draw up a parts list for the numbered items. State the essential material properties for each part.

- Select a suitable material for each part that you consider to have the required properties, be of reasonable cost, be readily available and that can be processed using conventional equipment.
- Submit your recommendations in the form of a report. This should include records of any data sources which you consulted for information. It should also include the reasons for your selections as against possible alternatives.

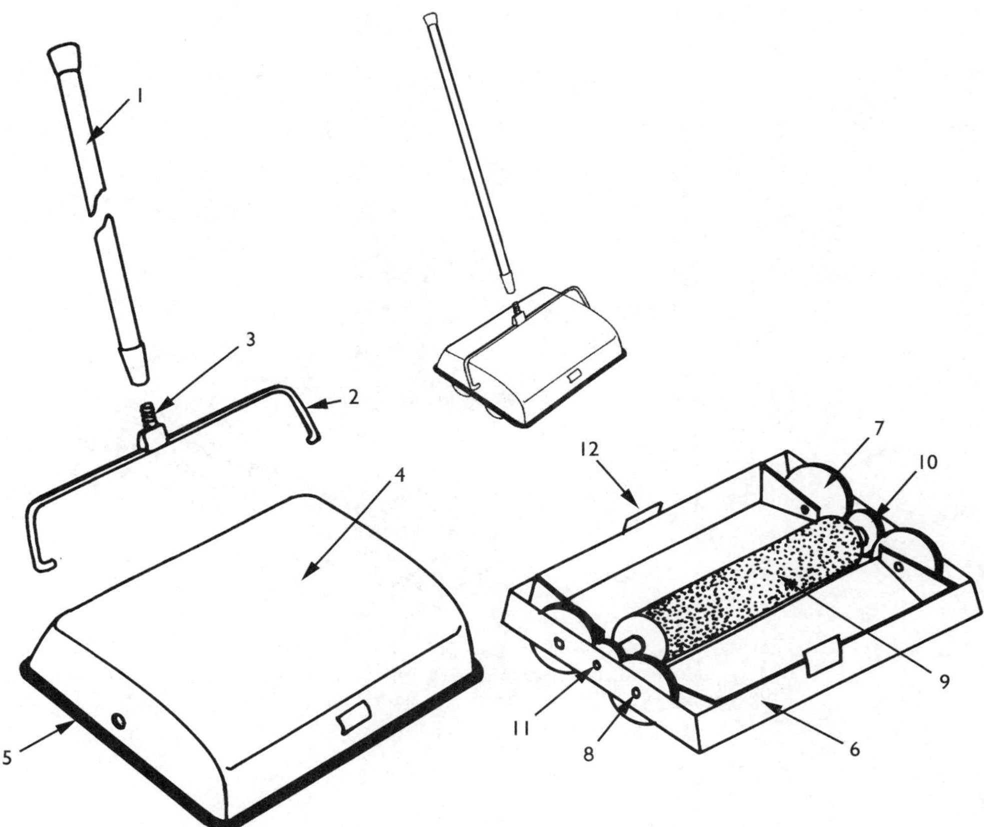

1, handle; 2, handle pivot; 3, screwed connection; 4, upper case; 5, flexible beading; 6, lower case; 7, wheel; 8, wheel pin; 9, brush; 10, brush drive roller; 11, brush spindle; 12, securing clip.

Figure 8.1 Carpet sweeper

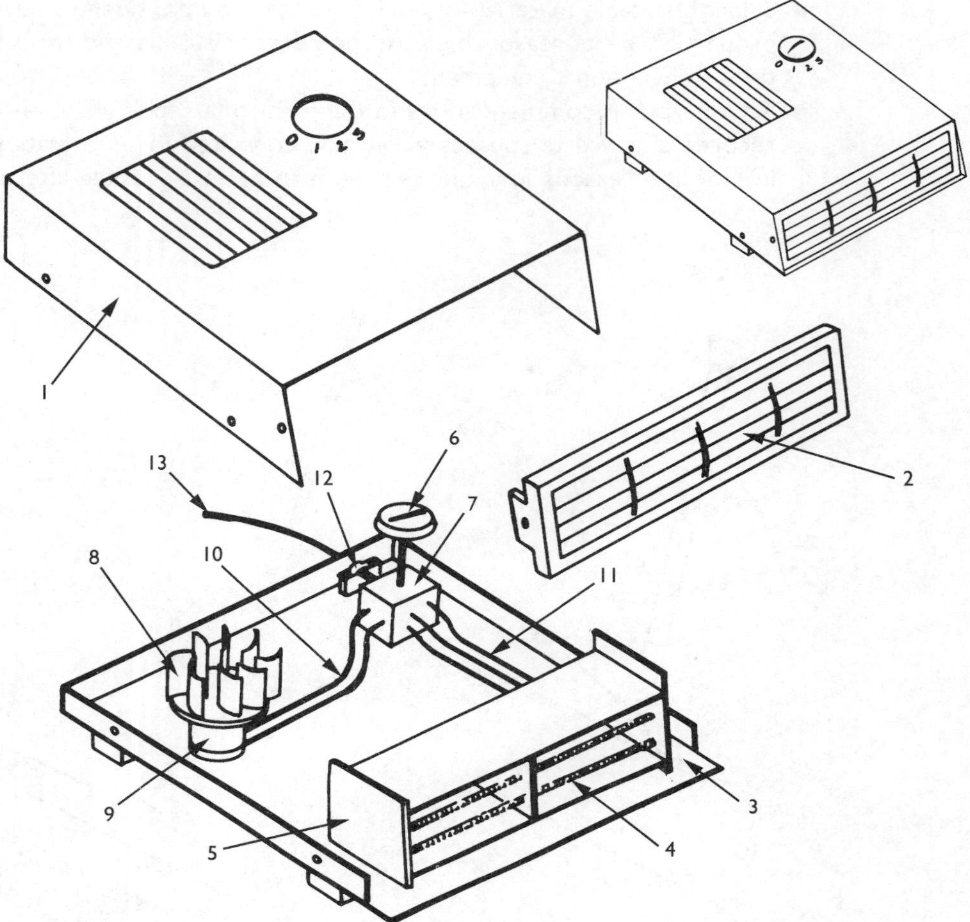

1, upper cover; 2, grille; 3, base; 4, heating element; 5, heating element support frame; 6, control knob; 7, 3-position switch; 8, fan motor; 9, fan motor; 10, motor connecting wires; 11, heating element connecting wires; 12, cable clamp; 13, flexible mains cable.

Figure 8.2 Fan heater

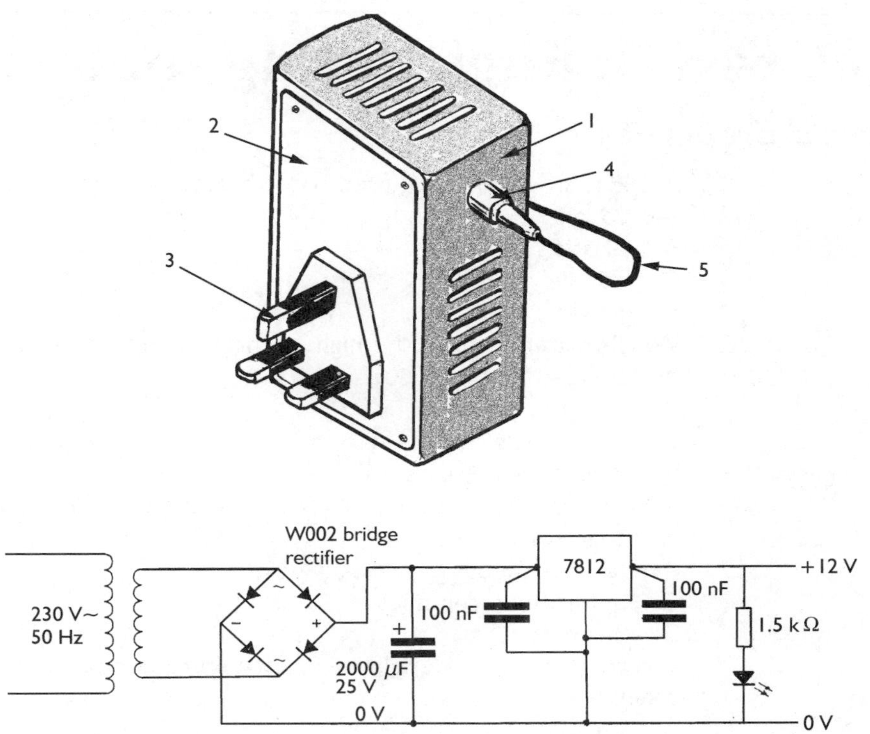

1, outer case; 2, rear panel; 3, mains socket pins; 4, output jack plug; 5, output lead.

Not shown: 6, printed circuit board; 7, internal connecting wire.

Figure 8.3 Low-voltage power supply unit

Unit test 3: Engineering materials

Material properties

1 A material which is required to withstand wear and abrasion must be
A tough
B elastic
C ductile
D hard

2 Which materials have the highest thermal and electrical conductivity?
A bonded ceramics
B thermoplastics
C alloy steels
D thermosetting plastics

3 Which materials are most likely to suffer attack from solvents?
A thermoplastics
B ferrous metals
C amorphous ceramics
D cermets

4 Which group of materials have a structure composed of long molecular chains?
A metals
B rubbers
C glasses
D semiconductors

5 Thermosetting plastics have a structure that is
A totally crystalline
B cross-linked polymer
C fully amorphous
D granular

6 The most malleable and ductile metals have a crystal lattice structure that is
A face-centred cubic
B tetrahedral
C body-centred cubic
D close-packed hexagonal

7 The polymers in plastic materials are largely composed of linked atoms of
A iron
B carbon
C silicon
D oxygen

8 The introduction of impurity atoms in silicon to make p-type and n-type semiconductor materials is known as
A alloying
B cross-linking
C doping
D degrading

9 A cermet is a composite material made up of
A a ceramic and a polymer
B a polymer and a metal

C a p-type and an n-type semiconductor
D a metal and a ceramic

10 An alloy formed when the atoms of one material occupy the spaces between the atoms in the crystal lattice of another is called
A a substitutional alloy
B an intermetallic compound
C an interstitial alloy
D a non-ferrous alloy

Processing materials

11 The cold forming of metals results in
A work hardness
B cross-linking
C recrystallisation
D grain growth

12 The effects of cold working can be removed by the process of
A carburising
B tempering
C annealing
D sintering

13 The process used to remove internal stresses and refine the grain size of steel components is
A anodising
B normalising
C oil quenching
D extrusion

14 Precipitation hardening is a process used to modify the properties of
A cast iron
B phosphor bronze
C mild steel
D duralumin

15 The process used to form ceramic carbide cutting tools is
A sintering
B die casting
C vacuum forming
D forging

16 Blow moulding is a process that is used for forming
A bonded ceramics
B semiconductors
C thermoplastics
D cast iron

17 The grain direction follows the contours of a forged component resulting in increased
A strength
B corrosion resistance
C ductility
D brittleness

18 Rivets should be made of the same materials as the parts being joined to reduce the risk of

A work hardening
B electrolytic corrosion
C grain growth
D solvent attack

Select materials

19 Which material can withstand high temperatures, resist atmospheric corrosion and is also a good electrical insulator?
A stainless steel
B porcelain
C PVC
D aluminium

20 What grade of plain-carbon steel would be the most suitable for the blade of a craft knife?
A medium-carbon steel
B mild steel
C high-carbon steel
D dead mild steel

21 The cups, saucers and plates used in caravans and for camping need to be light, tough and hygienic. A suitable thermosetting plastic material is
A epoxy resin
B bakelite
C polyester resin
D Melamine

22 The most suitable thermoplastic material for a light-duty gear wheel is
A nylon
B PVC
C Perspex
D polysyrene

23 The composite material most suitable for a boat hull is
A glass-reinforced plastic
B plastic-covered chipboard
C textile-reinforced rubber
D ceramic carbide

24 The prime consideration when selecting a material for a safety harness must be its
A cost
B properties
C availability
D processability

25 BS and ISO specifications for structural steels contain details of the
A composition of the materials
B material suppliers
C cost of the materials
D material availability

26 The cost of raw materials can best be obtained from
A British Standard specifications
B a directory of suppliers
C trade journals
D suppliers' catalogues

PART FOUR: DESIGN DEVELOPMENT

The purpose of engineering is to provide physical systems that meet human needs and in this design is a crucially important activity. The design brief sets an initial framework and specifies all requirements that must be met by the product or service. The design is then generated by selecting as many feasible alternative systems as possible. These different system solutions are then evaluated and the optimum solution selected. The chosen design solution is then subjected to thorough checks and tests to ensure that it will give safe and trouble-free service. This final design is then communicated by means of detailed specifications and drawings so that it can be manufactured or implemented.

This part has much in common with Part one, since one definition of design is 'the choice of suitable systems'. All three chapters in this part should be considered as a single topic, which shows how designs are initiated and realised. It is to be hoped that many interesting and innovative designs can be generated as the course progresses. One important lesson to be learned is that all designs must be produced within the constraints of whatever resources are available.

Chapter 9 Producing design briefs

This chapter covers:
Element 5.1: Produce design briefs for an engineered product and an engineering service.

. . . and is divided into the following sections:
- Design briefs
- Understanding customer requirements
- Standards and legislation
- Technical and operational constraints
- Design briefs for products and services.

Designing is one of the most creative, interesting and important jobs an engineer can do. Virtually everything we use today has been designed and produced by engineers. Our modern world is only made possible by engineers taking scientific discoveries and turning them into feasible and economic designs for typical products and services such as air, road and sea transport, telecommunications, computers, electricity and printing.

Activity 9.1

List four more products or services designed by engineers.

Design briefs

The starting point for any design is the design brief. This is an initial specification of the customer's needs, particularly of function. It also takes into account the overall framework in which the designer and customer operate. Thus, other very important considerations in the design brief are:

- safety
- conformity to the various rules, regulations and laws existing in the United Kingdom, Europe and the rest of the world
- environmental factors including both the effect of the product on the environment and the effect of the environment on the product
- timescale for realisation
- price and cost
- availability of materials, components, production capacity and suitable workforce.

Case study

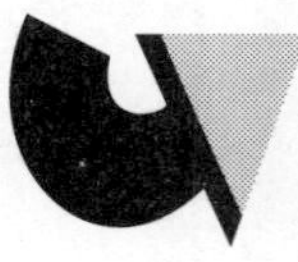

Channel Tunnel project

It is important to understand that the design brief is an initial plan, which sets a goal to be achieved within the constraints of society and of the marketplace. It will also state the criteria by which the finished outcome will be judged. Figure 9.1 shows this process in general terms and as applied to a practical example, in this case the Channel Tunnel project. Note that, in this case, a bridge or a tunnel for either rail or road (or both) could have been equally valid solutions. A brief was prepared for each, and they were tested against the constraints and criteria. As we know in this case, the final brief was for a rail tunnel. In the light of the disastrous lorry fire on the shuttle in November 1996, this solution is not without problems. The brief was perhaps not tested rigorously enough against the safety criteria.

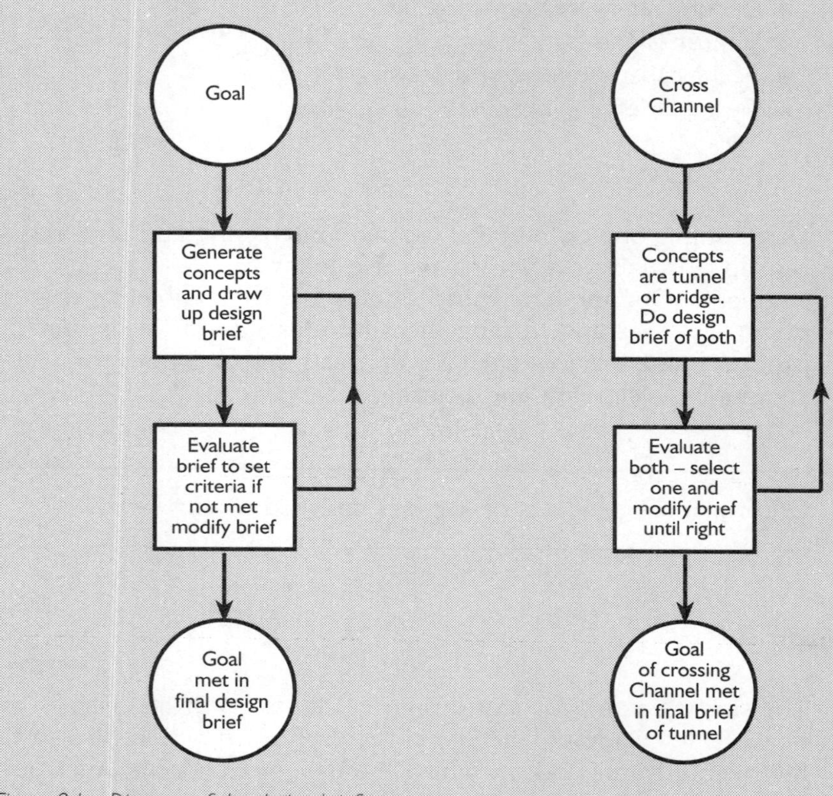

Figure 9.1 Diagram of the design brief process

Understanding customer requirements

This section deals with how to understand fully what the customer's needs are. Products or services can never be successful unless they really meet a customer's requirements and expectations.

Function and technical specification

Function has to be one of the most important requirements to be met. By stating functional needs and fully detailing a technical specification, the correct functioning of a product or service can be assured.

Problem statement Any design activity will start with a statement of the 'problem'. This first starting point of the design brief is usually a broad statement of what task or function is required to be achieved. This type of approach is called 'top down' in the designer's jargon. The designer will be given the general details of the function, in which the essential features are stated. This functional brief can be given by a client, a customer or the company management. Initially, the design brief can be very brief, as in the following case studies.

Case study

Short design briefs

An extremely short design brief was the stated goal in the early 1960s of the US President, John Kennedy: 'before the end of the decade, land a person on the moon and bring them back safely' (Figure 9.2). Another example is the one used in Figure 9.1 for the Channel between England and France to be spanned. In this case, there were two possible functional solutions, a tunnel or a bridge (Figure 9.3). In the case of the moon problem, the only feasible solution was by rocket. When the designer is just given basic aims, it means that maximum scope is given to search for the optimum solution.

Figure 9.2 Moon landing

Figure 9.3 The 'Chunnel' between England and France

Activity 9.2

Note down five examples of design solutions that started with basic design briefs. Say why the problem was solved in the way that it was.

Technical specification At other times, the initial functional requirement can be much more detailed and can actually be a technical specification. This is often true for engineering services, such as a network-linking computers, machines and processes in a factory. In this case, you cannot just state 'cable up the system'. Detailed specifications of things, such as transmission rates and 'handshaking' protocols, are required between the different pieces of kit to be connected up. This approach is nearer to what the designer calls 'bottom up'. This is because the solution is tightly defined and will be realised mainly by putting together standard components that are readily available. A similar example in an engineered product is the specification of a personal computer. This is 'top down' in the sense that the customer specifies things like speed of operation, size of RAM, etc., and this is the initial design brief. But the computer is designed and built by simply taking standard parts 'off the shelf' and assembling them, and so is inherently specified 'bottom up'.

Case study

Desktop computer system

For this case study, I have chosen to give a design brief for a good-quality desktop computer system for early 1997 – computers develop so fast that this brief will soon be out of date:

- 200 MHz 64-bit processor: 64 MB RAM: 512 k cache memory
- 3.1 GB hard disk: internal modem 3 3600 bits per second: 12 speed CD-ROM
- 17 inch 0.28 dp FST digital colour monitor: HP or equivalent laser colour printer
- 'cherry type' touch-positive keyboard: Microsoft mouse.

These technical requirements are, in this case, what the customer would consider essential, whereas requirements like size, shape, colour and finish would be more likely to be listed as desirable. Figure 9.4 shows this assembled system.

Figure 9.4 Assembled computer system

As can be seen, the design brief specifies most of the functional requirements, such as speed of operation, memory, input devices and output devices. However, all the specified items are available as standard items and the system can be assembled to the customer's requirements. In this case, the design brief is done bottom up since the items are picked from a standard menu.

Activity 9.3

Prepare a short design brief, which is a technical specification, for a product or service assembled from mainly standard items. Divide the requirements into essential and desirable.

Case study

A mass-produced car

This is an engineered product where the design brief is open as to possible solutions, but where standard items can also be used. This has a 'top down' design brief in which broad essential parameters, such as speed, economy, cost,

reliability and safety, are set for a target customer. Desirable features such as size, shape, colours, ergonomics and comfort, are also set. It should be noted that, in the case of many consumer products like cars and televisions, the desirable features will often help to sell the product. The functional specification given here would form an initial design brief for the car designer.
Functional specification for a new 'Eurocar':

- petrol engine using unleaded fuel
- top speed 180 km/h
- average fuel consumption at constant 90 km/h to be 18 km/l or less
- routine service intervals 24 000 km
- carrying capacity five adults with luggage
- body style 5-door hatchback
- front wheel drive with independent front and rear suspension
- optional manual or automatic transmission
- overall dimensions to be within length 4 m, width 1.5 m and height 1.4 m.

It can be seen that this specification sets out the essential targets to be met in terms of how the car is to function. It is also likely that other features, such as cost, shape and styling, will also be specified at an early stage. Thus, our written specification is likely to be accompanied by sketches, as in Figure 9.5, of possible shapes, which would have been tested on potential buyers.

Figure 9.5 Sketches for proposed Eurocar

Activity 9.4

a Prepare brief technical specifications for one engineered product and one engineering service in which the requirements are set 'top down', but where standard components could be used if required.

b Identify five essential functional requirements of a small electrically powered hoist to load and unload tools and workpieces for machine tools.

Ergonomics

Ergonomics is a concept of design that takes into account the safety, comfort and effectiveness of human users of products and services. The term is derived from the Greek words *ergon* meaning work and *nomos* meaning law. In the USA, it is called 'human factors engineering', which aptly summarises the function of ergonomics. It is an extremely important concept, since the end user of any product or service has to be able to use it easily, safely and without undue strain. A design has to be considered as a system or subsystem with a human controller. The person in control uses their senses to monitor what is happening, then actuates the controls to do a specific function. Some examples of these human–machine systems are:

- a bicycle being ridden
- a car being driven
- a manually controlled lathe being operated
- a nuclear power plant being operated.

Activity 9.5

List five other human–machine systems.

Case study

Bicycle

If one considers the case of a bicycle, it is essentially a system for personal transport:

- the cranks, chain ring, chain, sprockets and wheel are a lever subsystem whereby the downward pressure of the rider's legs is converted into forward motion
- steering and balancing are another subsystem in which the rider makes small adjustments of riding position and handlebar orientation to stay upright and travel in a straight line or around corners.

Figure 9.6 shows this rider/bicycle ergonomic system.

Figure 9.6 Rider/bicycle ergonomic system

Ergonomics and anthropometric data The design brief has to specify clearly the way the human will fit with the hardware part of the system. Human beings come in all shapes, sizes and capabilities, so the design has to take account of this:

- For a bicycle, this is done by varying the sizes of frame, wheels, cranks, etc. to suit different-sized riders and, within one size of bicycle, the positions and height of the saddle and the handlebars can be adjusted
- Where equipment cannot be adjusted, it has to be designed to suit as wide a range of users as is possible. This is done by referring to standard sets of anthropometric data for the target client group, which in many cases is the whole population

- Designers also have to consider the requirements of people with special needs. For example, a cooker suitable for blind people would need to have temperature controls that gave an audible or tactile output of temperature setting.

Activity 9.6

Take a typical human machine system and describe:
a how the system operates
b how the system accommodates different people.

Case study

Telephone system

A good example of a system in which ergonomics has been extensively applied is the telephone system. In particular, the telephone handset (Figure 9.7) has been extensively human engineered. It may look a deceptively simple piece of design, but this solution only came about as a result of much research and analysis:

- the keys are laid out in four rows of three push buttons after alternatives, such as two rows of six keys, circular and two columns of six keys, were tested and rejected. The numbers increase from top to bottom, which is faster and less error prone than the opposite arrangement on the calculator keyboard (Figure 9.8)
- the handset was also designed to be light and to suit the majority of hands with the microphone/earpiece spacings able to fit nearly all of the users.

Figure 9.7 Telephone handset

Figure 9.8 Calculator keyboard

Activity 9.7

Choose a product or service that, like the telephone handset, has been designed for ease of use. Sketch it and describe its ergonomic features.

Aesthetics

This aspect of design deals with the general appearance and attractiveness of a product. It is thus concerned with the look or feel of a product from the point of view of shape, proportions, surface and colour. It is where the work of the engineer overlaps with that of an artist or stylist. Sometimes, stylists and industrial designers are brought in as specialists to work purely on the aesthetic aspects of the design. However, it often happens that, when a product is designed simply and for optimum functionality, it also turns out to be aesthetically beautiful as well.

Case study

Good design

It is a reasonable rule of thumb when designing that if it looks and feels right it is right. This is only true if it is functional and safe as well.

- The supersonic jet liner *Concorde*, shown in Figure 9.9, is superb functionally and also has very graceful lines
- Another good example is the bicycle, shown in Figure 9.10, which Chris Boardman rode when he won the pursuit championship at the Barcelona Olympics in 1992.

Figure 9.9 Concorde

Figure 9.10 Chris Boardman's bicycle

Activity 9.8

Choose, and sketch, two engineered products that have been designed to carry out a specific function and that have naturally good aesthetics.

Case study

Car design

In some products, of which the car is a good example, the aesthetic aspects of the design brief are specified concurrently with the functional aspects. Thus, although wind tunnel tests may determine the best functional shape for a car, each manufacturer will wish to have a distinctive shape that appeals to the customer. Colour and finish are also very important to the customer. The more the reliable functioning of product or service can be relied upon, the more important things like aesthetics become. Figures 9.11 and 9.12 show two different, but equally good, aesthetic solutions to car design. In specialised products like Formula One racing cars, the result of using sophisticated design techniques by all manufacturers means that all the cars end up being almost identical. Aesthetics still come in here, with the team cars being differentiated by colour and the use of advertising logos.

Figure 9.11 Peugeot 406

Figure 9.12 Rover 400

Shape It is often said that 'beauty is in the eye of the beholder', which implies that any aesthetic judgements are subjective. This may be true, but there are certain fundamental principles that apply. On shape, it can be observed that in nature there are very few straight lines. Most natural shapes are curved, because this is generally the strongest and lightest structure, as well as being aesthetically pleasing.

Case study

Computer mouse

A good example of a design using curves is the computer mouse. This could be a plain box-like shape but how much more pleasing to the eye as well as to handle and use is the curved shape of the mouse shown in Figure 9.13. Microsoft are said to have spent several million dollars developing this design, which is an indication of the time and care required to develop outstanding designs successfully.

Figure 9.13 Microsoft mouse

Activity 9.9

Sketch two engineered products that use curves rather than straight lines and suggest why they are designed that way.

If rectangular shapes cannot be avoided, then designers often use the well-known classical Greek form called the golden section. This is a rectangle with sides in the approximate ratio 3:5, which gives aesthetically very pleasing shapes. Figure 9.14a

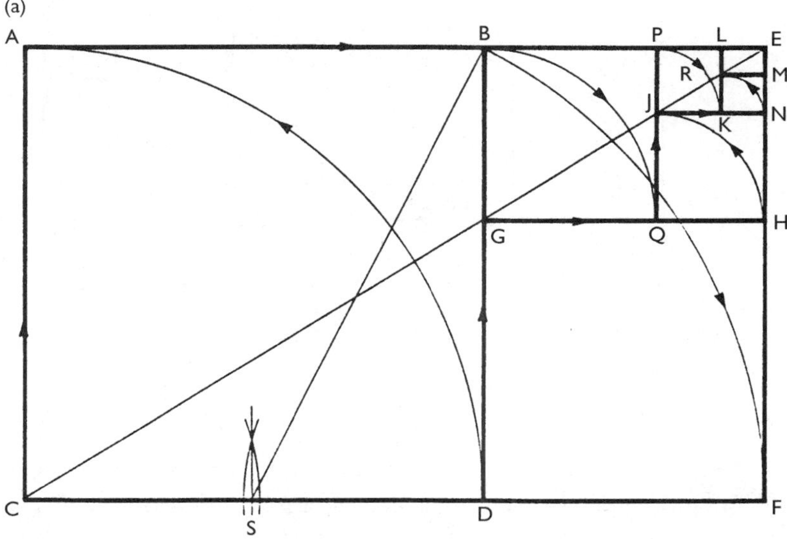

Figure 9.14a Method of constructing the Golden Section

shows the method of constructing the golden section starting with the square ABCD and drawing an arc of centre S and radius SB down to F. This gives one side of the rectangle, such that the ratio CD/CF = DF/CD = 0.618. The rectangles AEFC, BEDF, BGHE, PEQH, PJNE, LEKN and LERM are all golden sections. Figure 9.14b shows some examples of the use of the golden section.

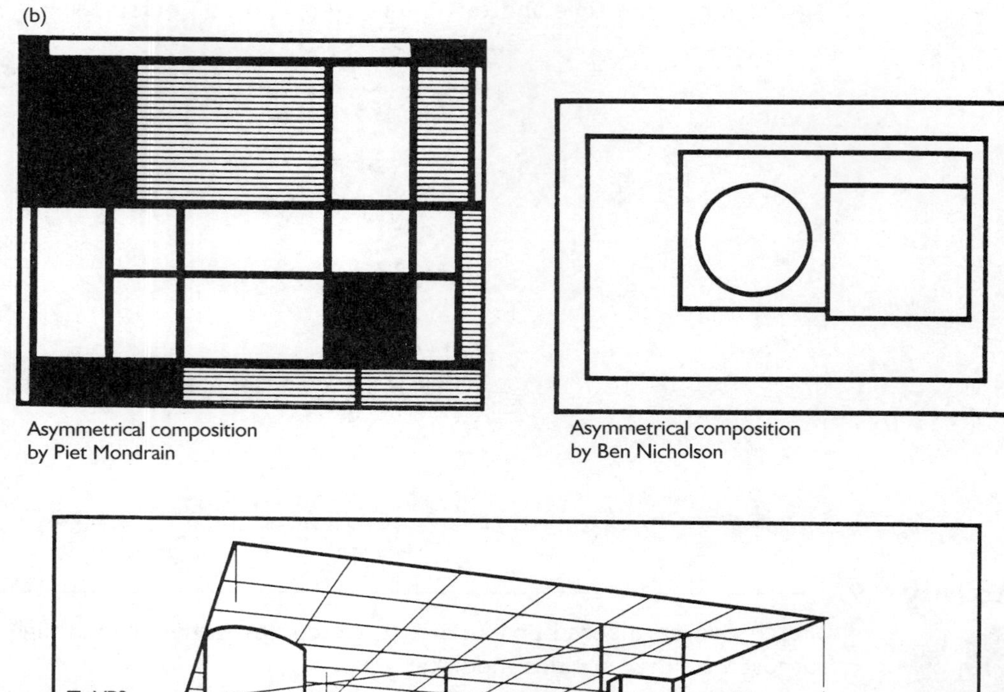

Asymmetrical composition by Piet Mondrain

Asymmetrical composition by Ben Nicholson

1930 building exhibition stand for Venesta Ltd by Le Corbusier (Olympia)

Figure 9.14b some examples of the use of the Golden Section

Activity 9.10

Find an engineered product, e.g. a machine tool, car, TV set, etc. that uses mainly straight lines. Measure and comment on its proportions and aesthetics.

Colour This is a vitally important factor in aesthetics.

- all colours can be constructed from the primary colours of red, yellow and blue. Mixing these primaries gives the secondary colours of orange (red + yellow), violet (red + blue) and green (blue + yellow)
- the most striking aesthetic effect of colours is when maximum contrast is obtained by matching a secondary colour with its complementary colour, which is the primary colour not used to make the secondary. Thus, orange is complementary to blue, violet to yellow and red to green
- for maximum contrast, complementaries are juxtaposed and, for a more harmonious relationship, yellow is matched to orange or green, blue to violet or green and red to orange or violet.

Activity 9.11

The robot arm in Figure 9.15 is attached to a lathe headstock as shown. Try various colour combinations from primary and secondary colours for the two and select:

a the one that gives the greatest contrast for both natural and artificial light
b the one that gives the closest match under the same conditions.

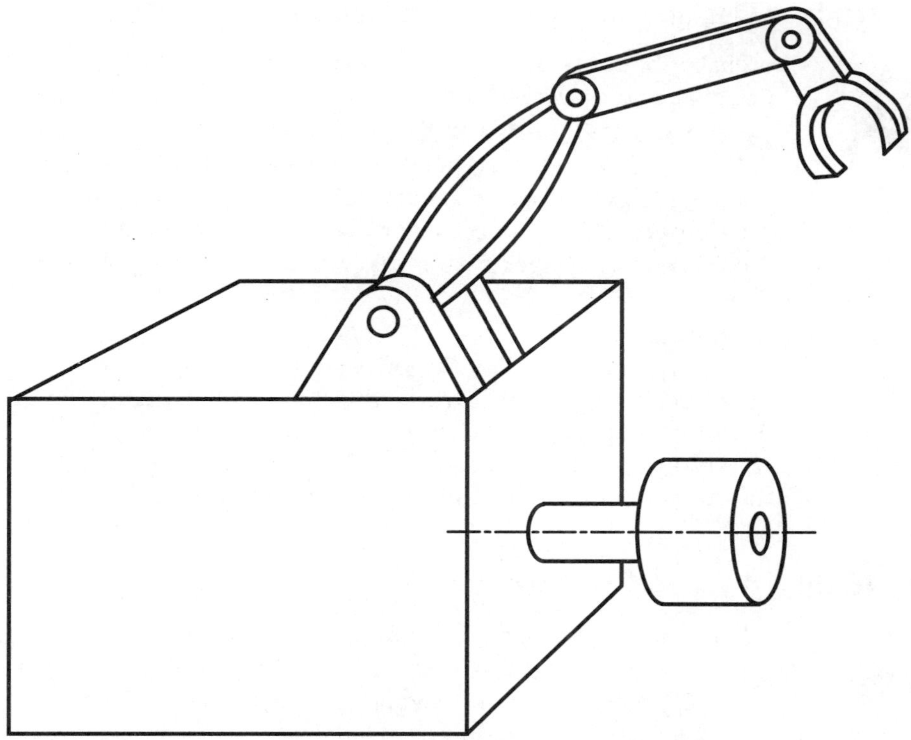

Figure 9.15 Headstock and robot arm

Quality

The quality of design is the 'level of fitness for purpose' of the proposed product or service. It should not be confused with the quality of conformance, which is the product or service being of acceptable quality if it conforms to specification. Thus, the design brief is very important in setting the quality levels of the product or service, the specified values of the customer requirements actually setting the quality levels:

- high quality of design sets customer requirements at high levels
- low quality of design sets customer requirements at low levels.

Case study

Different qualities of design

This case study looks at products with different qualities of design:

- a laser printer of 1200 dots/inch at a speed of 14 pages/minute and a mean failure time of 4000 hours has a high quality of design
- a dot matrix printer of 360 dots/inch at a speed of 1/2 page/minute and a mean failure time of 1000 hours has a lower quality of design.

Both printers may fit the purpose of their intended customers, but the quality of design is clearly different in each case. They are intended for different types of customers and are priced very differently too.

Case study

Design quality versus tolerance

Quality of design should not be confused with tolerance (see page 317). It may be that putting very low tolerances ensures better functioning, but it is not always the case. In World War Two, the German soldiers found that their very precisely made weapons worked excellently in normal conditions, but would jam up and not work either in very cold conditions or in the sandy environment of the desert. The more crudely made Russian weapons, with their larger tolerances and bigger clearances, worked adequately well under all conditions.

Materials

Naturally, when the quality of design is being set in the design brief, the materials specified will be greatly affected by the quality required. Materials have differing properties of strength, hardness, toughness, density, finish, conductivity, corrosion resistance, toxicity, etc. The determining factors in material selection are the functionality, reliability and price determined by the client or customer.

Case study

Types of bicycle

A bicycle can be made to a low, medium or high quality of design.

- a low quality of design would be a 'sit up and beg' machine retailing at under £100 (1997 prices).This would be made mainly of medium-carbon steel and would probably weigh around 20 kg
- a medium quality of design could be a touring bicycle retailing at around £300, which uses better and lighter materials and would weigh about 10 kg

- a high quality of design is a specialist racing cycle, as used by professionals in races like the Tour de France, which can easily cost £4000. This cycle would use more exotic and expensive materials such as titanium, carbon fibre, aluminium and magnesium alloys, etc. These materials have much superior strength:weight ratios, and the weight could be reduced down to as little as 5 kg without loss of strength or stiffness. These materials are also inherently more corrosion resistant, and so the cycles would last longer under normal use. Under racing conditions, they may not last as long, because the reliability and longevity of a product is dependent on the type of use it gets.

Activity 9.12

Choose an engineered product or service and produce a very short design brief for what you would consider to be:

a a high quality of design
b a medium quality of design
c a low quality of design.

Include in this brief possible materials for each design.

Tolerance

This is simply defined as a deviation from a specified value. This value can be a measured variable like length, voltage, time, etc. or an attribute like appearance or taste. A designer would like any value that is specified to be produced exactly to this value. It is, of course, impossible to produce anything exactly to value and so a permitted deviation, or tolerance, is also specified:

- the tolerance is usually specified in terms of an upper and a lower limit, with the tolerance being the difference between these two values
- all manufacturing processes have their own natural tolerances and the designer can use International Standards to specify these
- these tolerances must be achievable by the processes available
- size deviations and tolerances to achieve particular mechanical fits and ensure universal interchangeability are given in BS 4500, which is the British version of the ISO (International Organisation for Standardisation) system of fits and limits.

The three classes of fit specified are:

- clearance for when parts are moving with respect to each other, e.g. a bearing or sliding fit
- transition or location fit when parts need to be assembled and disassembled with accuracy
- interference or force fit when parts are permanently assembled using heat or force.

The actual setting of tolerances is also determined by the expected life of the product. As products are used, they will start to deteriorate, however imperceptibly, and change as a result. Examples of this are wear in bearings, corrosion of surfaces, fatigue and drift of electrical/electronic parameters such as resistance and capacitance.

What will happen in practice is that a tighter tolerance is specified than is actually required for correct functioning. Then, as wear and drift occur in service, the product will still be functional.

Typical examples of tolerancing with upper and lower limits are:

- cylinder:piston clearance 0.01 mm to 0.05; tolerance = 0.04 mm
- resistance 10 ohms ± 0.1 ohms; tolerance = 0.2 ohms
- regulated voltage 220 ± 0.05 volts; tolerance = 0.1 volts
- output power in kilowatts 1000 + 5 + 15; tolerance = 10 kW
- time to repair fault 0.5–2 hours; tolerance = 1.5 hours.

Activity 9.13

a select two components, e.g. a resistor and a screwed fastener. Find out from catalogues, data sheets or standards what the tolerances on these components are

b refer to data sheets BS 4500 A and B – Selected ISO fits: hole basis and shaft basis. Find the limits for a hole and shaft of 25 mm diameter for the following fits:

1. hole basis precision clearance (H7 – g6)
2. hole basis light transition (H7 – k6)
3. shaft basis maximum interference (S7 – h6).

Cost and quantity

Cost is an important element in the requirements of a customer. Indeed, one of the definitions of an engineer is 'someone who can make for a dollar something which any fool can make for five dollars'. The total cost is the sum of the individual costs of:

- research and development
- design and marketing
- materials and manufacturing
- services, administration, finance, etc.

Case study

High-technology products

For some high-technology products, the research and development and design costs are very large. For a totally new car, these costs can easily be around £300 million. If sales of the car are 5 million, the cost per car of this is £300 million/5 million = £60. On the other hand, for a battlefield helicopter, these costs could be around £200 million. On sales of 200, the cost per helicopter is £200 million/200 = £1 million.

Thus, the economics of production are directly related to quantity. The car can be made with expensive tooling using a lot of robotics and automation with production rates of typically 500 000 per year. The helicopter production costs are high because of the low volume, and the production rate may be as low as 20 per year. Tooling and equipment may be as expensive as that for the car but has to be covered by these much lower volumes.

Correct functioning is of greater importance than absolute cost. A very low cost achieved at the expense of function and quality is not worth having because the customer may end up by being dissatisfied. A better measure than cost is value, which relates the cost to the level of function specified. Value analysis is a useful technique, which involves a team of people looking at a part, product or service and improving its value. Value can be improved by cost reduction or by enhancing things such as life, function and aesthetics. A useful technique for a value analysis team is brainstorming (see Chapter 10). Typical questions asked in a value analysis exercise are:

- Can it be simplified or made with fewer parts?
- Can it be lighter and made of a different material?
- Can cheaper standard parts be used?
- Is it necessary at all?

Value is closely related to the 'whole life' (or 'life cycle') cost of a product or service. This 'whole life' cost is the total cost over the predicted life of the product or service. A cheap product that is unreliable and wears out quickly may cost more in total than a more expensive product that lasts longer and is more reliable.

Activity 9.14

a Choose an engineered product or service, e.g. a mobile telephone service, a motorcycle, a video recorder, etc. Find out (or get a reliable estimate) of:
  1 the cost of research and development and design
  2 the likely number of customers.

From this work out how much these activities will cost the customer over an estimated life of the product or service.

You should use a spreadsheet to model this exercise (core skills in IT). If you do this, you can easily modify the values to show the effect of varying costs and quantities.

b Choose a very simple product or service, e.g. a stapler, a three-pin plug, a factory maintenance service. Do a value analysis on the product or service in a team and improve its value.

Size

Designs can all be of a similar size, e.g. watches, laptop computers, or can be available in a range of sizes. This may be for ergonomic reasons (refer back to the section on ergonomics on page 307) or for cost, functional or performance requirements.

Case study

Effects of size

- A range of bicycles of different frame and wheel sizes can suit a range of customers from a 2-year-old child up to an adult of height 2 m
- Cars and televisions are made in different sizes to suit the pockets and needs of the customers. Small-screen televisions and mini-cars are not made just for small people!

When products are extremely small or very large, they obviously require different techniques for their specification and manufacture. Microelectronics is one industry that requires very specialised techniques of design and manufacture. A modern very large-scale integration (VLSI) chip can put around 10 000 transistors onto the surface of a silicon wafer of 2 or 3 square millimetres. The circuits for these chips are themselves designed by computer with any graphical displays being of a very large magnification.

At the other extreme are the larger scale products, such as ships and aeroplanes. These products have to be designed on a small scale, which is then enlarged up for manufacturing purposes. Large complex products like these require an initial simple design brief to be greatly enlarged to cover briefs for the multiplicity of systems and subsystems involved.

Tolerances and size Sizes over 500 mm cause difficulty in manufacturing and measurement, and large changes of dimension as a result of temperature changes have to be taken into account. Interchangeable manufacture is impossible or not required on large sizes. Components are individually matched to suit the functional requirements. This technique is referred to as 'matched fits'. BS 4500: Part three should be consulted for tolerances on large sizes.

Weight

Achievement of minimum weight is striven for when designing many products. The aerospace industry immediately springs to mind because every kilogram saved means an extra kilogram of payload that can be carried. Similar considerations apply to cars, where reduced weight means better performance and fuel economy, and to sports equipment, such as skis, where lightness and stiffness enhances speed and handling. Weight can be saved by using stronger lighter materials or by optimal design techniques such as finite element analysis or, most likely, by a combination of the two. Designers have been successful in reducing weight but still keeping a very strong structure by imitating nature. The honeycomb produced by bees has been copied and is often made with a cardboard honeycomb covered by a carbon fibre skin (Figure 9.16). In general, hollow structures like this are invariably stronger and stiffer than a solid structure of similar weight.

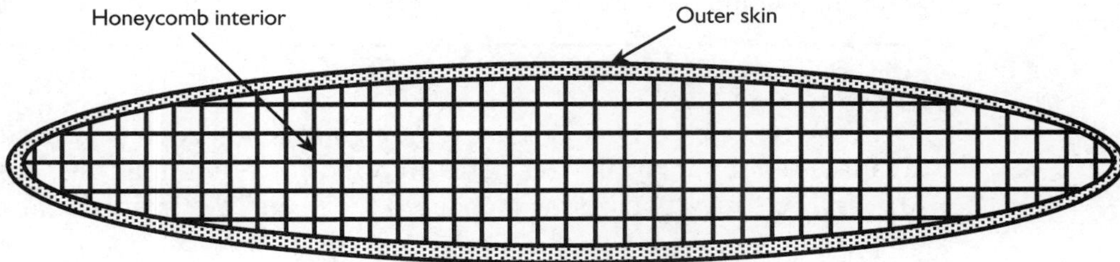

Figure 9.16 Section of honeycomb material

Timescale

Any customer or client needs to know when the product or service will be produced. The overall time between the inception of a design project and the delivery to a customer is referred to as the lead time. A schedule has to be planned and drawn up. As the project proceeds, progress is checked against this schedule

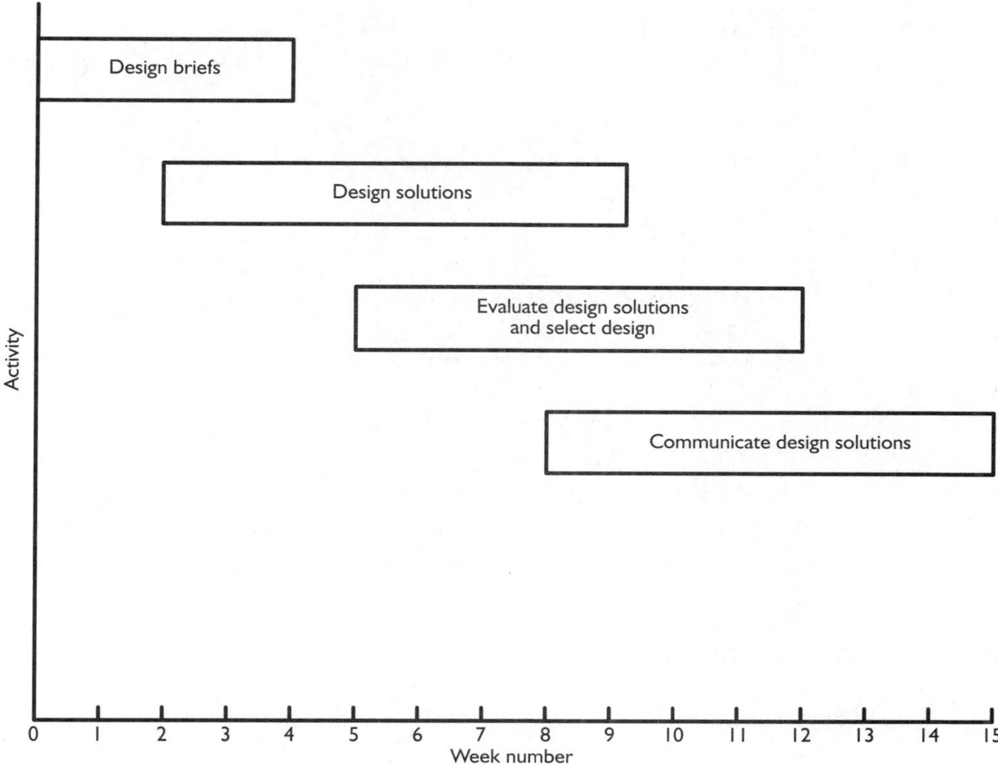

Figure 9.17 Gantt chart

and corrective action taken if slippage is occurring. The simplest type of design progress schedule is the Gantt chart, which is just a horizontal bar chart of the time of each activity. Figure 9.17 shows a Gantt chart for designing a product.

For very complex projects, there are techniques available such as critical path analysis and performance evaluation and review technique (PERT). These methods are beyond the scope of this book. They are more complex than the simple Gantt chart in that they link the activities together in a network.

They use computer programs to determine the 'critical path'. This is the path of activities, which is critical to the project being finished on time. Once this path is determined, more effort can be put into ensuring that there is no time slippage on this path.

Activity 9.15

Sketch a Gantt chart to plot the progress of a typical GNVQ engineering student.

Progress check

1. What is a design brief?
2. Explain what is meant by the technical and functional requirements of a design.
3. Define the term ergonomics and explain its importance in design.
4. What are the complementary colours? Explain why these colours give such a vivid contrast.
5. Define quality of design.
6. An estimated 100 000 products are to be made at a cost of £400 each. What are the total design and development costs if they are 5% of overall costs?
7. Describe two particular problems with:
   a very large products (over 500 mm)
   b very small products (under 0.1 mm)
8. What are Gantt charts used for?
9. Why are tolerances necessary?
10. What is meant by universal interchangeability? When is universal interchangeability not applicable?

Standards and legislation

A standard is a document that specifies agreed values to which everyone operates. They enable designers to specify components or materials by quoting the relevant standard. Typical examples are:

- precision ISO metric bolts to BS 3692
- steel to be supplied in conformity to BSEN 1020
- resistors to conform to quality standards of BSC EEC 4000
- design management to be to BS 7000.

Often the whole design of a product or service will be based on satisfying the requirements of particular standards. For example, manufacturers who wish to export vehicles to the USA must satisfy the US Federal Standards on vehicle emissions. Standards like these are invariably backed up by legislation to enforce them.

Legislation provides a mandatory framework within which designers must operate. Laws exist in all parts of the world to ensure that products and services are safe and cause minimum environmental damage. Consumer legislation also tries to ensure that customers have rights with respect to the quality and correct functioning of a product or service over a period of time.

Health and safety standards and legislation

In our modern and complex world, it is necessary to protect everyone from exposure to undue risk. There is no such thing as total freedom from risk, but properly designed products and services that are correctly made, installed and maintained should minimise risks. Since many people spend a large proportion of time at work, there is legislation to protect them. The Health and Safety at Work Act is general legislation for the UK covering all workplaces. This will include:

- factories, workshops and other industrial premises
- warehouses, offices and shops

- restaurants, cafes and hotels
- leisure facilities, e.g. cinemas, clubs, pubs, swimming pools, etc.

The Act directly affects anyone connected with plant and equipment:

- designers, manufacturers, suppliers and importers
- installers, maintainers, users and owners.

For designers, manufacturers and suppliers, there are strict guidelines laid down as regards the safety features of products and services. Section 6 of the Act requires that:

- any plant, machinery, equipment or appliance is so designed and constructed as to be safe and without risk to health when used
- inspection and testing must be done to ensure compliance with the law
- full and clear information and instructions for installation and safe use are provided.

The Health and Safety Executive (HSE) is the statutory body which is responsible for inspecting factories, workshops and industrial premises (all other types of premises are inspected by the local authority) and providing information and advice to anyone who requires it.

Designers need to consult HSE publications and British and International standards appropriate to what they are working on. The HSE also issues codes of practice (COPs) giving specific practices to be followed. Typical examples of standards and codes of practice are:

- BSEN 292: 1991 on safety of machinery, which covers the basic concepts and general principles of design for safety
- COP 37 on the safety of pressure systems as laid down in the Pressure Systems and Transportable Gas Containers Regulations 1989
- Control of Substances Hazardous to Health (COSHH) 1988.

COSHH governs the use of any substance which might cause harm, although lead and asbestos have their own special regulations. Therefore, the designer of any plant or equipment that uses any potentially harmful substances must consult these regulations, which are available from the HSE. The regulations cover:

- the safe storage, handling and use of any harmful substances and the safeguards to be built into any process using them
- safety equipment and protective clothing to be specified and provided when working with hazardous substances.

Safety engineering

Safety engineering is concerned with designing and operating plant and equipment so as to minimise or eliminate accidents and health hazards. It covers guarding and protective systems and the use of design techniques like ergonomics (see page 307) to produce inherently safe machines.

The use of safety engineering means that:

- all hazardous parts of the machine are covered by guards
- the machine cannot be operated if a guard is removed
- the control and operating system of the machine is made to suit the capabilities of a human operator and is also made 'fail-safe'.

' Fail-safe' means that, if a vital component fails or the operator makes a mistake, the machine will shut off or remain in a safe condition. One way of achieving this is through 'redundancy'. This is the provision of an alternative or 'back-up' system should the main system fail.

Case study

Aircraft controls

Connections from the pilot's controls to the flying surfaces of an aircraft can be electrical, hydraulic or mechanical. Thus, if the 'fly-by-wire' electrical connections fail, the hydraulic system takes over and, if this fails, the mechanical system comes into use. The only problem occurs if they are all routed through the same section of aircraft. A fatal crash occurred to a DC8 aircraft flying out of Paris Orly when a badly secured door came loose, resulting in a gross deformation of part of the underside of the aircraft. This in itself would not have been serious, but all the connections to the flying surfaces were severed as a result. Since that crash, aircraft control systems have been specified so that the routes of the alternative connections are separated. Another way of designing a 'fail-safe' system is that often used for brakes on heavy vehicle wheels or on machine spindles, as shown in Figure 9.18. In normal operation, the brake pads are held off the disc, usually by hydraulic or pneumatic pressure. When required, the brake is applied by releasing the pressure, which allows the pads to apply a force to stop the disc rotating. Thus, if the braking system fails, the brakes will automatically be applied.

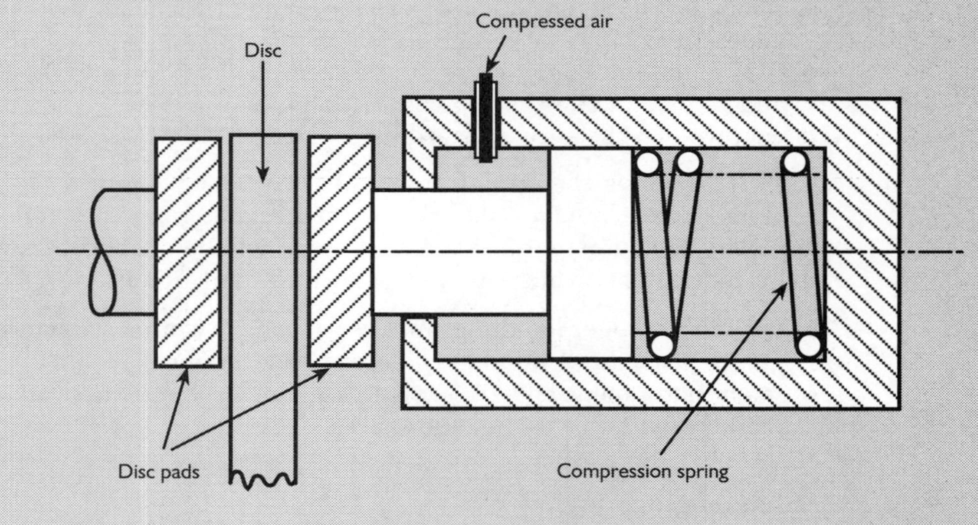

Figure 9.18 Fail-safe braking system

Activity 9.16

Describe another example of a 'fail-safe' device.

Progress check

1 What is the function of legislation that deals with design of goods and services? Give a typical example of one of these laws.
2 What does consumer legislation try to do?
3 What is the purpose of a standard? Give an example of a standard for a product or a service.
4 Outline the main provisions of the Health and Safety at Work Act.
5 What is COSSH? How might it affect a design?
6 What is a code of practice? Give a typical safety COP.
7 Outline the scope of safety engineering.
8 Describe what is meant by a 'fail-safe' system.

Technical and operational constraints

All designs are constrained by many factors. Generally, the available technology limits the range of possible solutions. However, sometimes the technology is not available and has to be developed. The example of space travel given earlier in this chapter is evidence of this (Figure 9.2). Availability of resources, such as labour, materials, plant and equipment, will be another restriction in the choice of a particular design solution. Increasingly exacting environmental factors also mean that designers must produce designs that minimise nuisances, such as noise and pollution. Finally, as already stated earlier in the chapter under customer requirements, most designs must be realised to strict cost limits.

Exceptionally, very large projects, such as the moon landing operation, have virtually unlimited allocations of kinds of resources. However, this is unusual. Most designs are in the 'real world' and have to be achieved to very tight limits of all kinds.

Technology

Technology can be defined in two ways as:

- the application of all kinds of knowledge, combined with scientific and pragmatic principles, to the solution of human problems
- organising people to work together in a systematic way with materials and machines.

Thus, for a particular technology to work effectively, the organisation of people with specialist skills is as important as the availability of knowledge and machines. So, when a designer is constrained by technology, it is often as much a problem of the people to build and maintain the new design as it is the actual techniques available to solve the problem. This is one reason why technology solutions supplied to Third World countries are often unsuccessful.

Case study

Appropriate technology

An African village is supplied with a modern tractor to help with its agriculture. This is fine for a while but then problems develop. It will not run on the local fuel and, when it breaks down, there are no spares or trained personnel to repair it.

What is taken for granted in a modern economy like the USA and Europe is that there is an infrastructure that can supply spares and trained people. Technology must be appropriate to the culture in which it is operating.

Today, the designer is constrained less and less by technology, since progress has been extremely fast in the latter half of the twentieth century. One only has to think of computers and electronics, which, allied with telecommunications, have produced the hybrid technology that we call information technology. The contents of this book can be communicated anywhere in the world for the cost of a local telephone call. Computers can transmit information at speeds of 33 600 bits/second or faster. I can send all the words in this chapter (approximately 90 000 bytes or 720 000 bits) in 720 000/33 600 = 21.43 seconds.

But technology moves on and, when fibre optic transmission rather than copper wire becomes available, data transfer speeds will easily be 100 times or more faster than now. It is beyond the scope of this book to survey all available technologies comprehensively. Although initial design briefs do not necessarily have to consider technology in detail, a typical case will be studied.

Case study

Burglar alarm technology

Burglar alarm technology brief for sensing technology:

- sensing to be contact and non-contact
- contact sensors will be selected from standard capacitive, magnetic, reed, tilt or vibration devices
- non-contact sensing will be realised by infrared, ultrasonic or photoelectric means.

In this brief, it is for a design that can be realised using standard parts and so is a 'bottom-up' approach. The way in which the whole system is put together is done 'top down' by the designer (see the earlier section in this chapter on

Activity 9.17

Write the technology part of a design brief for either:

a a device to change the direction of a driveshaft by 90°; or

b a robotic arm to pick up and stack components of up to 10 kg.

This activity can be helped by going into workshops or laboratories and examining drive systems on machines and robot arms. It would also be helpful to study manufacturers' catalogues.

Resources

These are the means the designer has available to complete the project. Before any design can be realised, the design brief needs to take account of how it can be done. The actual technology used to implement the solution will be greatly influenced by the resources at the designer's disposal. If you look back to the previous section on technology, the way technology is defined embraces the use of the resources of people, materials and machines.

Human resources (in the unit specification as labour resources) As regards human resources, the designer will have to take into account the type of people available for a product or service who can:

- design
- make
- install
- maintain.

Thus, totally different solutions will emerge for different scenarios. A situation in which unskilled labour is cheap and abundant but with skilled technicians in short supply will require a technologically simple design with mainly manual operation.

But when skilled technicians are available but expensive, and unskilled labour is scarce and also expensive, a better solution is a highly automatic and technologically advanced design.

Often, design briefs take account of the fact that people will have to be trained to do the designing, manufacturing, installation and maintenance.

Training programmes would be specified and designed along with manuals and software covering every aspect of the design. So, a design brief for an automatic meat deboning plant would specify the type of people required to build and maintain this plant. Since the plant uses robots and other automatic machines, the kinds of skilled people required would include:

- roboticists
- control engineers
- electronic engineers
- mechanical handling experts
- waste disposal specialists
- cutting tool engineers.

Activity 9.18

Investigate in a school, college or other organisation:

a the telephone system

b the heating and air-conditioning system.

List the type of skilled people required to build, install and maintain these systems.

Materials resources Since every product made or service delivered must use materials, the materials selected will affect most aspects of:

- function
- appearance
- cost: safety
- environmental impact.

Therefore, one aspect cannot be considered in isolation when selecting a material. Since materials have to be manufactured into finished products, their ability to be readily and economically processed is of prime importance. This is certainly true when cost is a major constraint. A good example of this is the use of low-carbon steel for the manufacture of car bodies. Although it is not ideal functionally, since it corrodes easily, it is widely used because it is readily available at low cost as well as being easily formed, machined and joined. Sometimes cost and ease of forming are secondary to function. This is the case for specialist products such as:

- carbon fibre racing cycle frames
- single-crystal metal gas turbine blades
- ceramic cooker hobs
- gold connectors for silicon chips.

Activity 9.19

Find as many examples as you can of products or services using materials which are chosen for:

a low cost and ease of processing
b function, irrespective of cost or ease of processing.

Designers are often constrained by the non-availability of materials for a variety of political and economic reasons. These can include wars, blockades, natural disasters and strikes, etc. In cases like this, 'substitution' is resorted to. This is where an alternative material, other than the one normally used, is specified.

Case study

Examples of substitution

1 One example of this was the very successful British wooden aeroplane of the Second World War, the Mosquito, which was made of plywood because of the shortage of aluminium (Figure 9.19). There were few problems flying it in Europe, but there were some difficulties when operating the aircraft in the conditions of high temperature and humidity encountered in the Tropics. Plywood is made by bonding thin layers of wood with adhesives. Unfortunately, in tropical conditions, some of these adhesives started to break down.

Figure 9.19 The 'plywood' Mosquito fighter-bomber

2 Another example was the use, in the 1960s, of steel pipe to connect up heating systems. This was because copper became scarce and expensive as a result of strikes and unrest in a major supply country, Zambia. The pipes functioned correctly, but there were major problems caused by corrosion at the steel–copper interfaces because of the differing electrochemical potentials of steel and copper. Both these cases show the extra care needed in design when 'substitution' becomes necessary.

Activity 9.20

Choose an example of a product or service in which 'substitution' of materials has been used. State whether any problems have been caused by this.

Often the choice of material is not clear cut, and acceptable solutions are available from designs in different materials. Indeed, the availability of competitive materials will often lead to better designs at lower cost. This is well illustrated by the cans used in the soft drinks industry. These are mostly made from either stainless steel or aluminium, both of which are functionally acceptable. The correct choice is not clear cut and will be based mainly on economics and customer preference.

Activity 9.21

Find two products that are usually made in two or more different materials with similar costs and functional properties.

There are an enormous range and variety of materials to choose from now, and most design solutions can be created without the need to develop new ones.

It is normally only in very complex and specialised projects that new materials have to be developed and tested. Some of these specially developed materials have come from the space programmes. These include the non-stick plastic polytetrafluoroethylene (PTFE), which is commonly sold under trade names, such as Teflon or Fluon. This is a totally inert material with an extremely low coefficient of friction, which was originally developed as a bearing material for use in spacecraft. Now, designers make good use of its properties in products such as non-stick coatings, valve seatings and low-friction bearings. The special ceramic tiles used to stop the space shuttle burning up on re-entry is another good example. The single-crystal material for gas turbine blades previously mentioned was developed specifically to prevent blade failure caused by a multicrystal structure.

Part three of this book deals specifically with materials and should be referred to for information when preparing your design briefs and producing designs.

Plant resources All materials specified for a product or service have to be turned into the end product. This is done by a combination of processes for:

- preparing
- shaping
- finishing

- assembling
- testing.

The actual plant and equipment available for doing this will influence the design. The manufacturing capability will clearly affect choice of:

- materials
- shapes
- processes
- tolerances.

When designs are specified, they may well be restricted by what is available as manufacturing facilities. Designers in industry will generally not be as restricted as those in education. This is particularly true for mass production where the large volumes justify the use of expensive machines and tooling dedicated to one product or process. In smaller scale manufacture such as the aircraft industry, most plant and tooling has to be adaptable to be able to produce many different parts on the same machine.

Case study

Producing components

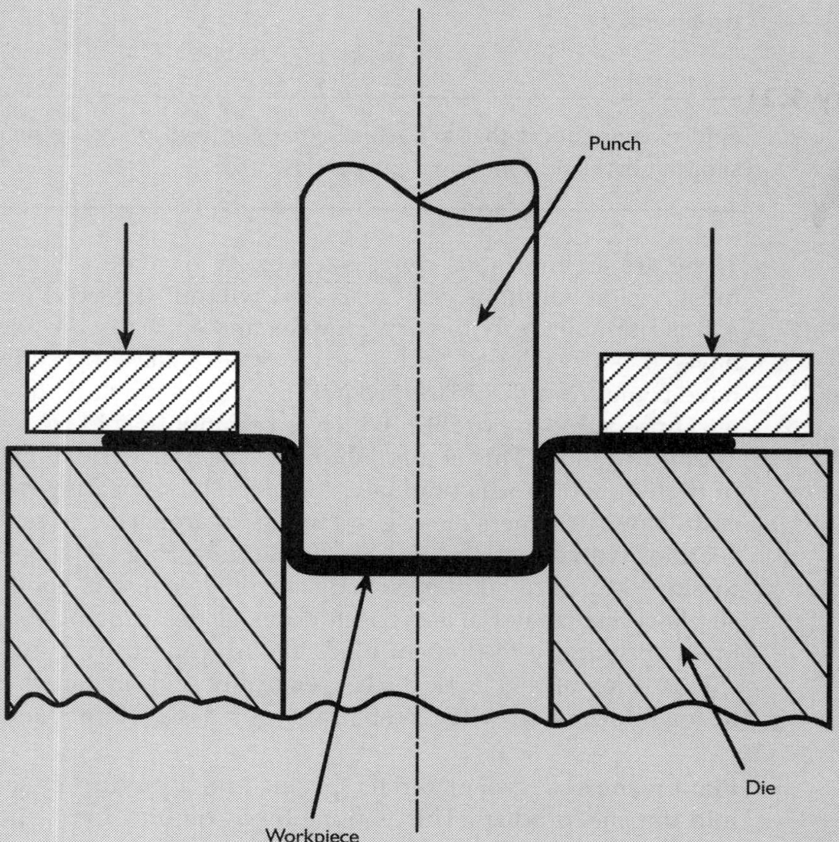

Figure 9.20 Deep drawing

A stainless steel sink unit produced in millions would be produced on a press by deep drawing (Figure 9.20). This process produces the sink in a single stroke of the press using complex male and female press tools and requiring very high pressures. A similar-shaped piece for an aircraft would most likely be produced by stretch forming (Figure 9.21), which involves stretching the metal over a former made to the internal shape of the piece.

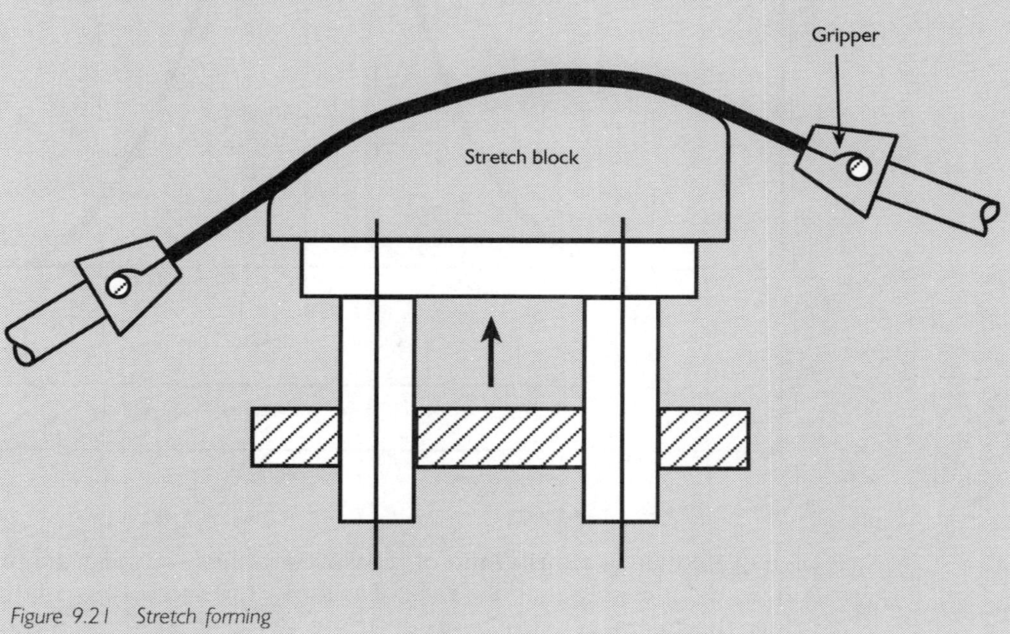

Figure 9.21 Stretch forming

The type and condition of plant available will also influence the specification of tolerances and the type of assembly configuration that can be used. Any machine or process will have what is called a 'process capability', which is the normal variability of the machine or process. This capability is the normal tolerance range that the process is capable of producing and will vary according to the age and quality of the machine. Thus, if a designer specifies tolerances tighter than this natural process capability, all parts must be inspected, and some will have to be rejected. So, the aim would be to specify tolerances that are within the capability of the machine. This capability or variability is determined by measuring the output of the machine or process and finding the standard amount of deviation from an average value. This standard value of deviation is called the standard deviation and is given the symbol σ. In statistical terms, the process capability is defined as 6 σ. The value of σ is equal to:

$$[1/n \Sigma (x - X)^2]^{1/2}$$

where, for values taken from a normal process, x = an individual value, X = the mean of all the individual values, n = the total number of individual values and Σ means 'the sum of all the values of the variable'.

The value of σ can readily be found by measuring items as they are produced and hence the process capability can be determined. Since it is a statistical measurement, the sample normally needs to be at least 50. Figure 9.22 shows a

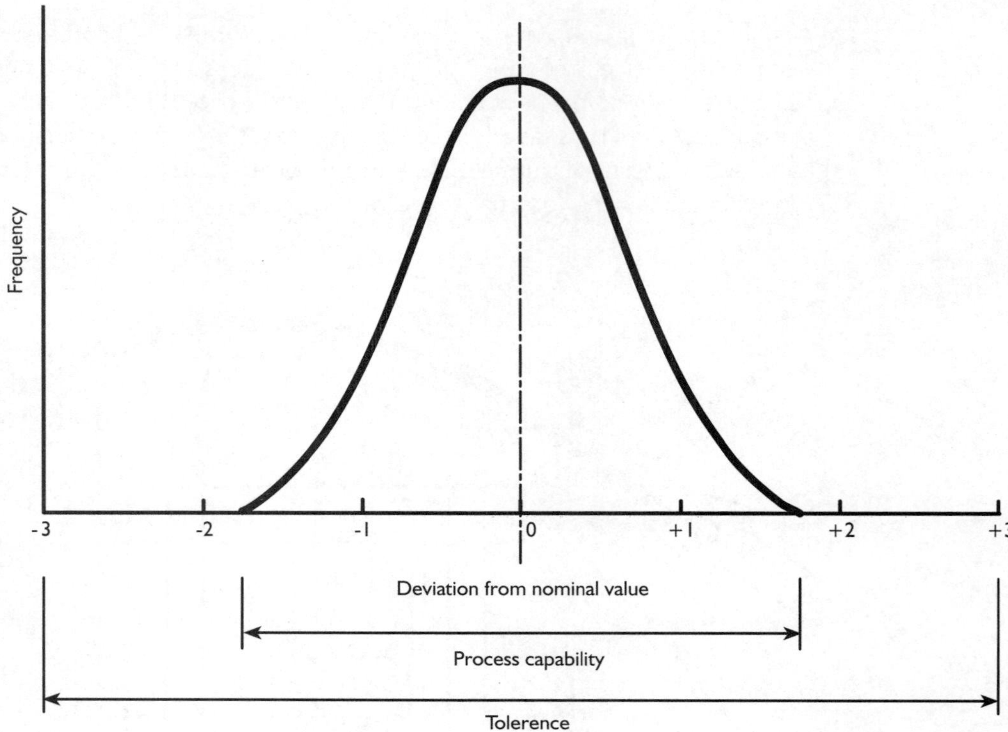

Figure 9.22 Process capability and tolerance

normal frequency distribution of measured values showing the process capability and its relationship to tolerance. To allow for some drift of the central mean value, the tolerance ideally needs to be twice as large as the capability. If the tolerance is twice the process capability, the process is said to have 'high relative precision'.

The following values were recorded in a process capability study ($n = 50$):

10.4 9.7 10.1 10.2 10.6 10.1 10.5 10.4 9.9 10.3

10.6 10.5 9.6 10.4 10.7 10.1 10.2 10.3 9.9 10.6

10.3 10.7 10.1 10.3 9.8 10.2 10.0 10.5 10.4 10.2

10.7 10.2 9.8 10.0 10.4 9.7 10.0 10.3 10.8 10.1

10.2 10.4 10.5 10.1 10.3 10.2 9.9 10.5 10.0 10.2.

$X = [1/n\ (\Sigma x)] = 1/50\ (500 + 14.3 - 1.7) = 1/50\ (512.6)$

$= 10.3$ (to three significant digits).

$\sigma = 1/n\ [\Sigma(x - X)^2]^{1/2} = [1/50\ (3.93)]^{1/2} = (0.079)^{1/2} = 0.28.$

Process capability $= 6\sigma = 6 \times 0.28 = 1.68$. So, if the process is set to an average of 10.3:

- the process capability is $10.3 \pm 3\sigma = 10.3 \pm 0.84$
- the upper capability limit is $10.3 + 0.84 = 11.14$
- the lower capability limit is $10.3 - 0.84 = 9.46$.

Therefore, for high relative precision, the tolerances would be:

- upper limit = 10.3 + 1.68 = 11.98
- lower limit = 10.3 – 1.68 = 8.62.

This gives a tolerance of 11.98 – 8.62 = 3.36 = 2 × 1.68 = 2 × process capability.

Activity 9.22

Select a workshop process and produce 50 items or do 50 operations/measurements e.g.:

- turn 50 parts on a lathe
- drill 50 holes
- mould 50 parts
- measure the resistance of 50 resistors of the same nominal value, etc.

Using the formula given, establish σ for the chosen variable and, hence, find the process capability and the tolerance for high relative precision. (This activity will be a numeracy core skill activity.)

Effect of environment on design of products and services

Any product has to function correctly in the range of environments likely to be encountered. It must also not be affected by the environment in which it is made, stored and transported.

Temperature For worldwide applications, the ambient temperature can go from –50°C in Siberia to 45°C in Arizona. This means that any equipment that is used outdoors must function satisfactorily within this range. It may mean that special procedures and service needs will be required, e.g.:

- special lubricants
- defrosting
- air conditioning.

For equipment that is designed to be operated within a building, the ambient range specified is normally much narrower. For example, a computer monitor handbook specifies a temperature range of:

- 10°C to 40°C for operating
- –20°C to 60°C for storage.

In special cases, the temperatures in which equipment has to operate are well outside the normal ambient range. This will apply to products operating at high temperatures such as:

- internal combustion engines
- gas turbines
- furnaces.

At the other end of the range are products used in low-temperature applications such as:

- refrigeration
- handling of liquid gases
- space satellites.

The main thing to consider for satisfactory functioning in a range of temperatures is the effect of temperature on material properties.

Size changes with temperature, and this change is unique to each material, so an assembly of parts using different materials must take account of any possible changes. Electrical properties change with temperature; therefore a circuit must be designed to operate correctly within the specified temperature range. All materials are subject to creep, i.e. a change of dimension under an applied load. At high temperatures, materials will creep at very much higher rates, at stresses well below their yield stress. Special creep-resistant alloys must be used for high-temperature applications. As temperatures decrease, most materials become more brittle and can change from ductility to brittleness quite abruptly at a certain temperature, called the glass transition temperature.

Case study

Liberty ships

An interesting example of low-temperature embrittlement occurred in World War Two when American 'Liberty' ships were produced as all-welded structures. Some of these ships failed catastrophically when they were in Arctic waters. The low temperatures induced brittleness, which caused cracking. Because there were no riveted joints to arrest these cracks, they propagated very rapidly and the ships broke in two. These problems were solved by a redesign involving the introduction of some riveted joints at critical points to resist crack propagation and the use of special steels that were not as prone to cracking.

Humidity and atmospheric conditions The conditions in which equipment operates can include:

- dry dusty desert
- wet tropical rainforest
- humid temperate climates like the British Isles.

This means that a designer has to consider the use of special materials and protection in certain environments. In a very dry atmosphere, materials do not need protection against corrosion but in a wet environment most metals must have corrosion treatments, such as phosphating, anodising and plating.

Resistance to chemicals In many design applications, typically in the food processing and chemical industries, materials must be used that are inert to chemical attack. This also applies to components in engines that can be subject to attack by carbon monoxide, sulphur dioxide and nitrous oxide. The effects of this can be seen in the rapid disintegration of mild steel car exhaust systems. In building and architectural applications, account must be taken of the chemicals present in the air as a result of atmospheric pollution.

Loading conditions In service, a product can be subjected to different kinds of loading. This loading can be:

- tensile
- compressive
- shear.

It can also be a steady load or a suddenly applied load. Parts can also fail at stresses below the normal breaking stress owing to fatigue. Fatigue failure occurs when a part is subjected to repeated alternating stresses, as when a steel paper clip is bent backwards and forwards a few times until it breaks. In some applications, this failure may not come until after a few million cycles of stress.

Pressure and pressure variations Safety regulations closely control the design, construction and regular testing of pressure vessels, pressure pipes and tubing. Special techniques are also necessary to seal the connections to these vessels. Another problem occurs when pressure changes, as with a pressurised aircraft. On the ground, the internal and external pressures are equal but, at altitude, the pressure inside is considerably greater than the pressure outside. This continual change in pressure can lead to a fatigue failure in the aircraft skin. The De Havilland Comet aircraft shown in Figure 9.23 failed in this way, largely because of the stress concentrations around its square windows.

Figure 9.23 De Havilland Comet

Vibration This can occur as a result of externally excited vibration when a product is in a vibrating environment, such as an audio player/radio in a vehicle. It can also happen by self-excitation as in the case of a rotating shaft reaching its critical whirling speed or a wheel being out of balance. Problems caused by vibration can be cured by design that prevents the out-of-balance forces that cause vibration. It can also be remedied by damping with energy-absorbent material or shock absorbers. Active damping is sometimes used where an equal and opposite force is automatically generated to counteract the out-of-balance force. When vibration is likely to cause problems with screwed fasteners becoming unscrewed, this has to be prevented by the use of shakeproof fasteners, split pins, lock nuts, serrated washers, etc. Some of these devices are shown in Figure 9.24.

Effect of living creatures and organisms Damage can be caused by bacteria, moulds, fungi, insects, vermin, etc. If the environment cannot be controlled to prevent attack, then the product must be suitably protected or encapsulated.

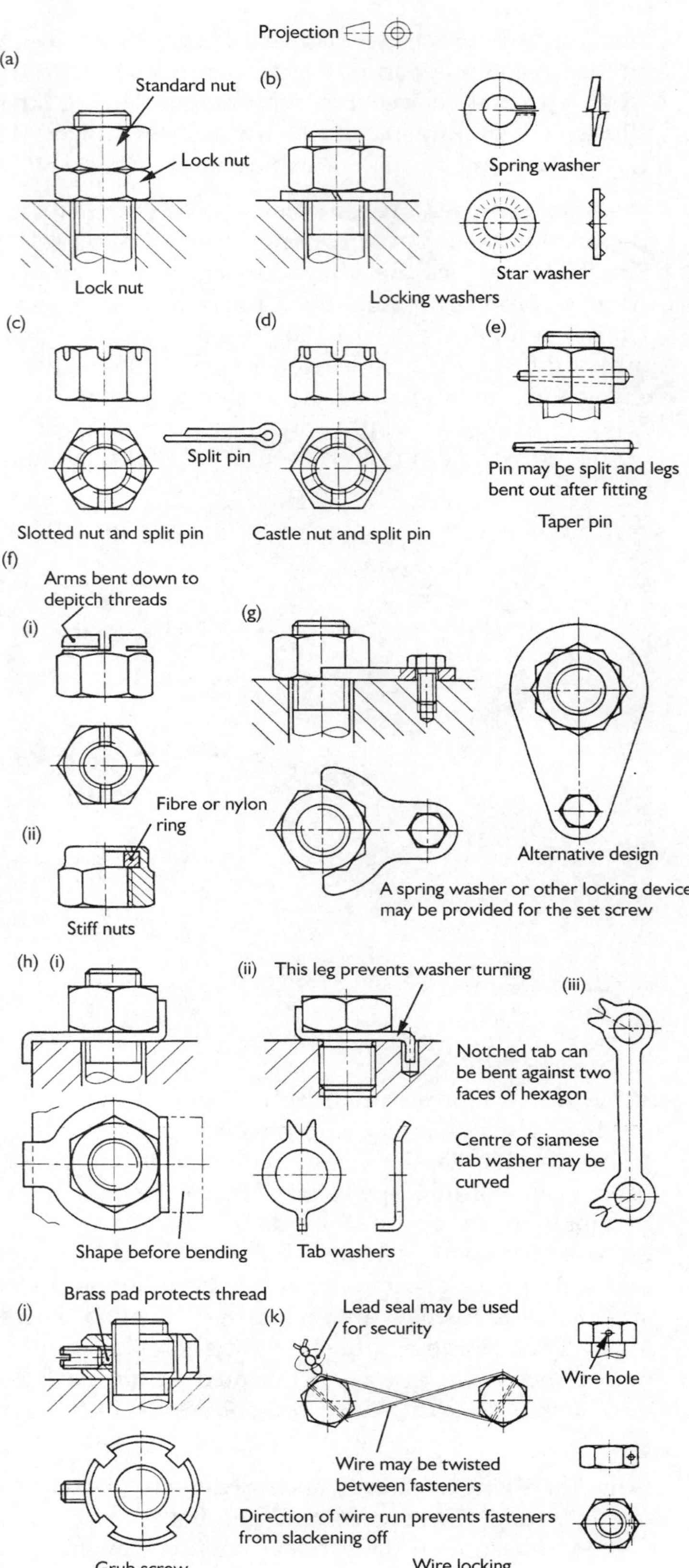

Figure 9.24
Shakeproof devices for screwed fasteners

Human use Since most products and services are for human use, they have to be designed in the most appropriate way. This is not just from the ergonomic view to make them 'user-friendly' but covers other aspects, such as vandal-proofing, fool-proofing and the effects of incorrect use.

Radiation and electromagnetic waves The deleterious effects of X-rays, gamma-rays and ultraviolet waves are well known. Similarly, electrical products, such as drills, can interfere with television signals and digital telephones can affect hearing aids. So any design should try to protect, usually by shielding, any product from the effects of these rays and waves.

Effect on the environment of a product or service
However benign, any product or service will have some effect on the environment. Even if a product or service has no effect on the environment in operation, it will initially have consumed resources in its creation. Any hardware will also have to be disposed of, or recycled, at the end of its useful life. An extreme example of this problem of disposal is a nuclear power station. It can be argued that, except for unforeseen accidents like the one at Chernobyl, it has minimal effect on the environment during its lifetime of operation. It consumes no non-renewable fossil fuels and emits no fumes or smoke. However, at the end of the 30-odd years, it has to be decommissioned. This process is almost as costly as building the station, and somewhere has to be found to store the radioactive materials for a very long period. Many products can, of course, be reused or recycled. It is plainly a sensible design strategy to make the product as totally reusable or recyclable as possible. In the case of a product like a car, it is difficult because a modern vehicle contains parts made from around 600 different materials. The metal content is around 66% so, if the car is just baled for metal scrap, a third of the materials will just be incinerated, which itself causes environmental damage.

Case study

Recyclable car design

Companies such as BMW and Volkswagen are working on car design so that most of the materials in a scrapped vehicle can easily be identified for reuse or recycling. Volkswagen are now designing their cars for automatic disassembly with recycling of all materials. As well as consuming resources in its creation, the operation of any product or service will normally have some adverse effect on the environment. It will consume resources while it is being operated. This is why it is important to consider the 'life cycle cost' of a product or service. For example, an aluminium-bodied car may initially cost £500 more than a steel-bodied car at 1997 prices. The Audi A8 aluminium space frame car is some 35–40% lighter than the equivalent car in steel. This weight saving, by reducing fuel consumption by 15%, can save the owner £1800 over a 10-year lifecycle. This is in addition to reducing the environmental impact, since most vehicles produce four times their own weight annually of carbon dioxide and other gases.

The use of any product will release emissions into the environment. Some, as in the case of vehicles, is direct. Using other power sources, such as electricity, merely shifts the problems back to the power station. Environmental nuisance like noise and pollution are strictly controlled by legislation and by directives issued by bodies such as the European Union and the US Federal Government. Therefore, any design brief must take full account of legislation and directives. Many electrical and electronic devices will cause interference to other devices, and there are regulations controlling this and giving details of the shielding/suppression that is required.

Activity 9.23

Select a product, e.g. bicycle, computer, television, vacuum cleaner, robot, tractor. Assess:

a the effect different environmental conditions likely to be encountered in service will have on the functioning of the product
b the effect on the environment of the use of the product.

Cost

This has already been mentioned under customer requirements, and it is certainly true that any design is constrained by what a customer is willing or able to pay. Thus, cost is one of the more important criteria against which any design will be judged. It is naturally difficult at the design brief stage to be absolutely precise about costs, but nevertheless a target figure must be set. The cost of any product or service is made up of both direct and indirect costs, and the indirect costs will include:

- research and development
- design and testing
- marketing.

All these costs have to be recovered from unit sales. The other indirect costs are all the costs incurred other than those directly attributable to making the product or delivering the service, e.g.:

- building costs, services, rent and property taxes
- maintenance of plant and buildings.

Direct costs will be made up of the materials used plus the cost of converting those materials into a finished product. In the case of a service, the direct costs are the cost of the materials and labour involved in the direct delivery of the service. For materials costs, it is important to include the cost of scrap and wastage as well as any possible disposal or recycling costs. Other direct costs for product manufacture and service delivery will include the direct costs of:

- labour
- energy
- plant and equipment.

Progress check

1 State an acceptable definition of technology.
2 Explain the possible reasons why technology may fail.
3 Which types of engineer are needed to design, build and maintain:

a an automatic cash dispenser for a bank
b portable electric screwdrivers?

4 Name five important factors to consider when selecting materials.
5 What are the processing stages when turning materials into finished products?
6 Define process capability. What is the ratio of tolerance to process capability to achieve high relative precision?
7 What effect can temperature have on:
a physical properties of materials
b fits between mating parts?
8 Explain the possible effect on components or assemblies of:
a different loading conditions
b pressure and pressure changes
c vibration
d human use.
In each case, give an example of how designs can overcome these effects.
9 State the possible effects on the environment of a product of its:
a raw materials
b manufacture
c use
d disposal.
In each case, how might these effects be minimised?
10 a explain the difference between direct and indirect costs.
b under which category are design costs put and how are design costs recovered?

Design briefs for products and services

This section covers the course requirements to produce two design briefs. Design briefs have to be produced for:

- an electromechanical product
- services for an electromechanical product or system, e.g. install, maintain, repair.

The first three parts of this chapter deal with each aspect of the design brief in detail and should be referred to when preparing the brief. Remember that when doing the brief you should:

- consult the 'customer' or 'client'. This should be the person or organisation for whom you are preparing the brief
- get the customer or client to sign the brief as being acceptable
- work as a team (this covers some core skills in communication)
- present it as a word-processed document (core skills in IT)
- use drawings and graphics if applicable (core skills in IT if you use CAD)
- identify clearly all references to standards and legislation, etc.

Your choice of a product or service must be restricted to those that have an electromechanical content. It is not acceptable to choose solely electrical/electronic or solely mechanical. While you can prepare a brief for something totally new, it is also acceptable to propose a redesign of a product or part of a product. It will be in your interests to choose something that can be carried forward to the Design Solutions and Technical Drawing elements. Use the next case study to help you formulate your own design briefs.

Case study

Design brief for an automatically opening garage door

- All designs to be done in conformity to BS 7000 Design management systems. All company procedures for design and manufacturing to conform to ISO 9000.

Essential functional requirements

- Fit all standard garages and normally operated from outside by remote control from owners own security device.
- Resist attacks by unauthorised entrants to standards of BS 3621.
- Operation to be satisfactory in all ambient temperatures between the two polar circles.
- Electrically operated from 220 V or 110 V AC supply.
- Manual override to be available for opening from garage interior or opening externally by key.
- Open or close in 15 seconds or less.
- Maximum range of remote control device to be 3000 mm.

Desirable functional requirements

- Fit any non-standard size garage door.
- Open or close in 7 seconds or less.
- Programmable to open on recognition of car number.
- Maximum range of remote control device to be 5000 mm.

Essential ergonomic requirements

- In manual mode, must be capable of being operated by the central 90% of the UK adult population with a maximum applied force of 5 kgf.
- Remote control device must fit comfortably into the hands of the central 90% of the UK adult population.
- Tactile devices, such as push buttons, to be capable of being operated one-handed with a maximum force of 1 N.

Desirable ergonomic requirements

- In manual mode, door to be capable of operation by the central 90% of the world's adult population with a maximum force of 3 kgf.
- Remote control device to fit comfortably into the hands of the central 90% of the world's adult population.

Essential aesthetic requirements

- Door surface to be free from visible defects and to be capable of accepting a painted finish.

- Available in all primary and secondary colours as well as black and white
- Automatic opening unit to be of pleasing form and shape and/or housed in a visually attractive enclosure.
- Any handles or attachments must match or complement door colours.

Desirable aesthetic requirements

- Door to be a mirror finish and capable of being plated or painted.
- Available in any colour specified by the customer.
- Company logo to be clearly visible and attached in such a way as to enhance the overall appearance.

Essential quality requirements

- Minimum life for door and automatic opening unit to be 15 years.
- Mean time to failure of automatic opening unit and door bearings to be 55 000 cycles of operation. One cycle of operation is defined as an opening followed by a closing.
- Five-year unconditional guarantee.
- All design, manufacturing, testing and quality systems to be in conformity to ISO 9000 or equivalent.
- All materials used should be specified and tested to the appropriate British or International Standards.

Desirable quality requirements

- Minimum life for door and automatic opening unit to be 30 years.
- Mean time to failure of automatic opening unit and door bearings to be 110 000 cycles of operation.
- Ten-year unconditional guarantee.

Essential tolerance requirements

- All assembled and bearing parts to use ISO fits and limits as detailed in BS 4500. All tolerances on assembled and bearing parts (other than fit of door to aperture) to be tolerance grade IT 9 or less.
- Fits to remain functional over the likely range of operating temperatures (see environmental constraints).
- Tolerances on door sizes to conform to BS 4500: Part 3 medium series which, for sizes between 1000 and 2000 mm, is ± 1.2 mm.
- All non-standard door sizes to be a 'matched' fit to the customer's aperture sizes.

Desirable tolerance requirements

- All tolerances on assembled and bearing parts to be grade IT 6 or less.

Essential cost and quantity requirements

- Target selling price for a standard automatic door of approximately 2200 mm by 2000 mm will be £1000. If the margin for wholesaler and retailer is 50% and the profit margin is 33%, then:

Maximum product cost = selling price/(1.33 × 1.5) = £1000/2 = £500.
This product cost will be approximately split into:

- materials, components and subassemblies 50% = £250
- manufacturing and assembly 30% = £150
- general overheads 10% = £50
- sales and marketing 5% = £25.
- design and development 5% = £25.

So, the allowable design and development costs, which will be recovered by 5 years' sales @ 25 000/year = 125 000 doors are £(125 000 × 500 × 5%) = £3 125 000.
Other cost and quantity requirements include:

- Product to sell 25 000 units/year over a 5-year period.
- Whole-life cost to be a maximum of the selling price + 10%.
- Whole-life cost to be minimised by the specifying assemblies, which are sealed for life and need no maintenance. Any parts that might possibly fail in service to be easily accessible and replaceable. Standard parts to be used if they meet the specification.
- Electric motors to be energy efficient and maintenance free.

Desirable cost and quantity requirements

- Product to sell 50 000/year over a 10-year period.

Essential size requirements

- As stated in the essential functional requirements, the doors must fit standard apertures. This will mean sizes for height and width of 2000 mm or more.
- The automatic opening unit to be contained within a size envelope of 1000 mm by 200 mm by 200 mm.

Desirable size requirements

- As stated in the desirable functional requirements, the doors should be capable of being made to fit any non-standard size of garage door. This can mean sizes of less than 2000 mm and of more than 2000 mm up to 5000 mm.
- The automatic opening unit to be contained in a size envelope of 750 mm by 150 mm by 150 mm.

Essential weight requirements

- Weight to be kept as low as possible for ease of manual and automatic operation. Security and safety aspects must also be considered here. Any materials used for the door must be strong and impact resistant, as well as being as light as possible.

Essential time requirements

- Product to be designed and manufactured within 2 years from project initiation to delivery of first doors to customers.

Desirable time requirements

- Product to be designed and manufactured within 1 year from project initiation to delivery of first doors to customers.

Standards and codes of practice

- BS EN 60730 specification for automatic electrical controls, safety, operating values and operating times.
- BS 3621 specification for thief-resistant locks operated by keys with mechanical patterns. This will be applicable to the manual override of the automatic system.
- BS 7036 code of practice for provision and installation of safety devices for automatic power-operated pedestrian door systems. This code is only for doors up to 2000 mm wide but will be a good code to use since the majority of the doors will be only slightly in excess of 2000 mm.
- PP 7317 (British Standard Educational Publication) Ergonomics – Standards and Guidelines for Designers.
- ISO 9000 Quality Systems.
- BS 7000 Design Management Systems.

Technology

- Remote control of automatic opening device to be implemented using existing technology including infrared, laser, analogue or digital radio signals etc.
- Prime mover device for opening/closing door to be operated using existing electrical technology from an AC mains input of 110 or 220 V single-phase supply.
- Any mechanical parts, such as bearings, springs, gears, etc., to be sourced, as far as possible, from existing specialist manufacturers.
- All manufacturing processes to use technologies with known capabilities, such as rolling, pressing, extruding, moulding, welding, etc.

Human resources

- Detailed designing to ensure that product can be made and assembled with as much automation as possible and with the help of semiskilled operatives.
- Overall design concept to make servicing or replacement of the automatic operating unit possible by the customer.

Materials resources

- Materials to be sourced from range of existing materials that are available from stock. All materials suppliers to be ISO 9000 approved suppliers and all material's specifications to be to approved standards.

- All components and subassemblies to be sourced as standard items from ISO 9000 approved suppliers.
- All materials, components and subassemblies to have 'substitute' alternatives specified to cover any possible supply blockages.

Plant resources

The plant available consists of:

- Automatic machines for forming, joining and finishing sheet material (all capable of IT tolerance grade 12 or less).
- CNC machine tools, such as drilling machines, milling machines and lathes (capable of IT tolerance grade 8 or less).
- Assembly machines such as riveters, soldering machines and presses.
- Assembly benches for mechanical and electrical assemblies.
- Testing equipment and test rigs.
- A full range of measuring equipment.
- Packaging and mechanical handling equipment.

Temperature

- As specified in the functional specification, the door must function satisfactorily in all regions between the two polar circles. This will mean operating temperatures of –50°C to 45°C.

Humidity and atmospheric conditions

- Function must be satisfactory in conditions varying from 0% to 100% relative humidity.
- Door and operating unit may be subjected to dusty and sandy conditions. The design must either prevent entry of damaging particles or satisfactorily accommodate their presence.

Resistance to chemicals

- Doors may be installed in areas with heavy atmospheric pollution. Materials will have to be protected against attack by sulphur dioxide and other atmospheric chemicals.
- The operating unit and the inside of the door may be subject to exposure to oils and fuels, and this should be catered for in the design.

Loading conditions

Loading conditions on the door will be wind forces with speeds of up to 120 km/h. This loading could be suddenly applied as a gust of wind.

- The door may be subject to sudden impact by slow moving vehicles.
- The automatic opening unit will normally have a steadily applied load, but the external wind loading will have to be considered.
- Shock absorbers to be used where applicable to cushion contact as door is opened or closed.

Pressure and pressure variations

- See preceding section where wind loading will cause differences between pressure inside and outside of the garage.

Vibration

- Door to be designed so that wind loading will not cause significant vibration.
- Shakeproof fasteners to be fitted to minimise the effects of any vibration in operation.
- Any vibration in the automatic opening unit is to be designed out.

Effect of living creatures and organisms

- Materials and design to minimise the effect of rodents, insects, moulds, fungi, etc.

Human use

- Fail-safe condition of automatic operating unit to be with the door in the closed position. However, manual operation should override this so that the door can still be opened. General construction of the door and lock to be robust enough to withstand normal usage and attempted forced entry. See BS 3621 for strengths of locks.

Radiation and electromagnetic waves

- Automatic operating unit to be designed to be unaffected by any waves, signals or radiation. The unit should be tested for the effects of devices such as radio, TV, mobile phone, etc.

Effect on the environment

- Maximum noise level to be 40 PNdB.
- Door to be 100% recyclable with operating unit components to be 80% recyclable.
- Energy usage to be as low as possible with automatic electrical switch-offs to be incorporated at the fully open or fully closed positions.

Cost

- See previous section on customer requirements for detailed cost considerations.

A design brief for a service would be done in a similar way under the same headings.

Assignment 9

This assignment provides evidence for:
Element 5.1: Produce design briefs for an engineered product and an engineering service
and the following key skills:
Communication: 3.1, 3.2, 3.3, 3.4
Information Technology: 3.1, 3.2, 3.3
Application of Number: 3.1, 3.2, 3.3

Your tasks

Prepare design briefs, using the previous case study as a guide, for one electromechanical product and one electromechanical service of your choice. These design briefs will be developed into design solutions in element 5.2.

Typical examples are:
a an electrically assisted pedal-cycle
b an electrically driven cement mixer
c an automatic car-park barrier
d a drilling machine
e installing, maintaining and repairing a metal cutting lathe
f installing, maintaining and repairing a lift system in a tall building
g installing, maintaining and repairing a hydraulic car lifter in a garage
h installing, maintaining and repairing of door security locks.

Chapter 10 Producing and evaluating design solutions

This chapter covers:

Element 5.2: Produce and evaluate design solutions for an electromechanical engineered product and an engineering service.

. . . and is divided into the following sections:

- Production and evaluation of design solutions
- Understanding requirements from design briefs
- Obtaining technical information
- Generating design solutions
- Evaluating and selecting design solutions.

The activity of generating and evaluating design solutions is where the design engineer takes the customer's requirements and turns them into practical designs. It will be essentially a team operation involving many other specialist engineers in fields such as electrical and mechanical systems, materials, manufacturing, etc. A project team on a new design will normally be led by a senior design engineer who will lead, inspire and co-ordinate.

Case study

Famous design engineers

Examples of this leadership in Britain where successful designs were developed include:

- Sir Harry Ferguson – the Ferguson tractor was the first tractor to have hydraulically operated implements as an integral part of the tractor instead of being towed. This greatly improved the safety and versatility of tractors and has been widely imitated (Figure 10.1).
- Sir Alex Moulton – the Moulton cycle was the first 'small wheel' cycle that could also be folded and carried. One of the very few major innovations in cycle design and still the best. Most imitations are not of the same quality (Figure 10.2).
- Sir Alex Issigonis – the Austin/Morris Mini was the first modern compact car with modern design features such as front wheel drive and transverse engine. This was hugely successful worldwide and has been copied by many manufacturers (Figure 10.3).

- Ted Ciastula – the Westland Lynx helicopter holds the world speed record for helicopters and has been exported widely. It was a highly innovative design with a semi-rigid rotor head, which enabled it to fly very fast and do aerobatics (Figure 10.4).

The importance of trade in manufactured goods to Britain means that it is essential to keep on developing successful new designs.

Figure 10.1 Ferguson tractor

Figure 10.2 The Moulton bicycle

Figure 10.3 The Austin/Morris Mini

Figure 10.4 The Westland Lynx helicopter

Activity 10.1

Find another example of an innovative British design and name the designer.

Production and evaluation of design solutions

The stages involved will be:

- Identify the requirements such as performance, fitness for purpose, reliability and cost.
- Obtain the relevant technical information, which includes materials, serviceability, functionality and processing capability.
- Generate design solutions using individuals and groups and including techniques such as brainstorming.
- Evaluate design solutions in terms of processing capability, schedules/lead time, cost/market price and output characteristics such as safety, portability, serviceability and installation.

Once a design has been selected, the detailed design will start. This detailed design will then be communicated for manufacturing in the form of drawings. If the drawings are produced on a CAD system, then this can be interfaced directly to computers controlling the manufacturing processes.

Figure 10.5 shows the generation, evaluation and communication of designs in diagrammatic form. It is particularly important to stress the iterative loop of generation/evaluation. Initial designs are generated, evaluated and then modified as a result of the evaluation.

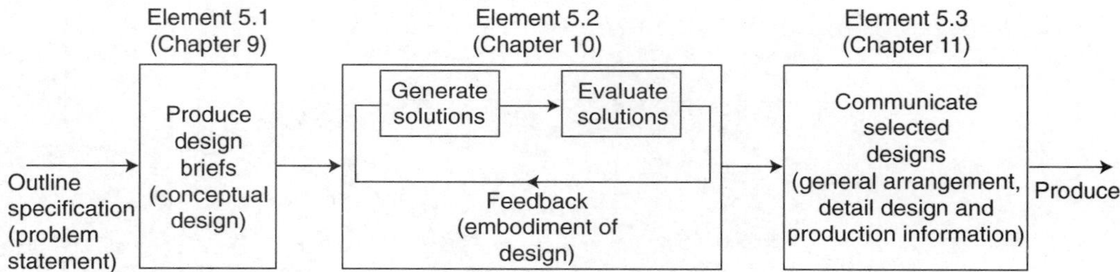

Figure 10.5 Generation, evaluation and communication of design solutions

Understanding requirements from design briefs

Design briefs will have to give all the likely requirements and analyse these requirements fully.

Performance

This is directly related to the functional customer requirements. These requirements have to be met in a way that is technically and commercially feasible within the given lead time. If it proves impossible to match the customer requirements, then the design brief may have to be reconsidered. Typical performance characteristics that have to be evaluated for a design solution are:

- speed
- accuracy
- reaction to changes in load, current, temperature, etc.
- range of operating parameters
- cost of operation
- control
- capacity.

Fitness for purpose

This is how well the product has been designed and covers all aspects of the design solution other than performance, reliability and cost. These other aspects have already been dealt with in Chapter 9 and include:

- ergonomics
- aesthetics
- quality
- size and weight
- compliance with standards and legislation
- maintainability, within agreed times, when in service.

Reliability

Reliability (R_t) is defined as the probability that the product continues to function, within the agreed performance parameters and in the specified operating environment, over a given timescale.

Mathematically, this is stated as:

R_t = number of items functioning after time t/total number of items in service.

For one particular case, R (1000 hours) = 0.975 (or 97.5%) = 975/1000.

So, 975 items are still functioning out of a total of 1000 after 1000 hours of operation.

What this means is that 25 items in 1000 will fail in 1000 hours of operation.

This gives a failure rate of $25/1000 \times 1000$ per hour = 25/1 000 000 per hour = 2.5×10^{-5} per hour.

A more meaningful measure is the mean time to failure (MTTF). This is the average time it takes for a single item to break down.

$$\text{MTTF} = (1/\text{failure rate})\text{ hours} = 1/2.5 \times 10^{-5}\text{ hours} = 40\,000\text{ hours.}$$

This becomes 40 000/24 = 1670 days = 1670/365 years = 4.58 years.

Summary of formulae

- reliability R_t = number of items functioning after time t hours/total number of items
- failure rate = $(1 - R_t)/t$
- mean time to failure (MTTF) = 1/failure rate = $1/[(1 - R_t)]/t$.

Case study

Quicksilver Communications

Quicksilver Communications has 20 million customers and keeps statistics on the breakdown of domestic telephone installations. Over a 1-year period, the number of reported breakdowns was 100 000.
Number of hours per year = $24 \times 365 = 8760$

$$\text{So, } R\text{ (8760)} = (20\,000\,000 - 100\,000)/2\,000\,000 = 19\,900\,000/20\,000\,000 = 0.995 \text{ or } 99.5\%.$$

So, the failure rate = 0.005/8760 per hour = 5.7×10^{-7} per hour
and hence MTTF = $1/5.7 \times 10^{-7}$ hours = 1 750 000 hours (200 years).

This figure shows how reliability figures have to be interpreted carefully. This figure of 200 years is the average time but, as the figures show, in a 1-year period 100 000 installations did break down. It is just that the probability of each individual one breaking down is very small. A good analogy is the National Lottery. As is well known, the probability of selecting the correct six numbers out of 49 is around 14 000 000 to 1. This is equal to a probability of 1/14 000 000 = 7.1×10^{-8}. However, even these daunting odds still mean that there are several winners each week because of the large number of entries. I would say that you have a slightly higher probability of your telephone breaking down than you have of winning the lottery!

Some ways of improving reliability are:

- Derating. This means having a safety margin. For example, a resistor rated for a current of 2 amps is run at 1.4 amps. The derating is 1.4/2 = 0.7. This is a typical figure for electronic parts.
- Parallel systems. This was mentioned in Chapter 9 under safety. What it means is that if one system fails another in parallel can take its place (Figure 10.6).
- Quality manufacturing with rigorous testing.

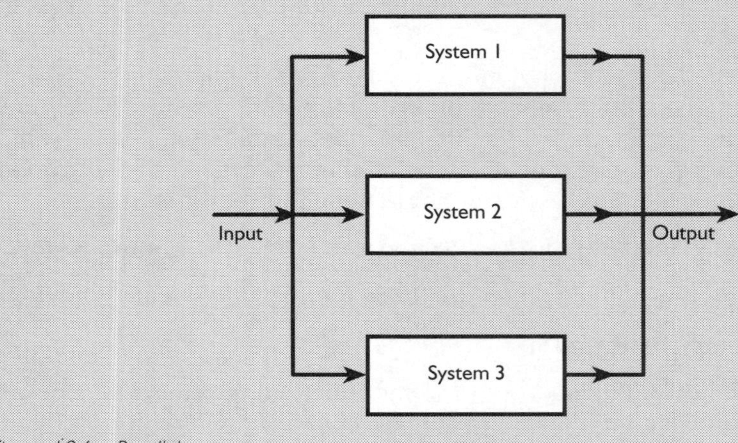

Figure 10.6 Parallel systems

Activity 10.2

Obtain a technical sales brochure or manual for an electromechanical product, e.g. an electric drill, washing machine, etc. Find out from this as much as you can on:

- performance
- fitness for purpose
- reliability.

Cost

Cost is a very important factor in design. It is not so much the cost of different qualities of design since these will vary. What is important is to be able to estimate the likely total cost of an individual design accurately.

Prototype cost Prototype originates from the Greek *proto* meaning first and *typos* meaning model. Thus, a prototype is the first model. For a large design, such as for an oil refinery, it can be a scale model. For smaller products, such as vehicles, electric motors and electronic products, the model is usually full size. Before a prototype is made, the design should be evaluated using mathematical modelling tools on computers. Some quite sophisticated mathematical modelling tools are available that give full 3D graphical displays in addition to mathematical evaluations. Typical examples of mathematical modelling are:

- gas flows, temperature distribution, stresses, power, fuel consumption, emissions, etc. in gas turbines and internal combustion engines
- kinematic modelling of linkages such as steering systems in vehicles and landing gear retraction mechanisms for aircraft
- evaluation of logic states and outputs from electronic circuits
- finite element modelling of stresses in parts such as car body shells, bolts and engine connecting rods
- simulation of sequence, speed, response, etc. of electromechanical, electro pneumatic and electrohydraulic systems.

Activity 10.3

Find as many examples as you can of commercially available software packages for mathematical modelling of electromechanical designs. Typical sources are engineering, materials, design and computing journals.

However good the mathematical modelling packages are, most designs will still require a prototype to be built for a fully realistic testing and evaluation. This is often to comply with legal requirements as well as to be tested under all environmental and loading conditions. For example:

- cars are crash tested
- vehicle engines are tested for noise and emissions
- aircraft have to undergo many safety tests to comply with civil aircraft licensing regulations
- consumer goods, such as washing machines and televisions, are life tested for reliability and safety.

Prototype costs are generally high because of the very small quantities involved. They often have to be hand built, resulting in an expensive labour-intensive process. This prototype cost will have to be recovered from the sale of the final design when it goes into production. In customised one-off production, the prototype is, in effect, the finished model. However, in this case, extensive testing will have been done on previous designs.

Production cost This has been mentioned in the previous chapter in the cost part of the section on technical and operational constraints. You may like to refer back to this. Production costs are all those incurred in producing the finished product or service. Direct costs incurred in production are:

- materials and bought-out components
- labour costs of those actually producing the product or delivering the service

- energy costs incurred in production
- plant and equipment including tools and moulds.

Indirect costs of production include:

- building costs, services, rent and property taxes
- maintenance of plant and buildings.

When evaluating alternative designs, it is the direct costs of production that should be focused on. It can be assumed that different designs being produced will incur similar indirect costs. Information on direct costs can be obtained from:

- catalogues and price lists of materials and component suppliers
- utilities such as electricity, gas and water companies
- suppliers of process plant, machine tools and equipment
- suppliers of cutting tools (for cutting speeds and feeds)
- manufacturing companies employing direct labour.

Case study

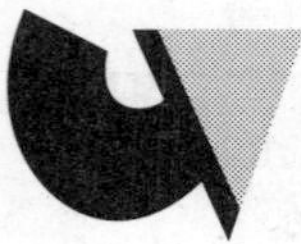

Valve assembly

Figure 10.7 shows the component parts of a valve assembly.
The component parts of the valve are:

- seating – injection-moulded ABS thermoplastic; faces machined, thread cut and counterbored after moulding
- spring guard – injection-moulded ABS thermoplastic; face machined and holes drilled after moulding
- stud – nylon; machined from ϕ 25 mm bar
- discs – stainless steel; bought-out item
- nut – stainless steel; bought-out item
- spring – stainless steel; bought-out item.

The discs in the pump valve are held against the seating by the spring in the spring guard. Pressure in the pump moves the discs upwards and opens the valve.
Costs in the company including all capital, labour and overhead costs are £40/hour.
Material costs are subject to variation and so are only roughly estimated here:

1 seating (moulding time 3 minutes) – cost = £(3/60) × 40 = £2.00
(machining time 2.5 minutes) – cost = £(2.5/60) × 40 = £1.67
plus material cost @ £0.50 total cost = £(2.00 + 1.67 + 0.50) = £4.17
2 spring guard (moulding time 4 minutes) – cost = £(4/60) × 40 = £2.67
(machining time 3 minutes) – cost = £(3/60) × 40 = £2.00
plus material cost @ £0.50 total cost = £(2.67 + 2.00 + 0.50) = £5.17
3 stud (machining time 5 minutes) – cost = £(5/60) × 40 = £3.33
plus material cost @ £0.60 total cost = £3.93
4 bought-out costs are 3 discs @ £2 each = £6.00
1 nut @ £0.50
1 spring @ £0.60
5 total cost of all items = £(4.17 + 5.17 + 3.93 + 6.00 + 0.50 + 0.60) = £20.37.

Assuming an assembly time of 4 minutes, the assembly cost = £(4/60) × 40 = £2.67.
Therefore, the total cost of one completed product = £(20.37 + 2.67) = £23.04.

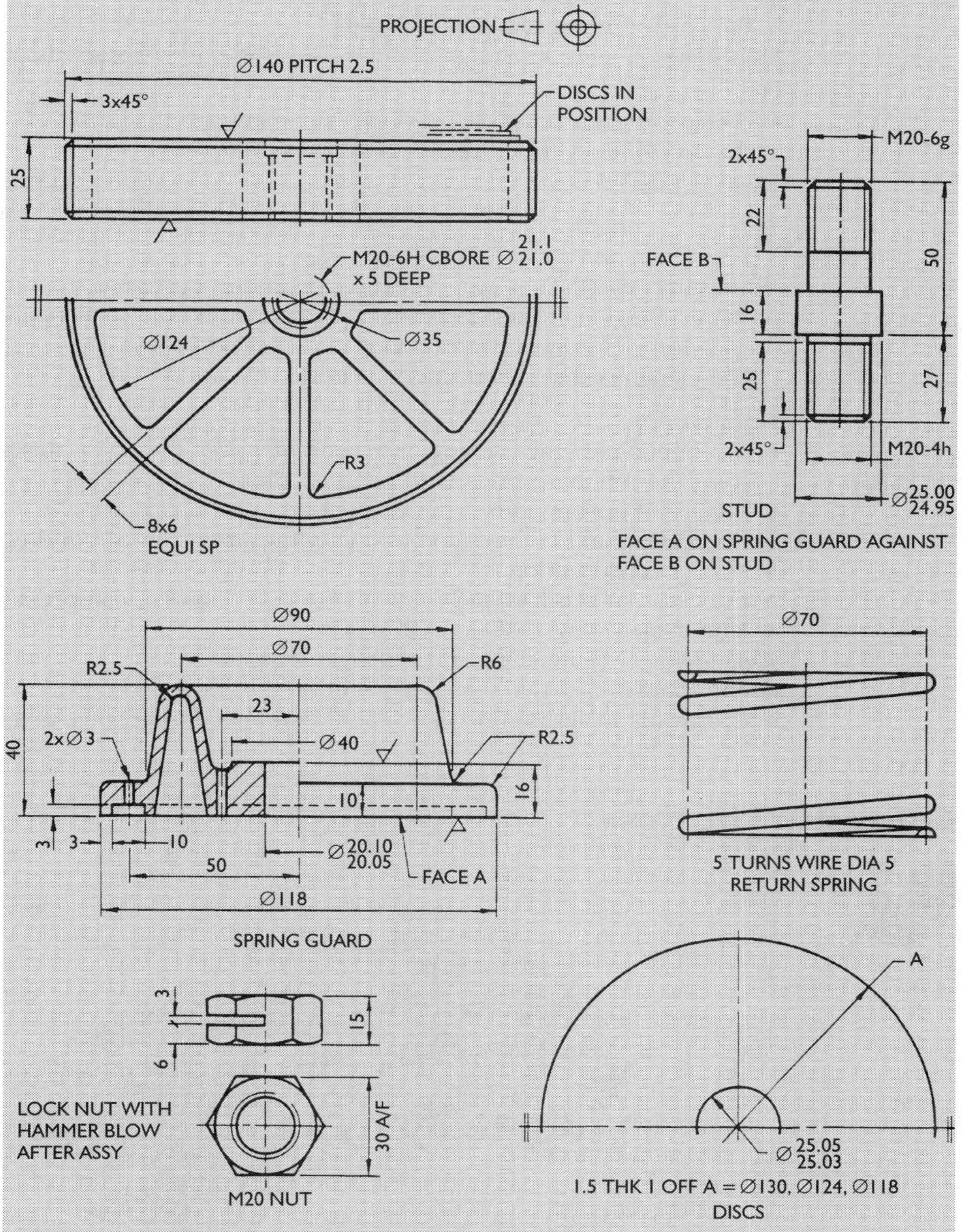

PROJECTION

Ø140 PITCH 2.5
3x45°
DISCS IN POSITION
25
M20-6H CBORE Ø $\frac{21.1}{21.0}$ x 5 DEEP
Ø124
Ø35
R3
8x6 EQUI SP

M20-6g
2x45°
22
FACE B
16
50
25
27
2x45°
M20-4h
Ø $\frac{25.00}{24.95}$
STUD
FACE A ON SPRING GUARD AGAINST FACE B ON STUD

Ø90
Ø70
R2.5
R6
23
2xØ3
Ø40
R2.5
40
16
10
3
3
10
50
Ø $\frac{20.10}{20.05}$
FACE A
Ø118
SPRING GUARD

Ø70
5 TURNS WIRE DIA 5
RETURN SPRING

3
15
6
30 A/F
LOCK NUT WITH HAMMER BLOW AFTER ASSY
M20 NUT

A
Ø $\frac{25.05}{25.03}$
1.5 THK 1 OFF A = Ø130, Ø124, Ø118
DISCS

The discs in the pump valve are held against the seating by the spring in the spring guard. Pressure in the pump moves the discs upwards and opens the valve.

Figure 10.7 Component parts of a valve assembly

Activity 10.4

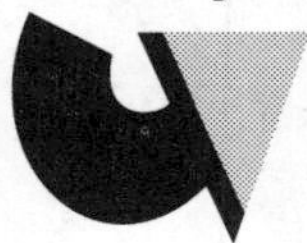

Select an assembly with about the same number of parts as in the previous case study.
Using materials suppliers' catalogues and materials data sheets find:
a the volume, weight and cost of all materials
b the cost of buying any standard parts.
Using data on feeds, speeds, cycle time, machine hourly costs, labour rates, etc. find:
a the cost of all production processes and materials
b the cost of bought-out items.

Whole-life cost This was mentioned under customer requirements in the previous chapter on design briefs. It is defined as the total cost of a product or service for a customer over a stated time period of use.

The elements that determine whole-life cost are:

- initial purchase price
- price obtained when selling at the end of a period of use – this price subtracted from the purchase price is the depreciation
- quality of design and of materials used
- installation and commissioning including provision of buildings
- energy consumption
- provision of services such as water, waste disposal, compressed air, etc.
- labour operating costs
- servicing, maintenance and repair
- environmental costs, e.g. anti-pollution measures
- overheads including insurance and insurance inspections, taxes, etc.

Case study

Ford Fiesta

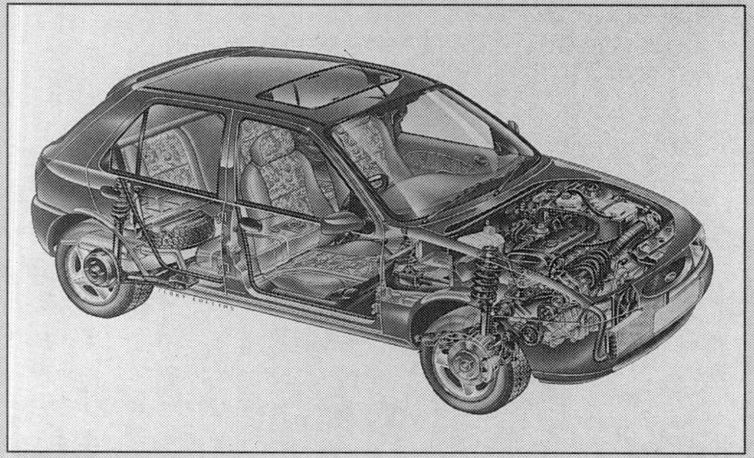

Figure 10.8 Ford Fiesta

The Ford Fiesta compact car (Figure 10.8) was one of the first cars to be designed to have minimum whole-life costs. It set the standard when it was first introduced and, according to Leasecontracts Business Fleet guide in 1996, the Fiesta 1.3 Encore still has the lowest overall costs per mile at 16.5 pence which, at 25 000 miles per year, is £4125 per year. These costs include the elements listed above such as fuel, maintenance and servicing, depreciation, taxes, insurance and membership of a motoring organisation for breakdown insurance.

Activity 10.5

Choose a product or service for which you can obtain information about whole-life costs. Typical examples could include:

- car
- motorcycle
- computer
- industrial equipment such as a lathe or a robot
- telephone service
- water service.

Establish as accurately as you can the whole-life costs based on a year's ownership or use. For vehicles, assume 25 000 miles/year. For equipment, assume it is used for 7500 hours/year.

Progress check

1 Draw a diagram showing the stages, with feedback loops, in the design process from initial concept to realisation.

2 In each case, state five typical performance characteristics for:
   a an electromechanical engineered product
   b an engineering service.

3 What is fitness for purpose? Give four examples of the types of things that have to be considered when designing for 'fitness for purpose'.

4 a How is reliability defined?
   b Describe two ways of improving reliability.

5 A building has four identical lifts. What is:
   a the reliability
   b the MTTF
   c the failure rate,
   if there are 100 breakdowns in a year (hint – work out the MTTF first – each lift will average 25 breakdowns/year or one every two weeks).

6 Define a prototype and explain why prototypes are generally expensive.

7 List all the elements making up the production cost.

8 What is whole-life cost and why is a customer particularly interested in it?

Obtaining technical information

A design engineer deals in information. If one thinks about it, the main function of a designer is to find, organise, improve and transmit information. There is an element of creativity, but the creativity is largely in the way the information is used and improved.

The important point about getting information is knowing where to go and which questions to ask. Sam Goldwyn, a famous Hollywood producer, was noted for his humorous misuse of words. He once said in a meeting: 'For your information, let me ask you a question', which aptly and wittily summarises the situation. Some ways of obtaining technical information on different subjects will now be dealt with.

Materials

There is nowadays a huge variety of materials to choose from, so it is important to be able to get accurate information on which to base design choices. This information can be obtained from:

- Textbooks. This includes Chapters 6, 7 and 8 of this book. Chapter 8 is particularly useful for materials selection. There are also many other specialist books on materials. A useful book for finding sources is Reynard, K.W., *UK Materials Information Sources*, Design Council, 1992.
- Journals on engineering, materials and design such as *Engineering Designer, Engineer, Engineering Materials and Design*, etc. These often contain up-to-date information on the use of material in a particular application.
- Manufacturers' catalogues. All suppliers of materials will provide technical information.
- Trade associations representing particular categories of materials. These can be found in a publication such as *Trade Associations and Professional Bodies of the United Kingdom*, Gale Research. Typical examples are the Aluminium Rolled Products Manufacturers' Association, the British Non-Ferrous Metals Federation and the British Stainless Steel Association.
- Professional and learned societies. Two applicable ones are *The Institute of Materials*, 1 Carlton Terrace, London SW1Y 5AF (0171 8394071) – student membership in 1997 £9/year, and *The Institution of Engineering Designers*, Courtleigh, Westbury Leigh, Wiltshire BA13 3TA (01373 822801).
- Software packages. These are available on disc or CD-ROM and allow one to search for a material for a particular application.

Libraries are a very important way of accessing the information on materials. They will normally be able to do a search for you on keywords. Typical keywords in materials selection would be:

- Materials. This would show up all the available information on all materials and so would generate a lot of information.
- Light alloys. This would generate information on all alloys such as aluminium alloys, magnesium alloys, titanium alloys, etc. This would also be a long list.
- Magnet materials. Here, one would only get those materials for this specific application. This would probably only show up a few suitable materials.

Therefore, in a search, the choice of key words is very important.

If you have access to a personal computer with a modem, it is possible to access information in various ways such as:

- sending an E-mail to manufacturers and suppliers
- via the Internet to access Web sites giving information on materials
- using an information service such as Bath Information and Data Services (BIDS). This will search its database of around 7500 journals worldwide and details of papers presented to 4000 conferences per year. You may find that your library is a subscriber to this service.

Activity 10.6

Select a design application requiring technical information on materials. Typical examples could include:

- casting alloy for die-cast cycle frame
- plastic casing for vacuum cleaner
- bearing shells for motor-cycle big end
- insulating grommets for electrical cables
- electric motor armatures.

Using as many as possible of the information sources suggested, obtain as much information as you can about suitable materials to use in your design application.

Serviceability

Serviceability has two meanings:

- ease and rapidity of servicing, maintenance and repair
- ability of the product to give service over a long period of time. This has much in common with the product's reliability, as discussed earlier in the chapter under requirements.

The stated aims of a design to fulfil the serviceability specifications can be achieved by:

- modularisation of design using proven standard components
- simplification of design to reduce number of parts
- use of good-quality materials and parts
- use of 'sealed for life' parts requiring no maintenance
- built-in devices to ensure regular services, such as oil and filter changes, are done
- use of information from service engineers and customers.

The main sources of technical information on serviceability will come from:

- manufacturers' catalogues and data sheets
- trade associations such as BEAMA – the Federation of British Electrotechnical and Allied Manufacturers' Association
- servicing data supplied by customers
- consumers' organisations – the Consumers' Association in this country supply data on consumer products like cars, televisions, washing machines, etc.

Functionality

This can be defined as the ability to perform a stated function.

For most engineering designs, this means that a full technical specification will state the way the system or component meets this function. Major sources of information on functionality include:

- British, European and international standards. If a standard is applicable, most suppliers will conform to the appropriate standard(s)
- catalogues from suppliers and manufacturers.

Case study

Hydraulic power presses

A company produces a range of hydraulic power presses. The electrically powered hydraulic power packs for these presses are bought from a specialist manufacturer. The functionality for these power packs includes:

- maximum pressure – this enables the correct valves, piping, etc. to be specified
- normal working pressure – this allows the capacity of the press to be calculated
- power – this determines the rating of the electrical supply and the size of electrical cable connections
- mechanical and electrical efficiency under various conditions – this helps determine the overall cost of operation
- flow rates of fluid – this governs how rapidly the press can operate.

Activity 10.7

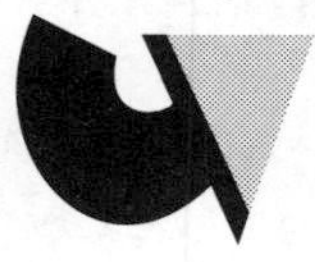

Select a functional engineered product such as:

a electric motor
b gearbox
c valve.

Using suitable information sources, write down a brief technical specification.

Processing capability

In the context of a design, processing capability can mean:

- response time of a control system. This is how fast a system can react to a change of input. The most fruitful sources here are likely to be systems text-books and data from installers and operators of systems
- speed of data processing or data transfer in an information technology system using computers and data links. There is ample information on this from manufacturers, suppliers and installers such as Microsoft, Apple, IBM, etc.
- ability of an installation, maintenance or servicing operation to process work. This would mean the number of units that could be installed, maintained or serviced per hour, day or week. Probably the best sources here would be local facilities such as garages, installers of industrial and domestic systems and equipment, electric motor reconditioning workshops, etc.

Activity 10.8

Selecting suitable examples, find out the processing capabilities of:

a a control system. Typical examples would be the ABS braking system on cars and a robot control system

b a data processing and data transfer system such as a personal computer system linked to the Internet

c an installation, maintenance or servicing facility such as lift installation and maintenance or a satellite TV installation and maintenance facility

d servicing of cycles, motor cycles or cars.

Progress check

1 Describe the role of a designer in using information.
2 List four sources of information on materials.
3 Define the serviceability of a product.
4 Suggest four ways of improving serviceability.
5 Define functionality and state its relationship to technical specification.
6 What are the best sources of information for product serviceability and functioning.
7 State three types of processing capability.
8 For each type, give the most suitable sources of information.

Generating design solutions

Once a design brief is established, the designer or design team will need to generate some solutions to the problem. These solutions can then be presented to the customer for a choice to be made. This process will essentially consist of:

- description of the problem
- finding potential types of solution
- selecting suitable types of model to use in solution generation
- generating actual solutions and design proposals.

Case study

Powered car jack

An engineering design company has been given a design brief that specifies 'a powered jack for a car'. This brief is vague and means that any kind of solution is possible. The stages are:

- description of problem – in this case to raise the car's wheels off the ground using any source of power
- finding potential types of solution – the jack could be powered by mains electricity, the car battery, high-pressure air or oil. Possible lifting mechanisms could be screw, piston, lever, airbag
- selecting model types: sketches/computer images (2D orthographic or 3D) and solid models, etc. (iconic models); models such as kinematic diagrams, flow charts and circuit diagrams, etc., which are analogous to the real situation (analogue models); mathematical formulae to evaluate power/forces/stresses; logic matrices showing digital inputs and outputs, etc. (symbolic models)

- generating actual solutions – this means testing the models and seeing whether they are initially feasible. In the first instance, this is best done with sketches and analogue/symbolic models. Testing of prototypes would come after the customer has selected from the alternatives.

Design ideas can be generated by individuals or groups. Both methods can be equally valid since most people in a group can contribute, working either alone or together. Care should be taken when working in a group that individuality is not stifled and that it is not a formal committee but a creative working group. It has been said 'a camel is an example of a horse which has been designed by committee'. On the other hand, there are some classic examples of failures, like the Sinclair C5 electrically powered tricycle (Figure 10.9), which are largely the product of one individual's ideas. Some methods of generating ideas will now be covered in a bit more detail.

Figure 10.9 Sinclair C5 electrically powered tricycle

Group methods

Since a lot of industrial design projects are large and complex, much of the work will have to be done in teams.

Case study

Aircraft design

Aerospace manufacturers need to generate designs in teams because of the size and complexity of aircraft. The design of a new aircraft would typically be split into teams such as:

- overall concept and co-ordination
- wings

- propulsion
- tailplane
- fuselage
- undercarriage
- systems and controls
- avionics.

This list is not meant to be exhaustive but is a guide to the types of team required. Each team will have a team leader who will carry the overall responsibility in each area. Each team will generate design solutions for consideration by the team responsible for the overall concept and co-ordination. Computers are making it possible to co-ordinate large projects more easily. The aircraft industry is using 3D models of the design, which are called 'electronic mock-ups'. Designers can actually work with this model to place their own parts and subassemblies. It can then be seen how the different parts actually fit together before the first prototype is built. It also saves the previous expense of having to build a full-size wooden 'mock-up'. Within the teams, a very important technique for generating ideas and possible solutions is brainstorming.

The aim of brainstorming is to get a group of people together to produce as many ideas and solutions as possible in a short period of time. The ground rules for conducting a brainstorming session are:

- no one is to be laughed at or criticised
- all ideas, however wild, are to be noted
- participants should be encouraged to produce lots of ideas
- other people's ideas can be taken and combined or improved
- all ideas should be recorded and later evaluated.

The whole point of brainstorming is to generate the maximum number of ideas. Sometimes, even the most implausible idea has proved to work well, despite the scepticism of so-called 'experts'. Sir Frank Whittle took his idea for a jet engine to the Institution of Aeronautical Engineers in the 1930s and was told that this form of propulsion for aircraft would not work. Need one say any more!

Activity 10.9

Get a group together and brainstorm as many ideas as you can on a particular problem.

Possible topics for brainstorming are:

a personal transport
b window cleaning devices
c anti-theft devices for home and vehicles.

Individual methods

Although group methods have to be widely used, there is still a place for the person working alone. Often designers specialise in one part of a design.

Case study

Car design specialisms

Major car manufacturers, such as Ford, Vauxhall and Rover, employ many designers who specialise in just one aspect of design. For example, in a car design group, there might be:

Engine design – within which could be:

- petrol engine design
- diesel engine design – within which could be:
- piston/con rod design
- cylinder head/valve design
- engine management systems design.

This specialism allows a designer to become really proficient in all aspects of a small area of design and is essential for a large design team.

Designers also often work as freelance consultants, designers and inventors. This type of individual work can be very rewarding, but it requires a special kind of person. Creativity is obviously important, but persistence is equally important, since possibly only one in a hundred of your ideas may be taken up.

Case study

James Dyson

One example of a very successful individual inventor, designer and entrepreneur is James Dyson. Dyson's first innovative idea was the ballbarrow, which uses a plastic ball instead of a wheel, and so can be used on any type of surface. Dyson then looked at existing vacuum cleaners and concluded that they were very inefficient and did not clean properly. His new and brilliant design was for a cleaner that did not lose power as it became full up with dust, was easy to empty and had a retractable hose. He worked on the idea for over 4 years and

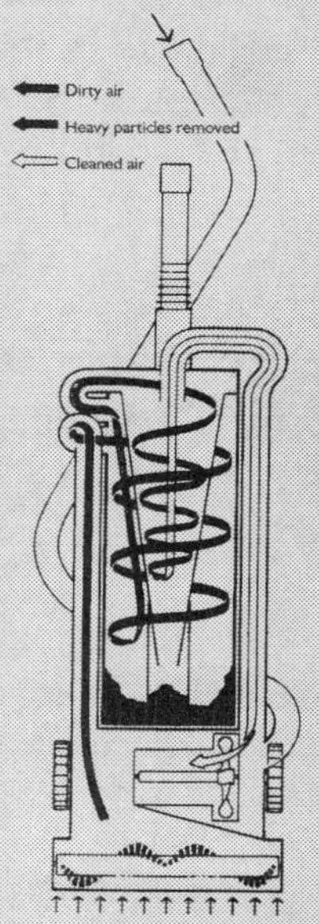

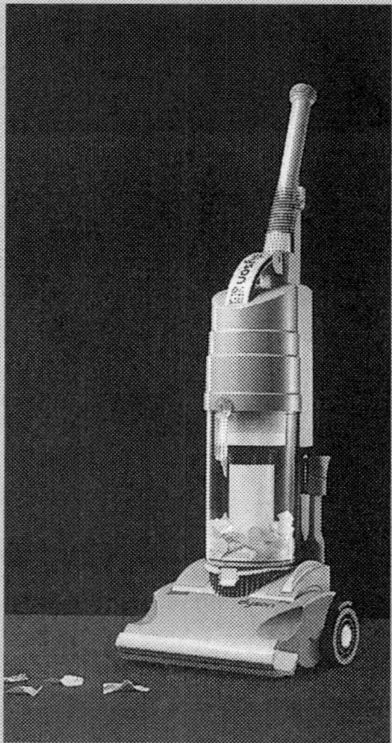

Figure 10.10 Dyson cleaner

produced around 5000 prototypes in the stables behind his house. His eventual solution was for a design using concentric cyclones, as shown in Figure 10.10. This product is now hugely successful and sold 16 000 in the first year. Now (1997) it is outselling every other model, with about 400 people being employed in Dyson's British factory.

Generation of ideas, whether by groups or individuals, can be helped by:

- getting off the tramlines – trying to break the usual mental habits and think of fresh approaches – reversing a problem can be useful, e.g. the tail wags the dog, the car drives the driver, etc.
- using comparisons and analogies – examples in nature can be used, such as bat navigation for radar and the hip joint of an animal for a universal joint
- trying to put yourself in the place of a design or part of a design, e.g. how would you feel if you were a piston in an engine?
- looking for similar problems that have already been solved
- building up a personal ideas inventory – be constantly curious and record ideas in notebooks
- collecting books, journals, data sheets, etc.
- brainstorming ideas (see group methods)
- researching previous ideas – go back as far as is possible – Leonardo Da Vinci had some good ideas, as did the Ancient Romans and Greeks! Also search for more recent ideas from books, journals and experts
- using classification systems, e.g. taking a grass cutter (or lawnmower) as an example, the different options will include power source: mains electricity, battery, two-stroke petrol engine, four-stroke petrol engine, diesel engine, push, pedal; cutter: cylindrical blade, rotary blade, cord, reciprocating blade; height adjustment: screw, lever, rack and pinion, hydraulic cylinder.

So, design solutions can be generated in many ways, either alone or in a group. The important thing is to generate plenty, so that a reasoned choice can be made. Do not always reject an idea because it does not appear to be the best at the time. Some ideas that did seem good at the time, like the motor car, now seem to be not so good. One has only to think of traffic congestion, road rage, pollution, destruction of countryside and the massive depletion of natural resources such as oil and metals, etc. On the other hand, as mentioned earlier in the brainstorming section, the jet engine was ridiculed when it was invented, but is now indispensable for world trade, communications and tourism.

Progress check

1 List four stages in the generation of design solutions.
2 What types of groups are used in generating design solutions?
3 What is a brainstorming session?
4 State five ground rules for successful brainstorming.
5 What is the desirable outcome of a brainstorming session?
6 Explain the role of specialisation in making use of individuals in design.
7 Describe six ways of assisting the generation of ideas by groups or individuals.

Evaluating and selecting design solutions

It is assumed that all the feasible design solutions that are generated will have been assessed as meeting the functional, fitness for purpose and reliability requirements in the design brief. The factors left for evaluation are:

- processing capability
- lead time and schedule
- economics, including total cost of design, development and production and selling price and size of market
- output characteristics, including safety, portability, serviceability and installation.

At this stage of design, finished drawings or prototypes will not normally be available. The evaluation will have to be done on the basis of outline proposals, with sufficient iconic, analogue and symbolic models to enable choices to be made.

Processing capability

This has been defined previously as:

- response time of a control system
- speed of data processing and transfer
- ability of a servicing unit to process work.

Suitable models to use here would be:

- analogue, e.g. flow charts enable inputs and outputs to be modelled
- symbolic, e.g. mathematical models can calculate speeds and response times.

If similar models can be applied to each proposal, then they can easily be compared.

Case study

Car park barrier

An automatic system opens a car park barrier as a vehicle drives up. The vehicle is sensed by an infrared detector, which switches on the motor operating the barrier arm. The motor runs at 1400 r.p.m. with a reduction of 50:1 to the shaft connected to the barrier arm. The shaft connected to the barrier arm has to rotate 90° to lift the barrier arm.

The symbolic model for this is:

- Speed of shaft = speed of motor/50 = 1400/50 r.p.m. = 28 r.p.m.
- Time for shaft to rotate one revolution = 1/28 min = 0.036 min = 2.2 s
- Time for barrier to lift = time for one revolution/4 = 2.2/4 s = 0.55 s.

So, if we assume the response time of the control system is 0.05 s, the response time of the automatic barrier = (0.55 + 0.05) s = 0.6 s.

The analogue system model is detector signal input → control system → motor/barrier system → output barrier up.

Activity 10.10

Select a control system, data processing/data transfer system or servicing system of your choice. Construct suitable models to describe it and work out the processing capability.

Lead time and schedule

The lead time is defined as the time taken from approval or selection of a design solution to the first customer taking delivery.

The schedule is a time plan of each stage that will add together to make the lead time.

The following stages will all be included in the lead time schedule:

- detailed design and communication
- prototype building
- testing
- production planning
- manufacturing.

The schedule shown assumes that concurrent engineering is used so that each stage overlaps. The advantages claimed for this are:

- shorter lead time
- better co-operation between specialists
- easier changes and modifications.

The alternative is the so-called 'over the wall' approach. In this, each stage is completely finished before the next one is started. This has the claimed disadvantages of:

- longer lead time
- it is harder for specialists to co-operate
- changes and modifications are more difficult – this is because, once the design has been finished and communicated, a lot of time and effort is required to change it.

It is very important that realistic lead times are calculated. Most design projects take longer than originally planned. If this happens, the customer will be dissatisfied, and the project will end up costing more. It is prudent to build in a contingency time of say 25% of the total lead time to take care of any unforeseen problems that may occur.

Economics

Economics are of crucial importance in design. No product or service can succeed unless it can be delivered at a price that will guarantee it a market within the quality of design set in the design brief.

Any economic evaluation of design solutions will have to assess:

- size of market
- selling price
- total cost of design, development and production.

This topic has been covered in Chapter 9 under customer requirements. The chosen quality of design will usually determine the size of market and the selling price. The cost of design, development and production will depend on how new and innovative the product is. If existing designs can be modified and standard components used, then costs will be a lot lower than for a totally new design with new components.

Case study

Vacuum cleaner

Two design solutions (A and B) are presented for an electrically driven vacuum cleaner for which the outline design brief was:

- comply with all relevant standards for function and safety
- be easily mass-producible
- sell to distributor for £100 or less (1997 prices).

Design solution A

- uses existing parts and technology in an attractively restyled casing with ergonomically designed controls
- market at target price estimated at 25 000 per year for 5 years
- cost of new tooling estimated at £300 000
- cost of production estimated at £50 each
- cost of redesign estimated at £200 000
- target selling price £80
- cost of design and tooling = £(300 000 + 200 000)/(5 × 25 000) per cleaner = £500 000/125 000 = £4 per cleaner
- total costs = £(50 + 4) = £54
  profit = £(80 – 54) = £26 per cleaner.

Design solution B

- is a total redesign using a technologically innovative solution
- market at target price estimated at 100 000 per year for 5 years
- cost of new tooling estimated at £1 000 000
- cost of production estimated at £60 per cleaner
- target selling price £100
- cost of design/development estimated at £1 500 000
  cost of design/development and tooling = £(1 000 000 + 1 500 000)/(5 × 100 000) = £2 500 000/500 000 per cleaner = £5 per cleaner
- total costs = £(60 + 5) = £65
  profit = £(100 – 65) = £35 per cleaner.

Solution A is taking the comparatively safe route of using tried and tested technology.

However, because it is in a very competitive market with lots of similar designs, it will sell at a lower price and the profit will be lower.

Solution B should take a larger market share because it is a technologically innovative design. It can also secure a higher selling price so, despite the higher costs, the profit will be higher. There will be a greater risk associated with solution B, but also greater rewards if it succeeds. In each case, a proper comparison can only be made if the costing is done on a realistic basis, with reasonable assumptions made about market share and price.

Activity 10.11

Select two different designs of the same product using:
a well-established technology
b innovative technology.
Find out the market share and selling price of each design. Estimate the profitability of each design based on your estimate of costs.

Output characteristics

The output characteristics are:

- safety – in most cases, safety will be evaluated to see if the design meets all standards and regulations. If the basic requirements can be bettered, then this will give a marketing advantage
- portability – all products have to be moved to their customer. Some will require to be moved in the course of their normal usage. Therefore, a typical evaluation will consist of measurement of size, weight, shock resistance, etc. If large, is transport practical? If not, can it be dismantled and reassembled? If it needs to be frequently moved in service, does it have wheels or other devices to make movement easier?
- serviceability – most products require routine servicing. Designs must try to make this as infrequent, quick and easy as possible. Evaluation will be on frequency of servicing, amount of servicing – look for parts that are 'sealed for life' and require no routine servicing, calibration and adjustments required – where possible use self-adjusting systems, time taken – take account of the total time required including dismantling and reassembling, condition-monitoring devices – automatically indicate when service/repair/replacement are necessary.
- installation – this evaluation will look into the ease with which the product can be installed and operated where it will be used. This will cover the provision of appropriate systems to cover available services. This includes type (AC/DC) and voltage of electricity, hydraulics, pneumatics, etc., adjustment devices to ensure product can be levelled, calibrated, etc., safe lifting points, designs of enclosures, foundations, etc.

Activity 10.12

Choose a suitable design and evaluate it for output characteristics.

Numerical evaluation of designs

In order to make a rational choice between different design solutions, a numerical method can be used. This involves rating the solutions on a scale from 1 to 10 so:

1 ____________________ 5 ____________________ 10

Poor | Average | Excellent

Each aspect is given a rating from 1 to 10:

- processing capability
- time
- economics
- output characteristics.

These ratings are then multiplied by a weighting factor, also from 1 to 10, which assigns the relative importance of each factor therefore:

1____________________5____________________10

Unimportant   Moderately important   Extremely important

The individual scores of (rating × weighting) for each factor are then added together. The design coming out with the highest score would then normally be selected.

Case study

Design weightings

The weightings assigned to a design problem are:

- processing capability – 10
- time – 5
- economics – 1
- output characteristics – 7.

Two designs are evaluated with the results shown in Table 10.1.

Table 10.1 Design weightings

	Design 1			Design 2		
	Rating	**Weighting**	**Score**	**Rating**	**Weighting**	**Score**
Processing capability	6	10	60	8	10	80
Time	10	5	50	5	5	25
Economics	9	1	9	1	1	1
Output characteristics	7	7	49	10	7	70
Total score			168			176

Note that a computer spreadsheet model is ideal to solve this problem.

So, on this basis, design 2 would normally be selected since it has the highest score. This assumes that both designs are equally good from the requirements of performance, fitness for purpose, reliability and cost. If required, a similar numerical analysis can be used to compare designs from the point of view of meeting requirements.

Activity 10.13

Take two design solutions – they could be your own or ones you have selected. Then:

a allocate weightings for each output factor

b analyse each design solution and rate them on each factor

c multiply ratings × weightings to give the score for all factors on both designs
d add all scores up for each design. The one with the highest score should then be selected as the optimum solution.

Progress check

1 Give three different definitions of processing capability.
2 Which types of model can be used when evaluating processing capability?
3 A computer is running at 133 MHz and is used for computer-aided design. The screen has a grid of (2048 × 3072) pixels. If each pixel requires 24 bits for a colour display, how long does it take the computer to process a full screen of graphics?
4 Define lead time and list the different elements that contribute to it.
5 State the advantages of concurrent engineering compared with over the wall methods.
6 A product costing £5 000 000 to design and develop has definite orders for 1000. The product takes 2000 person–hours to make. The material cost per item is £60 000, the direct labour rate is £30 per person–hour and the production overheads are £20 per person–hour. What is the total cost per item including design and development?
7 Describe a suitable method of evaluating the safety of a design.
8 What considerations apply when assessing portability?
9 State:
   a five factors to be considered when evaluating serviceability
   b three considerations to be used when looking at installation.
10 Two design solutions are assessed as shown in Table 10.2.
Score each design using weighted ratings and say which one is likely to be

Table 10.2 Design solution assessments

	Weighting	Rating design 1	Rating design 2
Processing capability	5	7	8
Time	7	10	5
Economics	10	8	7
Output characteristics	4	4	8

chosen. If possible, do question 10 using a computer spreadsheet model to gain core skills in IT.

Assignment 10

This assignment provides evidence for:
Element 5.2: Produce and evaluate design solutions for an electro-mechanical engineered product and an engineering service
and the following key skills:
Communication: 3.1, 3.2, 3.3, 3.4
Information Technology: 3.1, 3.2, 3.3
Application of Number: 3.1, 3.2, 3.3

Your tasks

Generate at least two design solutions for each of the two design brief assignments you chose for assignment 9. Evaluate your design solution as detailed in this chapter for:

a processing capability
b schedule and lead time
c cost and market size
d output characteristics including safety, portability, serviceability and installation.

Where possible, use the weighted rankings method, as suggested in the section on the numerical evaluation of designs on page 369, and select the final design based on your evaluation.
The project portfolio should include a project log and all evaluation notes, with reasons for the selection of the chosen designs. Properly written and presented notes, using word processing and spreadsheet models, will also gain key skills credits in Communication, Numeracy and IT.

Chapter 11 Technical drawing

This chapter covers:
Element 5.3: Use technical drawings to communicate designs for engineered products and engineering services

. . . and is divided into the following sections:
- Graphical methods.

All engineering designs need to be implemented in stages. Modelling is the first stage, when designs can be analysed and tested. These models can be mathematical, e.g. the finite element method of testing a design for stresses, simulation of the logic input and output states of an electronic circuit, etc. or a real-scale model of the design. The next stage is the building and testing of a prototype. Modelling and prototype testing should result in a final design that is suitable for the marketplace. The final design is then communicated to the manufacturing engineers for putting into production. For a service, the designs would include all the necessary information for installation and maintenance. Thus, design and engineering requires clear and precise communication at all stages. Much of this communication is done through the medium of drawing, either manual or computer assisted. Drawing is involved at all stages of a design from initial rough sketches through feasibility studies and into the drawing of the details and assembly for manufacture, installation and maintenance. It will also include drawings for publicity, brochures and manuals. This chapter concentrates on giving sufficient skills to communicate the finished designs from Element 5.2. Graphical methods are fully covered in Chapters 4, 5 and 6 of the Stanley Thornes' book for *Intermediate GNVQ in Engineering*. If Intermediate GNVQ has not been attempted, it will be worthwhile to consult the intermediate book.

Graphical methods

Manual and computer-aided methods

Both manual and computer-aided draughting are widely used in industry and commerce. It is essential first to master the basics of manual draughting so as to get a feel for sizes and scale. Also, rough sketches are a very useful initial design tool, and these are not so easy to do on a computer. Table 11.1 summarises some of the advantages and disadvantages of manual draughting.

Computer-aided draughting is becoming increasingly used and it can do many things that are not possible with manual draughting. Table 11.2 summarises some of the advantages and disadvantages of computer-aided draughting.

Table 11.1 Manual draughting

Advantages	Disadvantages
Low initial cost	Manual storage and slow retrieval of drawings
Initially fast and flexible	Modifications are slow and expensive
Can be done anywhere without electrical power	No linkages to manufacturing systems, etc.
Very suitable for initial sketches of concepts	Manipulation of images and data is difficult
Good for artistic impressions of designs	Animation is not really practical

Table 11.2 Computer-aided draughting

Advantages	Disadvantages
Modifications are easy to make	Initially not as fast as manual methods
All types of view easily displayed/plotted	Difficult to use for making rough sketches or artistic impressions
Multiple copies can be rapidly made	
Libraries of parts/assemblies can be stored on disc and easily retrieved	Possible loss of data if the system goes down. This can be prevented by making frequent back-up copies
Complex assemblies are easily managed by holding using a database of subassemblies and parts	
Manufacturing programmes can be produced directly from the drawing	
Parts can be tested for stressing using finite element meshes produced from the part drawing	
Simulations can be done of circuits, mechanisms, etc.	
Files of standard parts like springs, fasteners, electronic/electrical parts can be held on disk	

A mix of methods should be used for any of the drawings that communicate the chosen design solutions. Whatever method is used must, however, be done in conformity to all the appropriate standards and conventions.

Progress check

1 Why is it essential for a designer to be able to produce graphical design solutions?
2 List the different stages of design at which graphical methods would be used.
3 State why an ability to use manual methods is very important for anyone producing technical drawings.
4 Describe the advantages and disadvantages of manual drawing.
5 Describe the advantages and disadvantages of computer-aided drawing.

Fundamentals of drawing including projections, conventions and standards

Whatever graphical methods are used – sketches, schematic drawings, detail drawings, assembly drawings, etc. – a knowledge is required of:

- drawing sheets and types of drawings
- use of correct line and lettering types
- orthographic projection
- sectioning
- dimensioning and tolerancing
- pictorial projection
- conventions
- standards and symbols.

These items will be dealt with in turn.

The British Standards (BS) that should be consulted when doing the technical drawings of Element 5.2 designs are BS 308 Engineering drawing practice, BS 4500 Limits and fits, BS 3939 Graphical symbols for electrical power, telecommunications and electronic diagrams, BS 2917 Graphical symbols used on diagrams for fluid power systems and components, BS 1553 Graphical symbols for general engineering and BS 499 Welding symbols. It is worth buying the following BS publications: PP 7308 – an abridged version of BS 308 for schools and colleges; PP 7307 – graphical symbols for schools and colleges; BS 4500A and B – data sheets for selected ISO fits.

Drawing sheets and types of drawings Standardisation is essential here since drawings need to be copied for actual use. The originals are either kept in filing cabinets, on microfilm or on computer disk. It is possible to scan manually produced drawings into a computer system, and this greatly simplifies storage and retrieval. Figure 11.1 shows the sizes of, and relationships between, standard drawing sheets.

Figure 11.2 shows some typical examples of the different types of drawing, which are:

(a) design layout drawings that are preliminary sketches or drawings to show a designer's intentions
(b) detail drawings that give all the necessary information required to make a part. Note that tolerances are omitted in this drawing but would normally have to be on a working drawing (see page 392)
(c) tabular or parametric drawings that give details of geometrically similar parts to avoid having to draw all the possibilities
(d) assembly drawings that show parts, part numbers and the arrangement of the parts.

Once a drawing sheet has been selected, it must be given borders and title blocks. Most companies will have their own standard formats on preprinted sheets or computer disk. Figure 11.3 shows a typical drawing sheet with borders and title blocks.

Tables 11.3 and 11.4 give details on the basic and additional information required on all types of drawings. Table 11.5 gives examples of the specific type of information required for detail drawings, according to whether the part is made 'in-house' or purchased from a supplier. The compiling of parts and components lists are covered later on page 450.

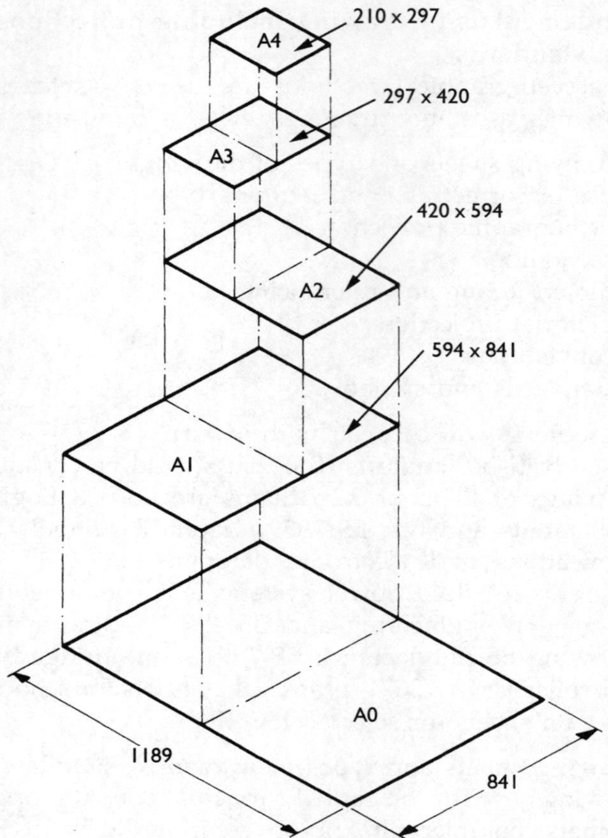

Figure 11.1 Sizes of, and relationships between, standard drawing sheets

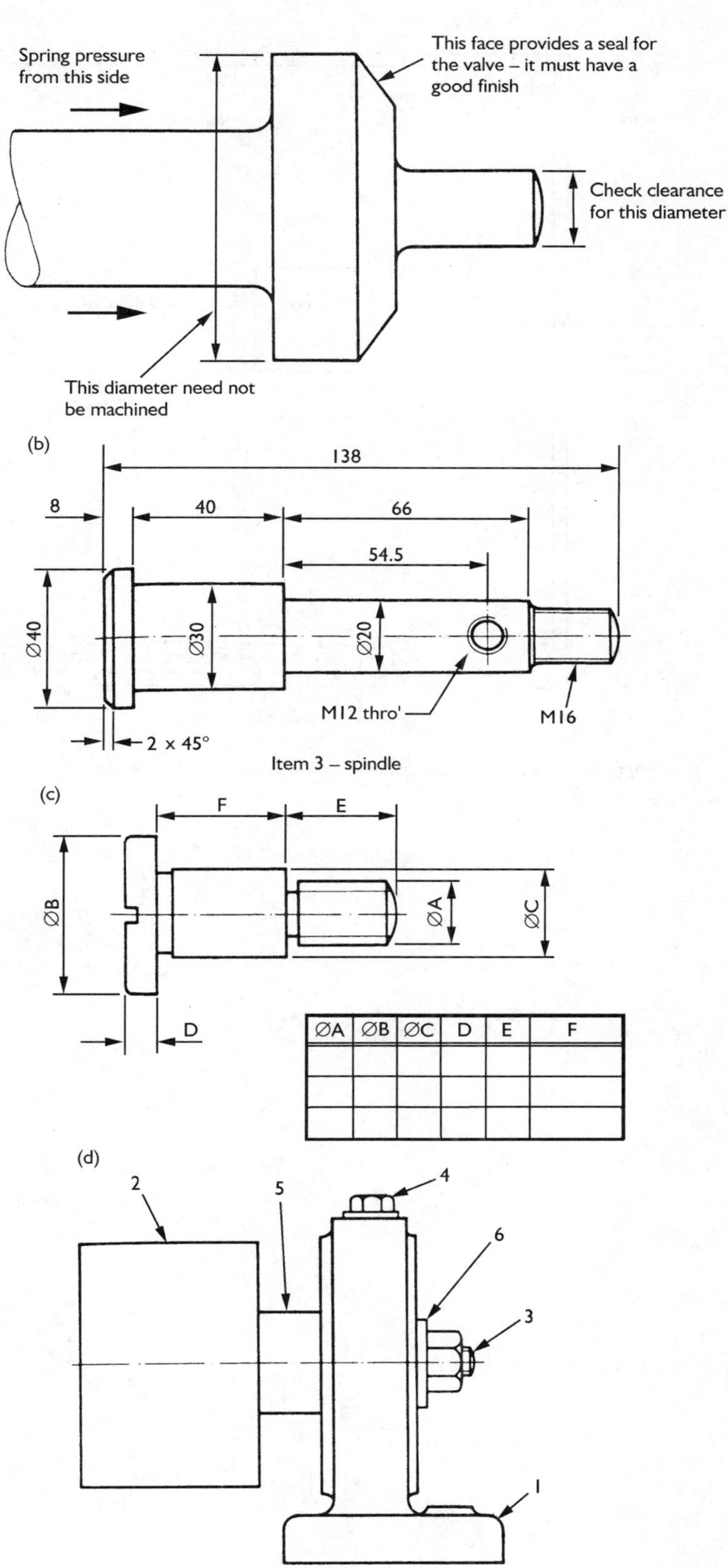

Figure 11.2 Examples of different types of drawing

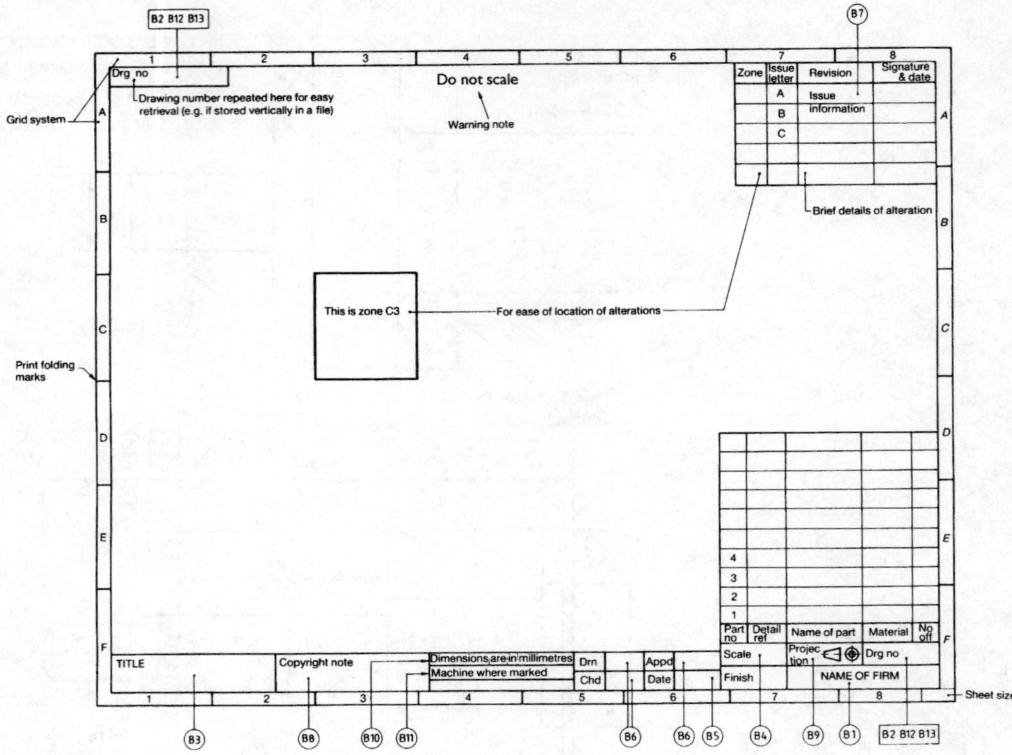

Figure 11.3 Typical drawing sheet with borders and title blocks

Table 11.3 Basic information on drawing sheets

Ref no	Basic information	What the user needs to know	Comment
B1	Name of firm	Who/where is this from?	Useful outside the place of origin
B2	Drawing number	Where the drawing can be kept and retrieved easily	Useful inside the place of origin
B3	Descriptive title of depicted part or assembly	What does it show?	This may be important if components are to be grouped for economic production
B4	Original scale	What is the original size of drawing: size of feature?	The size may be altered by subsequent reprographic process
B5	Date of the drawing	When was it drawn?	
B6	Signatures(s)	Who drew it? Checked it? Approved it?	
B7	'Issue' information	Is this the latest issue of this drawing?	Drawings are reissued when they are modified or updated
B8	Copyright clause	Is this information confidential? Secret? May it be copied?	Original ideas, patents need to be safeguarded
B9	Projection symbol	What type of projection has been used? First angle? Third angle?	
B10	Unit of measurment	Metric? Imperial?	
B11	Reference to drawing practice standards	Which standards have been used? Machining marks? Position tolerances?	
B12 B13	Sheet number Number of sheets	How many sheets are there related to this drawing?	Often in the same space as the drawing number. A simple statement '3 of 6' means that this is sheet number 3 from a total of 6 sheets

Table 11.4 Additional information on drawing sheets

Additional information	What the user needs to know	Comment
Material and specification	What is it made from?	
Related specifications	Identity of special requirements	Alloy steel? Concrete mix?
Treatment/hardness	Is the part acceptable as machined? Does it need any special treatment before it is used?	
Finish	Plated? Painted?	
Surface texture	Is it acceptable rough? smooth?	If so, how rough? How smooth?
General tolerances	How much variation from the general dimensions given can be tolerated?	This information will have a considerable influence on the manufacturing methods to be used
Screw thread forms	BS Whitworth? British Association? Metric? ISO?	
Sheet size	How big was the original drawing sheet?	This may have been increased or reduced when copied
First used on	Has this component/idea been used before	
Similar to	Does this component belong to a similar group of components?	These items help in the location of tools and equipment which aid manufacture
Equivalent part	Does an equivalent part exist?	
Supersedes or Superseded by	Is this drawing up-to-date?	Out-of-date reproductions are not always withdrawn from circulation quickly, but the fact that they are retained should be shown clearly
Tool reference	Are there any special tools to help with this job?	
Gauge references	Do I need any special gauges to help me to assess the quality of this job?	
Grid systems or zoning	How to locate particular features quickly	For instance if a drawing is being discussed by telephone
Warning notes e.g. DO NOT SCALE	Any special requirement to be noted	Drawings should never be scaled. Some are deliberately either increased or reduced in size
Print folding marks	Where to fold drawings so that important details do not become obscured	Drawings accumulate dirt along creases. Folds which obscure important features must be avoided

Table 11.5

Location of part	How many?	Part no.	Name	Raw material	Treatment required after machining
(a)					
1	6		Angle bracket	Cast iron casting	None
(b)					
17	32		Location pin	20 mm-diameter cold rolled D2% carbon steel	Case harden and grind
(c)					
2	4	XYZ SPEC No 20	Bearing bush	40 mm-diameter phosphor bronze bar	For details see maker's specification
(d)					
32	108	GKN16	Hex head screws	Not required as diameter, type of thread material etc. will be included in maker's detail or code	
(e)					
8	20	ABC 16/2	Cylindrical tension spring	Not required as wire diameter, coil diameter, shape of ends and heat treatment will be included in maker's detail or code	
(f)					
6	24	PQR 100	100 W lamp bayonet fitting	—	—

Typical entries in title blocks
(a) (b) and (c) manufactured items
(d) (e) and (f) bought-in items

Lines and lettering The types of line to be used are defined in BS 308: part 1 (1993) and are shown in Figure 11.4. Note the different uses of thick and thin lines, which must be used in a drawing in the ratio 2:1. The allowable line thicknesses are 0.25, 0.35, 0.5, 0.7, 1.0 and 2.0 (all in mm).

These lines can be produced by pen or pencil, but pencil and pen lines should not be used together on the same drawing. Pencils need sharpening to produce the correct width, while pens can be obtained to give the correct width automatically.

The application of the different line types is shown in Figure 11.5.

Examples of typical types of lettering are shown in Figure 11.6, and this shows that lettering is kept simple and is done in a thick line.

Orthographic projection This method of drawing projects views of different faces of the object onto different projection planes. First angle projection is shown in Figure 11.7a and third angle in Figure 11.7b.

Orthographic drawings are done full size, wherever possible, but the use of scale drawings is sometimes unavoidable. Scales are applied in multiples of 2, 5 and 10, and commonly used scales are:

- 1:1 – full size
- 1:2 – half full size

	Line	Description	Application
A		Continuous thick	A1 Visible outlines A2 Visible edges
B		Continuous thin	B1 Imaginary lines of intersection B2 Dimension lines B3 Projection lines B4 Leader lines B5 Hatching B6 Outlines of revolved sections B7 Short centre lines
C		Continuous thin irregular	C1 Limits of partial or interrupted views and sections, if the limit is not an axis
D		Continuous thin straight with zigzags	†D1 Limits of partial or interrupted views and sections, if the limit is not an axis
E		Dashed thick	E1 Hidden outlines E2 Hidden edges
F		Dashed thin‡	F1 Hidden outlines F2 Hidden edges
G		Chain thin	G1 Centre lines G2 Lines of symmetry G3 Trajectories and loci G4 Pitch lines and pitch circles
H		Chain thin, thick at end and changes of direction	H1 Cutting planes
J		Chain thick	J1 Indication of lines or surfaces to which a special requirement applies (drawn adjacent to surface)
K		Chain thin double dashed	K1 Outlines and adges of adjacent parts K2 Outlines and adges of alternative and extreme positions of movable parts K3 Centroidal lines K4 Initial outlines prior to forming §K5 Parts situated in front of a cutting plane K6 Bend lines on developed blanks or patterns

Note: The lengths of long dashes shown for lines G, H, J and K are not typical due to the confines of the space available.
†This type of line is suited for production of drawings by machines.
‡The thin F type line is more common in the UK, but on any one drawing only one type of dashed line should be used
§Included in ISO 128-1982 and used mainly in the building industry.

Figure 11.4 Types of line

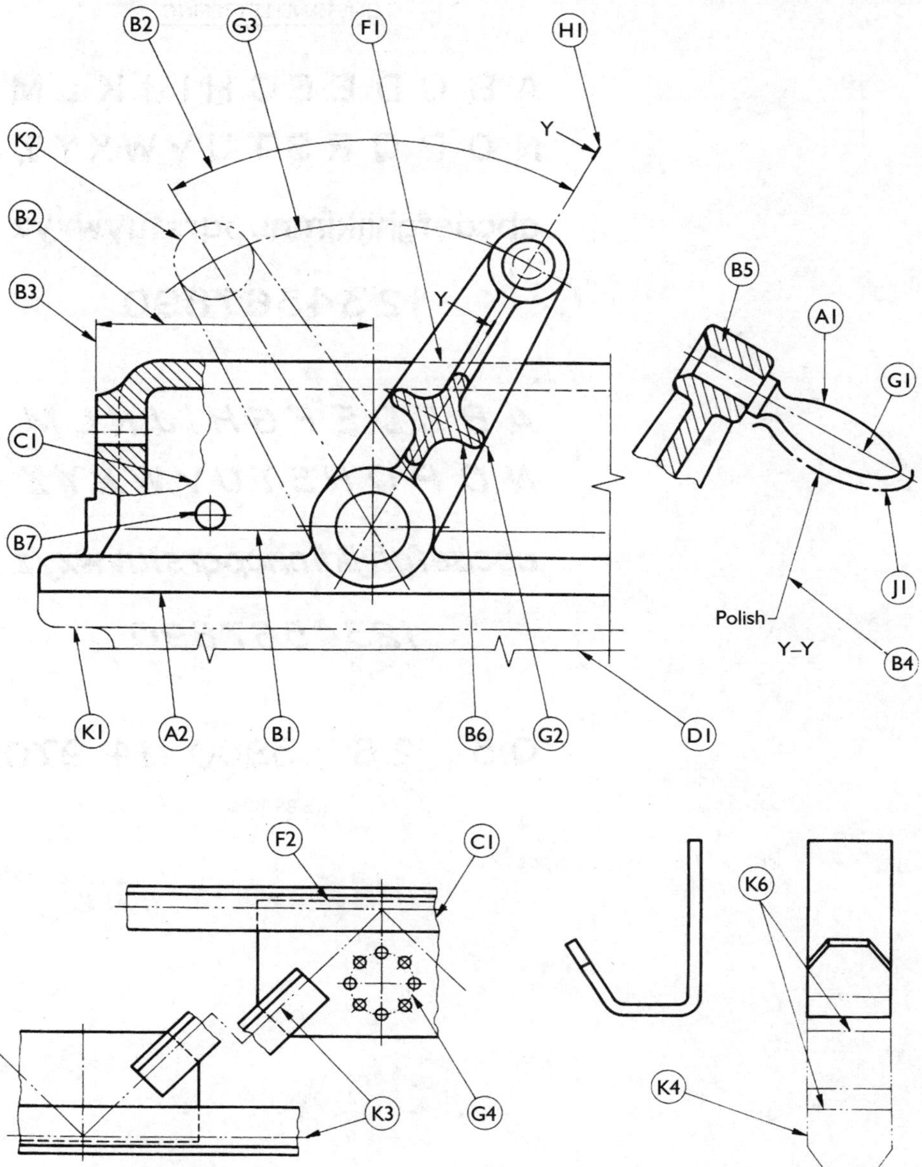

Figure 11.5 Application of the different line types

<u>LINES AND LETTERING</u>

A B C D E F G H I J K L M
N O P Q R S T U V W X Y Z

abcdefghijklmnopqrstuvwxyz

1234567890

A B C D E F G H I J K L M
N O P Q R S T U V W X Y Z

abcdefghijklmnopqrstuvwxyz

1234567890

0.5 2.6 3800 14 970

(to BS 308)

Figure 11.6 Lettering types.

Stage 1
The solid in a defined space.

25
Fold back the end view, i.e. the EVP
End vertical plane (EVP)
Front vertical plane (FVP)
View on arrow A
View on arrow C
View on arrow B
Horizontal plane (HP)
Fold down the horizontal plane with view B on it and see stage 2

Stage 2
The solid has been removed and the views folded back to form a single flat plane.

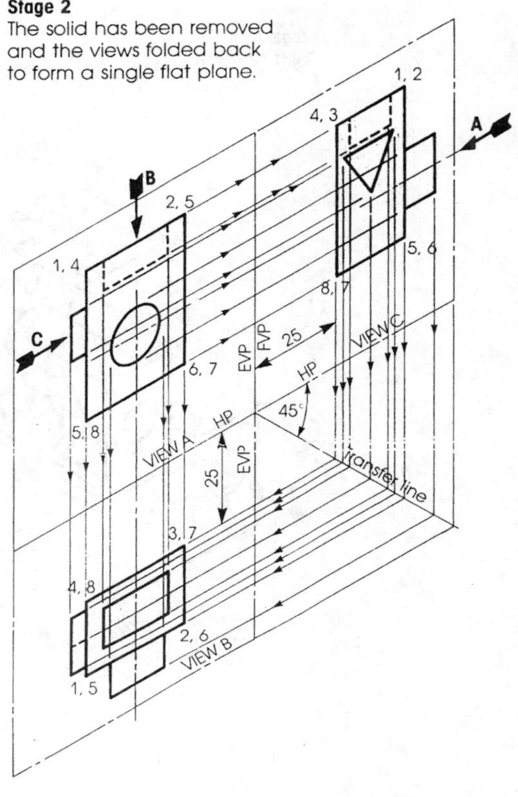

Stage 3
Produce the true shape of each feature as it appears in each view.

Note convention: When points or corners are numbered (e.g. 1,4) it means two points are in line, with point 1 being in front of point 4, that is, nearest the viewing position.

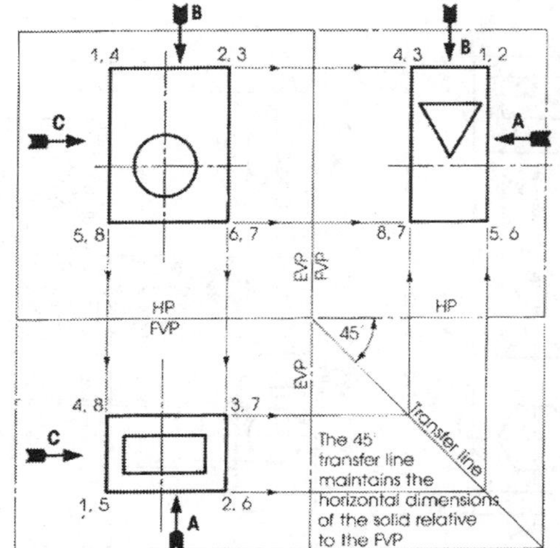

Stage 4
Project from the true shapes the apparent shapes in each view, i.e. stage 2 viewed direct.

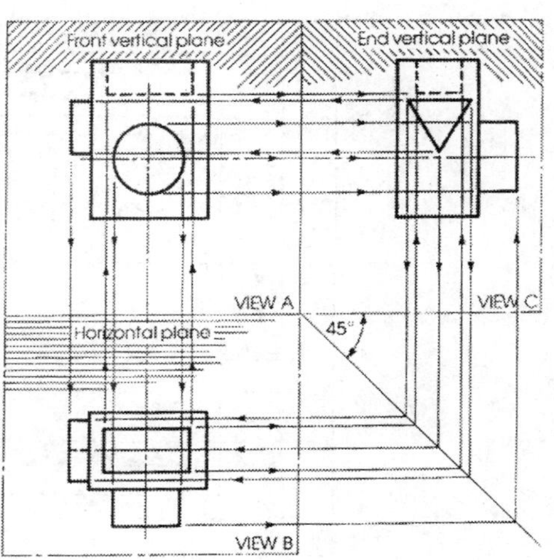

Figure 11.7a Orthographic projection. First angle projection

Stage 1
Solid defined.

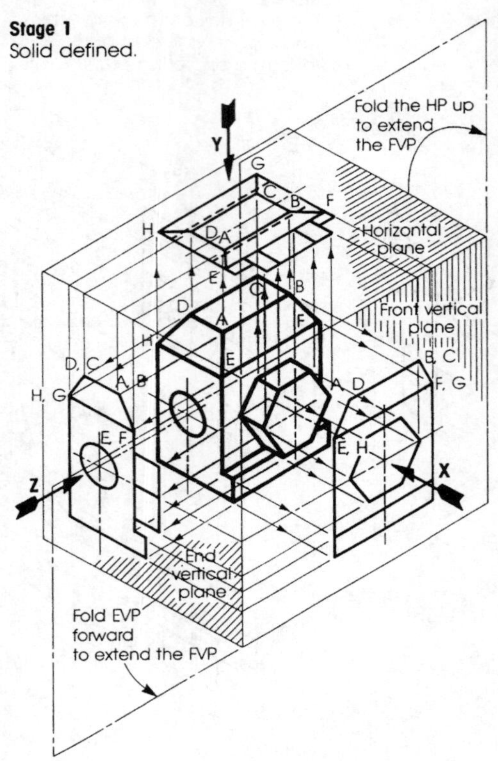

Stage 2
Projected views folded back to a single plane.

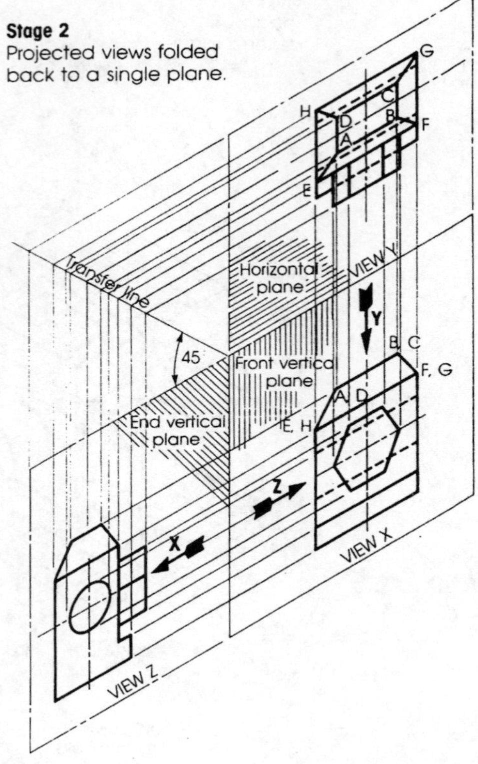

Stage 3
Work in all 3 views simultaneously and produce the shape of each feature which appears as a true shape.

Note: The true shape of a surface can only be seen when the viewing position is normal to the surface, i.e. at 90° to the surface.

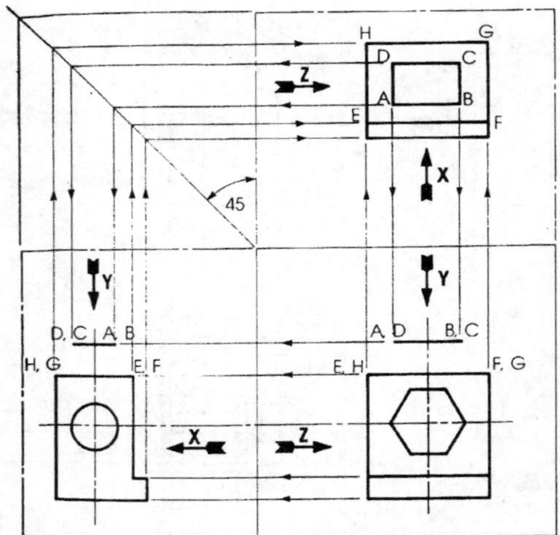

Stage 4
Direct view on stage 2. Join A to E, B to F, C to G and D to H to obtain oblique bevelled faces.

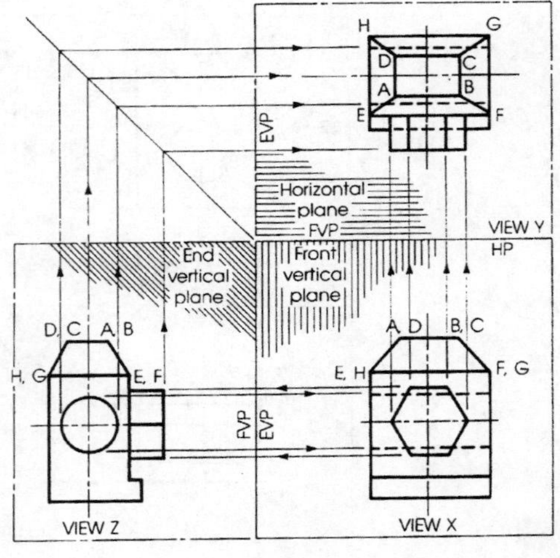

Figure 11.7b Orthographic projection. Third angle projection

- 1:5 – one-fifth full size
- 1:10 – one-tenth full size
- 2:1 – twice full size
- 5:1 – five times full size
- 10:1 – ten times full size.

One problem with orthographic projection is how to deal with hidden lines, such as internal features. Figure 11.8 shows the correct ways of showing hidden lines.

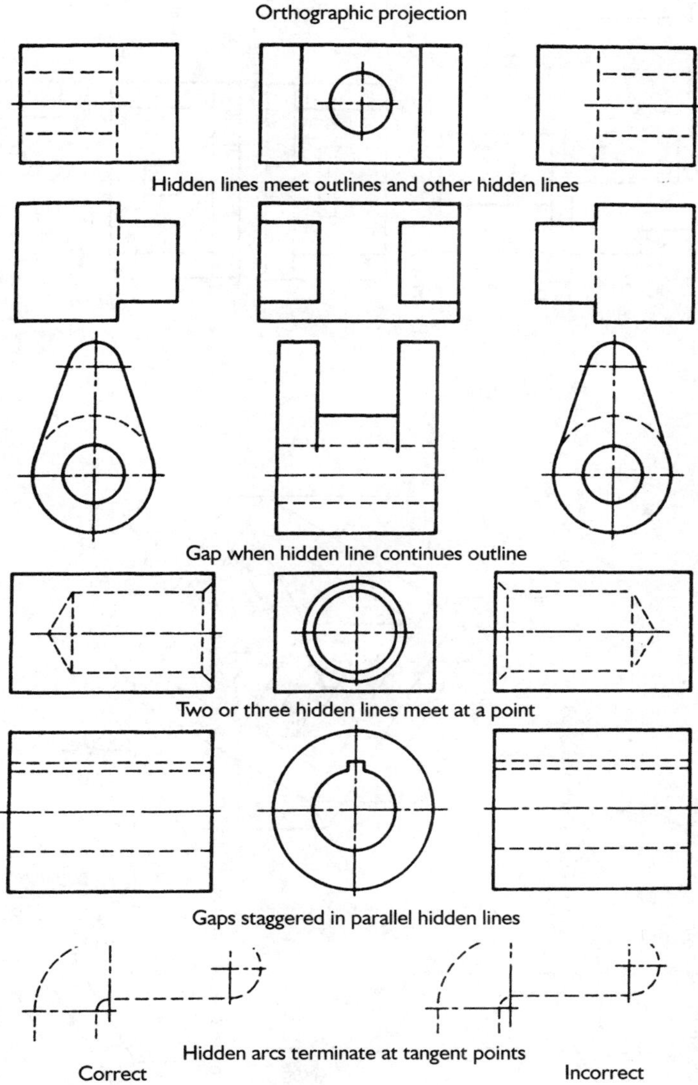

Figure 11.8 Hidden line technique

Case study

Orthographic projection

Typical examples of parts drawn in orthographic projection are shown in Figures 11.9 (first angle) and 11.10 (third angle).

Orthographic projection examples

Draw full size in First Angle projection the following views of the details shown.
(a) Elevation in direction of arrow T (b) End view in direction of arrow S
(c) Plan view projected from view (a)

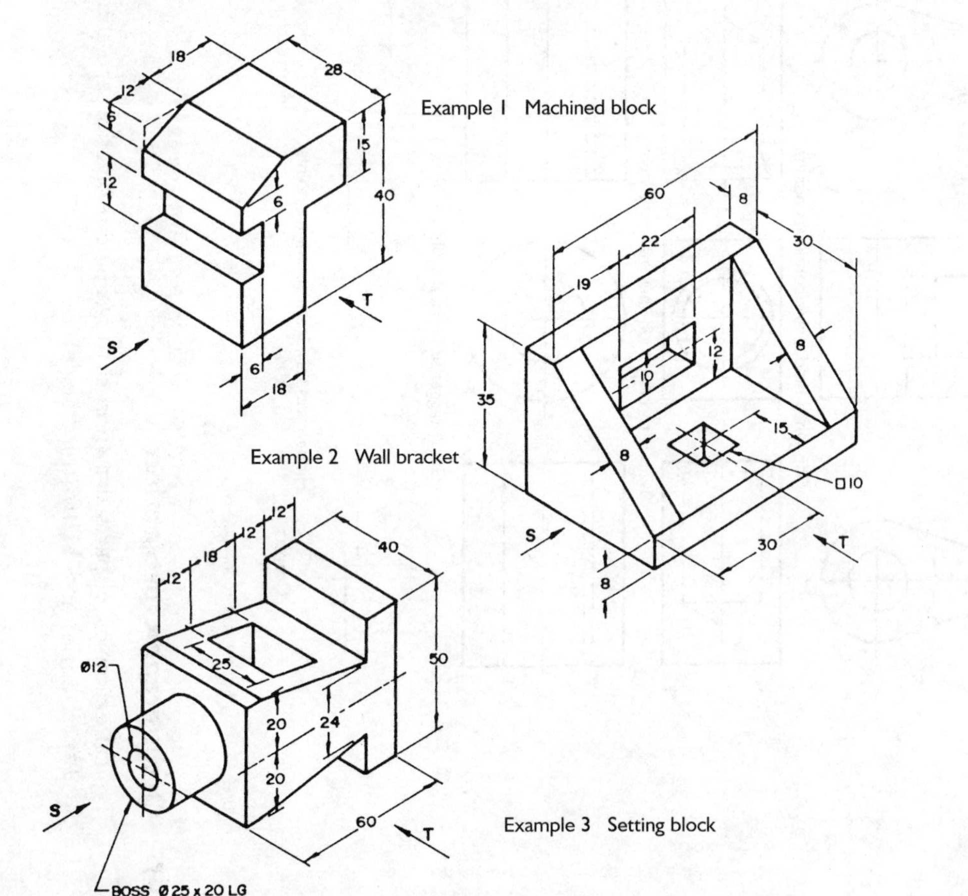

Orthographic projection examples

Example 1

Machined block

Example 2

Projection

Wall bracket

Example 3

Setting block

Figure 11.9 *Orthographic projection examples (first angle).*

Orthographic projection examples

Draw full size in Third Angle projection the following views of the details shown.
(a) Elevation in direction of arrow T (b) End view in direction of arrow S
(c) Plan view projected from view (a)

Orthographic projection examples

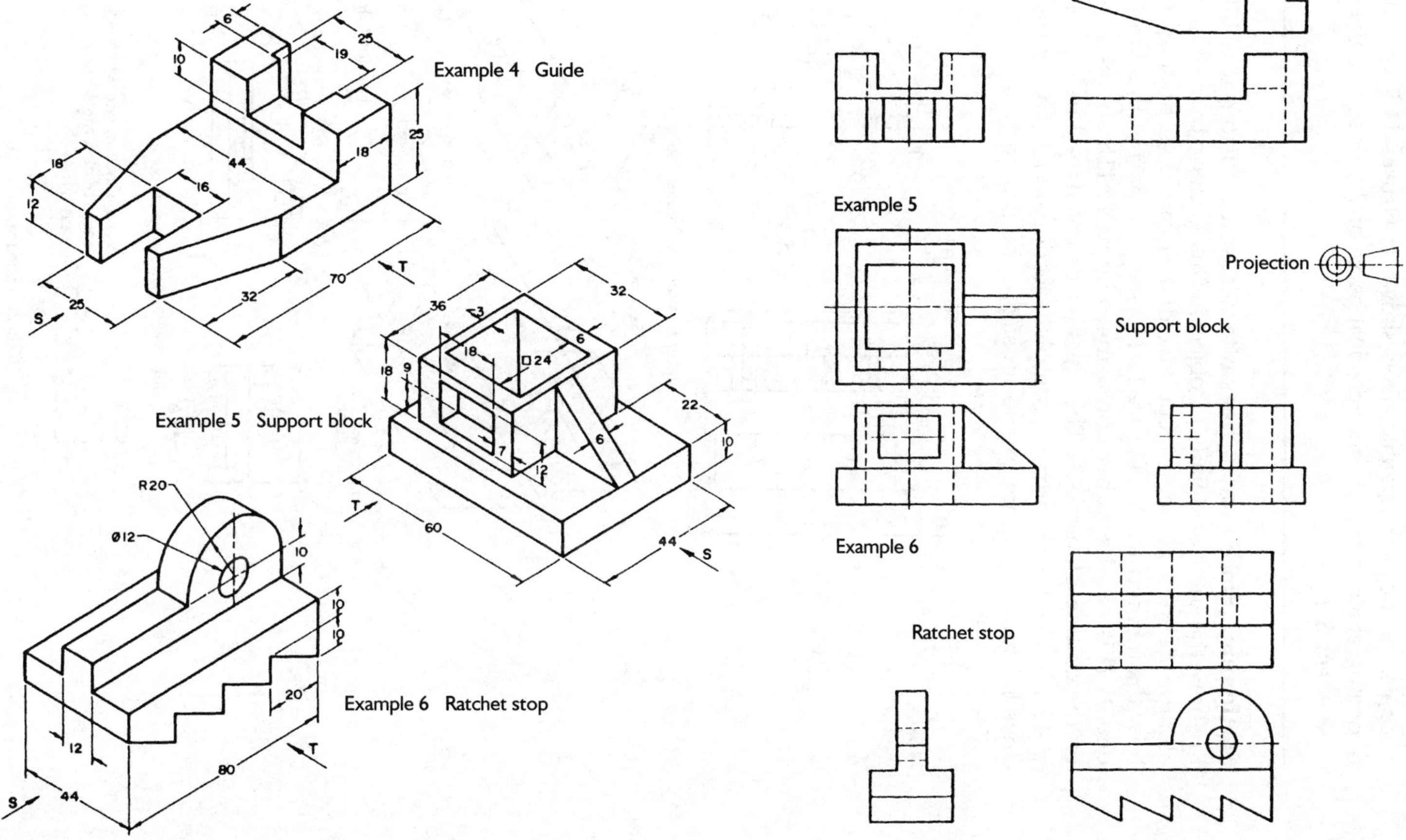

Figure 11.10 Orthographic projection examples (third angle)

Activity 11.1

Using both manual and computer-aided methods, produce:

a copies of the orthographic views shown in Figures 11.9 and 11.10
b orthographic views of the individual parts of the design solution assemblies from Element 5.2.

Sectioning Since hidden lines often cause confusion, it is important to be able to show the interior details of objects. This is done by cutting through the object along a defined plane, a technique that is called sectioning. Figure 11.11 shows how the method works. Figure 11.11a shows a part with a large amount of hidden detail. This can be rather confusing. Figure 11.11b shows a part fully sectioned by the cutting plane A–A. The cut surface is hatched by equally spaced thin lines at 45°. Notice the incorrect incomplete sectional view in Figure 11.11c. Any visible lines behind the cutting plane must also be drawn.

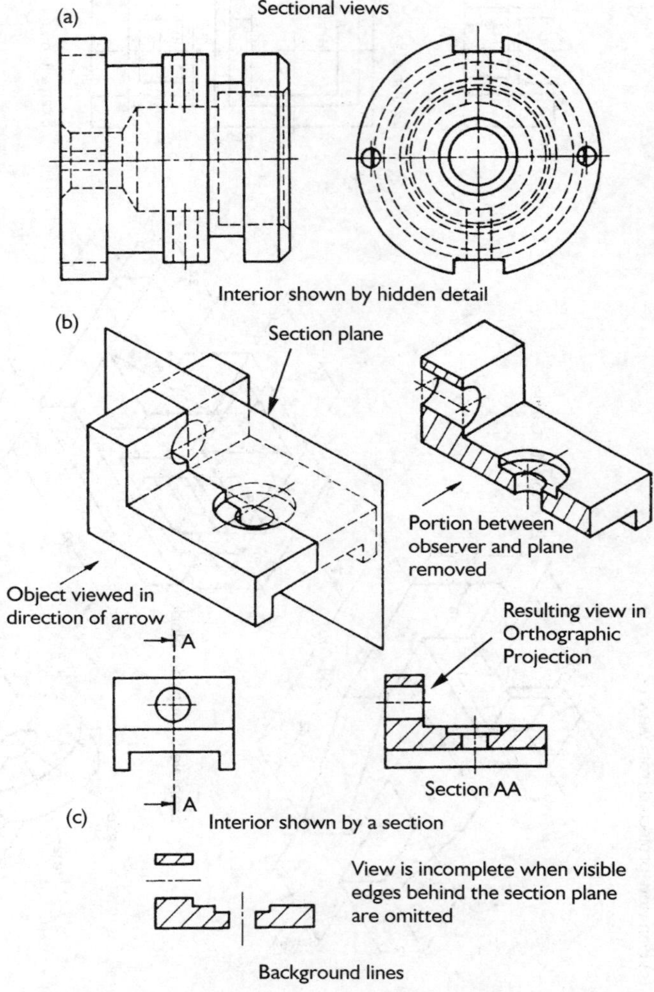

Figure 11.11 Sectioning. A part with a large amount of hidden detail (a); a part fully sectioned by the cutting plane A-A (b); and an incorrect incomplete sectional view (c)

Figure 11.12 shows correct methods of sectioning and the application of sectioning to different situations.

Certain features are not usually sectioned and these include:

- ribs and webs
- fasteners, e.g. screws, bolts, rivets, pins and keys
- solid shafts, gears, ball bearings and roller bearings.

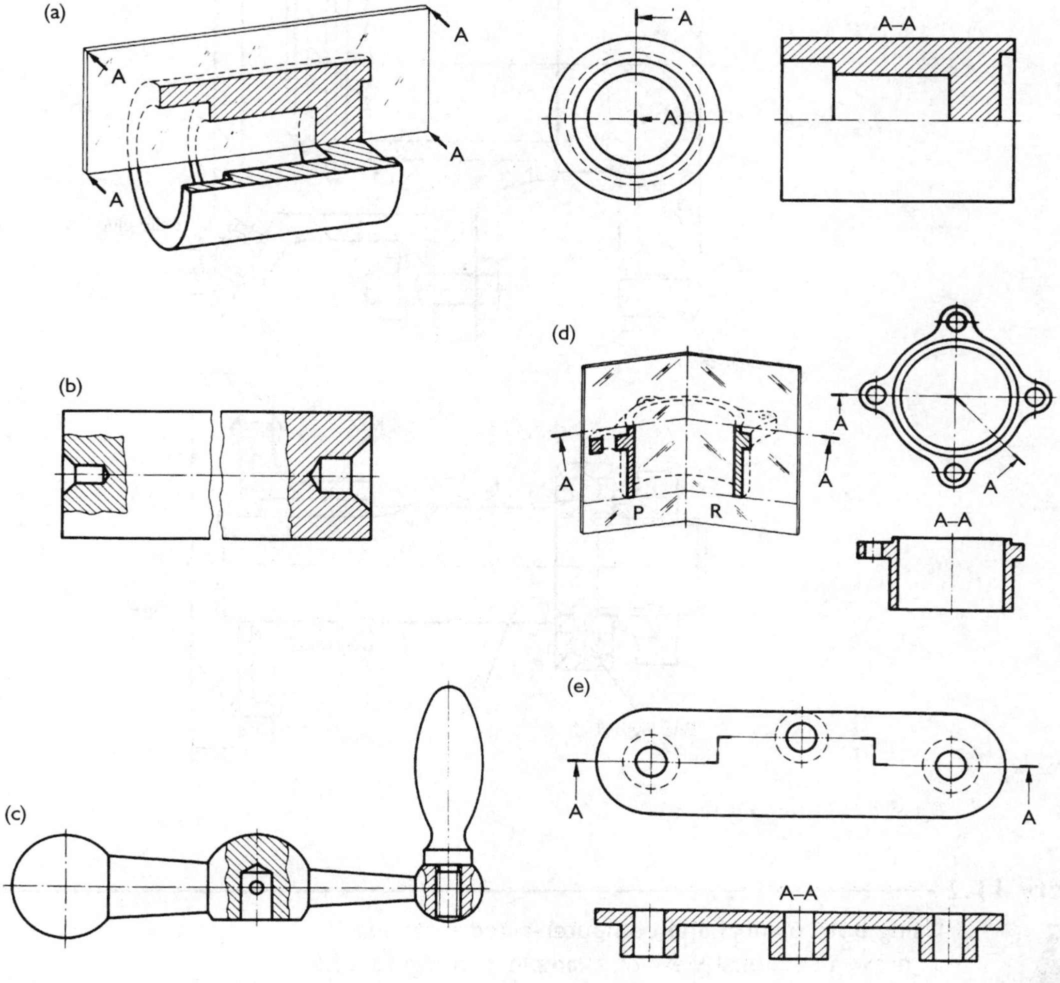

Figure 11.12 Correct methods of sectioning and the application of sectioning to different situations

Case study

Assembly drawings

The application of these principles in assembly drawings is shown in Figure 11.13. The unsectioned parts are clearly identified.

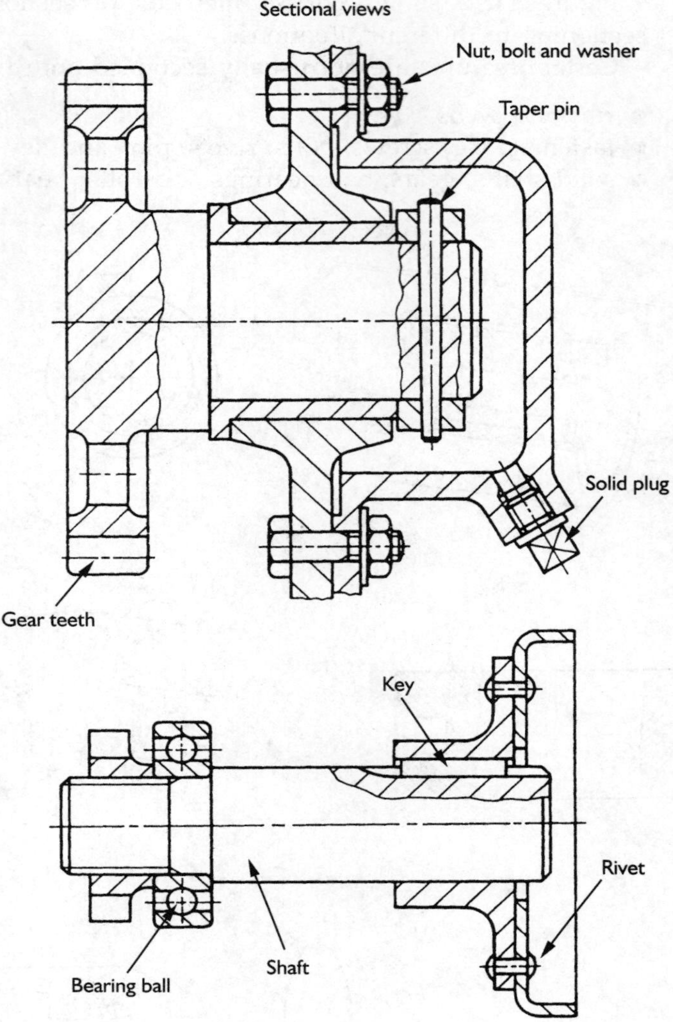

Figure 11.13 Sectional views

Activity 11.2

Using both manual and computer-aided methods:

a draw a sectional view of example 3 in Figure 11.9
b copy Figure 11.13
c draw a sectional view of at least one of the detailed parts from each of your design solutions from Element 5.2.

Dimensioning and tolerancing Dimensions and tolerances are obviously necessary in a drawing if it is to be interpreted for practical use in analysis and manufacture. Dimensions are classified as functional, non-functional and auxiliary. Figure 11.14 shows the application of all three types of dimension to an assembly, with functional dimensions being taken from a datum.

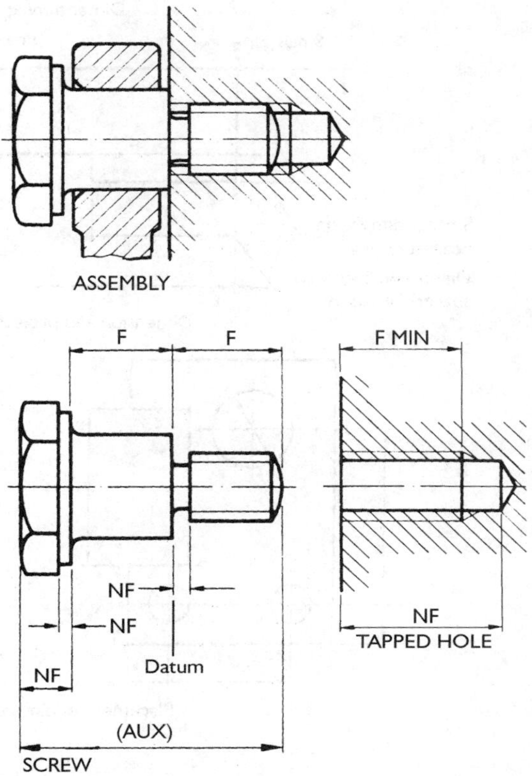

Figure 11.14 Application of all three types of dimensions to an assembly, with functional dimensions being taken from a datum

The general techniques and application of dimensioning to different situations is shown in Figure 11.15.

Tolerances are necessary so that parts and assemblies can be specified for correct functioning, interchangeability and economical manufacture. Chapter 9 has a section on tolerances, limits and fits and should be referred to (BS data sheets 4500A and 4500B for selected ISO fits should also be consulted). The BS 4500 data sheets give a reasonable selection of the types of fits. which are:

- clearance – for when parts are moving with respect to each other, e.g. a bearing or sliding fit
- transition or location fit – when parts need to be assembled and disassembled with accuracy
- interference or force fit – when parts are permanently assembled using heat or force.

The data sheets give the actual limits, and hence tolerances, for the full range of sizes from zero up to 500 mm. For sizes over 500 mm, special considerations apply – see Chapter 9 on tolerance and size. Figure 11.16 shows the basic types of fit as they apply in:

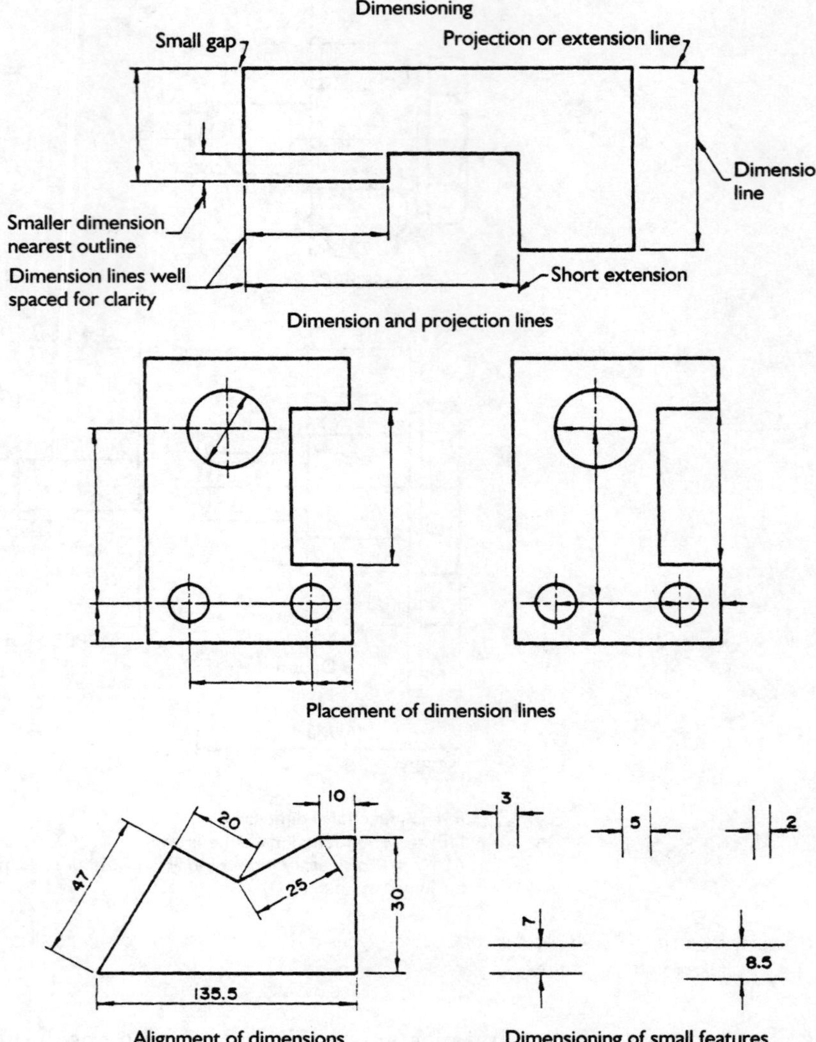

Figure 11.15 General techniques and application of dimensioning to different situations

- a hole basis system – this is generally preferred because of the availability of standard tools and gauges for holes. It is generally cheaper and easier to vary the size of shafts from the nominal size
- a shaft basis system – this is sometimes used when one long shaft of nominal size has to accommodate clearance, transition and interference fits

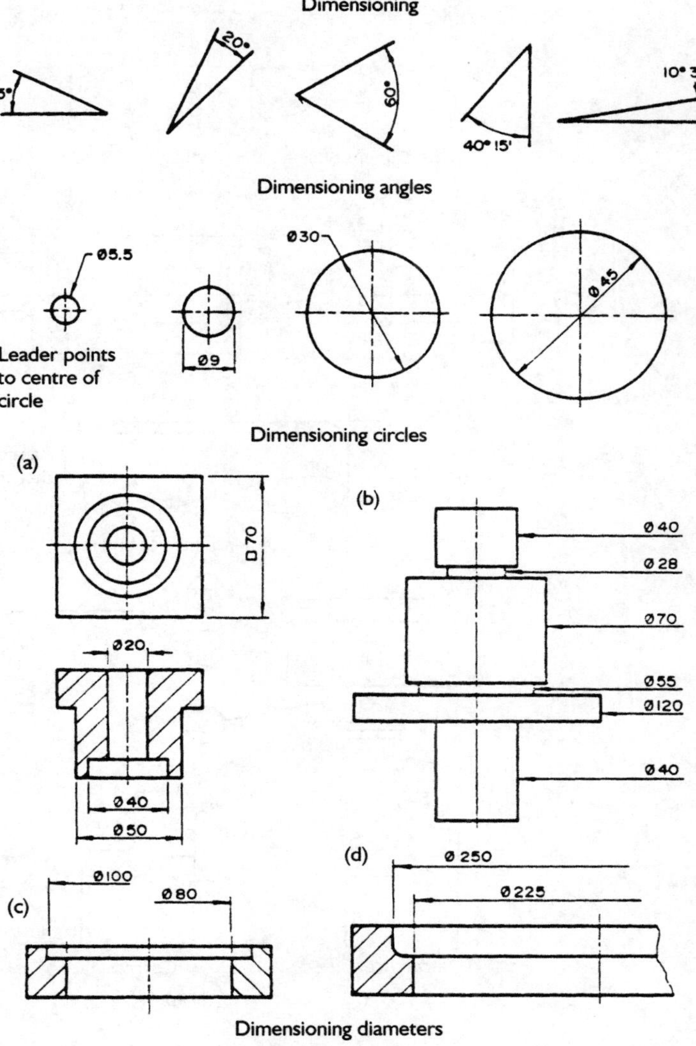

Figure 11.15(2) General techniques and application of dimensioning to different situations

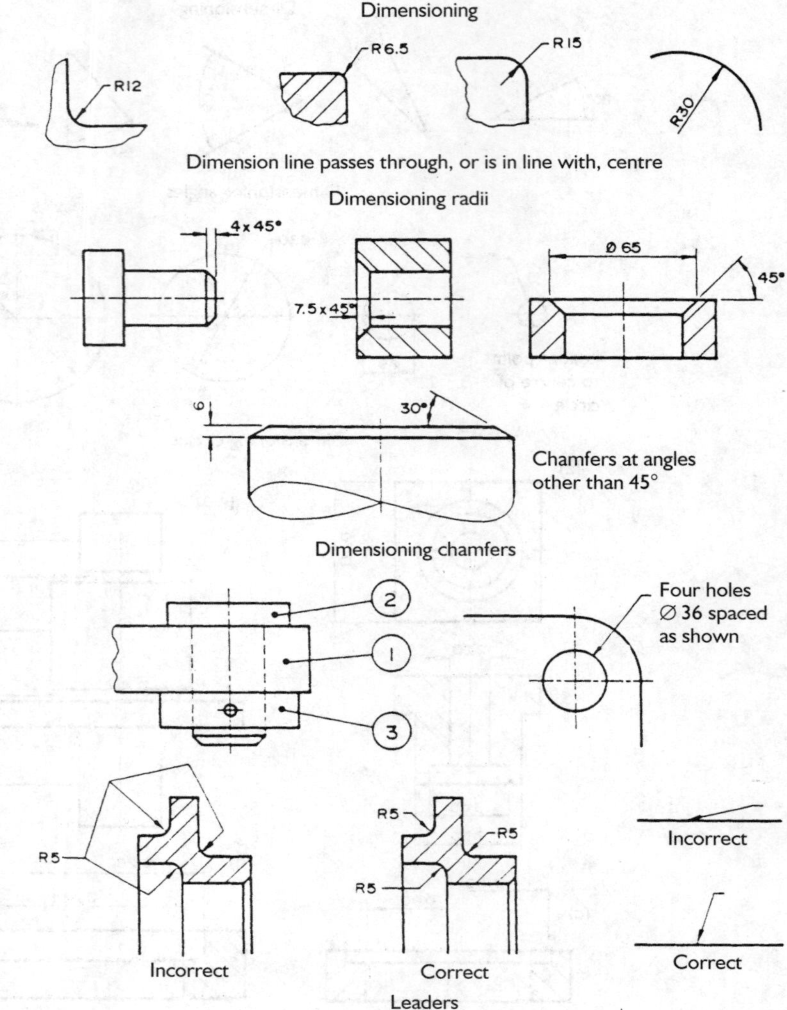

Figure 11.15(3) General techniques and application of dimensioning to different situations

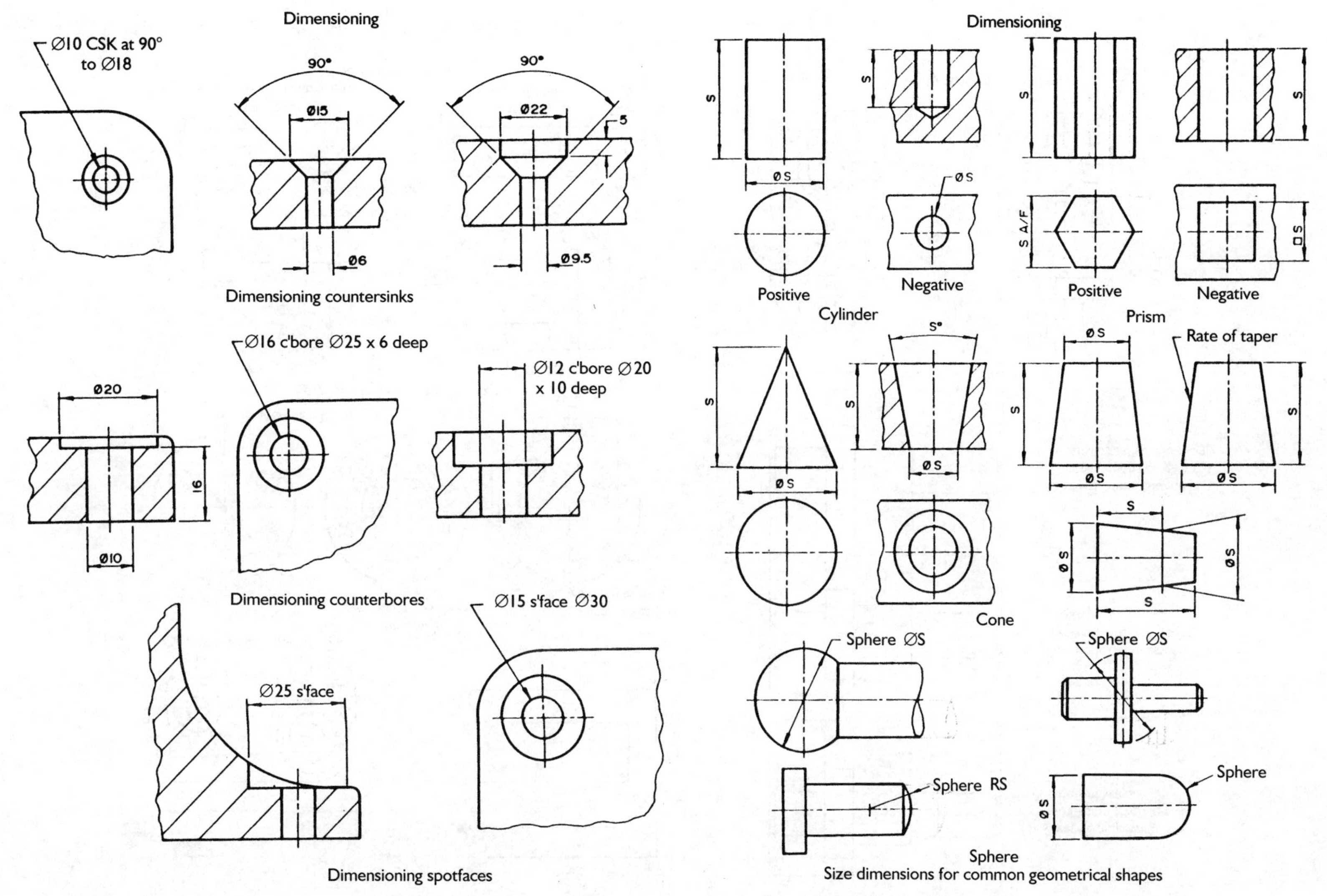

Figure 11.15(4) General techniques and application of dimensioning to different situations

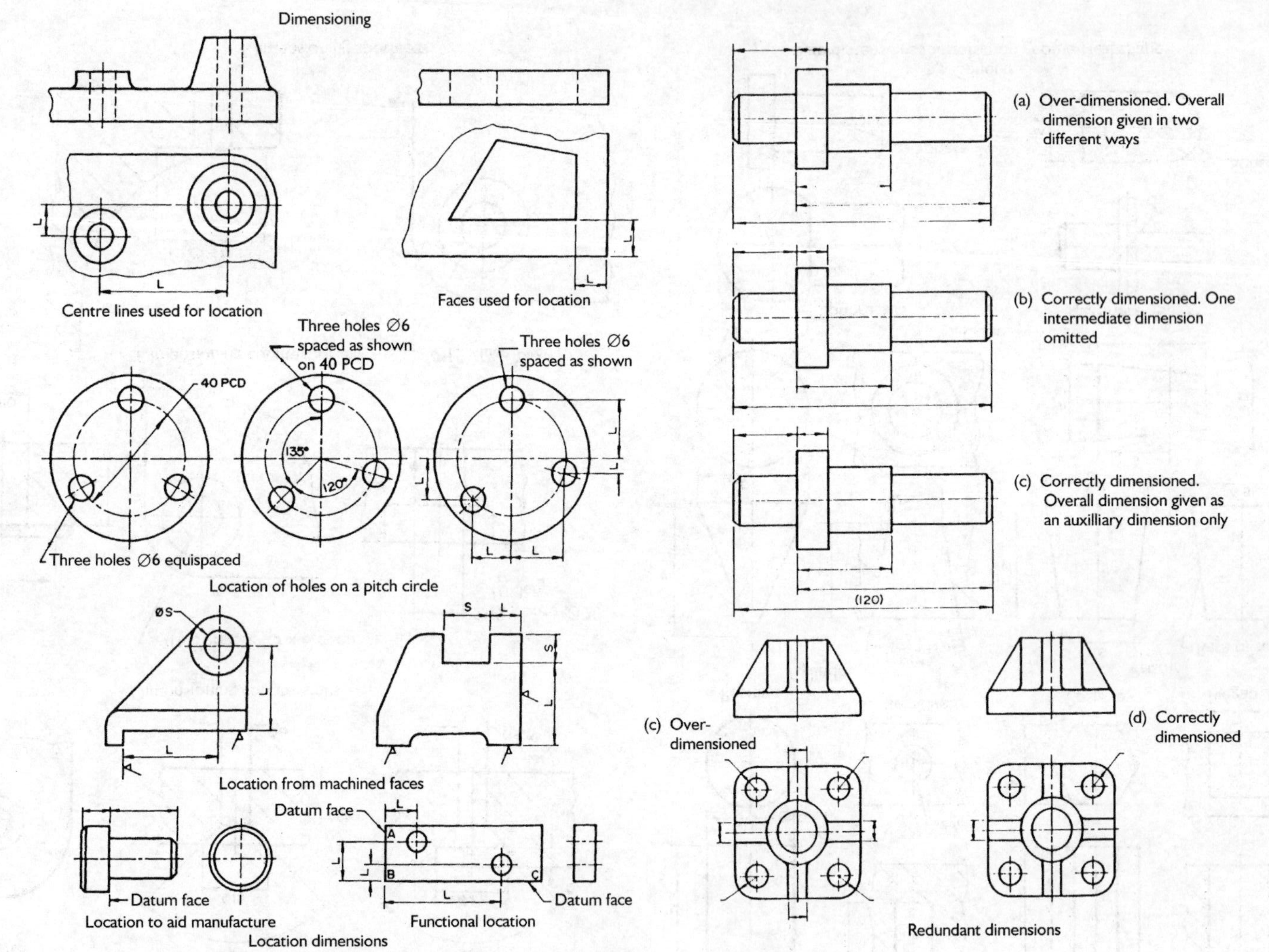

Figure 11.15(5) *General techniques and application of dimensioning to different situations*

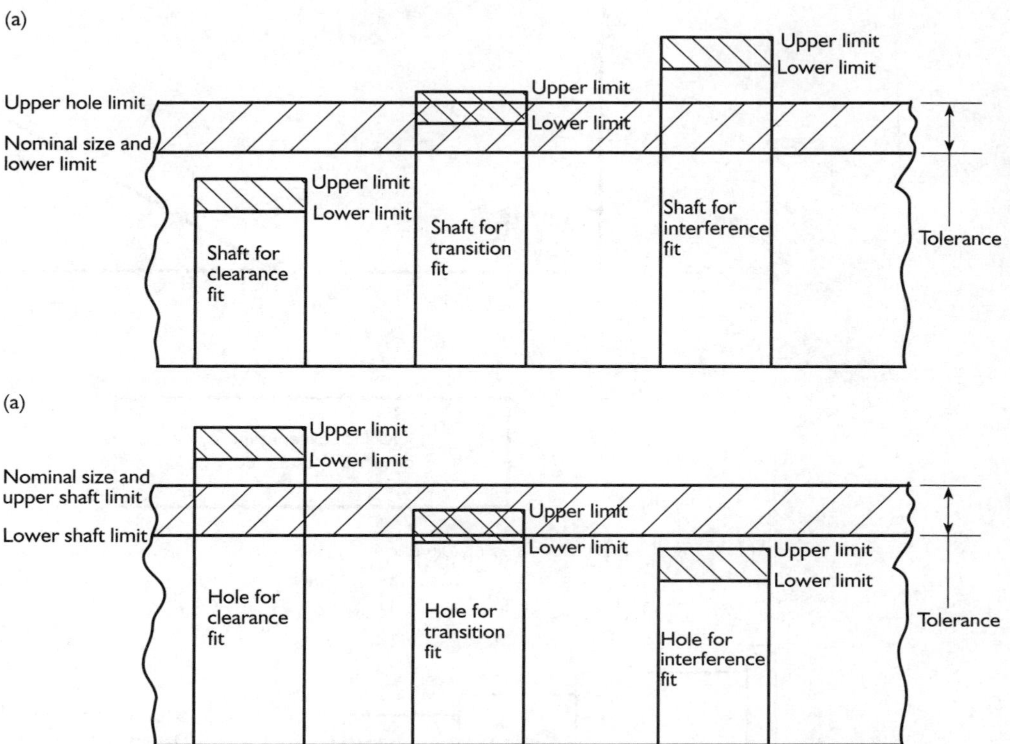

Figure 11.16 Basic types of fit: (a) hole basis fit and (b) shaft basis fit

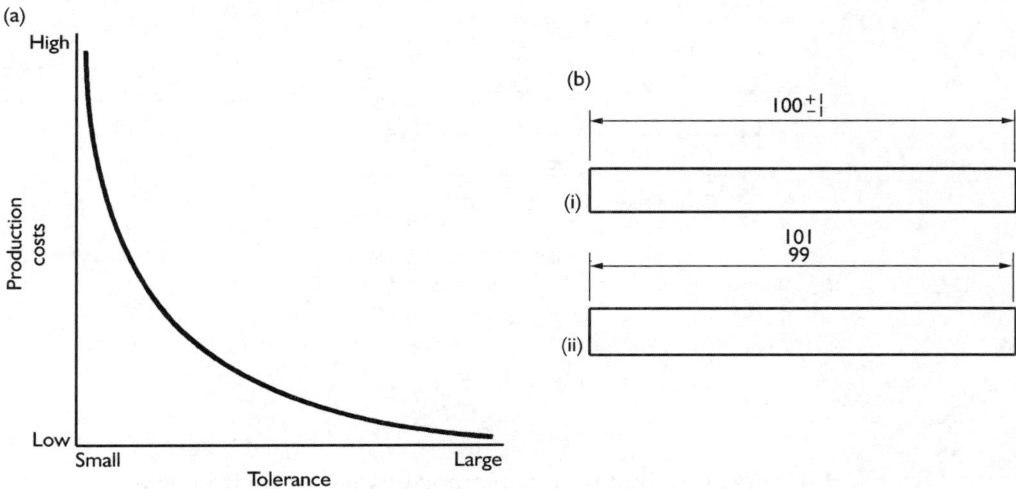

Figure 11.17a & b Basic principles of tolerancing

The tolerance in each case, as shown in Figure 11.16, is the difference between the upper and lower limit.

Tolerances should only be put on individual dimensions if strictly necessary for function. A general tolerance can be put on the drawing to apply to all dimensions unless otherwise stated. Figure 11.17 shows some of the basic principles to be followed when tolerancing such as:

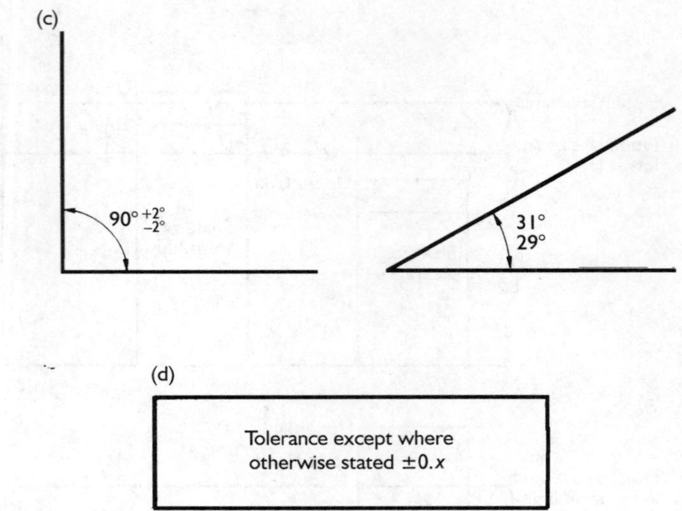

Figure 11.17c & d Basic principles of tolerancing

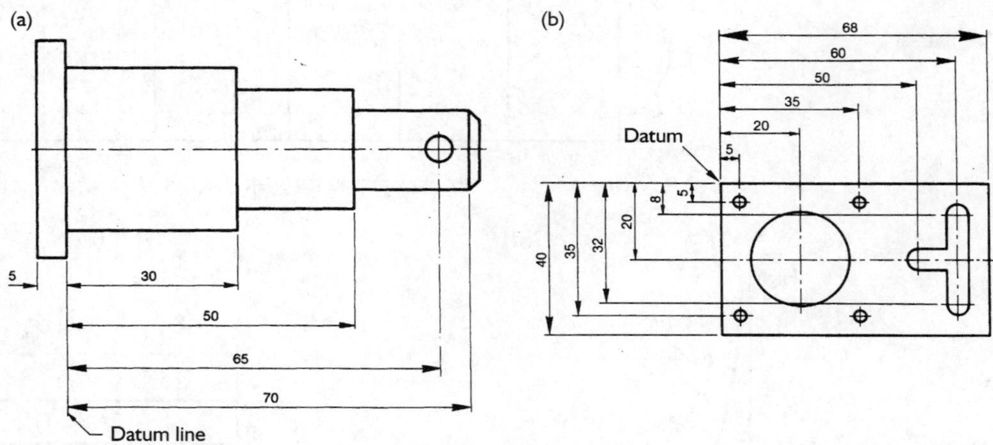

Figure 11.18 Application of datums and dimensioning methods: (a) dimensioning from a common datum line; (b) dimensioning several features on a component from one datum

(a) allocating maximum tolerance consistent with correct functioning, since costs increase rapidly as tolerances get smaller
(b) specifying the two limits of the dimension – here it is shown done in two different ways
(c) specifying angular tolerances in a similar way to linear tolerances
(d) correctly stating the general tolerance.

The application of datums and dimensioning methods is shown in Figure 11.18, which shows a shaft and a plate dimensioned from appropriate datums. The use of common datums ensures that there is not an excessive build up of tolerances.

Case study

Cumulative build-up of tolerances

Figure 11.19 shows an example of an assembly in which the cumulative build-up of tolerances cannot be avoided. This assembly consists of three rollers of different diameters, which are assembled onto a pin. If all the rollers are made to the maximum positive tolerance, the overall length of the assembly will be 20.2 + 20.2 + 20.2 = 60.6 mm.
If all the rollers are made to the maximum negative tolerance, the overall length of the assembly will be 19.8 + 19.8 + 19.8 = 59.4 mm.
This gives a possible total variation of 60.6 – 59.4 = 1.2 mm.
If this is unacceptably large, then the individual tolerances will have to be reduced, e.g. a reduction of tolerance to 0.2 on each roller will give a maximum total variation of $3 \times 0.2 = 0.6$ mm.

Case study

Dimensioned and toleranced detail drawings

Some examples of typical dimensioned and toleranced detail drawings are shown in Figure 11.20. Notice that centre lines can also be used as datums.

Activity 11.3

a fully dimension and tolerance the part shown in example 4, Figure 11.10
b fully dimension and tolerance the orthographic drawings of the detail parts from the design solution that were drawn in Activity 11.1.

Pictorial projection

Pictorial projection is a drawing technique to represent the object in three dimensions. There are two main methods of pictorial projection used in engineering drawing – isometric projection and oblique projection.

The basic principles of isometric projection are shown in Figure 11.21. Using the example of a cube, in isometric projection it can be seen that the cube is rotated relative to all three of the normal X, Y and Z orthographic planes. This rotation is shown by the 30° angle to the horizontal of lines EF and GF and the common angle of 120° of the lines meeting at the corner D.

The principles of oblique projection are shown in Figure 11.22. It can be seen that one of the faces of the cube is a normal front view with the other lines drawn at 45° to the horizontal.

Figure 11.23 shows the usual method of producing isometric and oblique drawings. A faint outline of the overall dimensions is drawn with a thin line. The outline of the shape is drawn firmly with a thick line. It is very convenient to use a standard grid when doing isometric and oblique drawings. Figure 11.24 shows suitable grids and also some sketches of parts drawn on the grid.

One of the isometric drawings in Figure 11.24 shows a part with circular features. In isometric projection, a circle appears as an ellipse. The grid enables the ordinates method to be used when drawing isometric circles. The ordinates construction method is shown in Figure 11.25, in which an isometric drawing of a cylinder is illustrated.

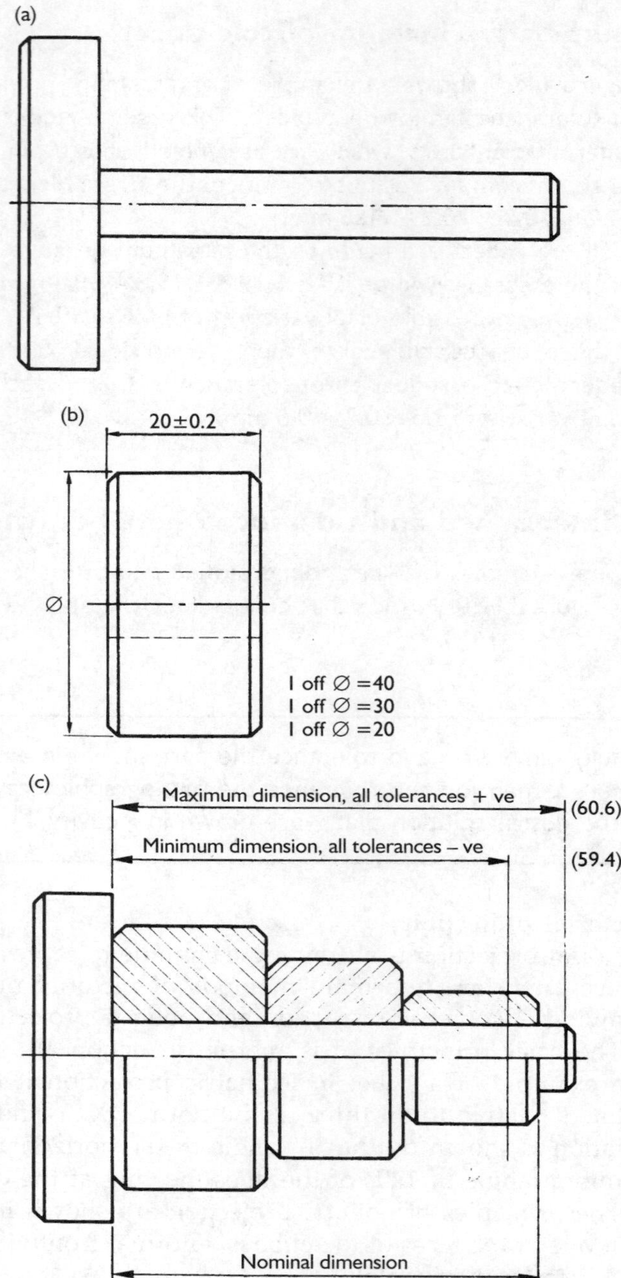

Figure 11.19 Assembly of rollers on a pin

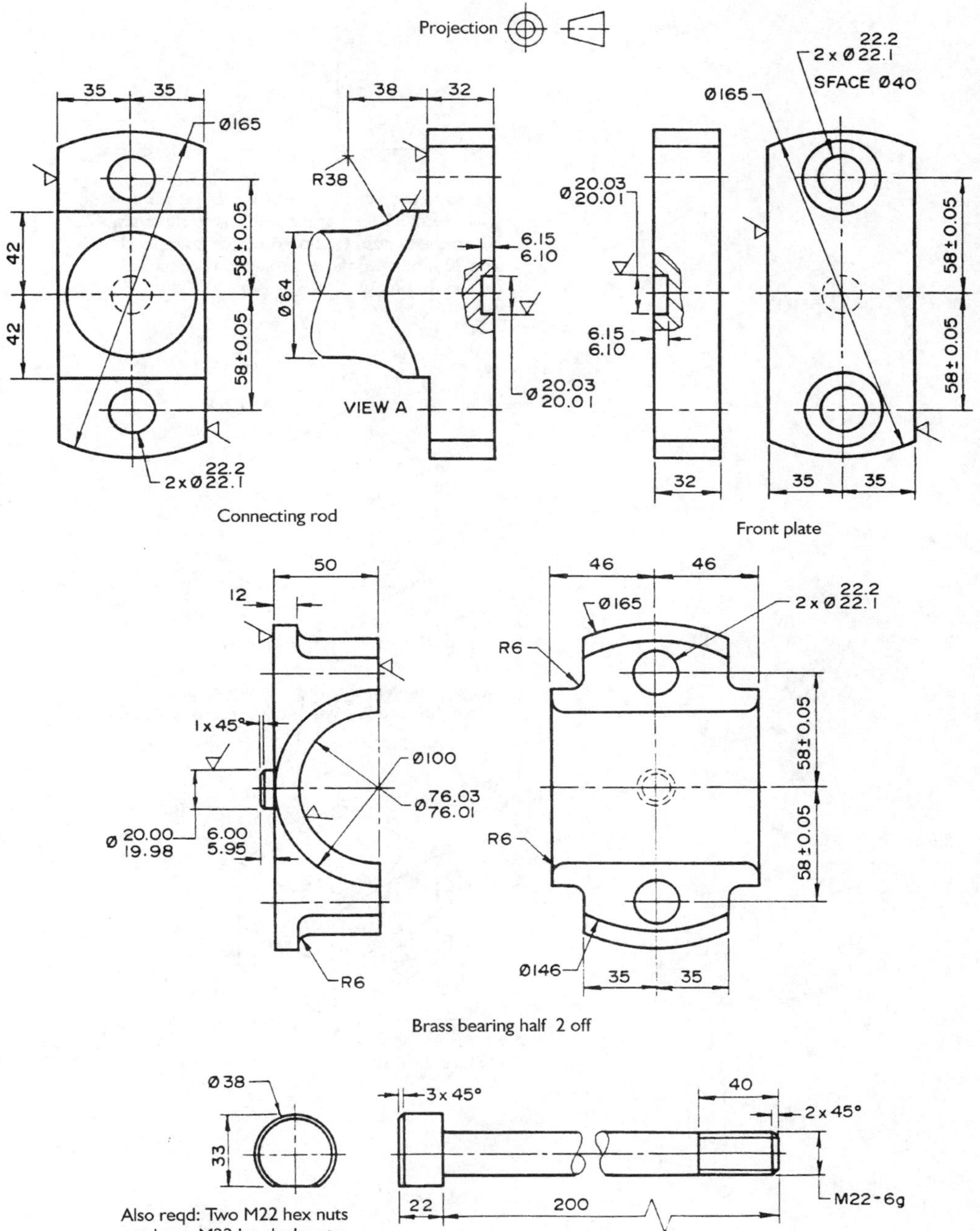

Figure 11.20 Examples of dimensioned and toleranced detail drawings

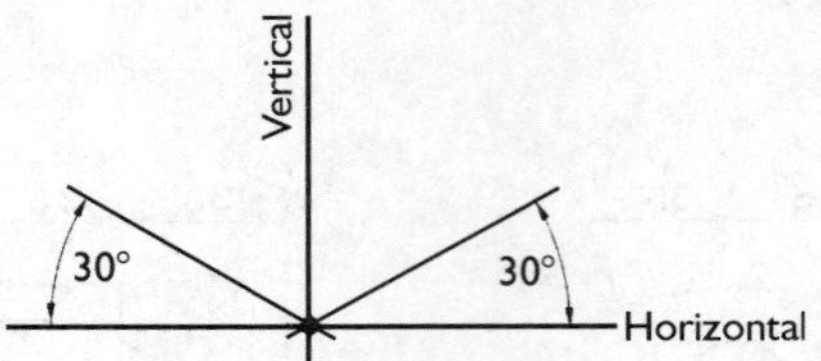

For isometric sketches vertical lines remain vertical while receding lines are shown to the same scale at 30° to the horizontal. So, to sketch a simple cube of side 20 mm in isometric projection proceed as in (a) and (b) below

(a)

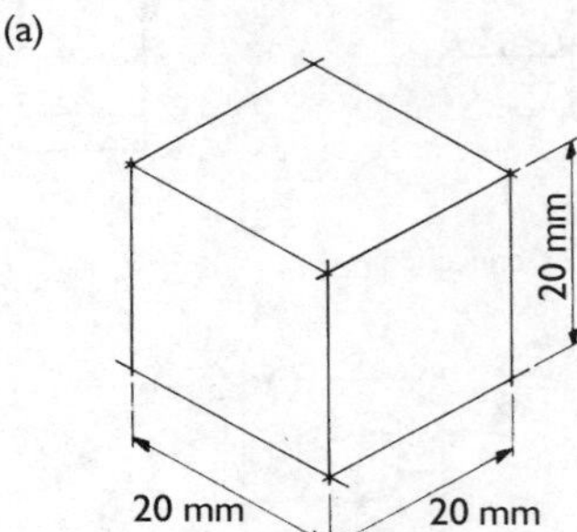

Stage 1: sketch, showing the simple basic shape (note the receding lines at 30° to the horizontal)

(b)

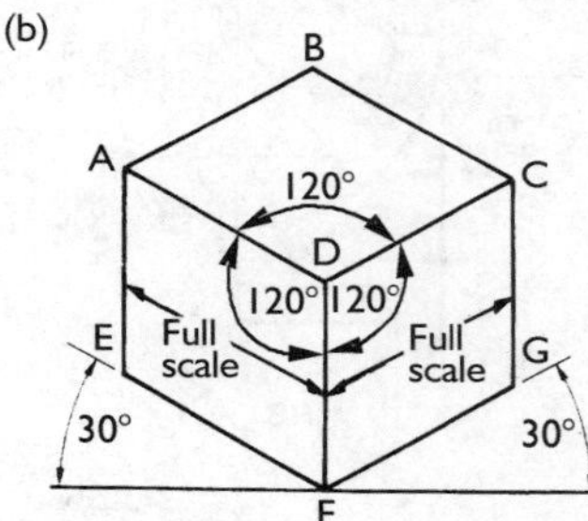

Stage 2: line it in

Figure 11.21 Isometric sketches

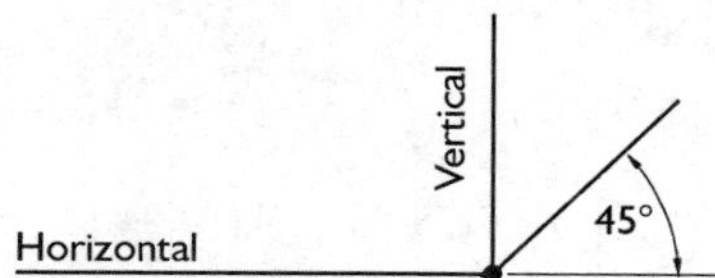

For oblique sketches the front face of a component is shown as it is; the receding lines are shown at an angle of 45° and to a reduced scale, to avoid a foreshortening effect. There are no fixed rules for the reduced scale of the receding lines, but sketches will appear in good proportion, if half scale is used. So, to sketch a simple cube of side 20 mm in oblique projection, proceed as in (a) and (b) below.

(a)

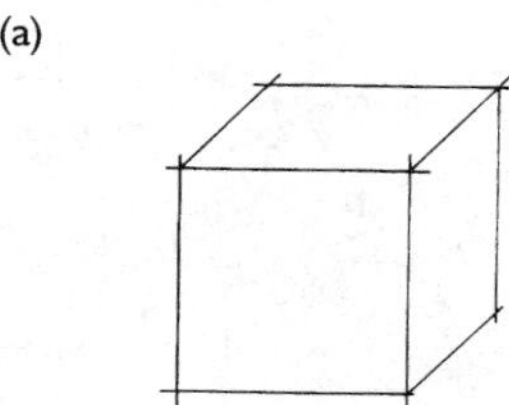

Stage 1: sketch, showing the simple basic shape (note the receding lines at 45° and 1/2 scale)

(b)

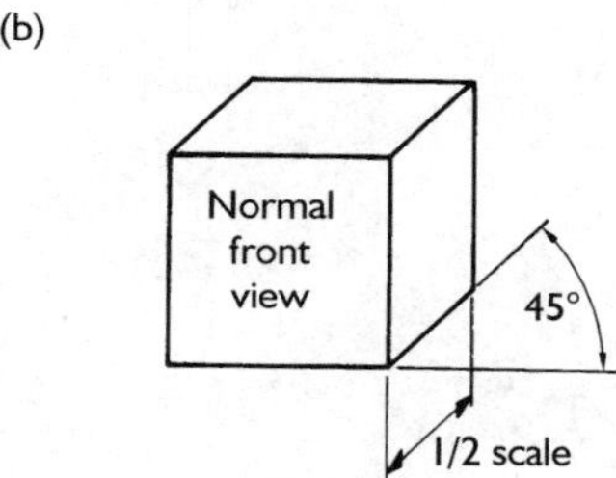

Stage 2: line it in

Figure 11.22 Oblique sketches

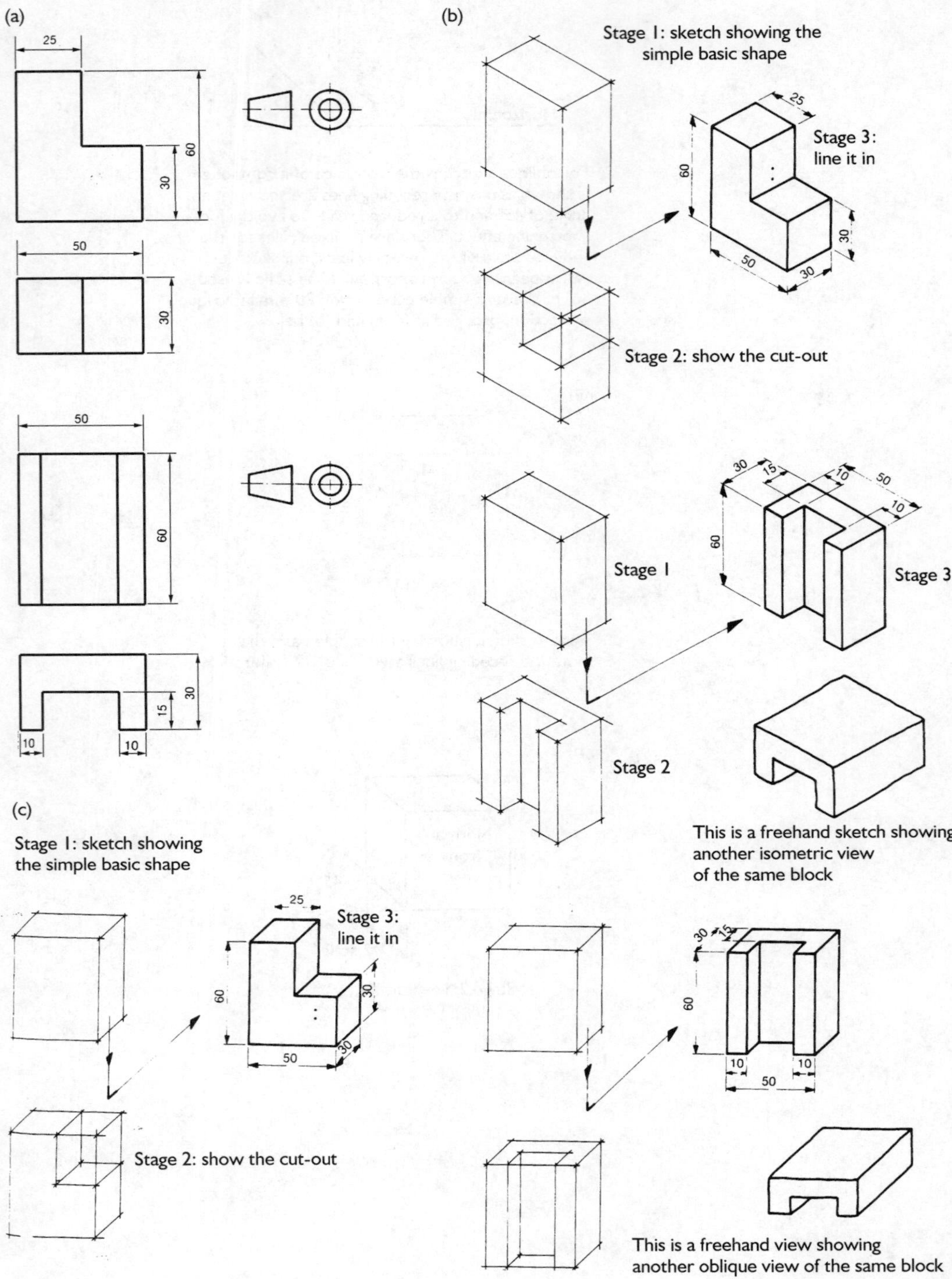

Figure 11.23 Usual method of producing isometric and oblique drawings

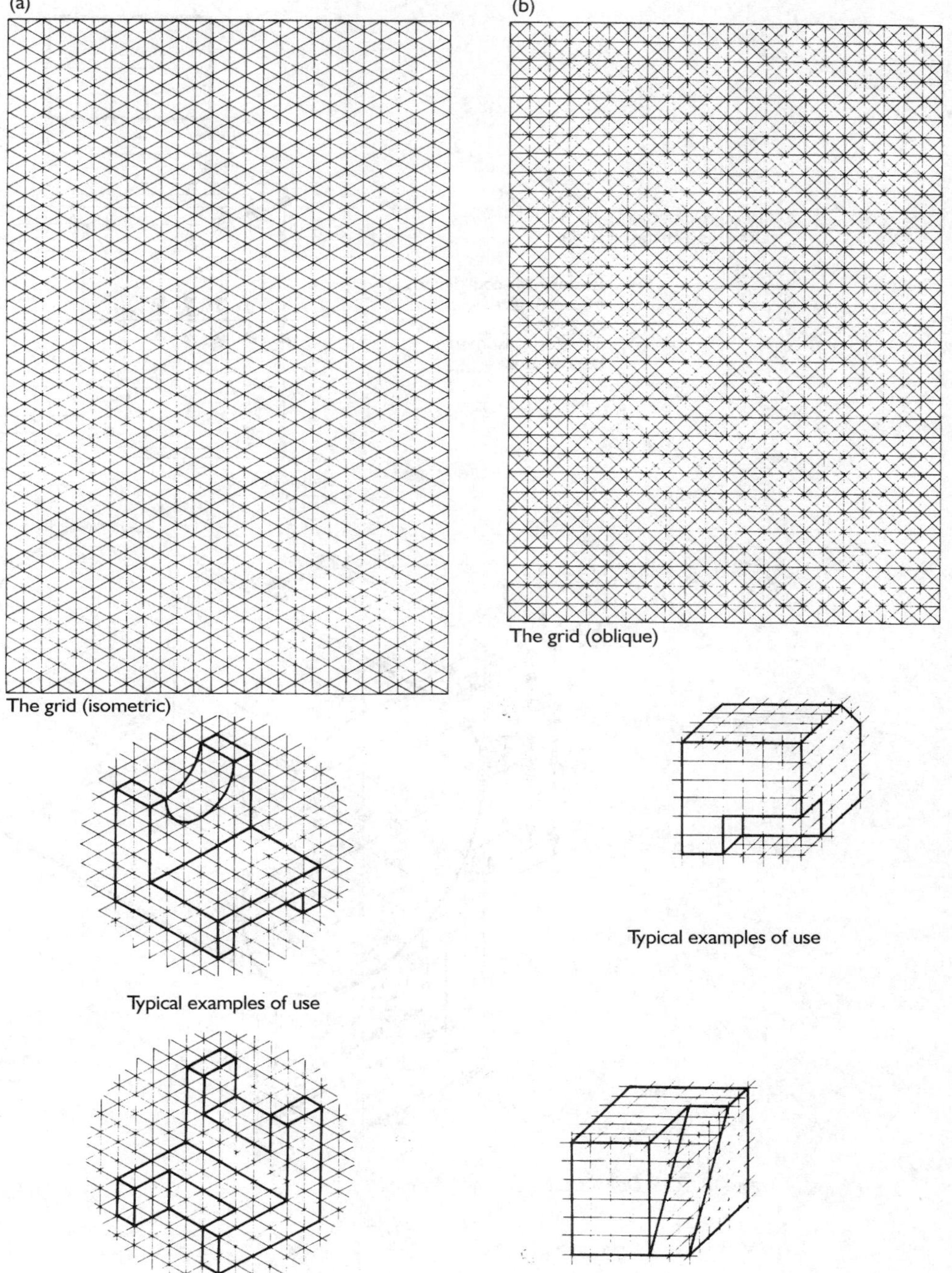

Figure 11.24 Grids for guided sketching: (a) isometric; (b) oblique

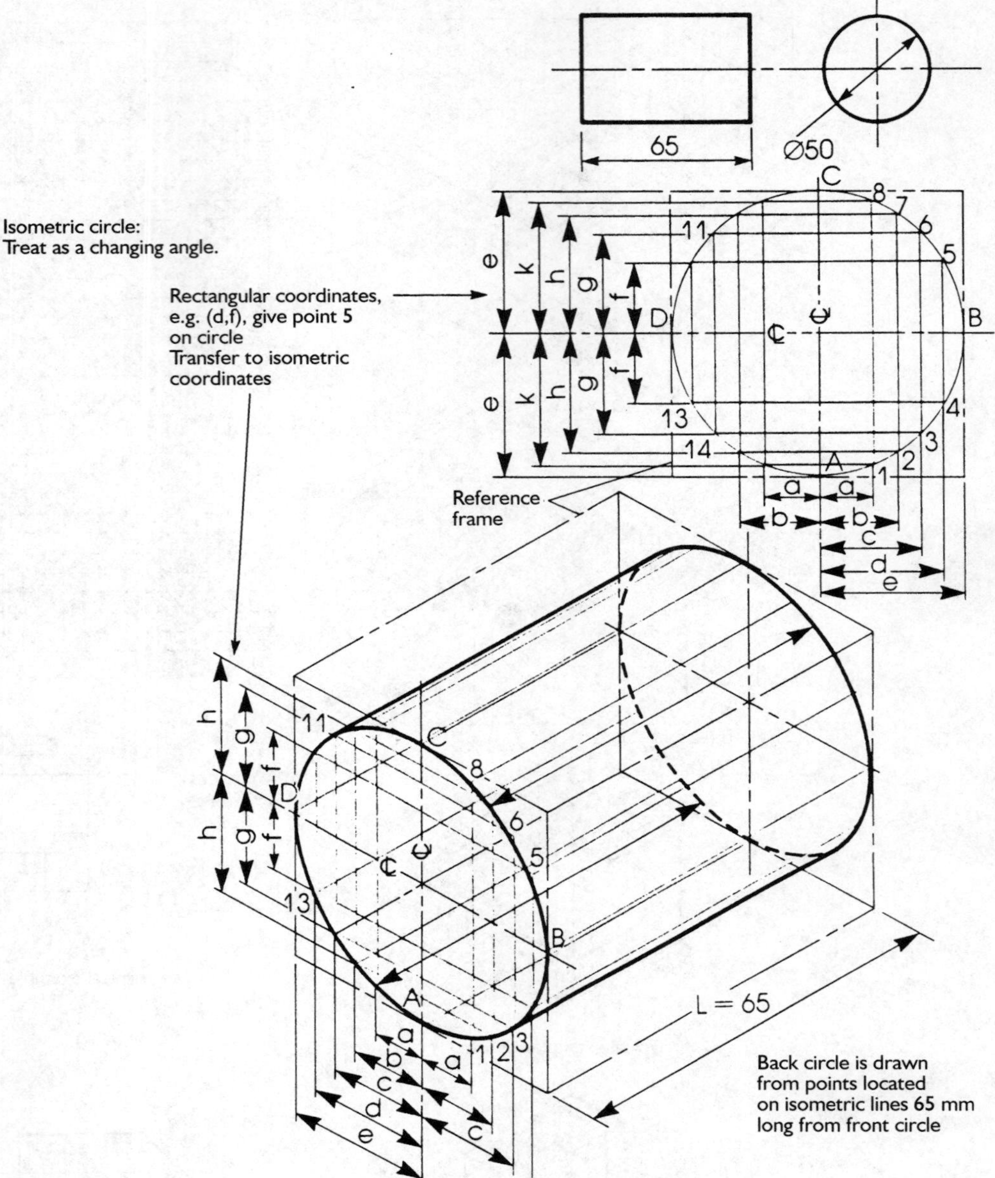

Figure 11.25 Ordinates construction method

Case study

Exploded diagrams

Exploded diagrams are particular examples of assembly drawings that show the parts of the assembly 'exploded'. This is used to show the method of assembly/disassembly, as well as being useful for spares and servicing. Figure 11.26 shows the parts of an assembly drawn in normal orthographic views. The parts are then drawn in isometric projection as an 'exploded' diagram. A parts list is also included. The method of drawing is given for guidance as to how to do this kind of drawing.
A typical example of a more complex isometric exploded diagram, in this case a car engine, is shown in Figure 11.27. Again, a comprehensive parts list is an essential part of the diagram to enable the parts to be located and identified.

Activity 11.4

Using both manual and computer-aided draughting methods:

- copy the isometric view of example 3 in Figure 11.9 and draw an oblique view of example 1 in Figure 11.9
- copy the exploded diagram in Figure 11.26
- draw at least one isometric and one oblique detail drawing from the design solution for Element 5.2
- draw an assembly exploded diagram for each of the designs from Element 5.2.

Conventions

Conventions are universally understood representations of features that ensure that anyone can understand the drawing. They also save a lot of time and expense in drawing. The conventions shown in Figure 11.28 are all from BS 308 Part 1 and show:

(a) to (m) screw threads
(n) to (r) interrupted views, repeated parts and repeated features
(s) to (u) plane faces, knurling and rolling bearings.

Representation of components and features using standard symbols Graphical symbols are all defined in the appropriate standards. As with other drawing conventions, standard symbols ensure that anyone can understand the drawing.

Figure 11.29 shows some spring types using the BS 308 representation.

Figure 11.30 shows the method of drawing hexagon-headed screw fasteners and studs and illustrates how they are used in assemblies.

Other types of screwed fastener are available and some of these are illustrated in Figure 11.31.

Once drawn in the correct form, the screw threads will need to be dimensioned and toleranced. Thread sizes and tolerances are always indicated by a note. The most commonly used thread in Europe is the ISO metric thread, which is specified in BS 3643. Threads are typically specified as follows:

- M12 × 1.75–6H for an external thread
- M12 × 1.75–6g for an internal thread.

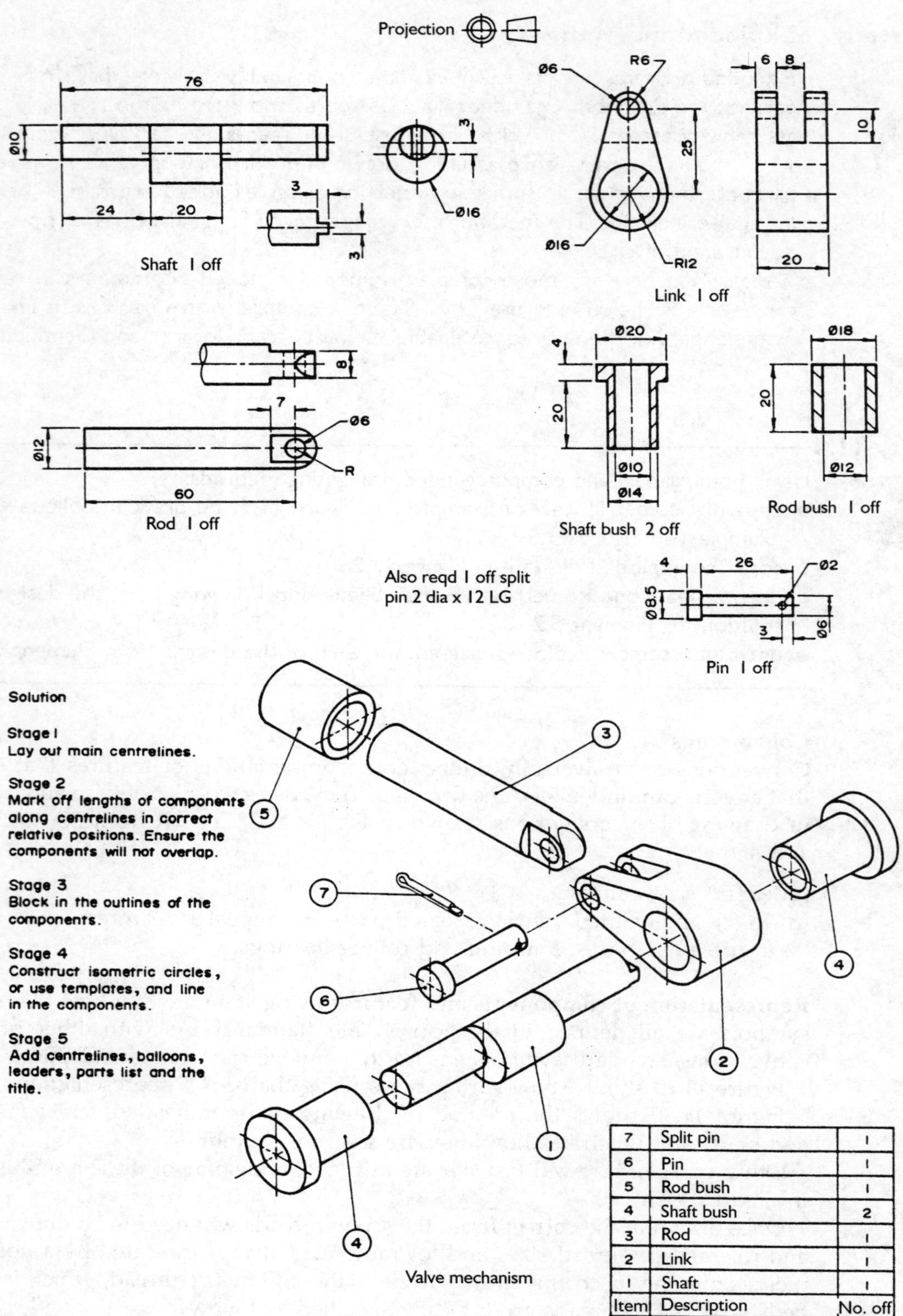

Item	Description	No. off
7	Split pin	1
6	Pin	1
5	Rod bush	1
4	Shaft bush	2
3	Rod	1
2	Link	1
1	Shaft	1

Figure 11.26 Parts of an assembly drawn in normal orthographic views

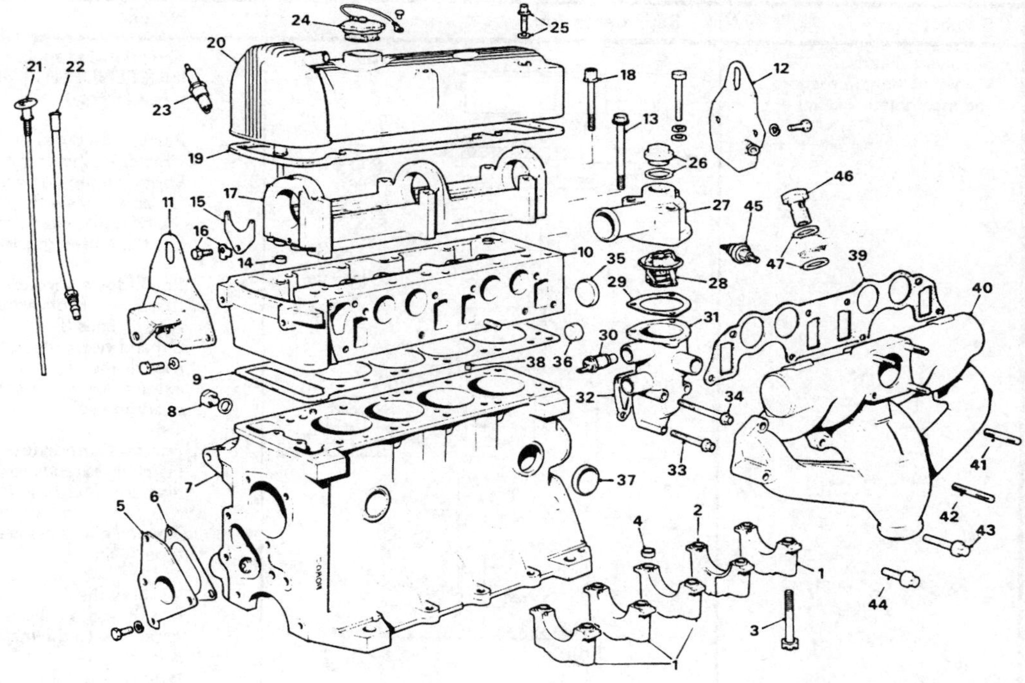

Key to the engine external components

No. Description

1. Main bearing caps
2. No. 4 thrust-main bearing cap
3. Set screw for main bearing cap
4. Ring dowel for main bearing cap
5. Engine front cover
6. Gasket for front cover
7. Cylinder block
8. Cylinder block drain plug and sealing washer
9. Gasket for cylinder head
10. Cylinder head
11. Engine lifting bracket-front
12. Engine lifting bracket-rear
13. Set screw for cylinder head
14. Ring dowel for camshaft carrier
15. Locating plate for camshaft
16. Set screw and lock washer
17. Camshaft carrier
18. Set screw-carrier to cylinder head
19. Gasket for cover
20. Cylinder head cover
21. Oil dipstick
22. Oil dipstick tube
23. Sparking plug
24. Oil filler cap
25. Set screw and 'O' ring seal-cover to cylinder head
26. Filler plug and 'O' ring seal
27. Water outlet pipe
28. Thermostat
29. Gasket for water outlet pipe
30. Thermal transmitter
31. Thermostat housing
32. Gasket for thermostat housing
33. Short set screw for thermostat housing
34. Long set screw for thermostat housing
35. Core plug for cylinder head
36. Plug for main oil gallery
37. Core plug for cylinder block
38. Dowel for cylinder head
39. Gasket for inlet and exhaust manifold
40. Inlet and exhaust manifold
41. Short stud-carburetter to manifold
42. Long stud-manifold to cylinder head
43. Long set screw
44. Short set screw
45. Oil pressure switch
46. Screw for brake servo banjo union
47. Washers for banjo union

Figure 11.27 An exploded drawing of a car engine

Subject	Representation	Notes
Screws threads Visible, hidden; in section and assembled	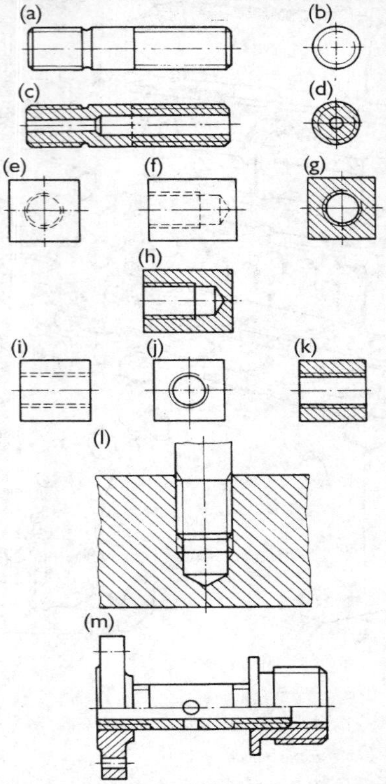 (a) (b) (c) (d) (e) (f) (g) (h) (i) (j) (k) (l) (m)	The following applies INDEPENDENTLY of the type of screws thread **For visible threads** Crests are shown by a type 'A' line Roots are shown by a type 'B' line On an end view the line showing the root diameter extends for not less than three quadrants (b) (j) **For hidden threads** Crests and roots are shown by type 'F' lines (f) (i) On end views the line showing the root diameter is also type 'F' and extends for not less than three quadrants (e) **For sectioned views** Hatching extends to the crest diameter (c) (d) (g) (h) (k) Tapping hole sizes are shown by a type 'A' line (f) To show the limits of useful length of a thread, a transverse line extending to the major diameters is used: type 'A' for visible threads (a) (h) type 'B' for hidden threads (c) (f) On sectional views where parts are shown assembled external parts are shown covering the internally threaded parts (l) (m)

Subject	Representation	Notes
Interrupted views	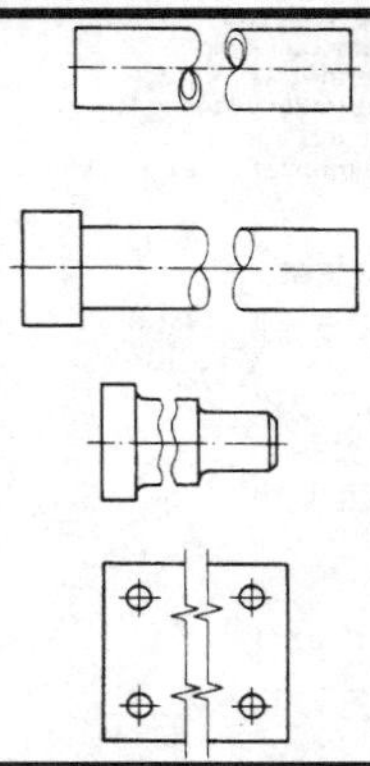	In order to save space only those parts of a large or long object which are sufficient for its definition are shown. Breaklines are shown using a type 'B' line and revealing the circular shape. The type 'D' breakline continues for a short distance beyond the outline.
Repeated parts and features	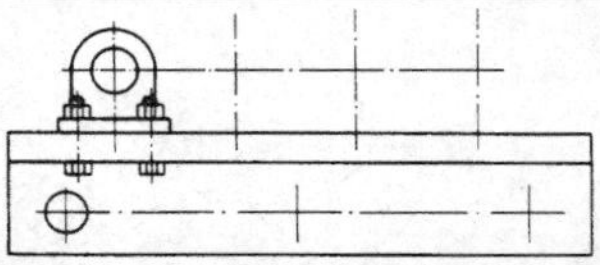	In order to avoid repeated illustrations of identical features or parts, they are shown once, the position of others being indicated by their centre lines.

Figure 11.28 Conventions from BS 308 Part 1

Subject	Representation	Notes
Patterns of repeated features	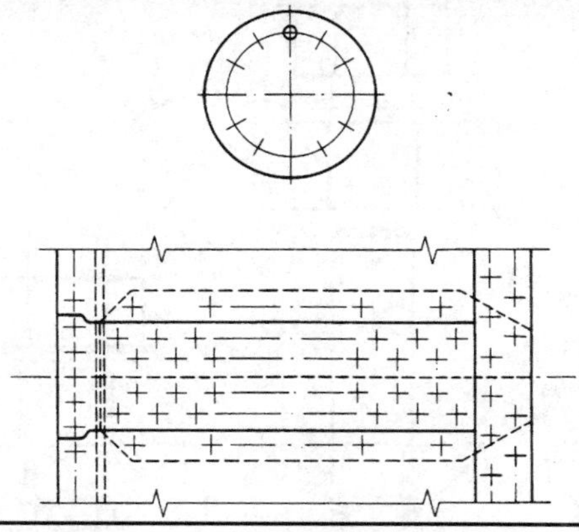	When several holes, bolts, rivets, slots,etc. are required in a regular pattern, only the locating centre lines to establish the pattern are shown. Other information is given in a note.
Plane faces on cylinder parts	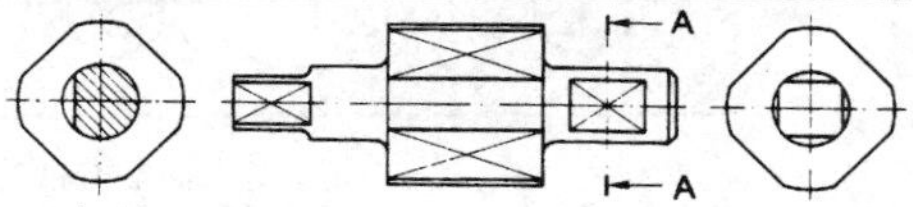 A–A; A; A	Flat surfaces e.g. squares, tapered squares and local flats are shown by thin, crossed, diagonal lines.
Knurling	Straight knurl; Diamond knurl	The type of knurling is shown by type 'B' lines on the surface to be knurled.
Rolling bearings	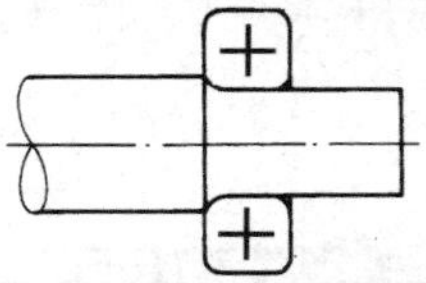	Rolling bearings are shown in this manner, without taking into account the type of bearing (e.g. roller of ball: radial or thrust).

Figure 11.28 Conventions from BS 308 Part I

Subject	Representation			Notes
	View	Section	Simplified	
Cylindrical helical *compression spring* of wire of circular cross section	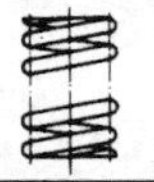	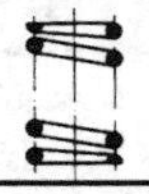		If necessary, an indication "Wound left' or 'Wound right' is included in an adjacent note.
Cylindrical helical *compression spring* of wire of rectangular cross section	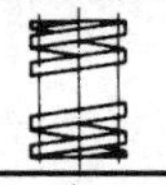	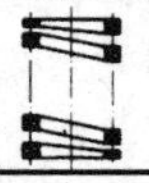	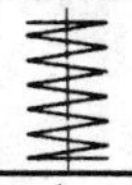	
Cylindrical *helical tension spring* of wire of circular cross section	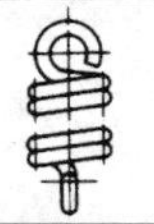			
Cylindrical *helical torsion spring* of wire of circular cross section (wound right-hand)	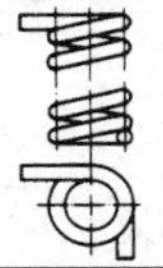		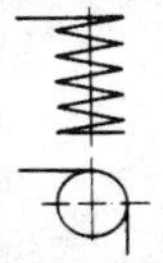	
Cup spring	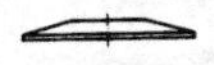		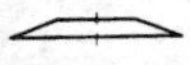	
	View	Simplified		
Semi-elliptic *leaf spring*			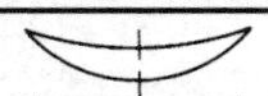	
Semi-elliptic *leaf spring* with eyes	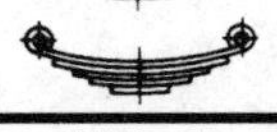		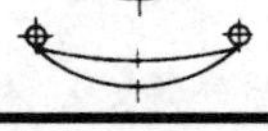	

Figure 11.29 Some spring types using the BS 308 representation

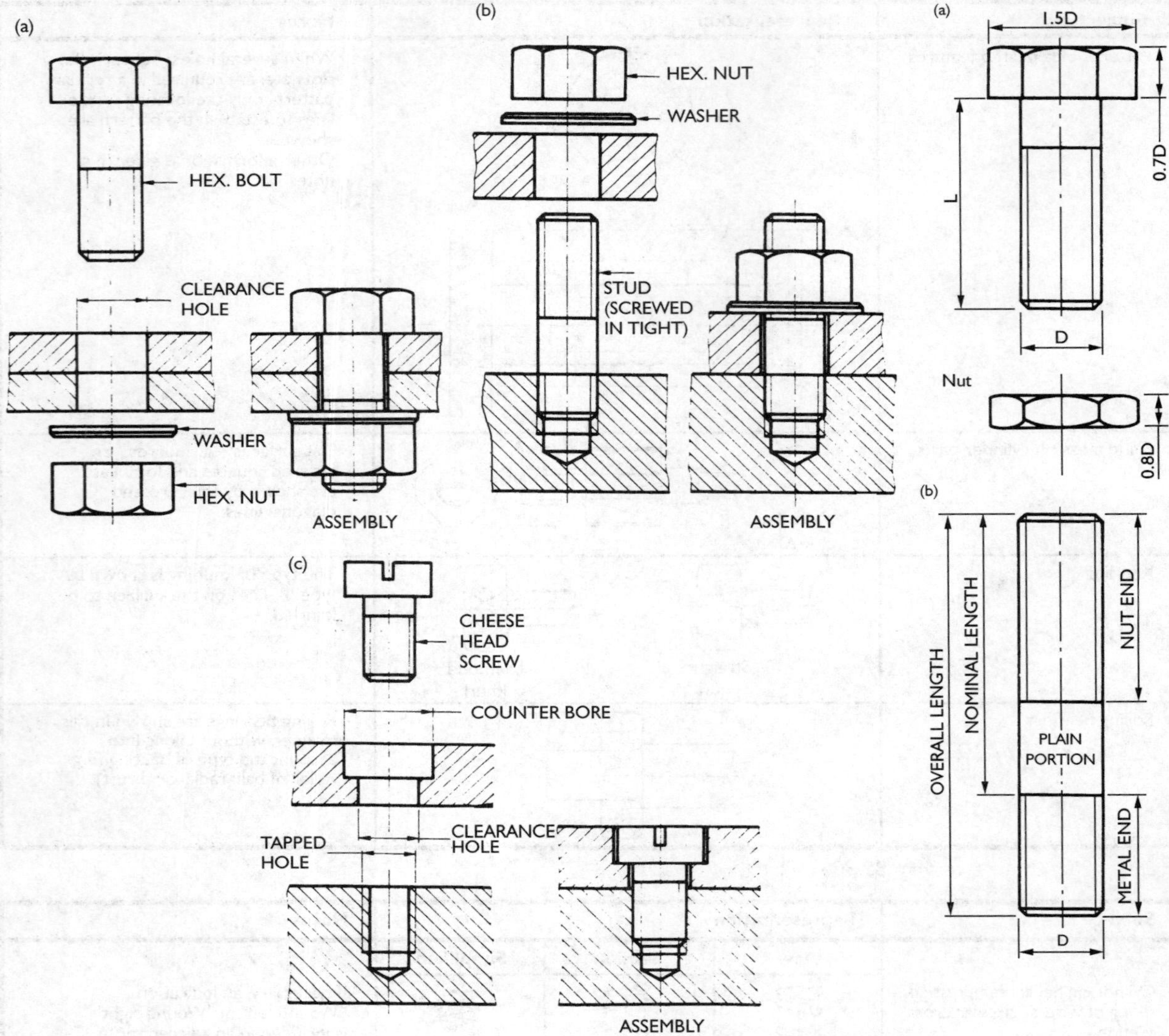

Figure 11.30(1) Method of drawing hexagon-headed screw fasteners and studs and how they are used in assemblies

Figure 11.30(2) Method of drawing hexagon-headed bolts and how they are used in assemblies

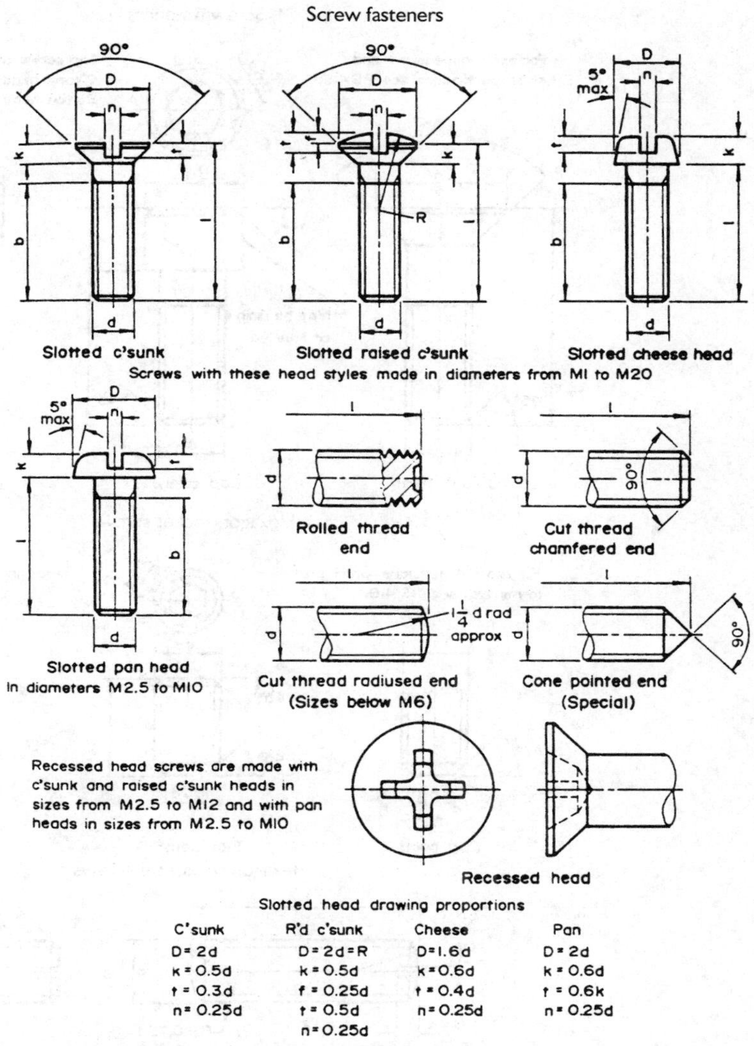

Figure 11.31(1) Other types of screwed fastener

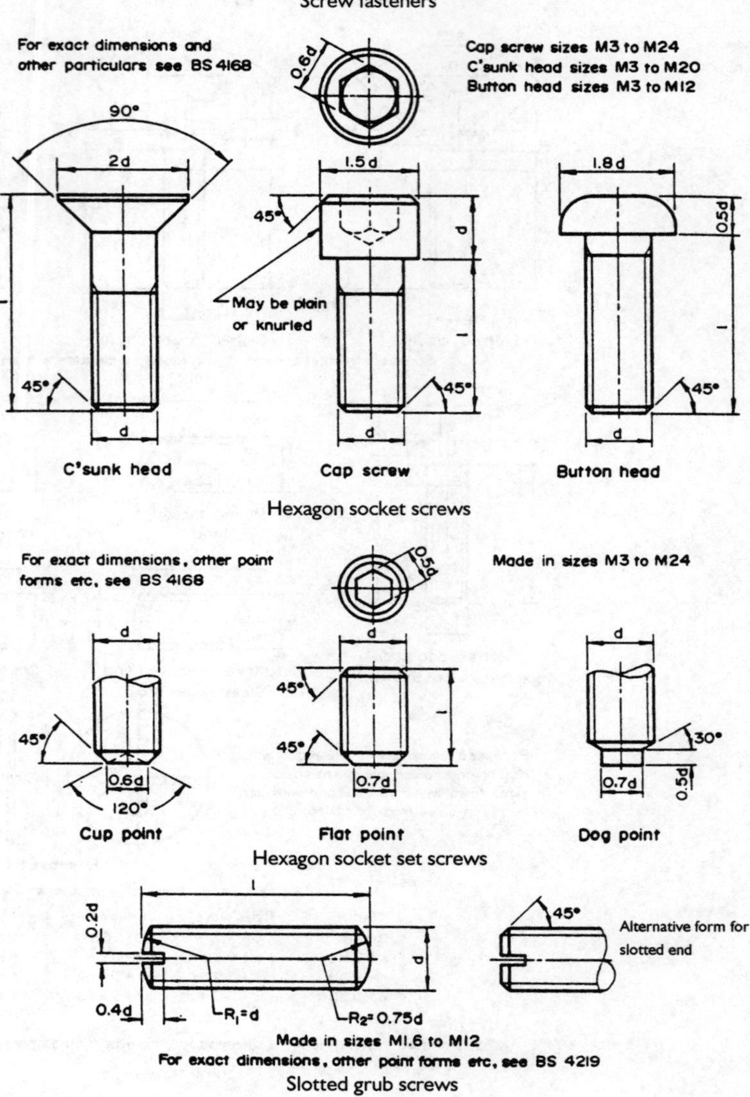

Figure 11.31(2) Other types of screwed fastener

In this system:

- M is the symbol for an ISO metric screw thread
- 12 is the nominal major diameter in millimetres
- 1.75 is the pitch in millimetres
- 6 is the international tolerance (IT) number, which specifies the thread tolerance
- H and g are the deviations from nominal size of the threads.

Some examples of screw thread dimensioning are shown in Figure 11.32. In particular, note that:

- (a) and (b) show the use of dimension notes to the circular view of the thread
- (c), (d) and (e) show dimensioning methods for internal threads
- (f) and (g) show how thread lengths are shown.

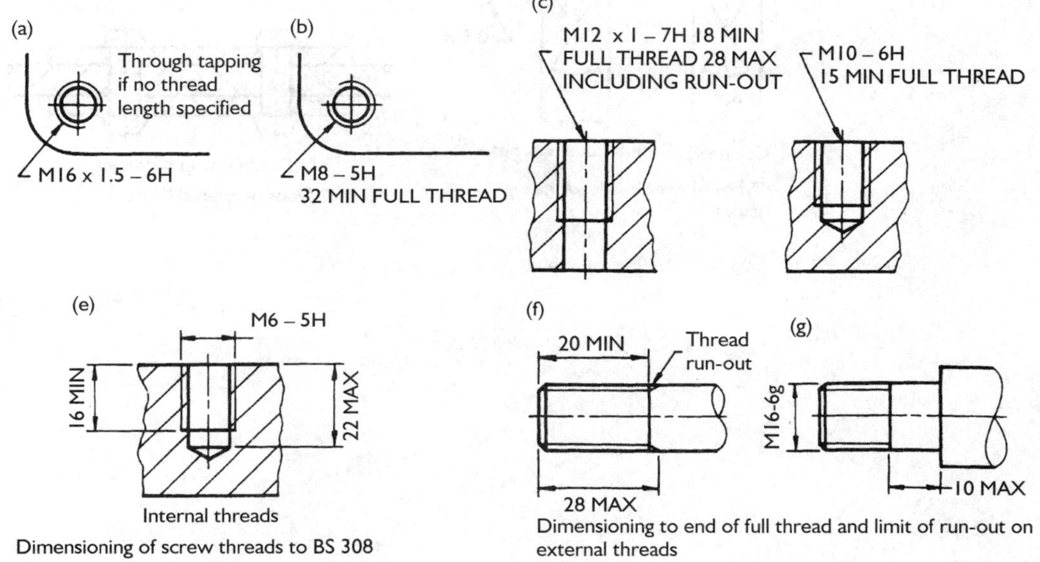

Figure 11.32 Dimensioning of screw threads

Rivets are used where a permanent joint is required. Figure 11.33 shows some typical rivets and types of riveted joints.

Keys, splines, serrations and cotters are mainly used to prevent rotation and/or axial movement between a shaft and a hub. Figure 11.34 shows methods of drawing and typical examples.

Gears are used to transmit power from a power source to an output. Figure 11.35 gives examples of the common types of gears and shows both schematic symbols and conventional representations. Gears are specified and dimensioned using the metric module system where pitch circle diameter (PCD) = module × number of teeth. So, a 20-tooth 4-mm module gear will have a PCD of (20×4) mm = 80 mm.

Bearings allow parts to rotate and/or slide with respect to each other. The conventional representation of rolling bearings has already been shown in Figure 11.28. Figure 11.36 shows schematic symbols for bearings and pivots.

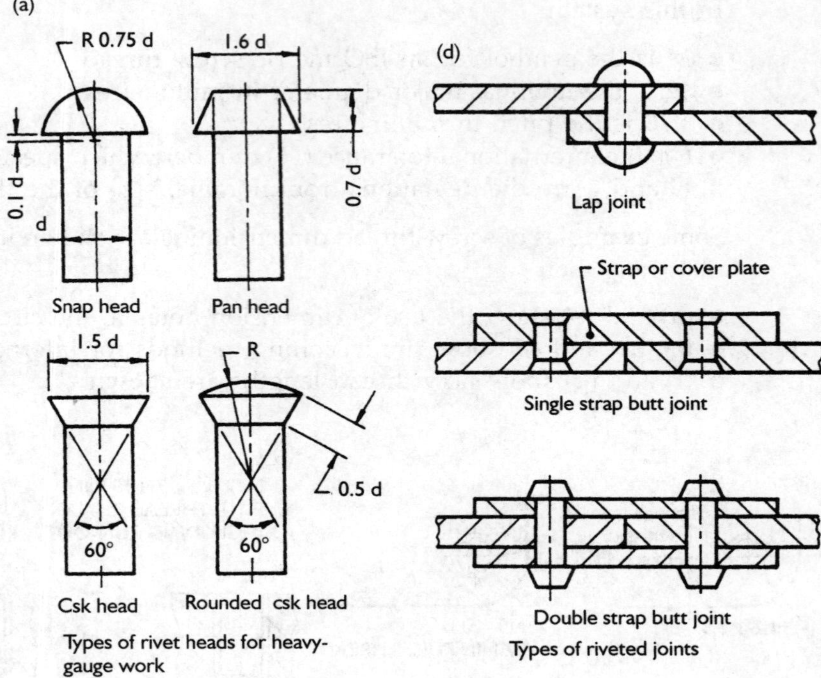

Figure 11.33(1) Rivets and types of riveted joints

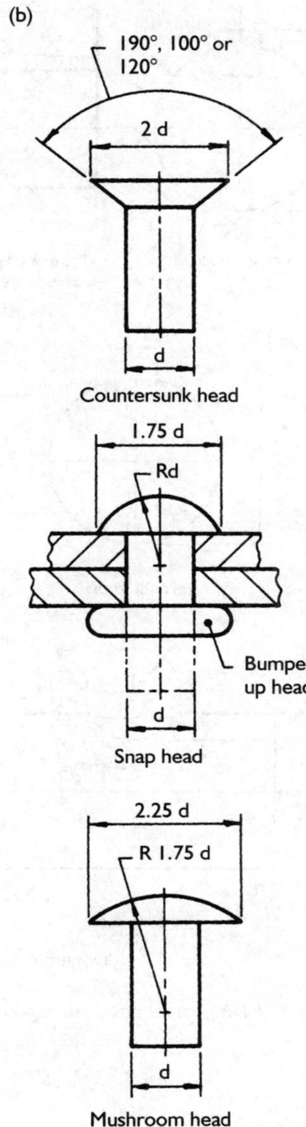

Figure 11.33(2) Rivets and types of riveted joints

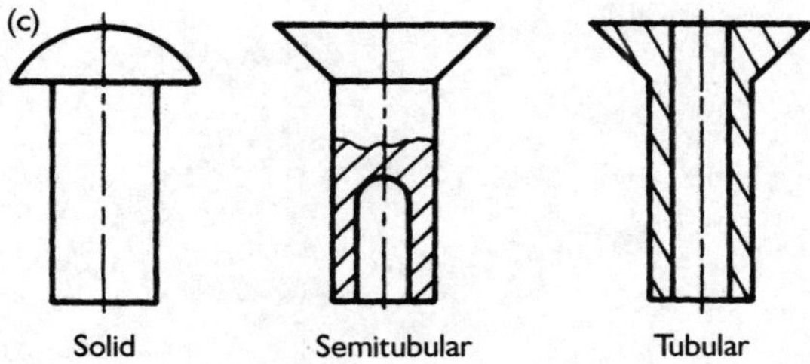

Figure 11.33(3) Rivets and types of riveted joints

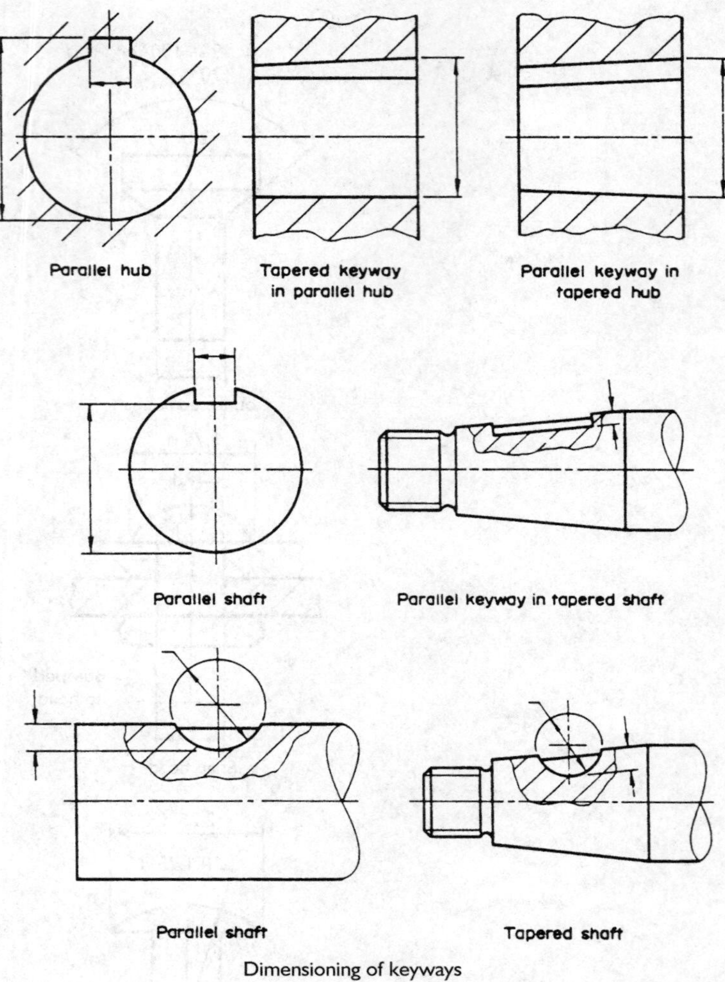

Figure 11.34(1) Keys, splines, serrations and cotters

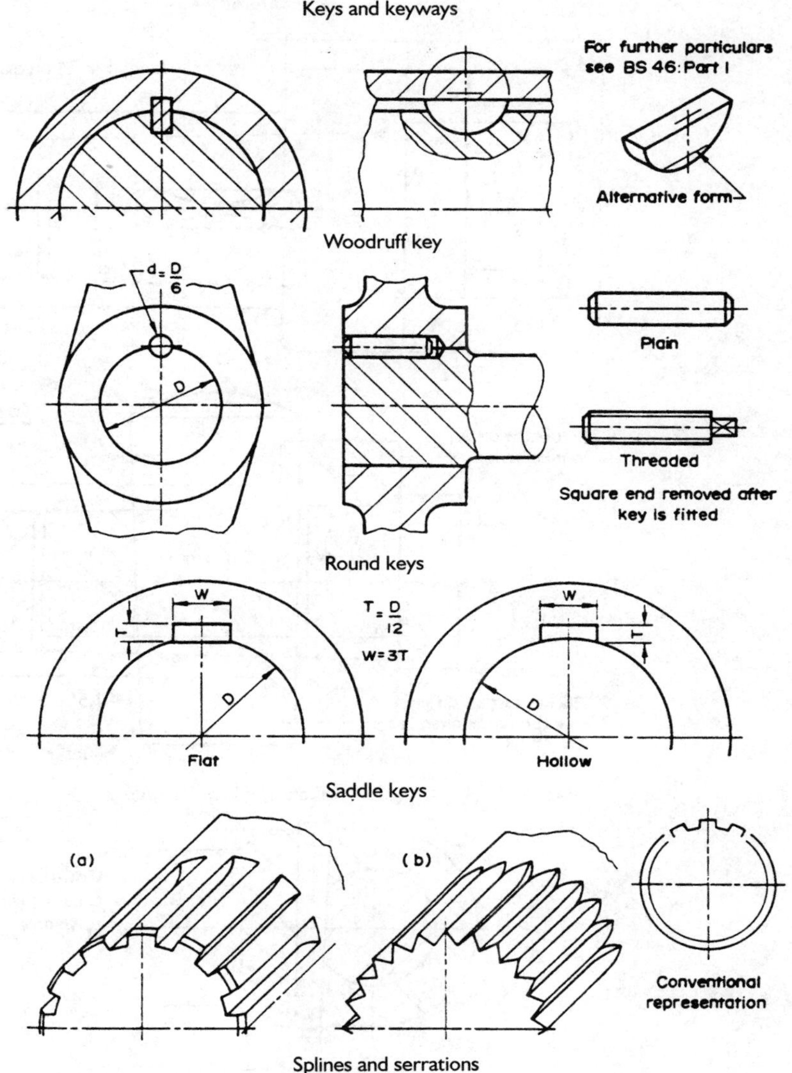

Figure 11.34(2) Keys, splines, serrations and cotters

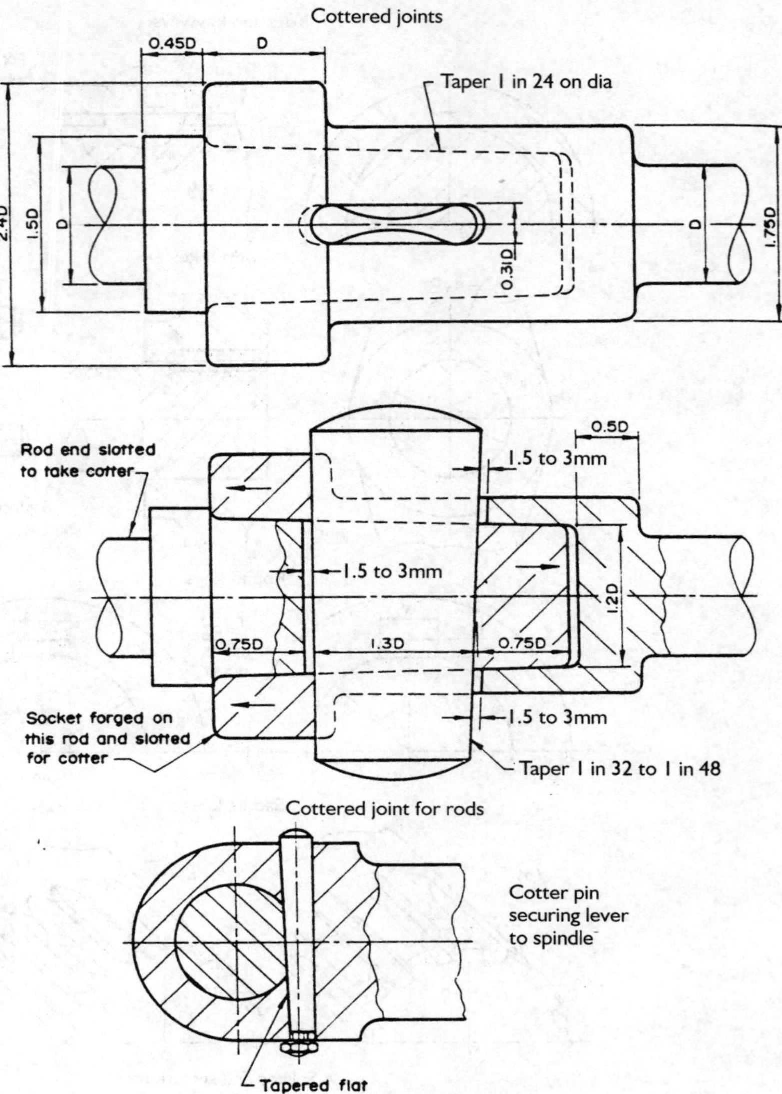

Figure 11.34(3) Keys, splines, serrations and cotters

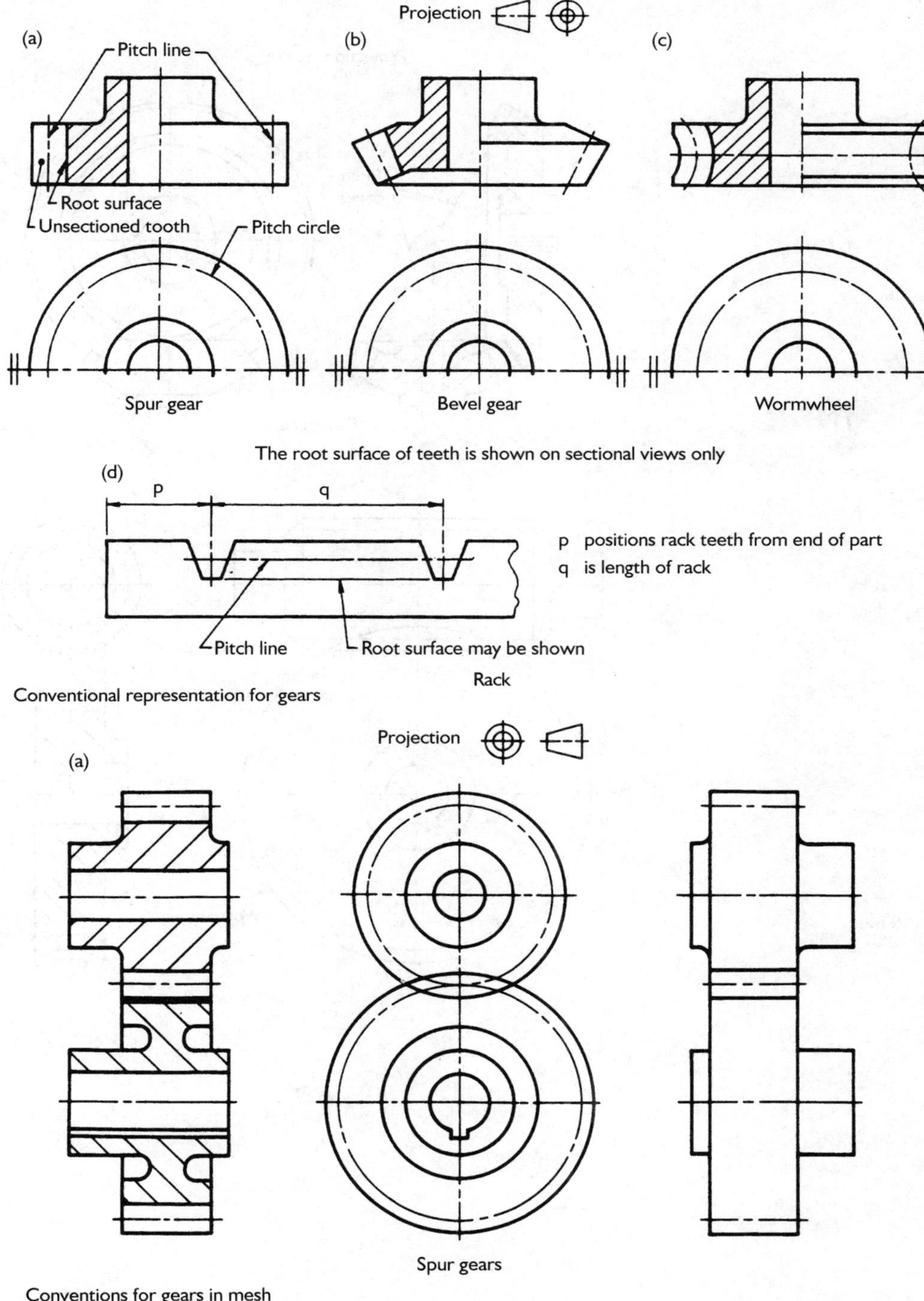

Figure 11.35(1) Gears

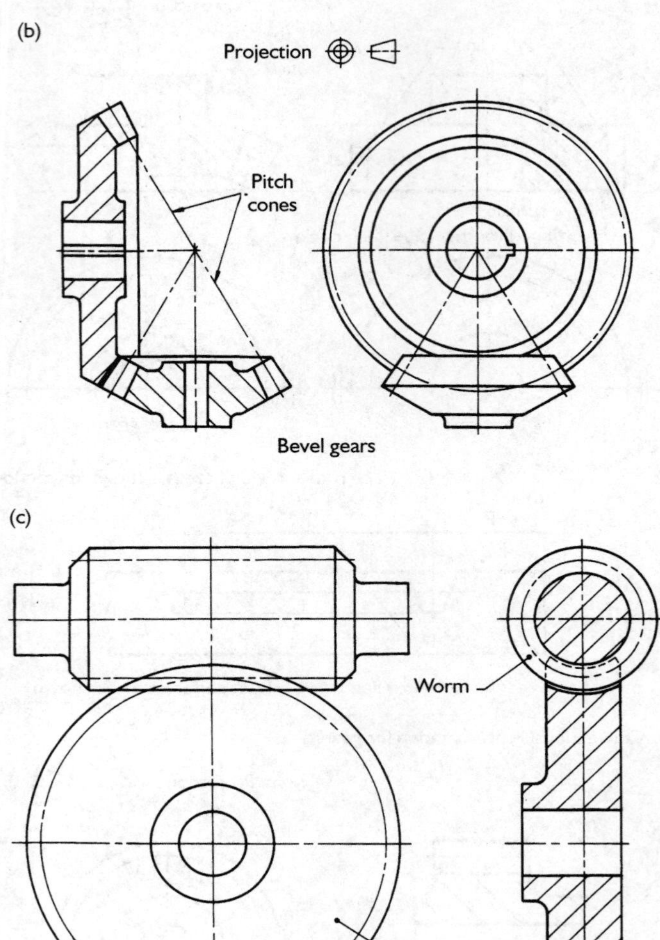

Figure 11.35(2) Gears

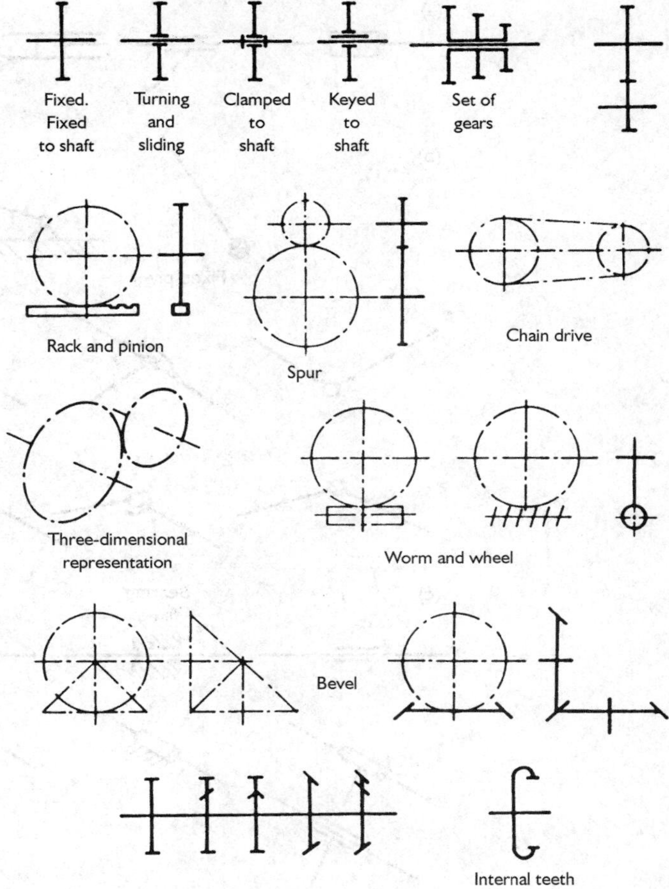

Figure 11.35(3) Gears

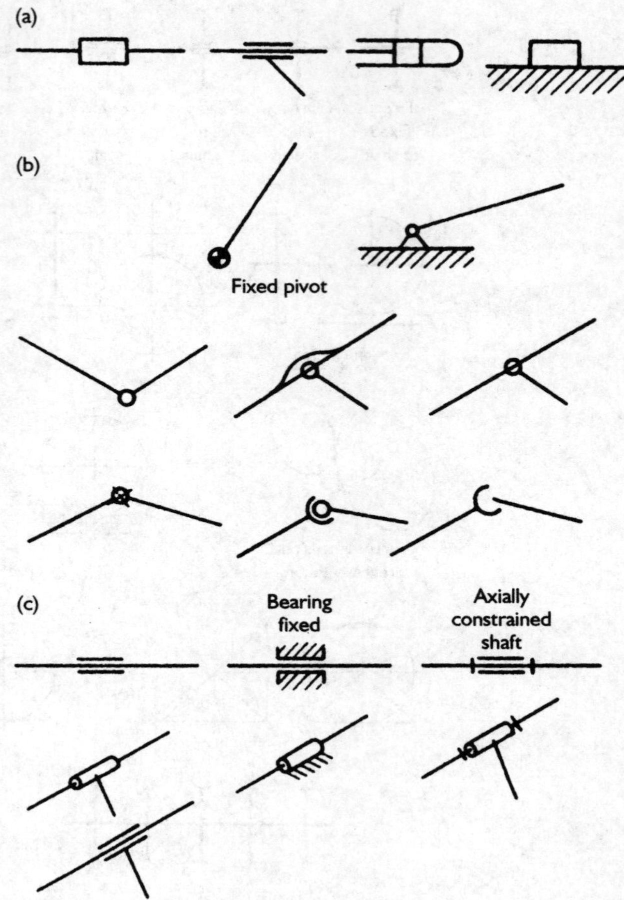

Figure 11.36 Schematic symbols for bearings and pivots: (a) sliding pairs; (b) pivots; (c) rotary/sliding bearings

Friction
Socket
Flexible
Ball and socket
Claw
Flange

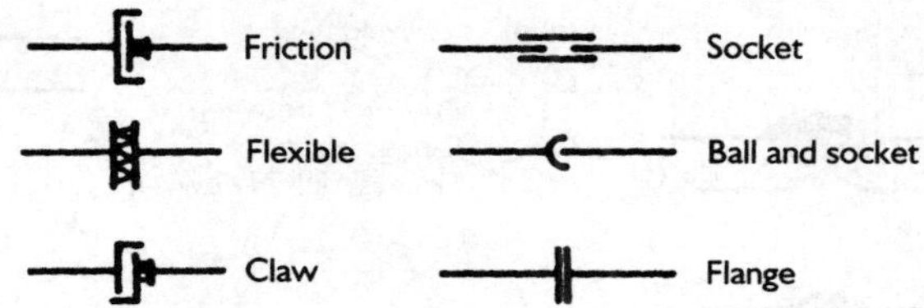

Figure 11.37 Symbols for different types of couplings

Couplings allow drives to be transmitted and to be connected or disconnected. Figure 11.37 shows the symbols for the different types of couplings.

Welding is used to make permanent joints, and symbols are used to simplify the drawing of the welded joints. Figure 11.38 shows the most common types of weld with the appropriate symbol to BS 499, and the application of the symbols on a drawing.

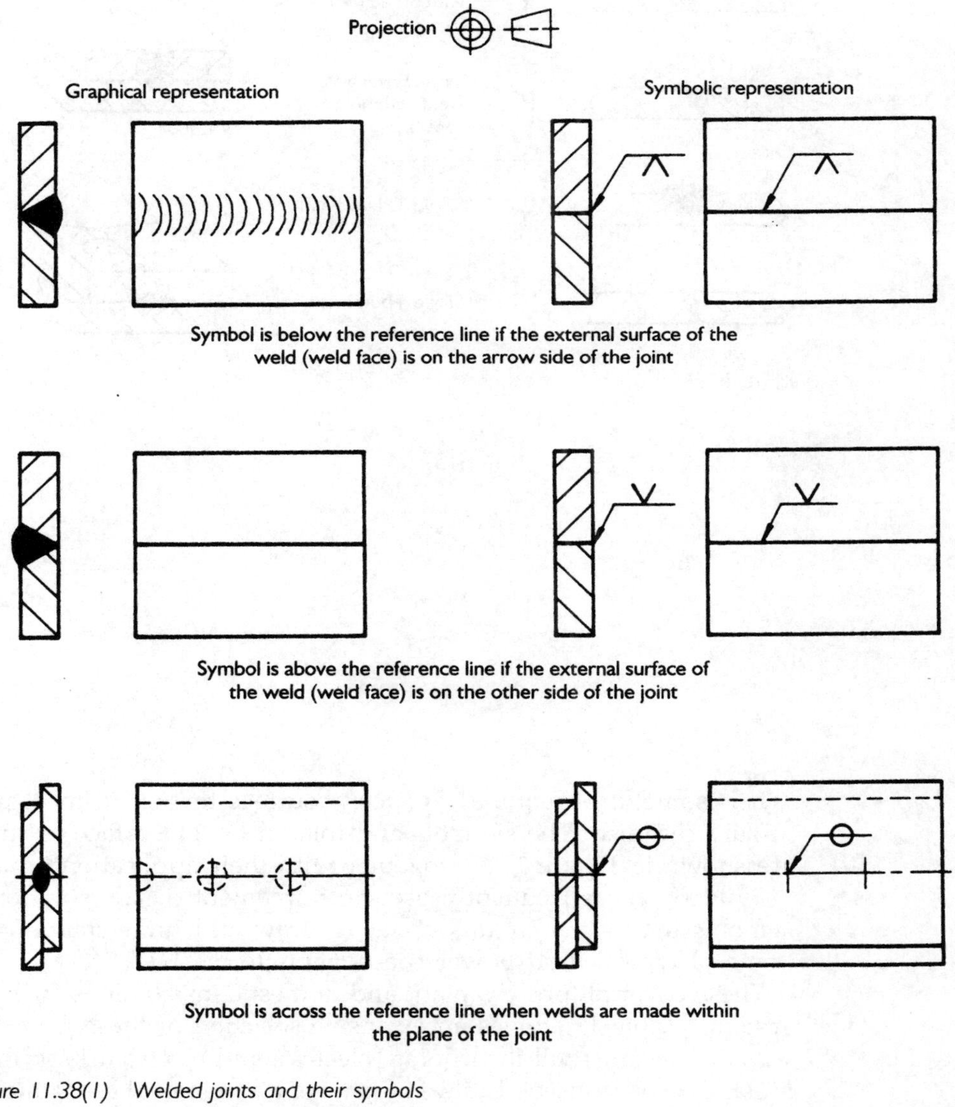

Projection
Graphical representation
Symbolic representation
Symbol is below the reference line if the external surface of the weld (weld face) is on the arrow side of the joint
Symbol is above the reference line if the external surface of the weld (weld face) is on the other side of the joint
Symbol is across the reference line when welds are made within the plane of the joint

Figure 11.38(1) Welded joints and their symbols

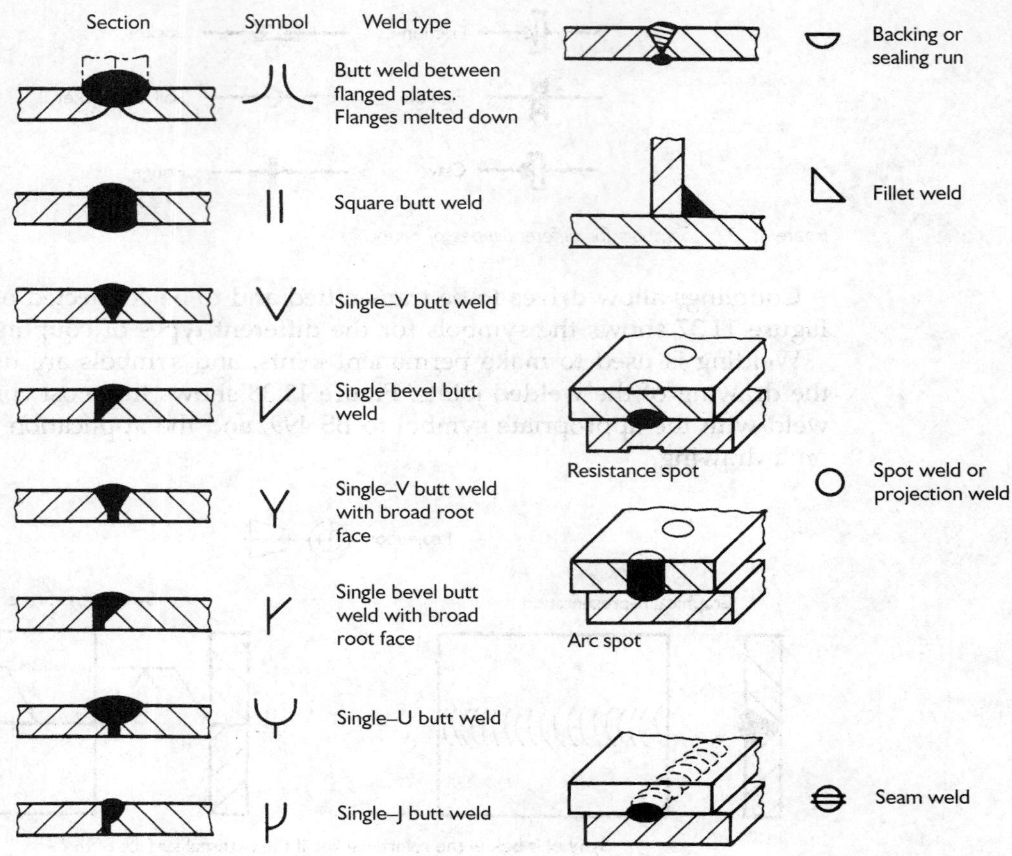

Figure 11.38(2) Welded joints and their symbols

It is sometimes required to specify surface texture to meet particular qualities of surface finish. The symbols and tolerances for surface texture and machining are shown in Figure 11.39, together with their application on a drawing.

Fluid power is frequently used to implement design solutions, and the use of symbols makes the circuits easier to draw and understand. Table 11.6 shows a range of typical fluid power components to BS 2917.

These cover all process plant and domestic installations such as central heating systems. Table 11.7 gives an extensive selection of these from BS 1553.

Many designs will incorporate electrics and electronics, so the correct symbols must always be used. Table 11.8 shows a range of these symbols to BS 3939.

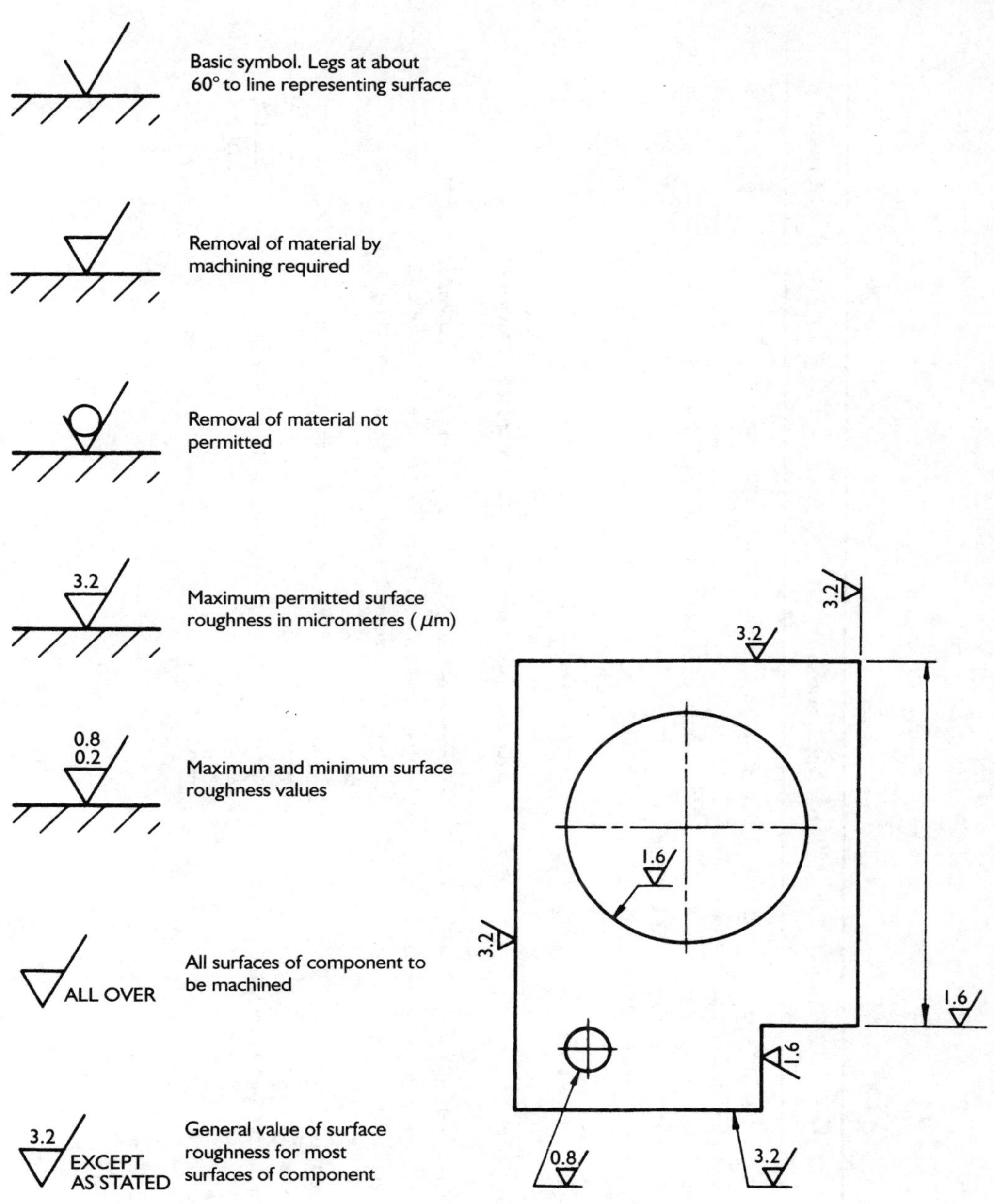

Figure 11.39 Symbols and tolerances for surface texture and machining

Table 11.6

Component	Purpose/function	Applications	Symbol
Pump	Converts electrical energy to kinetic energy of fluid flow at pressures up to 100 bar and flow rates up to 700 litres/min	Actuation of hydraulic motors and cylinders	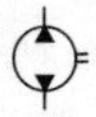
Compressor	Converts electrical energy to kinetic energy of air flow at pressures up to 175 bar and flow rates up to 4000 litres/min	Actuation of pneumatic motors and cylinders	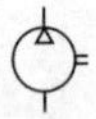
Motor	Converts energy of fluid kinetic energy of rotation	Rotation for machine operation etc.	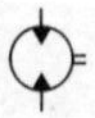
Accumulator	Stores fluid under pressure to act like a fluid spring	Supply of fluid to meet a sudden demand	
Reservoir	Container for hydraulic fluid	Source of fluid supply and place where fluid is returned to	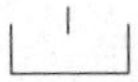
Single acting	Power actuation in one direction	Where one-directional motion is required with power actuation in one direction and return by gravity, e.g. hoist/jack	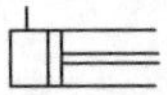
Single acting with spring return	Power actuation in one direction	Where one-directional motion is required with spring return e.g. clamp	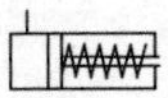
Double acting	Power actuation in both directions	Where powered motion is required in both directions, e.g. machine tool slide	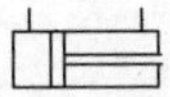
Air–oil actuator	Conversion of pneumatic pressure to hydraulic pressure or vice versa	Where change from air tool or oil to air is required	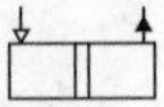
Push-button control mechanism	Manual valve actuation	Valve control	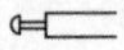
Lever control mechanism	Manual valve actuation	Valve control	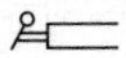
Pedal control mechanism	Manual valve actuation	Valve control	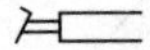
Plunger or tracer control mechanism	Mechanical valve actuation	Valve control	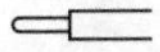

Spring control mechanism	Mechanical valve actuation	Valve control	
Roller control mechanism	Mechanical valve actuation	Valve control	
Solenoid control mechanism	Electrical valve control	Valve control	
Motor control mechanism	Electrical valve control	Valve control	
Pilot control mechanism	Indirect valve control by fluid pressure	Valve control	
Non-return valve	To open if inlet pressure is higher than outlet pressure	For clamps, jacks etc. to prevent failure if inlet pressure drops	
Shut-off valve	Prevention of any fluid flow	To shut down circuits rapidly	
Pressure control/ throttle valve	To control flow and pressure	Provision of suitable pressure/flow rates for circuits	
Pressure-relief valve (safety valve)	Opens the outlet to reservoir or atmosphere if working pressure is exceeded	Prevention of danger or damage by excessive pressure	
Two port–two position (2/2) directional control valve (DCV)	To control one inlet port and one outlet port	Control of single-acting cylinder, etc.	
Three port–two position (3/2) DCV	To control two outlet ports and one inlet port or vice versa	Control of two cylinders, etc.	
Five port–two position (5/2) DCV	To control two outlet ports, two inlet ports and one exhaust port to atmosphere or reservoir	Control of double acting cylinders etc.	

Table 11.7

Description	Symbol	No.
Pipe; major flow line		1
Pipe; minor flow line		2
Pipes connect, i.e. junction		3
4-way junction		4
Pipes cross but not connected		5
Butt welded joint		6
Soldered or joint made by adhesive		7
Screwed joint		8
Sleeve joint		9
Compression joint		10
Socket welded		11
Flanged and bolted		12
Flanges welded on		13
Flanges screwed on		14
Pipe bore change		15
End capped		16
End capped screwed		17
End blank flanged and bolted		18
Pipe guide		19
Pipe support		20
Pipe hanger		21
Soldered tee		22
Screwed tee		23
Flanged tee		24
Soldered bend or elbow		25
Screwed bend		26
Flanged bend		27
Open vent		28
Instrument; temperature gauge	T	29
Instrument; pressure gauge	P	30
Discharge to atmosphere		31
Trap; release or retention		32
Trapped vent		33
Sight flow indicator		34
Strainer or filter		35
Orifice plate		36
Valve (in line); general symbol		37
Parallel slide valve		38
Wedge gate valve		39
Globe valve		40
Butterfly valve		41
Diaphragm valve		42
Needle valve		43
Valve closed		44
Valve; hand operated		45
Valve; power operated		46
Motor operated valve	M	47
Pressure control valve		48
Power operated pressure control valve		49
Flanged, hand operated valve		50
Valve to open on failure		51
Flanged, power operated pressure control valve to close on failure		52
Valve to maintain its position on failure		53
Float operated valve e.g. cistern ball cock		54
Non–return valve flow direction		55
Plug or cock annotated T or L according to style of porting	L	56
3-way plug or cock with style of porting annotated T or L	T	57
Safety valve spring loaded		58
Safety valve; flanged and weight loaded		59
Angled valve		60
Angled valve; flanged and hand operated		61
Flanged, power operated, angled pressure control valve to open on failure		62
Radiator		63
Convector		64
Towel rail		65
Pump		66
Boiler	Blr	67
Hot water cylinder	HWC	68
Hot water tank	HWT	69
Feed and expansion tank	F&ET	70
Condensate tank	CT	71
Air vessei		72

Table 11.8 Schematic circuit, wiring and layout diagram symbols

Name	Symbol
Cell	+ −
Battery	
Battery (alternative)	6V
Polarity POSITIVE	+
Polarity NEGATIVE	−
Direct current DC	
Alternating current AC	~
Suitable for AC and DC	≂
Conductor	
Conductors not connected	
Conductors connected	
Separable contact	
Fixed or bolted contact	
Fuse	
Fuse	
Fuse-supply side indicated	
Winding	
Transformer	
Resistor	
Resistor (alternative)	
Variability	
Variable resistor	

Name	Symbol
Rheostat	
Potentiometer	
Impedance	Z
Capacitor	
p n Diode	
p n Diode on a "chip"	
Light sensitive *p n* Diode	
Thyristor	
p n p Transistor	
n p n Transistor	
Microphone	
Earphone	
Loudspeaker	
Headphones	
Conductors twisted	
Cable with 3 conductors	
1 line to mean 3 conductors	3
2 conductors in one line	
Link (separable)	
Link (fixed)	
Filament lamp	
Signal lamp	

Name	Symbol
Circuit breaker	
Make contactor	
Break contactor	
Winding with one tapping	
Transformer	
Single element relay	
Heater	
Heater alternative	
Mechanical coupling	
Relay coil	
Socket (female)	
Plug (male)	
Multiple pole plug and socket	
4-pole plug and socket	4
Coaxial socket	
Coaxial plug	
Jack sleeve and spring	
3-pole concentric jack and plug	
Ammeter	A
Voltmeter	V
Wattmeter	W

Name	Symbol
Ohmmeter	Ω
Generator	G
Motor	M
DC Generator	G
DC Motor Alternator	M
AC Generator	G
AC Motor	M
Oscilloscope	
Discharge lamp	
Clock-timer	
Make contact	
Break contact	
Push button make contact	
Push button break contact	
Mechanically coupled make contactors	
Change over contact	
Change over contact	
Change over diagonally	
Change over contact unit	
Make/break contact unit	
Mech. operated break pulse	

Name	Symbol
Electric bell	
Buzzer	
Chassis-ground	
Earth	
Pyrometer Thermometer	t°
Thermocouple	+ve −ve
Single stroke bell	

Installation symbols

Name	Symbol
Main control or supply	
Main switch	
Meter	kWh
Spotlight	
Wall light with built in switch	
Lighting point built in switch	
Wall mounted light point	
Single tube fluorescent	
Multiple lamp fluorescent	3 40W
Fluorescent lamp-multiple	
Multiple light point	3 40W

Name	Symbol
Lamp or lighting point	
Switch with a pilot light	
Cord operated one-way switch	
Single pole one-way switch	
Push button contact	
Illuminated push button	
Transformer	
2-pole one-way switch	
Two-way switch	
Electrical appliance	
Intermediate switch	
Rheostat or dimmer switch	
Socket outlet	
Switched socket	
Multiple socket outlet	2
Socket with pilot light	
Switched socket with pilot light	
Consumer's earth point	
Fan	
Thermostat	t°
Time switch	
Aerial	

Progress check

1 What are the standard sizes of drawing sheets? State the relationships between the different sizes of drawing sheets.
2 What printed or written information is put onto a drawing?
3 State the standard line widths and say what the ratio is between a thick line and a thin line.
4 What are the uses for:
   a thick lines
   b thin lines?
5 Describe what orthographic projection is and name the two different systems.
6 Why are drawings sectioned and which parts are not usually sectioned?
7 List the principles to be followed when dimensioning parts.
8 What is a datum and why is it used?
9 Why are tolerances necessary on dimensions?
10 Name the three types of standard fit between mating parts.
11 State the difference between hole-based and shaft-based systems of fits. How are tolerances and limits allocated in the two systems of fits?
12 Name and describe the two main types of pictorial projection.
13 What is an 'exploded' diagram and what is it used for?
14 Give the reasons for the use of drawing conventions for representing screw threads, repeated features, knurling, etc.
15 Which British Standard is referred to for:
   a drawing conventions
   b electrical and electronic symbols?
16 What are the advantages of using standard symbols on drawings?

Circuit diagrams

Circuit diagrams are used to show the interconnections between the individual components in an electrical, electronic or fluid power device. The components are represented by standard symbols, and some of these are shown in Tables 11.6–11.8.

Computer-aided methods are particularly useful for circuit design and drawing because:

- large libraries of standard symbols can be held on file
- the function of the circuit can be simulated
- the optimum layout of the circuit can be determined automatically.

When producing circuit diagrams, reference should be made to BS 3939, BS 2917, BS 1553 and BS publication PP7307.

Case study

Examples of circuit diagrams

Three examples of circuit diagrams are given in this case study.

Digital electric voltmeter

Figure 11.40 shows an example of an electronic device, in this case a digital electric voltmeter. Referring to Figure 11.40, the diagrams are:

- a block diagram, which is a system diagram of the operation of the voltmeter. The flow of signals is shown, with the input and feedback signals, which are compared in the comparator. Blocks A, B, C, D and E are represented by components on the left-hand side of the circuit diagram. Blocks F, G and H are represented by components on the right-hand side of the circuit
- a circuit diagram showing all components and connections. Note that the pin numbers on the ICs are non-sequential in order to simplify the circuit diagram. A full parts list is given with the specification of each component
- a layout of the tracks on the reverse of the printed circuit board (PCB). The left-hand and right-hand sides of the circuit are shown separately
- a general arrangement drawing, which shows the actual position of the components on the PCB. The left- and right-hand sides of the circuits are again shown separately.

Water supply and microbore heating system

Referring to Figure 11.41, the diagrams are:

- a circuit diagram showing the arrangement of a water supply and microbore heating system for a bungalow. All components and connections are identified
- an architect's plan of the bungalow, showing locations of heating and water services. Locations of electrical services are also specified.

Fluid power circuit

Figure 11.42 shows a continuously operating reciprocating pneumatic cylinder, and the two diagrams are:

- a circuit diagram, which shows a double-acting pneumatic cylinder, a lever-operated 2/2 DCV – valve number 1; a plunger-operated 2/2 DCV – valve number 2, a pressure-controlled 5/2 DCV feeding air into each end of the cylinder and also exhausting air from each end. It is controlled by valves 1 and 2
- a layout drawing, which shows the cylinders, valves and piping in their actual positions and to scale.

Activity 11.5

Using both manual and computer-aided methods:

a copy the circuit diagrams in Figures 11.40, 11.41 and 11.42

b produce circuit diagrams for any circuits used in the design solutions for Element 5.2.

Freehand sketching

Freehand sketching is a very valuable technique for a designer to acquire. It is used for modelling, communicating and recording. It can be applied to most kinds of drawing including detail drawings, assembly drawings, block and flow diagrams, circuit diagrams, conceptual diagrams, etc. A designer's sketchpad is very useful for recording ideas and suggestions. It can help clarify ideas during

(a)

The following diagrams describe a digital electronic voltmeter in different ways and are provided as examples of the various stages of illustrating an electronics network.

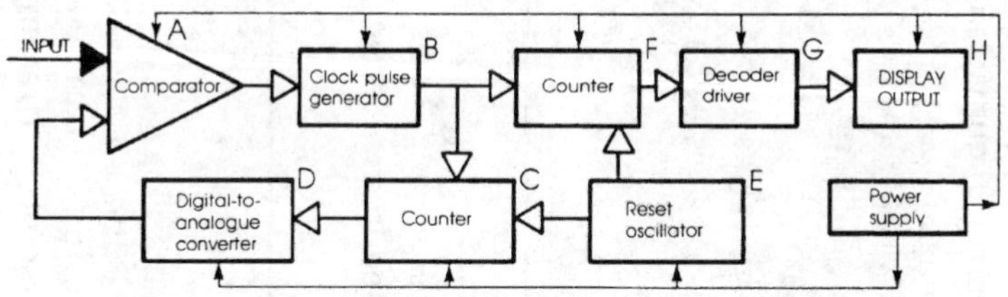

Block diagram. The block diagram elements A,B,C,D and E are represented in the left-hand side of the circuit diagram; F,G and H by the right-hand side.

Note the symbols shown on the integrated circuits (ICs), locate the pin number 1 with the other pins numbered consecutively around the IC:

IC   8 7 6 5 / 1 2 3 4

This consecutive numbering is not strictly necessary on the schematic circuit diagram

(b)

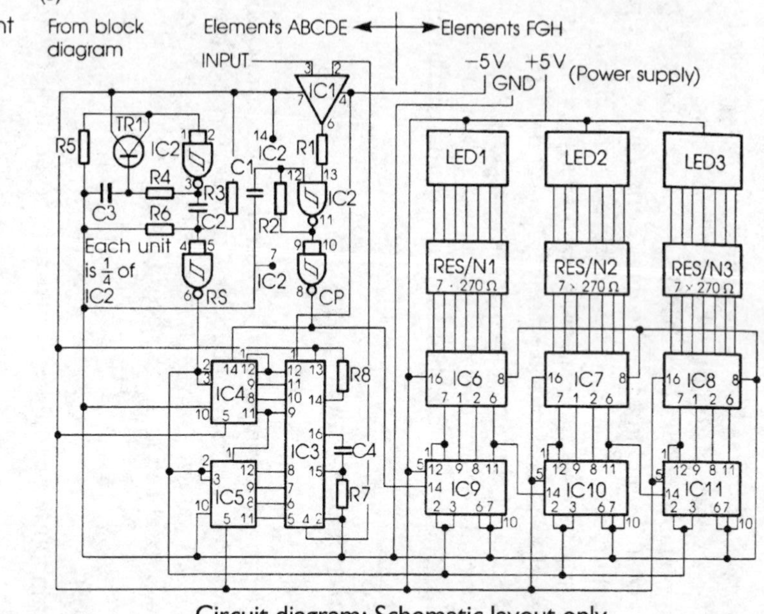

Circuit diagram: Schematic layout only

TR1	BC109	Transistor
IC1	741	Op. amp
IC2	74132	Quad Schmitt trigger
IC3	MC1408	8-bit DAC
IC4 and 5	7493	4-bit binary counters
IC6, 7 and 8	7447	Decoder driver
IC9, 10 and 11	7490	Decade counter
LED1, 2 and 3		Light emitting diode

RES/N1, 2 and 3	7 × 270 Ω
R1	2.2 kΩ
R2	330 Ω
R3	1 kΩ
R4	3.3 kΩ
R5	330 Ω
R6	2.2 kΩ
R7	2.7 kΩ
R8	2.7 kΩ

C1	10 nF
C2	10 nF
C3	10 μF polarised
C4	50 pF

Note: GND = ground or earth
CP = clock pulse
RS = reset
DAC = digital to analogue converter

Figure 11.40a&b Digital electric voltmeter circuit diagram

(c)

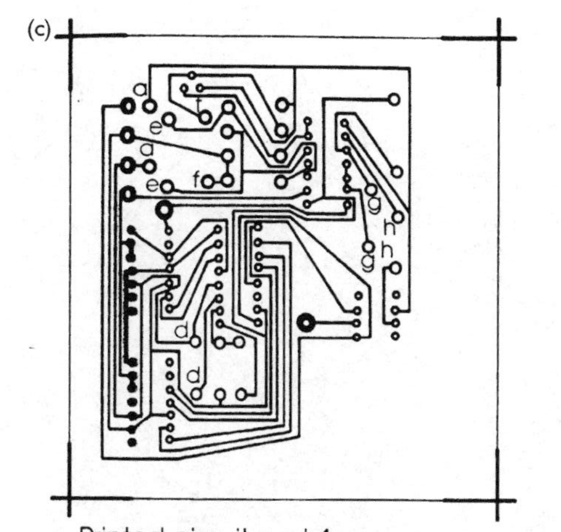

Printed circuit part 1.
The through board foil print as it is known, is the counterpart of the wiring diagram

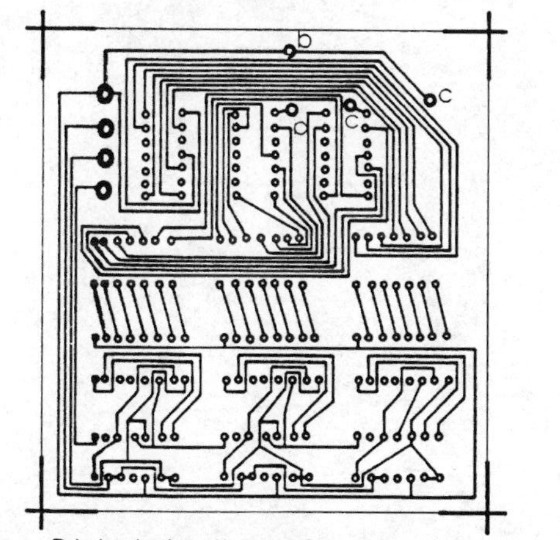

Printed circuit part 2.

(d)

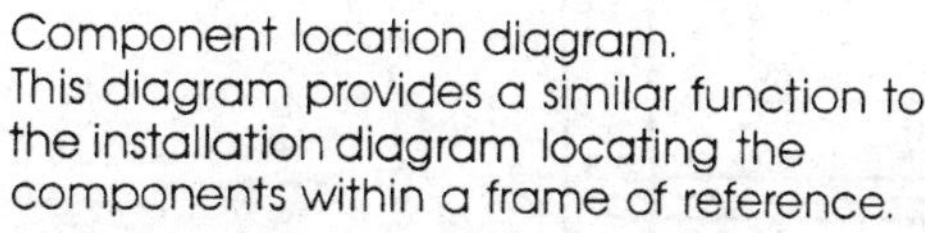
Component location diagram.
This diagram provides a similar function to the installation diagram locating the components within a frame of reference.

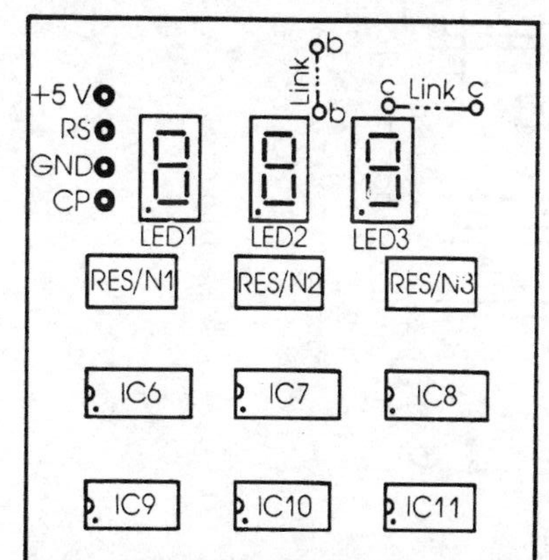

Component location diagram.

Wired connections are required between the two boards, CP to CP, RS to RS, GND to GND and 5V to +5V with an external supply provided to the +5V and -5V. Unlike the schematic, the conductors on the boards cannot cross but, where this is necessary, insulated wired links are provided, i.e. a to a, b to b, and c to c. The links have only been included to illustrate their use and could have been avoided.

Figure 11.40c&d Digital electric voltmeter circuit diagram

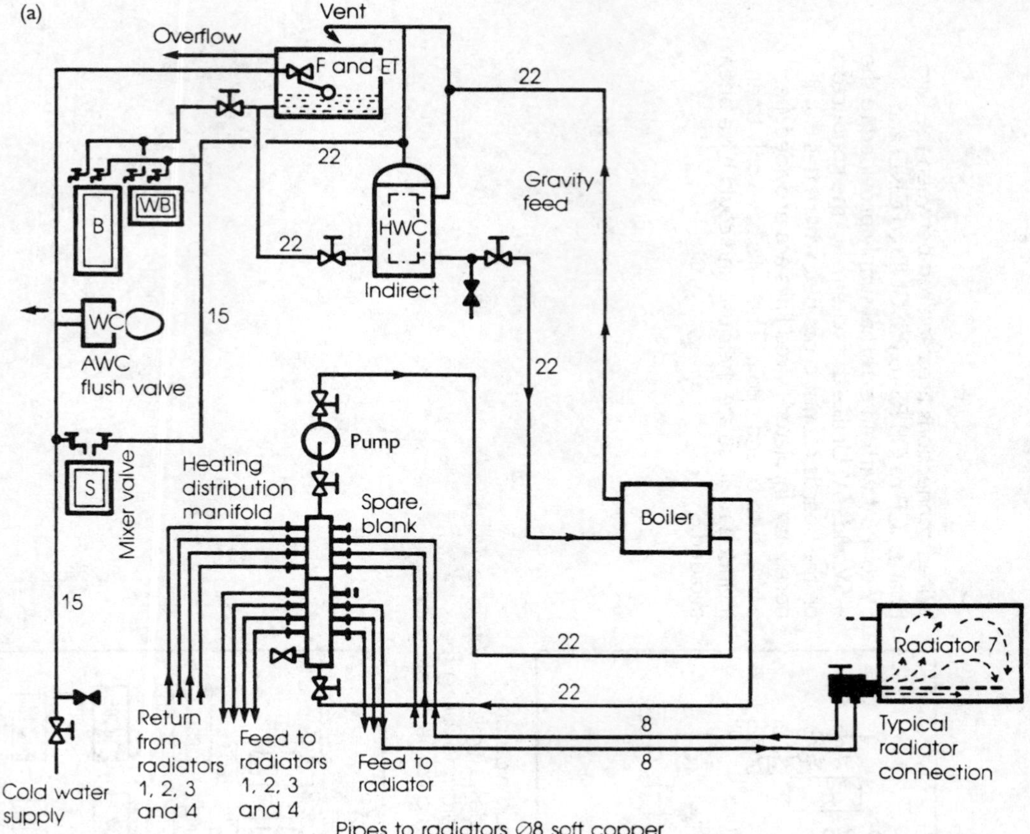

Figure 11.41a Water supply and microbore heating system circuit diagram

(b)

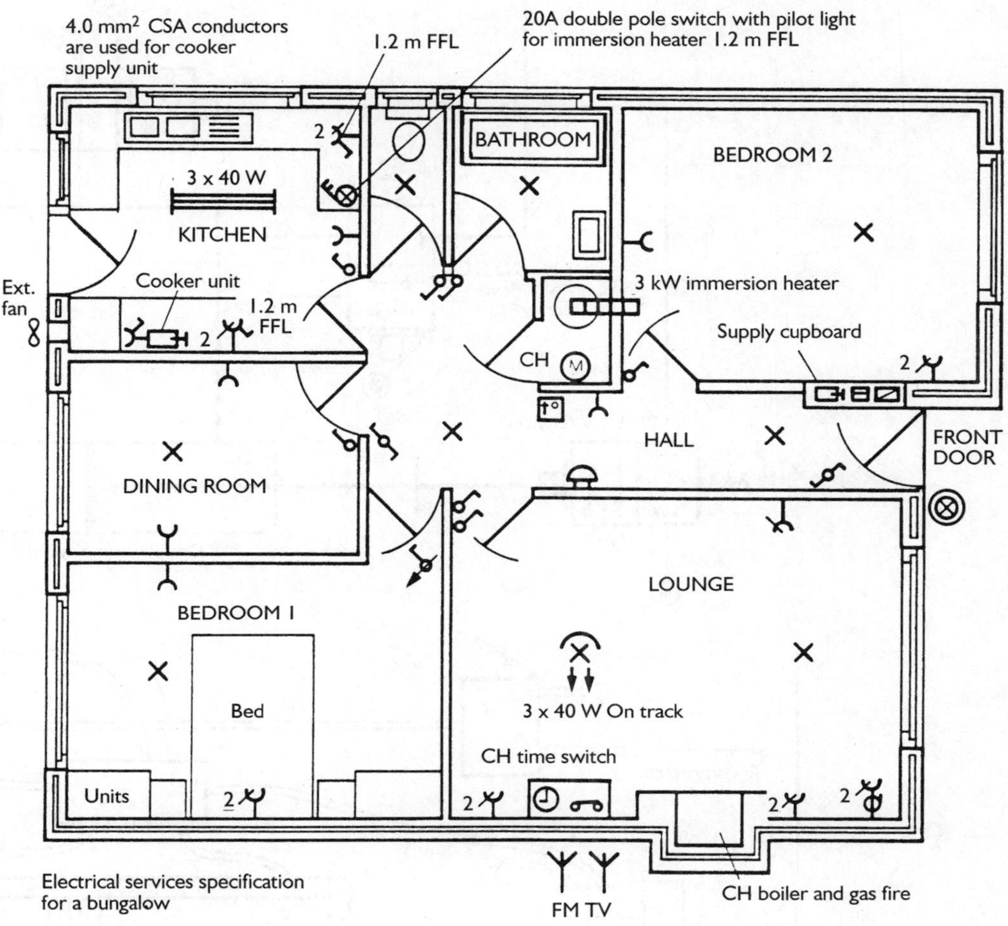

Figure 11.41b Water supply and microbore heating system circuit diagram

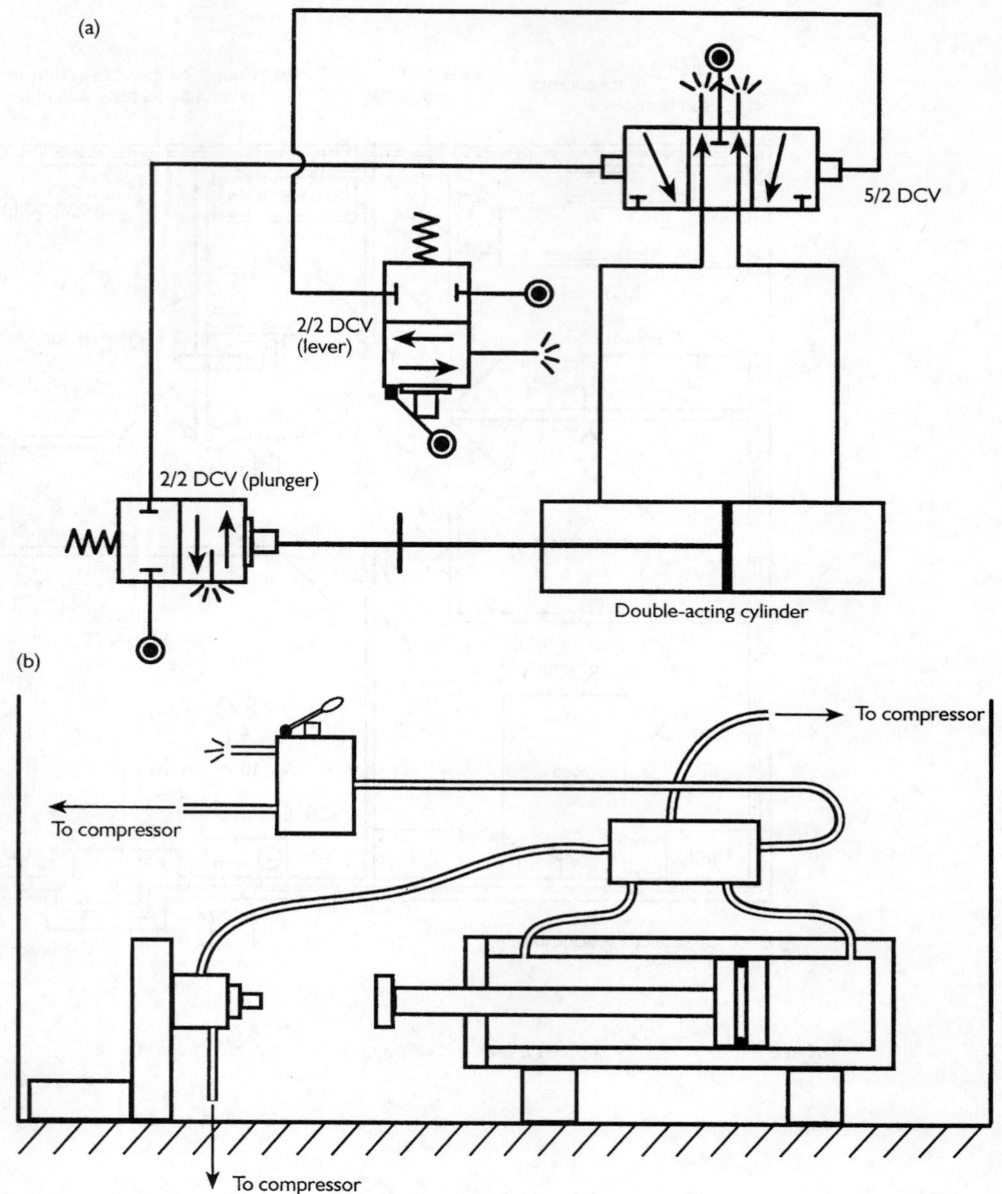

Figure 11.42 Fluid power circuit

a search for alternative solutions. One of the big advantages of freehand sketching is that it requires no equipment other than pencil or pen and paper. When producing a freehand sketch, it is a good idea to build up the drawing using guidelines drawn in thin lines, based on simple geometric shapes such as cubes, rectangular prisms, triangular prisms, pyramids, cylinders, cones and spheres. These basic principles applied to simple shapes are shown in Figure 11.43.

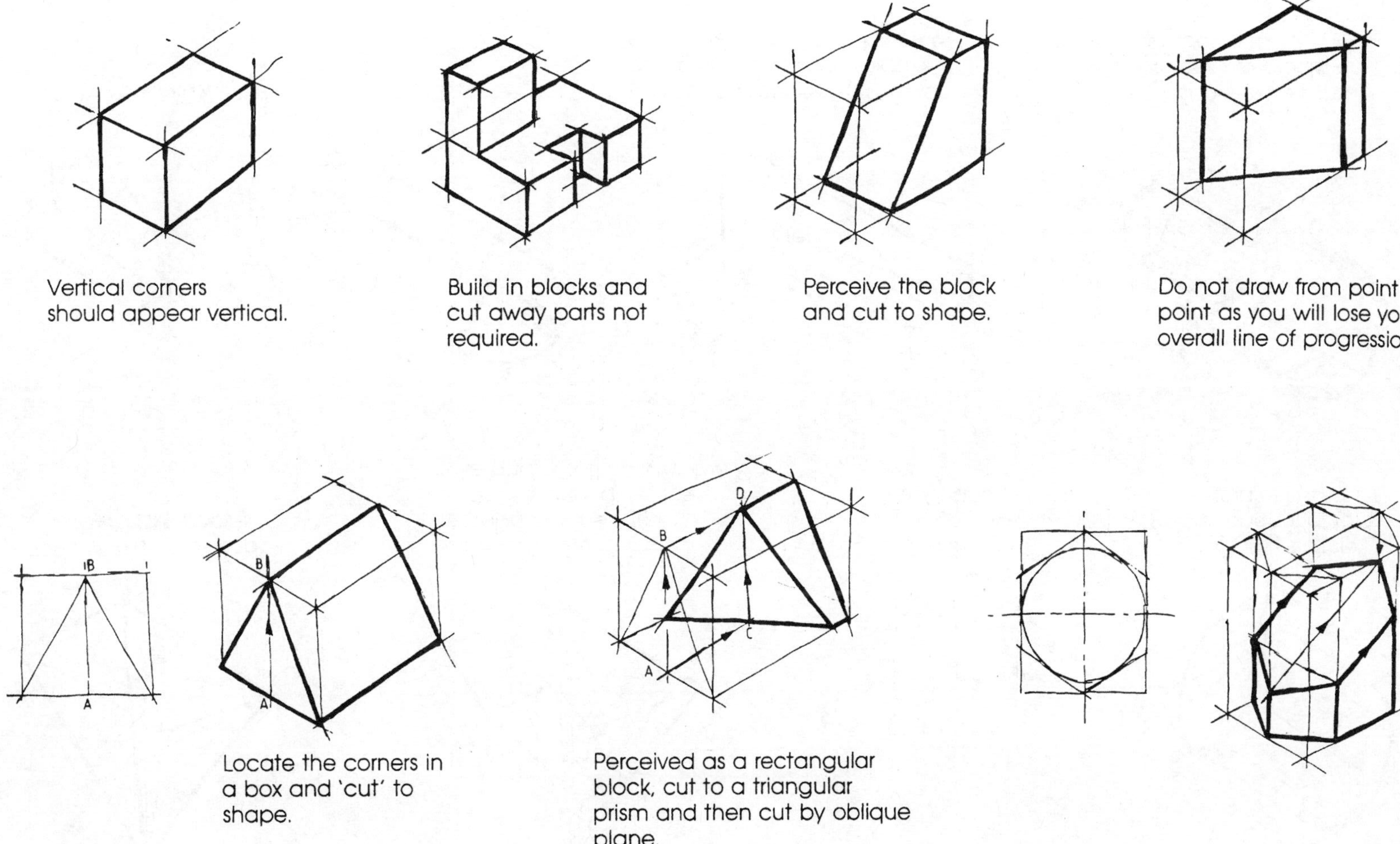

Figure 11.43(1) Freehand sketching

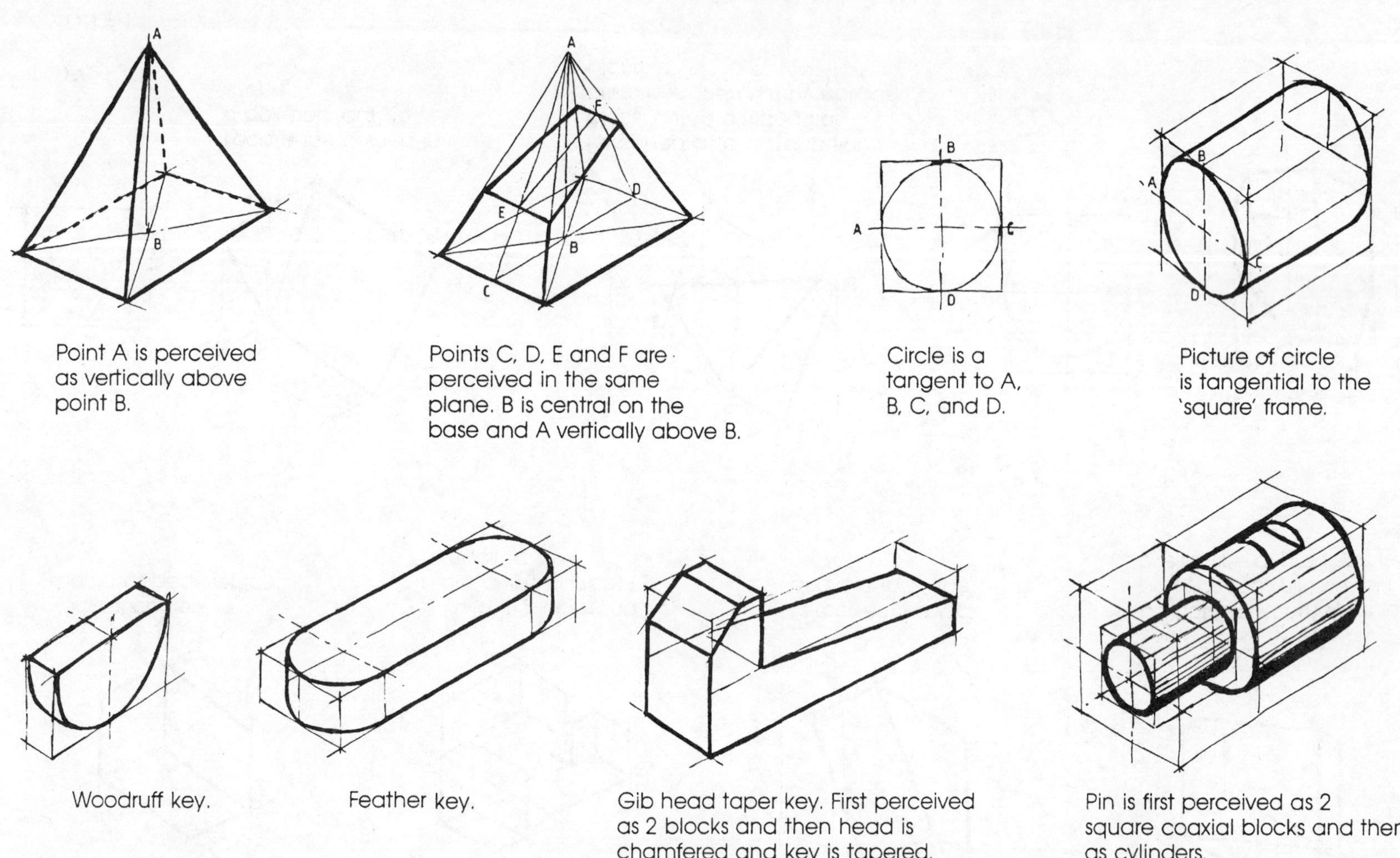

Figure 11.43(2) Freehand sketching

Case study

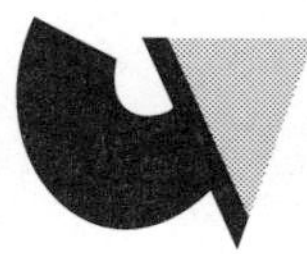

Freehand sketches

This case study shows the application of freehand sketches in different situations. Typical isometric drawings are shown in Figure 11.44, in which both a detail and an assembly drawing are featured.

Orthographic views can also be sketched freehand. An example of this is shown in Figure 11.45, along with the method of construction using centre lines and guidelines.

The use of shading to produce a more realistic impression is illustrated by the exploded piston and con rod assembly in Figure 11.46.

Activity 11.6

Draw orthographic and/or pictorial freehand sketches of:

a example 6 Figure 11.10

b at least one assembly and one detail drawing of the design solution from Element 5.2.

General arrangement drawings

General arrangement drawings are usually of two types:

- assembly drawings – these show the arrangement of the individual parts that make up the functioning assembly. A complex assembly would be further broken down into subassemblies. Assembly drawings can be used for the actual assembly process and for installation, servicing and maintenance. The exploded assembly drawing is widely used for servicing and for ordering parts (Figure 11.27)
- layout drawings of installations – these give all the details required to put together installations such as process plant, heating systems, etc.

Case study

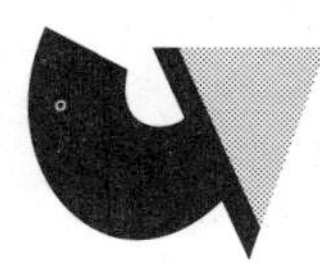

Assembly drawings

Three examples of typical assembly drawings are shown in this case study.

Figure 11.47 is a sectioned assembly drawing of a fuel pump for a car. This is operated by a cam rotating in contact with a rocker arm, which moves a diaphragm. A downward movement of the diaphragm sucks fuel from the inlet into a chamber. An upward movement of the diaphragm then pumps fuel to the outlet.

A slightly different kind of assembly drawing is shown in Figure 11.48. In this assembly drawing of a fishing reel, the main purpose is to describe the principle of operation. The handle turns gear A, which rotates the eccentric pin. As the pin rotates, it oscillates the spool with simple harmonic motion. The fishing line comes through the winding arm and is distributed over the whole surface of the spool.

Figure 11.49 is an example of a typical schematic layout drawing, in this case of a domestic water supply and heating system. Both plain and isometric schematics are shown. Note that the drawing shows the relationships between all component parts but not the exact location of things like radiators, boiler and tanks. All symbols used are from BS 1553.

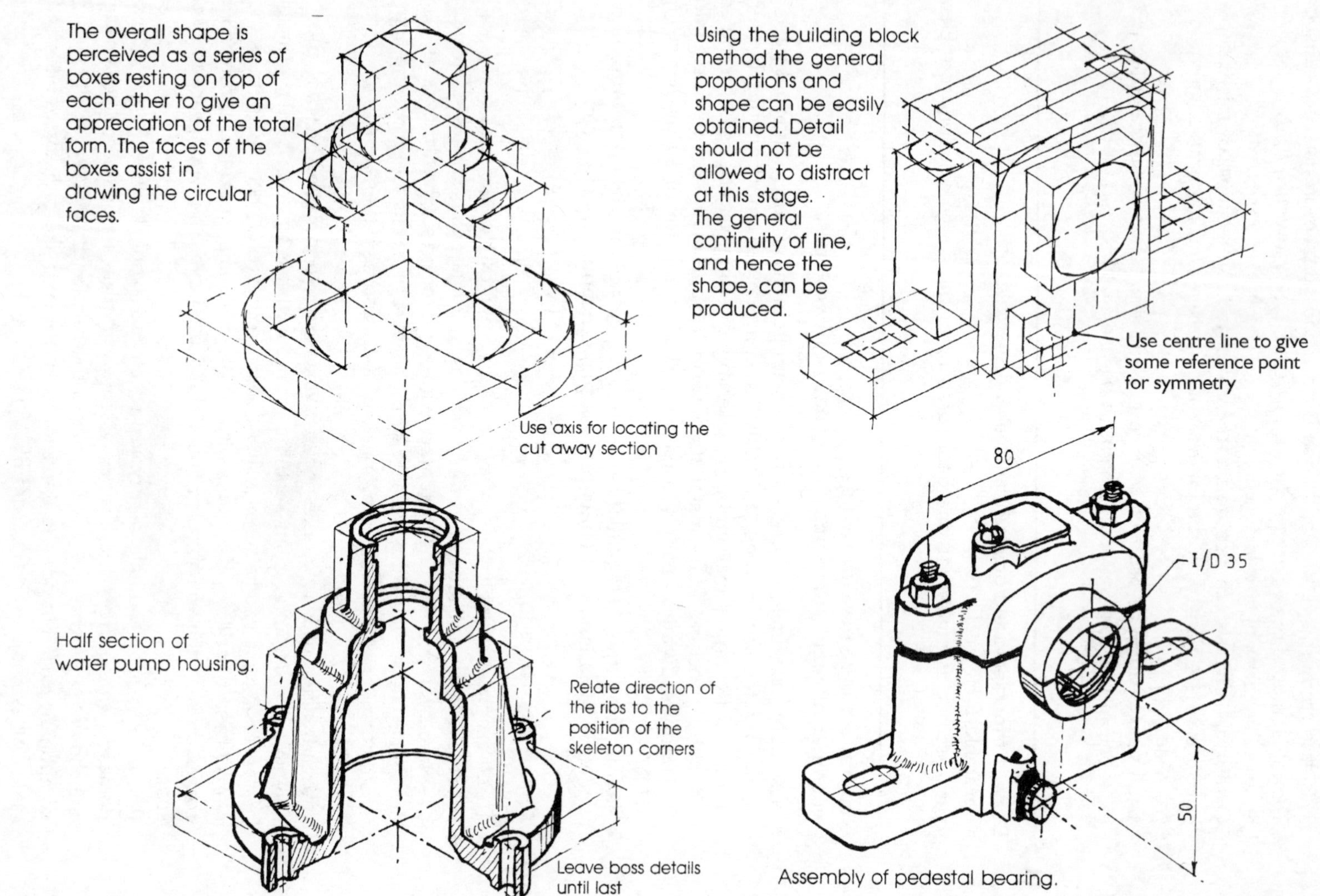

Figure 11.44 Freehand sketching of an isometric drawing

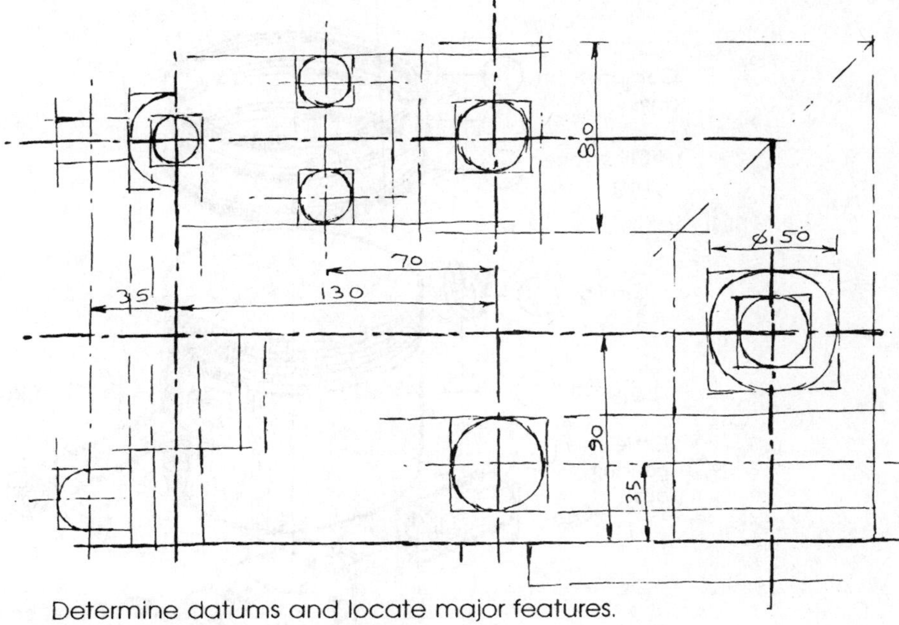

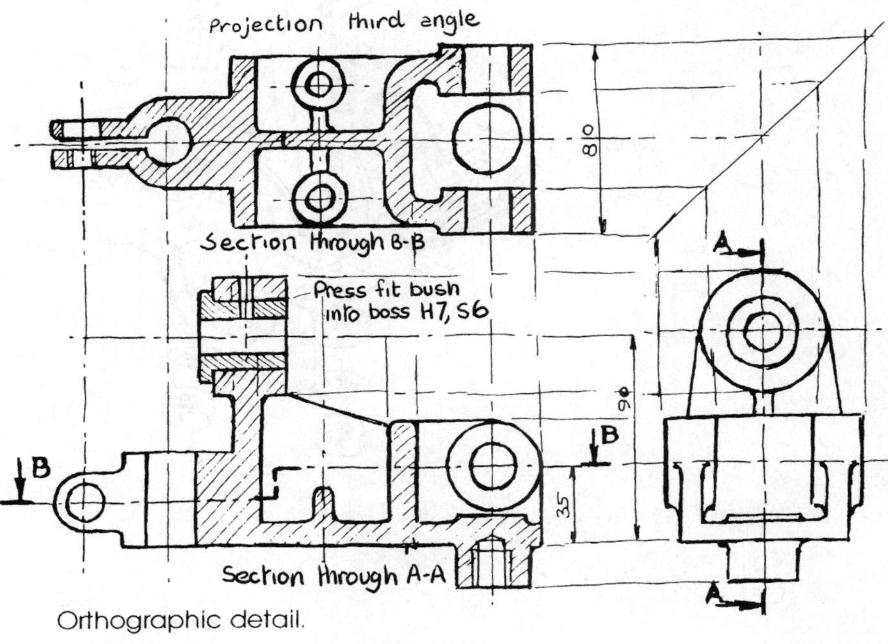

Figure 11.45 Freehand sketching of an orthographic drawing

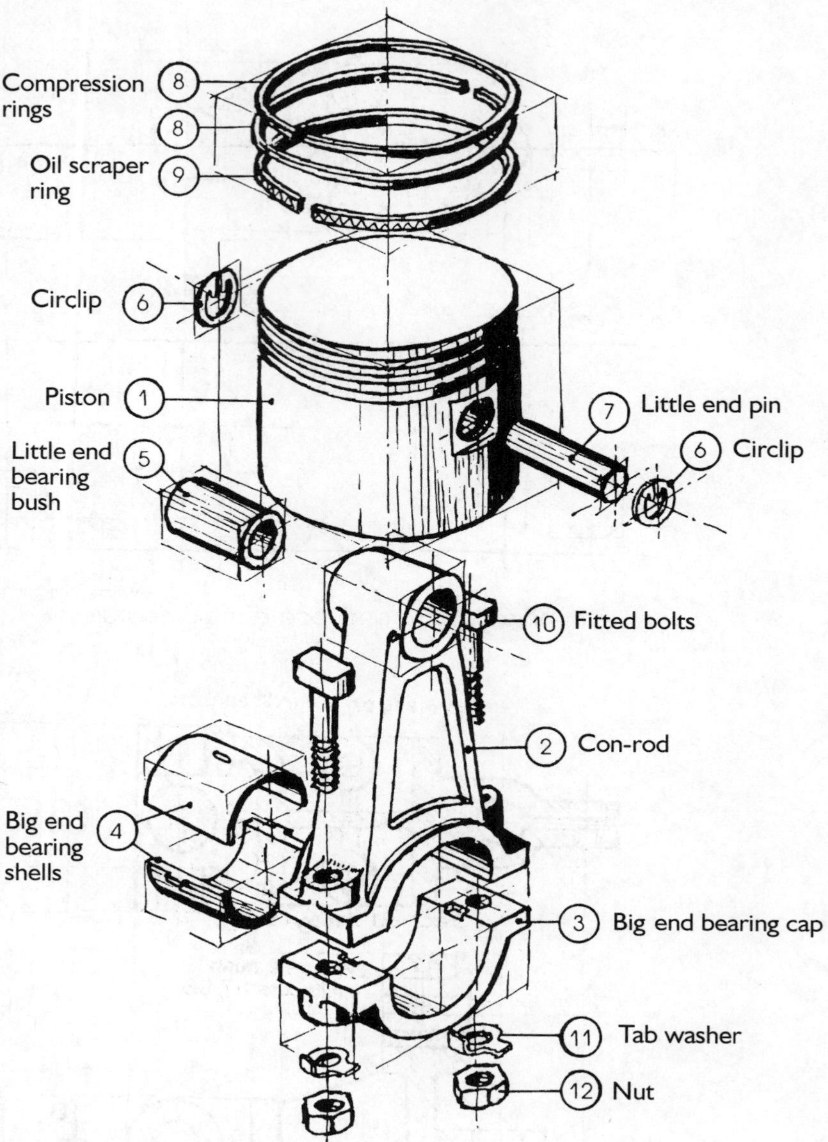

Figure 11.46 Freehand sketching using shading

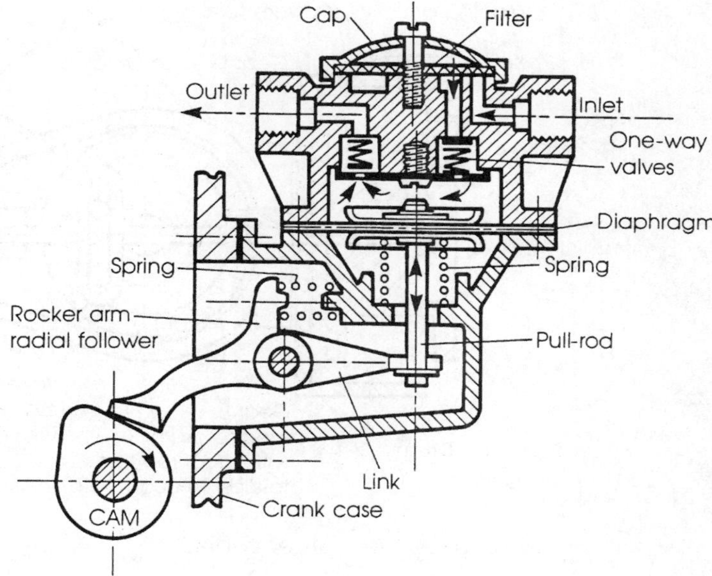

Cam operated diaphragm pump (car fuel pump).

Here the cam is used to produce a displacement in the pull-rod which displaces the diaphragm and sucks fuel into the chamber through the appropriate one-way valve. The diaphragm is then returned by the spring thus forcing the petrol through the other one-way valve and on to the carburettor.

Figure 11.47 Sectioned assembly drawing of a car fuel pump

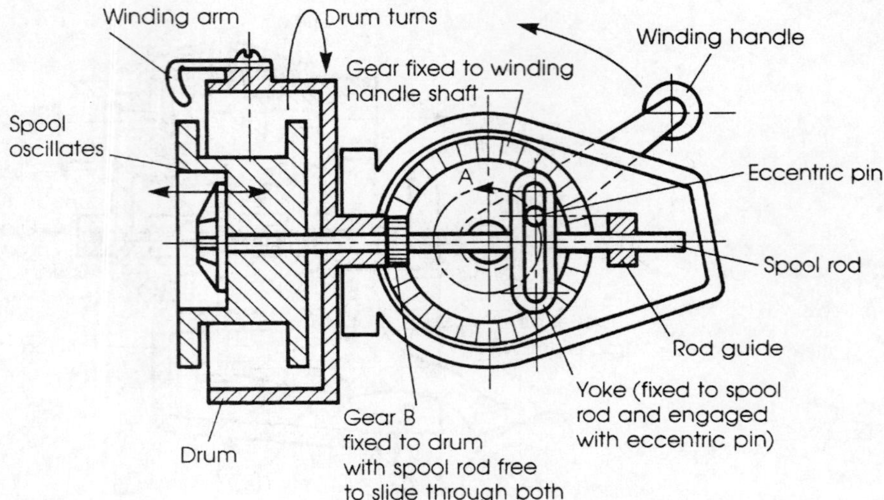

Fishing reel uses an eccentric.

The handle drives gear A to which the eccentric pin is fixed. The pin locates in the yoke attached to the spool rod and as the pin rotates with the gear A it causes the yoke and spool rod to oscillate with simple harmonic motion (SHM) when the handle turns with a constant speed. The rod is free to slide through the small gear B which rotates the drum and winding arm. As the spool oscillates it distributes the line being wound evenly along the spool.

Figure 11.48 Assembly drawing of a fishing reel

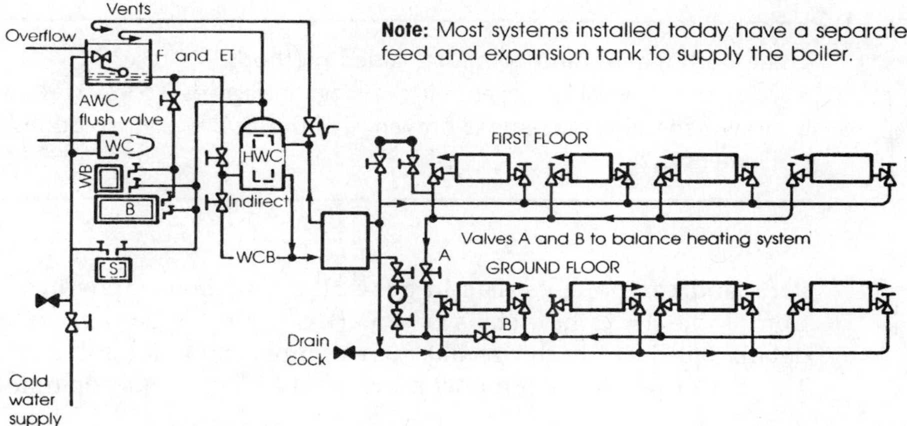

Plain schematic of domestic water supply and central heating system (2-pipe) for typical 3 bedroom house.

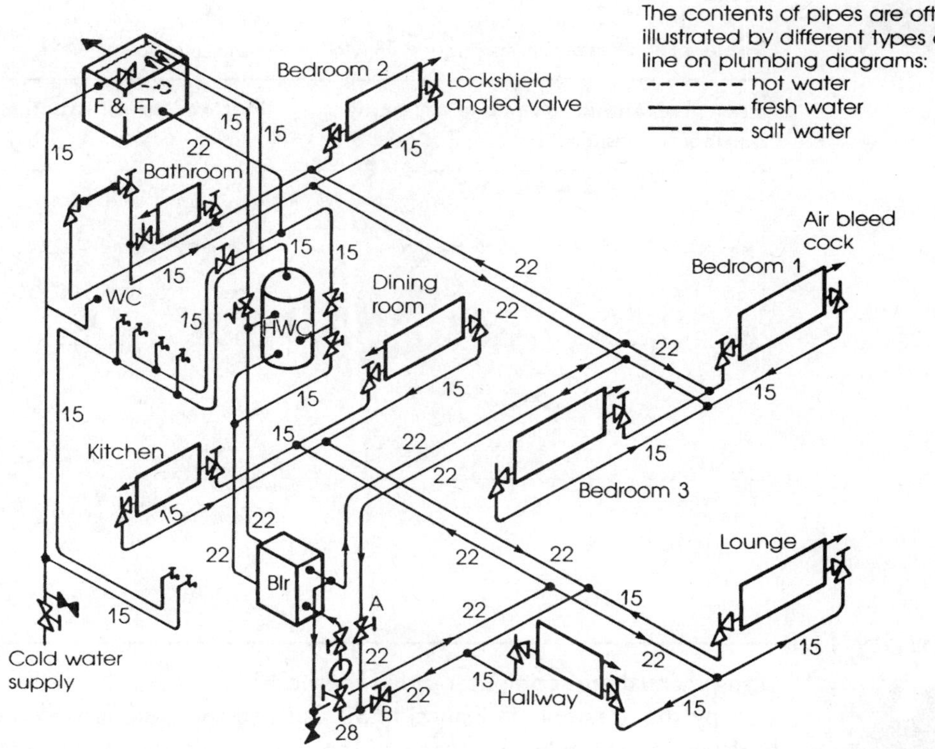

Figure 11.49 Typical schematic layout drawing

Activity 11.7

Using both manual and computer-aided methods:

a copy the general arrangement drawings in Figures 11.47, 11.48 and 11.49

b draw general arrangement drawings of the two design solutions from Element 5.2.

Parts and components lists in general arrangement drawings As already seen from Table 11.5, the details of a part can be put onto a drawing sheet. These details will include things like part number, material and standards applicable. This list of parts or components can also be done as a separate list.

Case study

Parts list

Figure 11.50 is a drawing of an assembly together with all the parts. The parts list is given in Table 11.9.

Table 11.9 Parts list for figure 11.50

Part number	Number off per assembly	Part name	Material specification	Treatment
Stop 1	1	Locking handle	Free-cutting steel 220M20	
Stop 2	1	Pivot pin	Free-cutting steel 220M20	
Stop 3	1	Square head bolt	Free-cutting steel 200M20	
Stop 4	1	Block	Free-cutting steel 220M20	Case harden to company spec. H35
Stop 5	1	Finger	Free-cutting steel 220M20	Case harden to company spec. H35

Activity 11.8

Using manual and computer-aided methods:

a copy the assembly in Figure 11.50 and add the parts list to the drawing

b draw an assembly from each of the design solutions for Element 5.2.

Detail drawings

As the name implies, these drawings are for showing the details of individual parts and items. Detail drawings are most frequently used for manufacturing purposes, so it is usual to specify things like dimensions, tolerances, material and any other properties of the part.

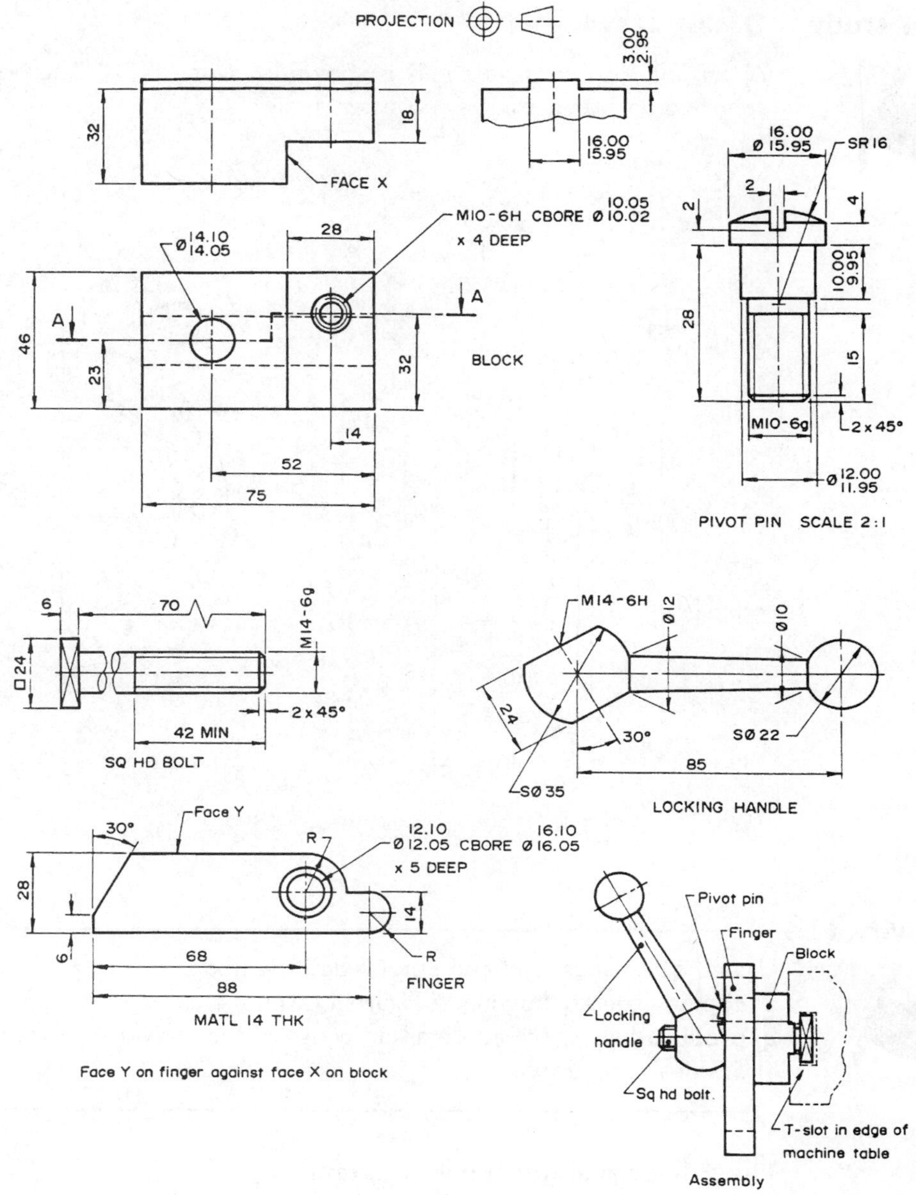

Figure 11.50 An assembly and all its parts

Case study

Detail drawing

A housing for a gear assembly is shown in Figure 11.51. The housing is fully specified so that it can be made correctly.

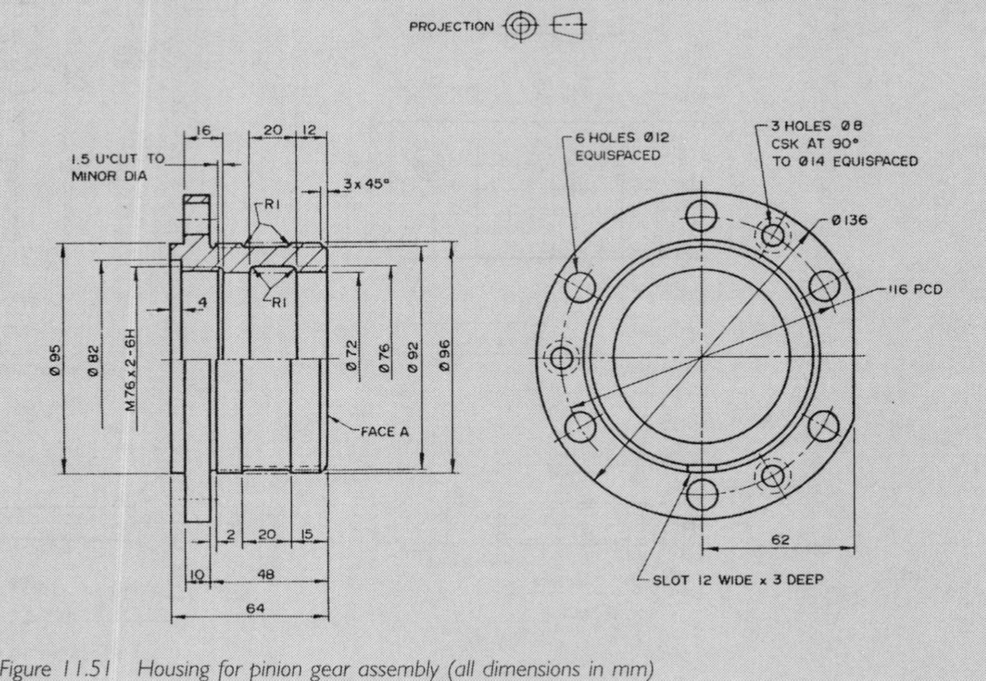

Figure 11.51 Housing for pinion gear assembly (all dimensions in mm)

Activity 11.9

Using both manual and computer-aided methods:

a copy the detail drawing in Figure 11.51

b produce fully specified detail drawings of the individual parts from the Element 5.2 design solution.

Flow diagrams and block diagrams

Flow diagrams, also called flow charts, are used whenever it is required to show a sequence of events. They have similarities to block diagrams (see chapters 1 and 2), but use a variety of operations. BS 4058 details all of the symbols for the different operations, and a selection of the most commonly used ones is given in Figure 11.52.

Case study

Domestic heating system

The case of a domestic heating system, which was partly shown in a previous case study, is further studied here. Figure 11.53a is a block diagram of this system with Figure 11.53b being a circuit diagram of the fail-safe system and pilot light ignition.

Symbol	Description	Symbol	Description
(flow line with arrow)	Flow lines — arrows used to increase clarity of flow direction.	(staggered intersecting flow lines)	Flow lines which intersect should be staggered. Arrows need only be shown to increase clarity, e.g. if flow is other than right to left or top to bottom.
(crossing lines)	Flow lines which do not join but cross should be avoided if possible.		
(rounded rectangle)	Start, stop, halt, delay.	(parallelogram)	Input or output to the flow chart or diagram.
(rectangle)	Process or data. Defines an operation or group of operations.	NO (diamond) YES	Decision — exit from symbol identified by a route specification.
3 (rectangle) 2	Process reference for identification, top left e.g. 3, cross-reference top right e.g. 2.	(hexagon)	Modification of an instruction, index register or starting a routine.
(rectangle with double sides)	Process which has been predefined by a lower order program.	To sheet 2	Exit or entry from another flow chart on another sheet.
(rectangle with bracket)	Annotation — addition of descriptive comments or explanatory notes.	To 4	Connector — entry to or exit from another part of the same flow chart.

Figure 11.52 Flow chart symbols

While Figure 11.53 gives a lot of information about the two systems, the block diagram and circuit diagram do not fully explain the operation of the systems. Flow diagrams will show the sequence of operations in all circumstances and are particularly useful for operating the systems and for trouble-shooting and fault-finding. Figures 11.54a and b are complete flow diagrams for the heating system and the fail-safe and pilot light ignition system.

Activity 11.10

Using both manual and computer-aided methods:

a copy the flow, block and circuit diagrams in Figures 11.53 and 11.54

b produce flow and block diagrams showing the operation of the chosen design solutions from Element 5.2.

(a)

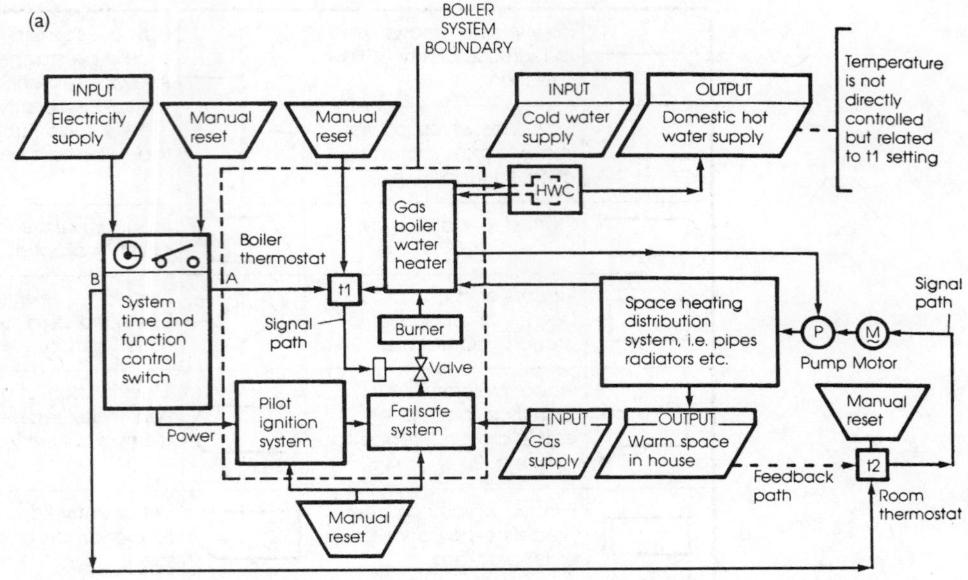

(b)

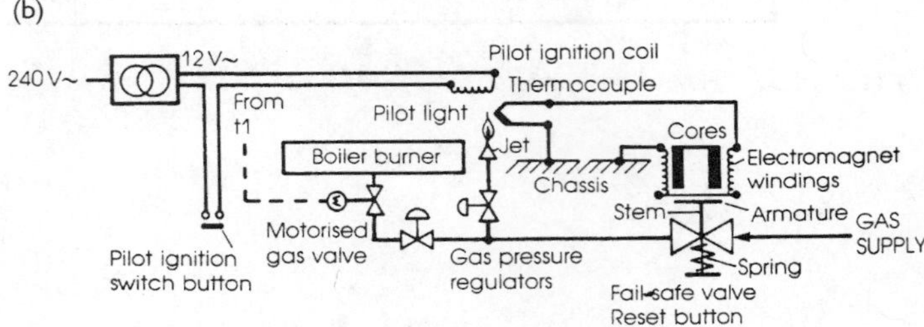

Figure 11.53 (a) Block diagram of complete domestic heating system; and (b) circuit diagram of fail-safe system and pilot light ignition

Progress check

1 What is the function of a circuit diagram?
2 Why are computer-aided methods very useful for producing circuit diagrams?
3 Which types of drawing are particularly suitable for being done as freehand sketches?
4 Describe the main types of general arrangement drawings and give their applications.
5 State the information that would be put on a list of parts or components.
6 What are the uses of a list of parts or components?
7 What is a detail drawing and what information must be put on such a drawing?
8 Describe flow and block diagrams and state their applications.

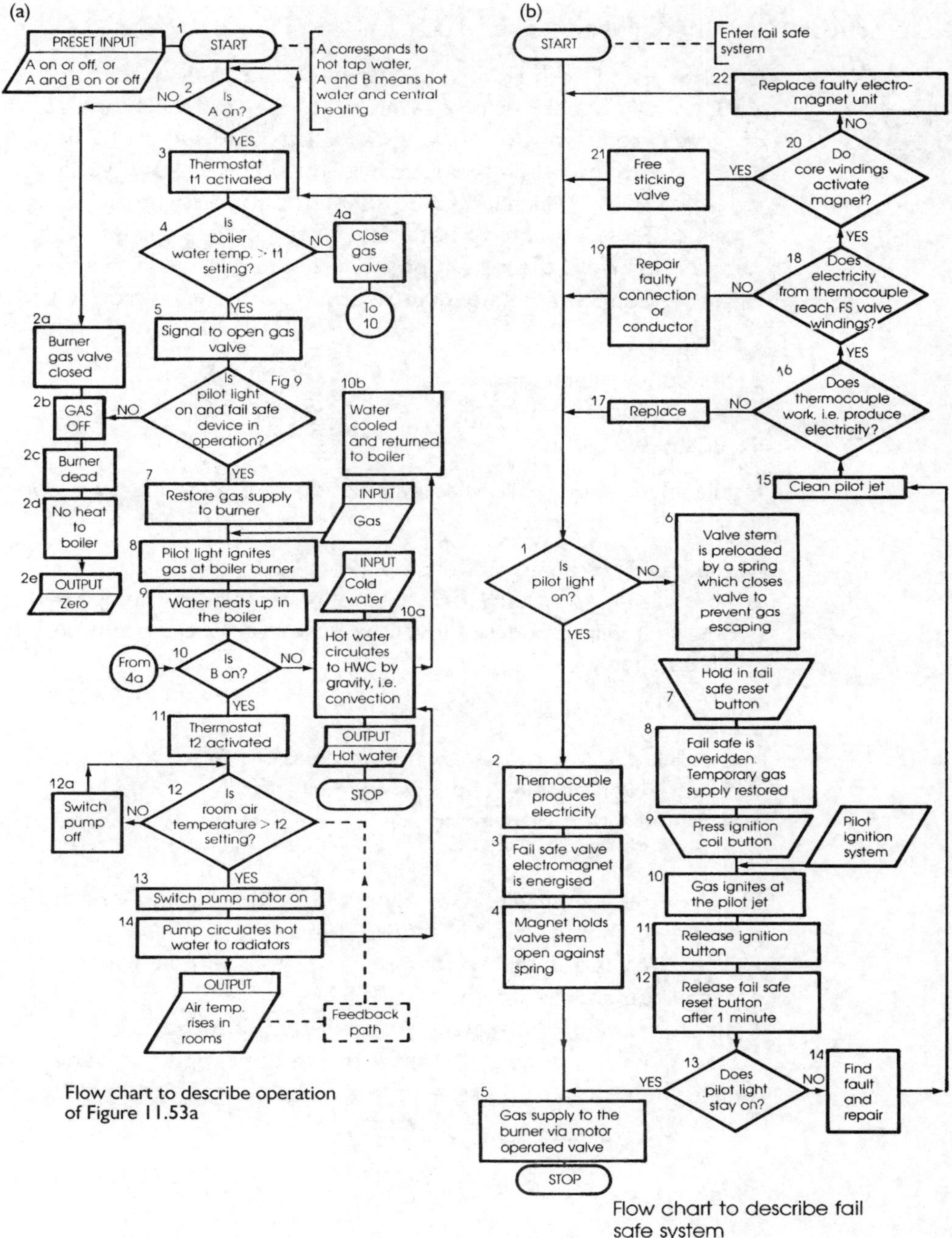

Figure 11.54 *(a) Flow chart to describe operation of Figure 11.53a. (b) Flow chart to describe fail-safe system*

Case study

An integrated case study

It will be useful to build on the previous case study in Chapter 9 concerning the automatically opening garage door. This case study will look at typical examples of the different types of drawings and sketches used in the design, from initial concept to the finished product. It is intended to show typical examples of the type of graphics that need to be submitted for Assignment 2.11.

Figure 11.55 is a graphical statement of the basic problem, which is to move a garage door from the closed position to the fully open position.

Figure 11.56 gives four possible solutions to the garage door problem:

1 vertical roller door
2 horizontal sliding door
3 hinged doors
4 up-and-over doors.

Figure 11.57 shows the selected solution of the up-and-over doors.

Figure 11.58 shows:

- the basic mechanism of the door
- possible types of automatic actuation including nut and screw, rack and pinion and hydraulic cylinder. All will be powered by electric motor from the normal mains supply.

Figure 11.59 shows:

- a block diagram of the open-loop control system for the automatic actuation of the door opening and closing mechanism
- a flow diagram showing the sequence of events when arriving with a car outside the garage door.

Figure 11.60 is a partial assembly drawing showing the motor with the rack and pinion assembly. A parts list is also given.

Figure 11.61 is a detail drawing of the pinion, with all tolerances, material specification and treatment.

Figure 11.62 shows a sketch for installation purposes.

Figure 11.63 is a circuit diagram showing the connections and components to enable the motor to be reversed.

Door
Opening

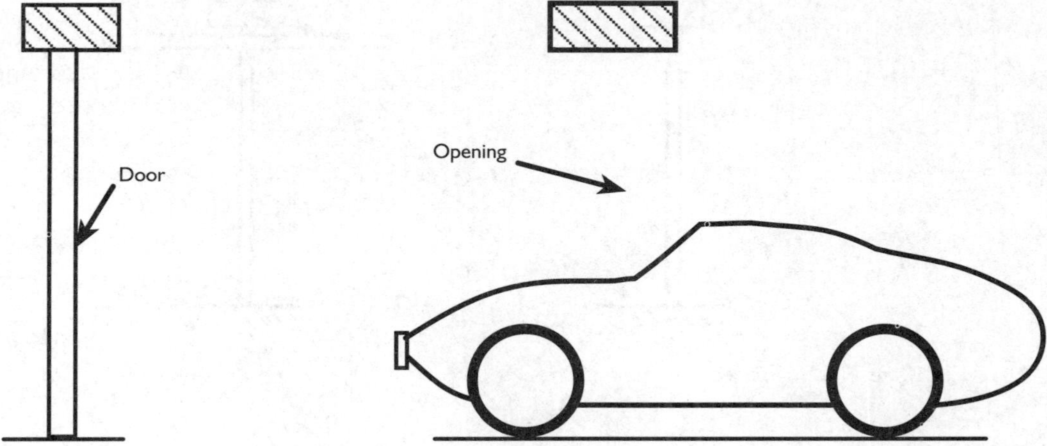

Figure 11.55 Graphical statement of the garage door problem

(a)
Vertical roller door
(b)
Horizontal sliding door
(c)
(d)
Up-and-over doors

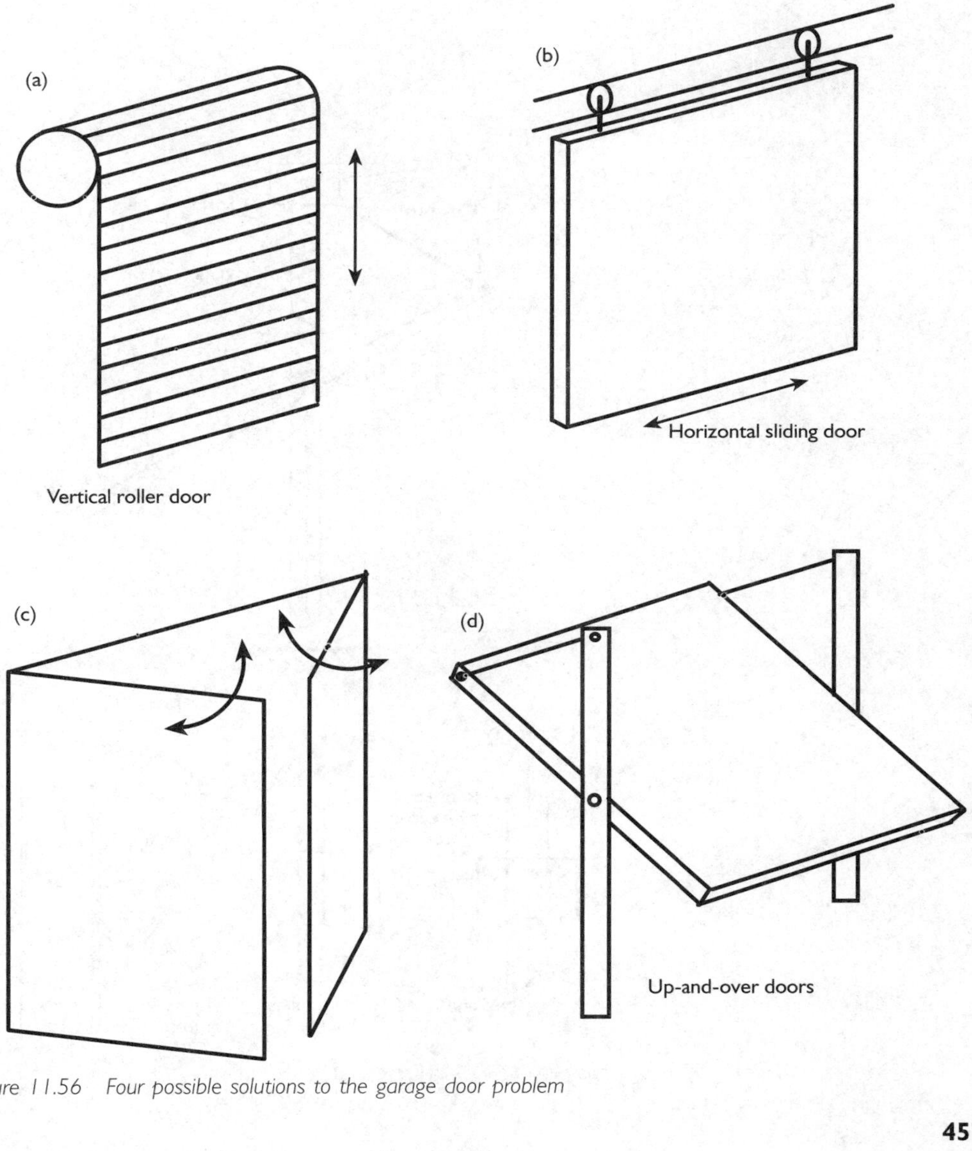

Figure 11.56 Four possible solutions to the garage door problem

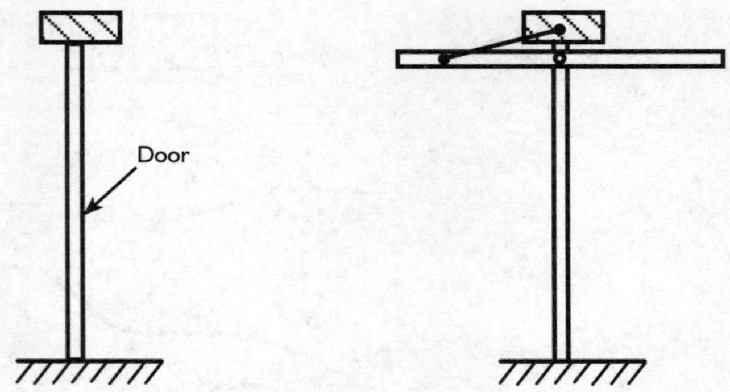

Figure 11.57 Selected solution of up-and-over doors

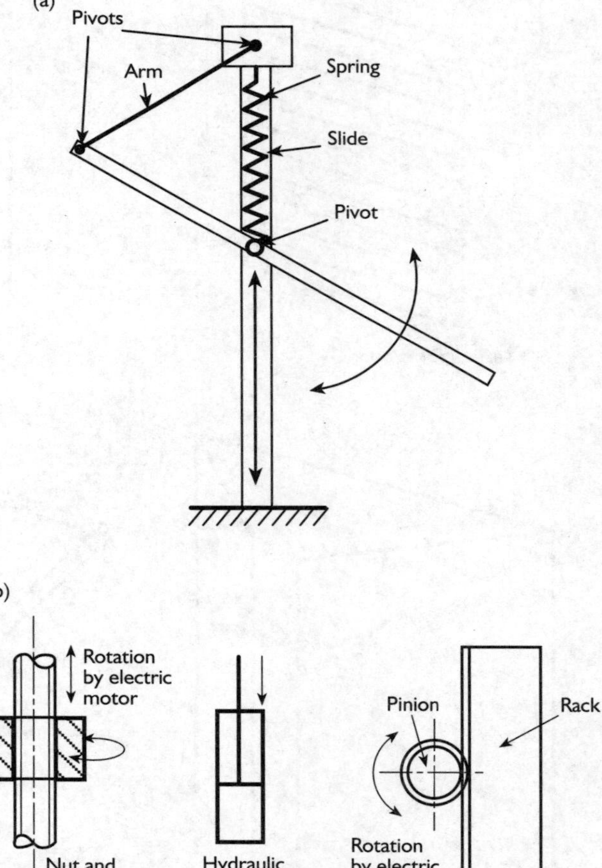

Figure 11.58 (a) Basic actuation; (b) types of automatic actuation

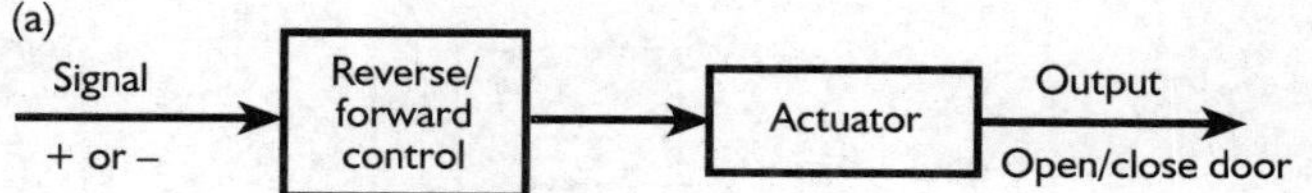

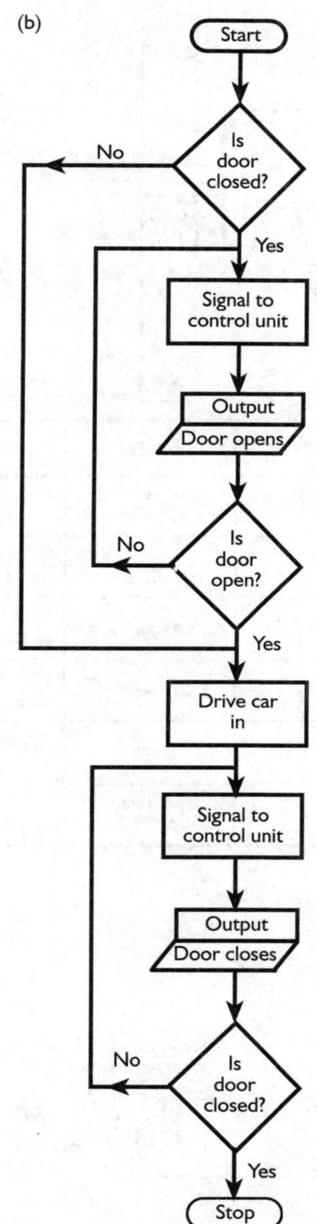

Figure 11.59 (a) Block diagram of control system; (b) flow diagram for door opening/closing

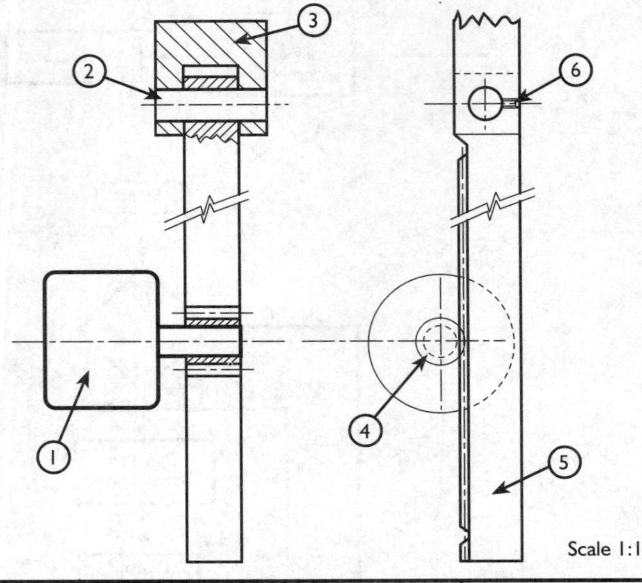

Part No	No. off	Name	Material	Treatment
1	1	24V 200W DC motor	Bought-out item	
2	1	Pin	070M20 steel to BS 970	Case harden to company spec. H20
3	1	Connecting block	070M20 steel to BS 970	Case harden to company spec. H20
4	1	Pinion	070M20 steel to BS 970	Case harden to company spec. H20
5	1	Rack	070M20 steel to BS 970	Case harden to company spec. H20
6	2	Socket head set Screw M4x5mm	Bought-out item	

Figure 11.60 Partial assembly drawing

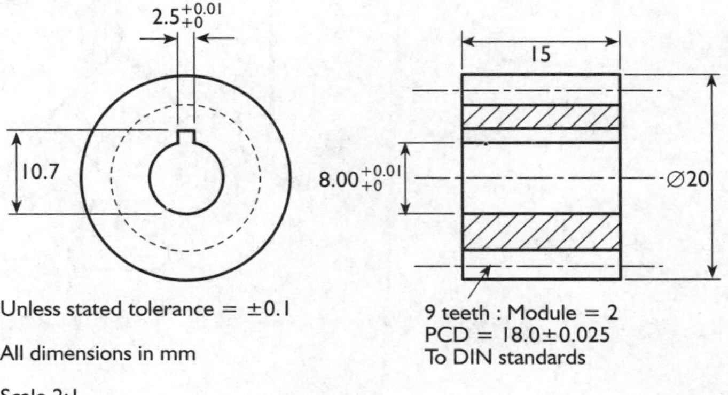

Pinion – Part No.4 Matl : 070M20 steel to BS970
Treatment : case harden to company spec. H20

Figure 11.61 Detailed drawing of the pinion with all tolerances, material specification and treatment

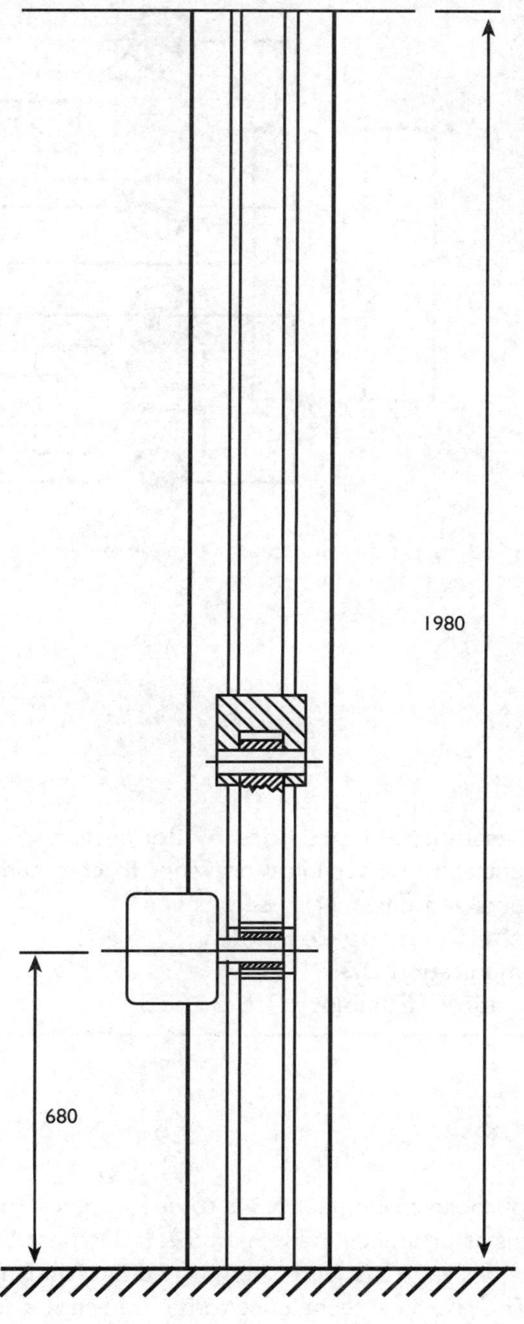

Figure 11.62 Sketch for installation purposes

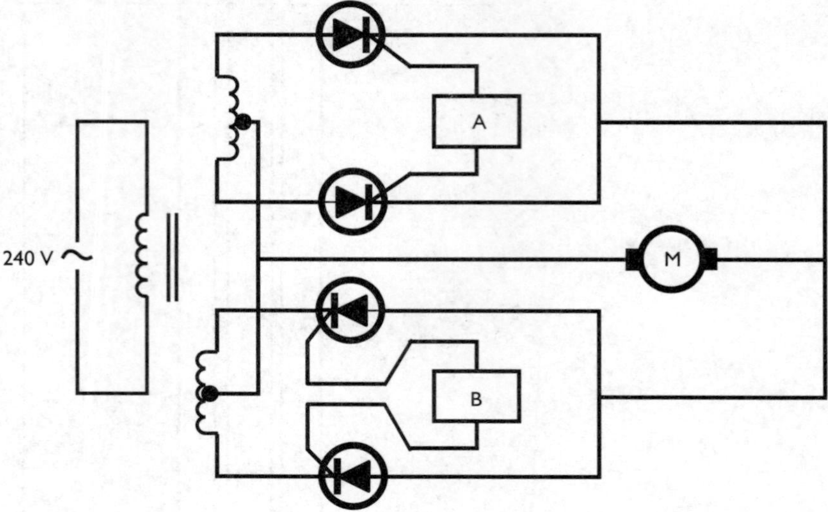

Figure 11.63 Circuit diagram showing the connections and components to enable the motor to be reversed. (a) Forward trigger; and (b) reverse trigger

Assignment 11

This assignment provides evidence for:
Element 5.3: Use technical drawings to communicate designs for engineered products and engineering services
and the following key skills:
Communication: 3.3
Information Technology: 3.1, 3.2, 3.3

Your tasks

The purpose of Element 5.3 is to use graphical methods to communicate the finished designs as produced in Element 5.2. In Element 5.1, the design briefs were formulated. These design briefs were then turned into design solutions and evaluated in Element 5.2. The graphical techniques to be used should include both manual and computer-aided methods. A full set of working drawings should be produced of the designs of both an electromechanical engineered product and an engineering service involving installation and/or maintenance. The portfolio should include:

- orthographic views in both first angle and third angle
- isometric and oblique pictorial views
- exploded views
- sketches
- general arrangement drawings with parts and components lists
- detail drawings with full specifications
- circuit diagrams
- flow diagrams and block diagrams.

As a minimum requirement the following standards should be correctly used where applicable:

- BS 308 Engineering drawing practice
- BS 3939 Graphical symbols for electrical power, telecommunications and electronic diagrams.

Note: This assignment can be turned into a fully integrated assignment by doing a full investigation of alternative materials and a study of the manufacturing methods and processes that could be used to implement the design.

Glossary

ABS (anti-lock braking system) A braking system, originally developed by Mercedes-Benz, which prevents the wheels of a vehicle locking when the brakes are applied

Accumulator A device for storing energy

Accuracy Deviation from an expected value

Actuating mechanism A mechanism for moving something

ADC (analogue to digital converter) A device for converting an analogue signal into a digital signal, i.e. a continuously varying quantity into a series of discrete values

Aesthetics Taste and beauty as applied to a design or artefact

Allotropic Material that can exist in different states with different crystal lattice structures

Alternator A generator that generates alternating current

Aluminium bronze Alloy whose main constituents are copper and aluminium

Ambient The temperature surrounding a particular system

Amorphous Material without any discernible crystalline or polymer structure; also used to describe the various grades of glass

Amplifier A device, usually electronic, for increasing the power of a signal

Amplitude The magnitude of a signal

Analogue An object, model or quantity used to represent another quantity, e.g. the dial on a clock used to represent time

Annealing Heat treatment technique to remove the effects of cold working in metals, and internal stresses in glass

Anodising A surface protection for aluminium which is used to protect from corrosion and also to harden and colour the surface

Anthropometric data Measurements relating to the human body

Arbor A tool-holding device used in machine tools, particularly milling machines

Armature The central rotating core of an electric motor

Assembly drawing A drawing showing all the different detail parts put together to form a complete system or subsystem

Atmosphere A measurement of pressure equal to normal atmospheric pressure at sea level

Automatic transmission A system of changing gear ratios automatically as load and speed change

Automation A general term applied to whenever a process is done automatically

Avionics All the electronic navigation and control systems in an aircraft or a helicopter

Bandwidth The total frequency spectrum of a signal

Bar A measurement of pressure. One bar is equal to normal atmospheric pressure

Barometric pressure Atmospheric pressure as measured by a barometer

Barstock Raw material in the form of bars

Bearing A component which is designed to support a shaft or other moving part. There is a clearance fit (see Clearance fit) between the bearing and shaft and also provision for lubrication to minimise friction

Blow moulding A forming technique widely used to make bottles and other similar containers from thermoplastic materials and glass

Bonded ceramics Products made from clay mixtures which have been moulded to shape and kiln-fired

Boundary The total functional area enclosed by a system. Anything outside the boundary is called the environment

Brainstorming A group technique for generating ideas to solve a problem

Brass An alloy whose main constituents are copper and zinc

BS (British Standard) A standard which applies to the United Kingdom but is often also applicable to Europe and the rest of the world

BSI (British Standards Institution) The institution which is responsible for generating, maintaining and publishing British Standards

Bus line A communication line which transfers data. The term is normally applied to links between subsystems in a computer, i.e. processor to memory

Cache memory A very fast-acting memory in which data that needs quick access is held in preference to the main RAM. This greatly speeds up many computer operations

Calibration A procedure in which the accuracy of a measuring or metering device is checked by comparing its output with a known input

Capacitance A measure of the amount of electricity an electrical capacitor can store

Case hardening Heat treatment technique in which a material is given a hard wear-resistant outer case while retaining a softer and tougher core

Cast iron An alloy made up of iron with 3.2% to 3.5% carbon and small amounts of other elements which remain from the smelting process

Casting A forming process in which molten metal solidifies to the required shape

Catalyst A substance that speeds up a chemical reaction without itself being changed by the reaction

Cavitation The erosion of surfaces by the flow of a fluid

CD-ROM (compact disc read-only memory) A method of storing large amounts of data for accessing by computer

Cermet A composite material made up of a ceramic and a metal

Circuit diagram An analogue diagram that shows the interconnection of different devices in a system. It is mostly used for electrical, electronic and fluidic systems

Clearance fit When there is always clearance between a shaft and a hole. This type of fit is used when parts slide or rotate relative to each other (see Bearing)

Closed loop Defines a system which has feedback

CNC (computer numerical control) A control system for machine tools such as lathes and machining centres that uses a computer to run the overall system

Comparator A measuring instrument that compares the size of a component with a standard of known size such as a stack of gauge blocks

Complementary colour The primary colour *not* used to make a secondary colour. Thus, orange is complementary to blue, violet to yellow and red to green.

Composite A combination of two or more materials bonded together in such a way as to improve their mechanical properties

Compound A chemical synthesis of two or more elements, e.g. sodium chloride, iron oxide

Compressor A device for increasing the pressure of a fluid or gas

Conceptual diagram A diagram showing the basic ideas or concepts in the initial phases of a design

Contact adhesive An adhesive that is allowed to dry on the joint surfaces before they are brought together to form an immediate bond

COP (code of practice) A set of rules and procedures usually concerned with the safe operation of machines, processes and systems

COSHH Regulations (The Control of Substances Hazardous to Health Regulations) A set of regulations which control the storage and use of hazardous substances

Counterbored In machining when another larger hole is bored concentric to the first one

Coupling A device for connecting different parts of a system, e.g. a constant-velocity joint is a coupling for connecting a drive shaft from a vehicle gearbox to its wheels

CPU (central processing unit) The part of a computer where arithmetic and logic operations are performed on data. Also just simply called a processor

Crimping The technique of compressing terminations onto the bared ends of electrical cables

Critical path analysis A method of analysing a complex project to find the critical factors affecting the progress of the project

Cross-linking Permanent links formed between the polymers in thermosetting plastics

Crystal The name given to the basic structural component of a material based on the regular arrangement of atoms

Crystalline Made up of crystals or 'grains'

Cupro-nickel An alloy whose main constituents are copper and nickel

DAC (digital to analogue converter) A device for converting data as discrete numbers into a continuously varying quantity

Damping The prevention of excessive oscillation in a system, i.e. by use of viscous or other friction

Datum A reference point or line from which measurements are taken

DCV(directional control valve) A fluidic (hydraulic or pneumatic) system valve that controls the direction of motion of an actuator (see Actuator)

Decommissioned Taken out of service

Deep drawing A process for forming deep shapes from sheet material, e.g. kitchen sinks

Depreciation The reduction in monetary value of an item over time

Derivative The rate of change of one value with respect to another

Detail drawing A drawing of an individual part which goes to make up an assembly or complete system (see Assembly drawing)

Diaphragm A thin disc that deforms when a force is applied to it

Dies Material-forming tools made from hard metal to the shape or cross-section of the product that is to be formed

Differentiation The determination of a derivative value (see Derivative)

Digital Any variable or quantity represented as a number

Disassembly The organised breaking down of an assembly to its constituent parts

Disturbance Undesirable or unplanned inputs to a system

Doping A process in which impurity atoms are introduced into semiconductor materials to modify their electrical properties

Drawing A forming technique used to produce wire, bar and tube by pulling the material through a series of dies to reduce its cross-section

Drilling A material removal technique for producing holes in a stationary workpiece using a rotating drill bit

Ductility The ability of a material to be drawn out in tension

Duralumin A light and strong aluminium alloy whose hardness can be adjusted by heat treatment and which is widely used in aircraft production

Eccentric pin A pin with two diameters not concentric to each other. It can be used to convert rotary motion to reciprocating motion

ECM (electrochemical machining) A machining process that uses the principles of electrolysis to remove material

Efficiency Usually applied to energy conversion systems, it is the ratio of output/input

Elastomers The name given to the long intertwined molecular chains of which rubbers are composed and also a general name for rubbers

Electrochemical potential The amount of electrical charge a material possesses as compared with a standard hydrogen electrode

Electroplating A chemical technique involving the use of electric current and an electrolyte to plate a component with a layer of another metal

Electrolysis The process of passing an electrical current through a liquid to produce a chemical reaction in the liquid

Electrolyte A conducting chemical solution used in electrochemical plating and material removal processes

Element A single substance (of which there are 92 naturally occurring) which cannot be separated into a simpler substance by chemical means, e.g. iron, carbon, hydrogen

Embrittlement Making a material brittle by the effects of temperature etc.

EMF (electromotive force) The force (in volts) that causes a current to flow in an electrical circuit and which is supplied by a battery or a generator

Emission The giving out of things like heat, light, smell, gas, waste, etc.

EMU (electronic mock-up) A solid modelling technique for producing a realistic solid model of a component or assembly on computer screen

Encapsulated Totally enclosed in a protective casing

Encoder A type of transducer that is normally used to convert linear or rotary motion into digital data

Endothermic reaction A chemical reaction which requires heat to be input

Ergonomics The science of analysing the operation of systems from the viewpoint of the user or operator

Etching A chemical technique in which a chemical solution is used to remove material

Exothermic reaction A chemical reaction which outputs heat

Exploded diagrams Diagrams showing the different detail parts in an assembly exploded apart from each other but still in the correct relationship to each other. They are frequently used in service manuals to show how to assemble and disassemble the parts in the assembly

Extrusion A forming process used to produce a variety of cross-sections by forcing

Fatigue The failure of a material below its normal ultimate strength owing to repeated alternations of stress

Feedback The connection of the output value of a system to the input value to form a closed-loop system. When output is subtracted from input it is negative feedback and inherently stable. When the output is added to the input it is positive feedback and inherently unstable.

Feedforward An anticipatory means of control that uses a model of the process to make changes in response to a load change before it actually happens

Ferrous metal A metal in which iron is a main constituent

Finite-element-analysis A technique for modelling the behaviour of a component or assembly when put under stress. This is done by breaking down the structure into small finite elements and simulating the effects of different loadings on each element

Fixturing The location and clamping of workpieces onto a machine

Flow chart A chart which shows the logical flow of operations in a procedure or process and which allows decisions at each stage with backward and forward looping as appropriate

Flux A substance that is used to clean and/or protect the joint surfaces from oxidation during soldering and welding processes

Fly-by-wire A technique used for controlling the flying surfaces of aircraft by using computers to remotely control each surface

Foolproofing A method of designing devices so that they cannot be improperly or unsafely used

Forging A forming technique in which a metal is worked to shape by hammering or shaped by compressing it between dies

Frequency The speed of a waveform variable expressed in cycles/s (hertz, Hz)

Frequency distribution The arrangement of a series of values or measurements so that the frequency of each number or group of numbers is displayed in graphical form

Fuel injection The propulsion of fuel, at high pressure, into a combustion chamber in an engine

Functionality The ability of a design to fulfil its intended purpose

Furnace An industrial oven for heating objects

Gantt chart A linear graph which displays the time taken to do each task during a project

Galvanising A surface finishing technique in which steel is coated with a protective layer of zinc

Generator A device for producing electric current but can also be applied to devices for blowing air (wind generators)

Grommet A sealing ring, normally made of plastic or rubber, which provides a seal between different components

Handshaking A technique for transmitting/receiving data between electronic devices using standard protocols (see Protocol)

Hard soldering A joining process in which metal parts are bonded together using brass or silver solder as the joining material

Hardness The ability of a material to withstand wear and abrasion

Hierarchy A method of arranging people, objects, systems, etc. in ordered relationships

Hole basis system A system of fits and limits based on keeping the hole size constant and varying the shaft to achieve the correct fit (see Clearance fit, Transition fit and Interference fit)

Honeycomb material A structure for a material based on the hollow honeycomb as produced by bees

HSE (Health and Safety Executive) The executive body responsible for all aspects of health and safety

Hysteresis The lagging of physical effect behind cause, e.g. a room will remain at a certain temperature even after the heating is switched off

Iconic A term used in modelling to denote a model which attempts to look like the object or value being modelled, e.g. a digital clock gives an iconic model of time

Indentation test A test in which a metal ball or a diamond is pressed into a material to assess its surface hardness

Inertia The tendency of a physical characteristic to remain the same unless deliberately changed, e.g. a car will continue moving forward at a certain speed unless the brakes are applied or will not move forward unless energy is used from the engine

Infrastructure The built environment of roads, railways, airports, telephone lines, etc.

Injection moulding Forming technique used widely with thermoplastics in which the heated material is injected into the cavity between metal dies

Innovative A term which relates to a design or product being new and original

Integral The sum between stated limits of many very small, usually varying, quantities

Interchangeability The ability of standard components to be readily replaced by others which could be made anywhere, e.g. any standard light bulb will fit any standard socket and so the bulbs are interchangeable

Interchangeable manufacture A system of specifying parts in an assembly so as to allow the replacement of one part by another made anywhere to the same specification as the original

Interfacing The connection of devices, usually electronic, so that data can be transmitted and exchanged

Interference fit A fit in which a shaft is always bigger than the hole into which it is fitting. It is used when the two parts are to be permanently fixed together but could be separated, if required, by force

Interlocking When one action is dependent upon another action taking place

Intermetallic A chemical compound formed by the constituents of an alloy

Internal combustion The burning of fuel at a rapid rate inside an internal space to rapidly heat and expand a gas

Interstitial alloy An alloy in which the atoms of one constituent occupy the spaces in the crystal lattice between the larger atoms of another constituent

ISO (International Organisation for Standardisation) The international body responsible for implementing world wide standardisation

Isometric projection A method of pictorial projection which shows the object using three axes at 120° to each other. All lines on the object are equally foreshortened hence the name isometric meaning equal measure.

IT (information technology) The technology which combines computing and telecommunications

JCB (J.C. Bamford) A self-propelled digger/loader. The name JCB is a trademark but is used to describe any digger, rather like a hoover being used as a general name for a vacuum cleaner or a biro for a ball-point pen

Kinematic diagrams Diagrams used to simulate the movement of a mechanism and also enable velocity and acceleration of parts of the mechanism to be found

Kinetic energy The energy possessed by something in motion, e.g. a moving car, a rotating flywheel, a waterfall, etc.

Knurling An indented hatched pattern normally applied to screws for ease of turning by hand. The knurling operation is done on a lathe using a standard knurling tool

Laminate A composite made up of bonded layers of material

Layout drawing A drawing showing how machines, furniture etc. are positioned relative to each other and the space they occupy

Leadscrew A rotating screw which converts rotary motion into linear motion

LED (light-emitting diode) A glass encapsulated device which emits light when a current is passed through it

Logarithmic scale A scale based on a base number raised to a power, e.g. base 10 logarithmic scales go

arithmetic scale	logarithmic scale
1	0
10	1
100	2
1000	3

etc., so a small increase in a logarithmic scale means a large increase in the corresponding arithmetic scale.

Logic matrix A diagram with rows and columns which shows the logic state between inputs and outputs in a digital system

Machining centre A multipurpose machine tool which has automatic control of all functions including axis movement, spindle speed and tool changing

Malfunction When a device or system fails to function normally

Malleability The ability of a material to be deformed by compressive forces

Mathematical model A representation of a physical situation by means of equations, e.g. a finite element stress model

Mean The simple average of a series of values, i.e. the sum of the values divided by the total number of values

Microbore Normally applied to mean a pipe which has a very small internal diameter

Microelectronics Electronic circuits using semiconductor devices, such as integrated circuits, which perform many functions using minimal space and power

Micronic filtration A method of cleaning hydraulic fluid by filtering out any solid particles bigger than a few microns (1 micron = 0.001 mm)

Milling A material removal technique in which the workpiece is fed past a rotating multitoothed cutter

Mixture A substance made up by intermingling two or more elements or compounds which are not chemically bonded together

Modem (modulator/demodulator) An interface device that is mostly used to connect computers to other computers via the analogue signals down a telephone line

Modularisation In design this means the breaking down of the design into smaller parts or modules. These modules are often standard components or subassemblies for low manufacturing and maintenance costs

Modulus of elasticity A measure of the elasticity of a material given by the ratio of induced stress and the resulting strain

MTTF (mean time to failure) The average time before a component or system fails

Network A set of linked systems or sub-systems

Neutral flame A flame used in oxy-acetylene welding in which acetylene and oxygen are burned in equal quantities.

Noise An undesired disturbance to a signal

Non-ferrous metal A metal which does not contain iron or in which iron is present in only small amounts to enhance its properties

Normalising A heat treatment technique used to remove the internal stresses from metal components and refine their grain size

Nozzle An orifice through which a fluid or gas is discharged

Nut and screw A screw thread assembly normally used to transmit motion or to clamp parts together

Oblique projection A method of pictorial projection using two axes at 90° to each other and another receding axis, usually at 30° or 45° to the horizontal. This system allows true lengths to be shown in two axes with only lines on the receding axis being foreshortened

Open-loop A control system without feedback, i.e. there is no connection between input and output

Optimum solution The best solution to a problem with regard to a specified parameter (see Parameter), e.g. cost, speed, reliability

Orthographic projection A system of projecting views of solids onto mutually perpendicular planes. It is the standard system used in engineering drawing

Overshoot In a system is when the desired value is exceeded

'Over the wall' The usual sequential sequence for a product of design, prototype building and testing, planning, tooling, material procurement, manufacture, assembly and test

Oxidised When a fluid becomes contaminated by excessive contact with the oxygen in air

Parameter Another word for factor or property (see Optimum solution)

Parametric A method of dimensioning solids of similar shape by defining a variable dimension for each feature which can be changed as required

Pasteurisation The process of destroying bacteria in food at a temperature low enough not to destroy any vitamins

Pattern Wooden or metal facsimile of an engineering component which is used to make the moulds for sand casting

Permeability A measure of the influence which a material has on a magnetic field

Permittivity A measure of the influence which a material has on an electric field

PERT (performance, evaluation and review technique) A graphical method of simulating and co-ordinating the activities in a project so as to plan the most efficient use of time, materials and people

Phase The relative position of two or more signals, i.e. if they are in phase they are together and if they are out of phase there is a time difference between their occurrence

Phosphating A surface treatment often applied to steel sheet to help prevent corrosion and aid paint adherence

Pickling A chemical process in which components are placed in a chemical solution to remove surface oxides and scale.

Pictorial projection A system of representation of a three-dimensional object on a flat screen or paper. The main types are isometric projection (se Isometric projection) and oblique projection (see Oblique projection)

Pilot light A small light in a gas boiler system which is permanently on and which ignites the main heating jets when required

Plain-carbon steel An alloy made up of iron and up to 1.5% carbon.

Plating An electrical technique for depositing metal onto the surface of a material

PLC (programmable logic controller) A dedicated process control computer with multiple interfaces to inputs and outputs

PNdB (perceived noise decibels) A scale for measuring the loudness of a noise. It is a logarithmic scale, i.e. a sound of 40 PNdB is 10 times louder than one of 30 PNdB

Polymerisation The process of forming a polymer from a monomer, e.g. polyethylene from ethylene

Polymers Atoms linked together to form long chain molecules; also a general name given to plastic materials

Polymorphic As allotropic

Power The rate of doing work in J/s (watts, W)

PPI (programmable peripheral interface) A programmable device for connecting peripherals such as printers, plotters and keyboards to a computer

Precipitation Heat treatment technique used to adjust hardness and hardening

Precipitator A device, usually in a chimney or flue, which can remove chemicals from solid solution, e.g. sulphur dioxide in solution in the flue vapour is precipitated as solid sulphur dioxide

Precision For a repeated series of values, the measure of deviation from the mean value (see Mean)

Pressure For a fluid or gas is the ratio of force/area in units of N/m^2

Prime mover The main means of power in a system, e.g. the prime mover in a diesel electric train is the diesel engine

Process capability A statistical measure of the overall spread of values obtained when using a particular process

Processor See CPU

Protocol A set of standard rules governing a procedure

Prototype The first full-size realisation of a product

PTFE (polytetrafluoroethylene) A polymer which is inert and has a very low coefficient of friction. It is frequently used to line 'non-stick' cooking utensils

Quality of conformance A measure of the ability of a process to produce items within the stated tolerance

Quality of design A measure of the excellence and 'fitness for purpose' of a product

Quench hardening A heat treatment technique in which medium- and high-carbon steel components are hardened by heating to a specified temperature and quenching in oil or water

Rack and pinion A circular gear meshings with a linear rack. It is used to convert rotary motion to linear motion or vice versa

RAM (random access memory) The main memory of a computer where data currently used by a program is held

Ramp change A continuously varying input to a system from zero to the desired value (see also Step change)

Recrystallisation The formation of new undeformed grains in cold worked metal when raised to a specified temperature.

Rectified Electrical current which has been changed from alternating to direct current

Refrigerant The fluid used in a refrigerator to extract heat from the interior

Relay A device for switching which allows a very small signal to be converted to a much larger one, e.g. a few millivolts into full mains power

Reliability In statistical terms is the probability of a device functioning for a specified period of time

Repeatability The ability of a device to repeat a desired value with accuracy

Resistance A measure of resistance to flow. In electrical terms resistance (ohms) = potential difference (volts)/current (amps)

Resistivity A measure of the electrical resistance of a material

Response rate The rate at which a system is able to respond to changes of input

Robot A device for manipulating objects in three dimensional space. Robots can be 'pick and place', when they just move an object from one position to another, or 'continuous path', when they can follow any programmed path

Roboticist A person who is an expert on robots and robot systems

Robotics The technology of robots and associated systems

Rolling A forming process in which material is passed through a succession of rolls to produce plate, sheet, bar, angle, channel and a variety of other structural sections

ROM (read-only memory) A computer memory where data is held which is constantly used for routines like booting up the computer. As the name suggests, data cannot normally be written into this memory

Scenario A description of a particular situation or 'scene'

Semiconductor A material such as silicon or germanium whose electrical conductivity varies widely with temperature

Sensitivity A measure of the smallest change which a system is able to respond to, e.g. a measuring device with a sensitivity of 1 micron will record changes of 1 micron or larger

Servomotor A fluid power or electric motor which is controlled by a negative feedback system to provide precise movement

Settling time The time it takes for a system to settle into its new setting after a change of input

Shaft basis A system of fits and limits based on keeping the shaft size constant and varying the size of the hole (see Clearance fit, Transition fit and Interference fit)

Simple harmonic motion A type of motion which is mathematically based on the sine wave and is produced by a mechanism such as a crank and a piston, i.e. if a crankshaft is rotating at constant speed the piston is reciprocating with simple harmonic motion

Simulation A means of artificially imitating the operation of a system, e.g. a flight simulator exactly imitates the behaviour of an aircraft in response to changes of control inputs or disturbances

Soft soldering A joining process in which metal parts are bonded together using a tin–lead alloy (soft solder) as the joining material

Solenoid An electrical device which normally gives a small linear motion when a current is applied to it. It is frequently used in switching devices for valves etc.

Soluble oil Oil which can be mixed with water for use as a coolant in material removal processes

Solvent A chemical substance which can cause a material to dissolve or degrade

Spool valve A fluid power valve which moves linearly to switch direction and position of fluid flow

Stability The ability of a system to remain at the set input despite disturbances such as change of load or current etc.

Standard deviation A statistical term which gives a measure of the spread of values from the mean (see Mean) value of a series of numbers

Step change A suddenly applied input change to a system, i.e. the input goes from 0 to the desired value almost instantaneously (see also Ramp)

Sterilisation The process of rendering clean and free of bacteria and viruses, usually by a heat or chemical treatment

Stiffness The ratio of stress/strain in a material

Strength–weight ratio The ratio of the strength of a material to its weight

Stress The ratio of force/area in units of N/m^2 for a solid material

Stylus A needle-like device usually made of a very hard material. It is normally used to follow closely a surface to reproduce sound or measure surface roughness

Substitution The technique of using alternative materials when the one normally used is not available

Super glue Cyanoacrylate adhesive which cures in the presence of moisture

Suppression The reduction or elimination of interference in an electrical or electronic circuit

Symbolic A term used to describe a mathematical model, algorithm or equation which describes a characteristic or relationship

Tachogenerator A device attached to a shaft to give velocity feedback

Telemetry The measurement or recording, by electronic means, of a distant event

Telephony Telecommunications system for the transmission of sound and other data

Tempering A heat treatment technique used to toughen quench hardened steel components by reheating them to a specified temperature and quenching again.

Thermal expansivity A measure of the effects of temperature change on the dimensions of a material

Thermodynamics The science of the relationship between heat and energy

Thermoplastic material A polymer material which can be softened by heating

Thermosetting material A polymer material which cannot be softened by heating

Thermostatic A term applied to devices which can maintain temperature at a set value

Thrust A propulsive force which can produce forward motion, e.g. the thrust of a jet engine will propel an aircraft or rocket

Time constant The time it takes, after a change of input, for the system to achieve approximately two-thirds (actually 63.2%) of the set value

Tin–bronze An alloy whose main constituents are copper and tin

Tinplate Mild steel sheet which has been given a protective coating of tin

Tolerance The band between the specified upper and lower limits of a dimension or other quality parameter of a product

Toughness The ability of a material to withstand shock loads

Toxicity The degree to which a substance can cause illness when ingested

Transducer A system for converting one form of energy to another, which is usually used as a measuring or feedback device, e.g. a linear transducer which converts linear movement to digital pulses

Transfer function The output/input ratio of a system

Transfer line A manufacturing system of machines and conveyers which processes parts through several operations in sequence

Transformed current Electrical current which has been changed in value by being passed through a transformer, e.g. mains current generated at the power station at over 100 000 volts is transmitted through the National Grid and transformed down in stages to 240 volts for supply to domestic and premises

Transition fit When a hole and shaft are nominally the same size. It is used when shaft and hole have to be in accurate location together and allows easy assembly and disassembly

Troubleshooting A systematic problem-solving technique

Turbine A machine which produces rotary motion when fluid or gas at high pressure is passed through it

Turning A material removal technique in which a single-point cutting tool removes material from a rotating workpiece

UART(universal asynchronous receiver/transmitter) A type of interface for peripheral devices. It converts a parallel stream of bits into a single serial stream of bits, to and from peripherals, which can only accept data in serial form

Ultimate tensile strength Tensile stress at which a material fractures

Undershoot In a system is when the desired value is not reached

Vacuum forming A forming process widely used with thermoplastics in which atmospheric pressure forces the heated sheet material into the evacuated cavity in a die.

Value analysis A technique for establishing and improving the value of a product or service

Velocity The rate of change of position; and is a vector quantity, i.e. it has both magnitude and direction

VGA (video graphics array) Computer graphics interface for colour and high-resolution graphics

VLSI (very large-scale integration) An integrated circuit technology which can put up to 10 000 components on a single chip. A typical application is a microprocessor chip as used in a personal computer

Welding A joining process in which the joint surfaces are heated and fuse together in the molten state

Wire wrapping A joining process used in electrical and electronic circuits in which a wire is terminated by wrapping it tightly around a terminal post.

Answers to unit tests

Answers to unit test 1

1. B (1)
2. D (1)
3. A (1)
4. 1-A, 2-C, 3-D (3)
5. 1-B, 2-C, 3-D (3)
6. D (1)
7. A (1)
8. B (1)
9. B (1)
10. D (1)
11. A (1)
12. C (1)
13. 1-B, 2-C, 3-A (3)
14. 1-D, 2-A, 3-B (3)

Total marks = 22

Answers to unit test 2

1. C
2. C
3. A
4. C
5. D
6. B
7. D
8. C
9. D
10. B
11. B
12. D
13. B
14. A
15. C
16. A

17. A
18. C
19. D
20. C
21. C
22. A
23. C
24. C
25. B
26. D
27. C

Answers to unit test 3

1. D
2. C
3. A
4. B
5. B
6. A
7. B
8. C
9. D
10. C
11. A
12. C
13. B
14. D
15. A
16. C
17. A
18. B
19. B
20. C
21. D
22. A
23. A
24. B
25. A
26. D

Index